6th International Conference on Electrical Engineering, Control, and Robotics (EECR 2020) and the Workshop of the 3rd International Conference on Intelligent Control and Computing (ICICC 2020)

IOP Conference Series: Materials Science and Engineering
Volume 853

Xiamen, China
10 - 12 January 2020

ISBN: 978-1-7138-3615-5
ISSN: 1757-8981

Printed from e-media with permission by:

Curran Associates, Inc.
57 Morehouse Lane
Red Hook, NY 12571

CURRAN ASSOCIATES INC. proceedings.com

Some format issues inherent in the e-media version may also appear in this print version.

Institute of Physics
Dirac House, Temple Back
Bristol BS1 6BE UK

Phone: 44 1 17 929 7481
Fax: 44 1 17 920 0979

techtracking@iop.org

Additional copies of this publication are available from:

Curran Associates, Inc.
57 Morehouse Lane
Red Hook, NY 12571 USA
Phone: 845-758-0400
Fax: 845-758-2633
Email: curran@proceedings.com
Web: www.proceedings.com

TABLE OF CONTENTS

HV TRANSMISSION AND INSULATION TECHNOLOGY

GIS PARTIAL DISCHARGE PATTERNS RECOGNITION WITH SPHERICAL CONVOLUTIONAL NEURAL NETWORK 1
Wei Yang, Guobao Zhang, Taiyun Zhu, Mengyi Cai, Hengyang Zhao, Jing Yan, Yanxin Wang

STUDY ON INSULATION CHARACTERISTICS OF SCRATCH DEFECTS IN 35KV XLPE CABLE TERMINALS 7
Wei Jiaxiang, Wang Chong, Gan JianWen, Lu Zhengxu, Liang Dawei, Li Lini

ANALYSIS OF THE INFLUENCE OF FLUX DETECTOR ON THE SUBSYNCHRONOUS OSCILLATION CHARACTERISTICS OF DFIG 14
Haiqiang Jin, Jun Wen, Tongtong Wang, Hanhua Zhang

COORDINATED GENERATION AND TRANSMISSION EXPANSION PLANNING APPROACH CONSIDERING PROBABILISTIC AVAILABLE TRANSFER CAPABILITY 20
Ping Zhou, Peng Kang, Ruiguo Chen, Hao Tian

COORDINATED GENERATION AND TRANSMISSION EXPANSION PLANNING APPROACH BASED ON CHANCE-CONSTRAINTS USING GENETIC ALGORITHM 28
Ping Zhou, Peng Kang, Yuan Zhu, Ruiguo Chen

SMART GRID AND MICRO GRID

A NOVEL METHOD FOR IMPLEMENTING HARMONIC COMPENSATION WITH MULTI-INVERTERS IN A MICRO-GRID 36
Xiang Shao

SECONDARY CONTROL OF ISLAND MICROGRID BASED ON DISTRIBUTED MULTI-AGENT SYSTEM UNDER SYMMETRIC TIME DELAY 45
Bai Shaotong

DOUBLE-LAYER ELECTRICITY MARKET TRADING STRATEGY FOR MICROGRIDS BASED ON MULTI-AGENT SYSTEM 54
Le Du, Xiaofei Zhu

SAFETY EVALUATION OF MICROGRID USING CHAOTIC TIME SERIES AND RBF NEURAL NETWORK 62
He Qin, Wenbo Tang

RESEARCH ON OPTIMIZING THE RECOVERY STRATEGY OF A DISTRIBUTION NETWORK COMMUNICATION SYSTEM UNDER EXTREME DISASTERS 70
Husheng Lin, Xia Zhou, Jianqiang Lu

A SECURITY ASSESSMENT OF DISTRIBUTION NETWORK CPS BASED ON ASSOCIATION MATRIX MODELING ANALYSIS 77
Zhou Yang, Xia Zhou, Ming Ni, Manli Li, Husheng Lin

THE MODIFICATION OF LINE PARAMETERS IN DISTRIBUTION NETWORK CONSIDERING THE ELECTRO-THERMAL COORDINATION 86
Fang Tianyu, Zhang Hengxu, Shi Fang, Hou Yanruo

LOW FREQUENCY OSCILLATION ONLINE IDENTIFICATION BASED ON SYNCHRONOUS PHASOR OF DISTRIBUTION NETWORK 93
Li Yi, Yan Fen, Wang Kun, Sun Chaohui, Zhang Wenwei

POWER ELECTRONICS AND TRANSMISSION TECHNOLOGY

CALCULATING THE WINTER BASE LOAD AND AC LOAD IN CHONGQING CONSIDERING THE CORRELATION OF TEMPERATURE AND POWER SYSTEM LOAD 100
Ruimiao Wang, Zheng Liao, Chunyan Li, Bo Hu, Kaigui Xie, Tao Niu, Qingsong Sun, Gang Wang

DEADBEAT PREDICTIVE CURRENT CONTROL WITH DISTURBANCE OBSERVER FOR PERMANENT MAGNET SYNCHRONOUS MOTOR 108
Zhang Chuntao, Dang Ruina, Wang Yang, Zhang Chengning, Zhang Shuo

THE REALIZATION OF INTELLIGENT KNOWLEDGE ADAPTIVE LEARNING METHOD IN THE FIELD OF SUBSTATION OPERATION AND MAINTENANCE 114
Jianlin Yin, Yali Liang, Qi Gao, Lu Liu, Yabin Lei, Xilan Zhao

DFIG MODEL SUITABLE FOR SSR RESEARCH 123
Hanhua Zhang, Haiqiang Jin, Tongtong Wang, Jun Wen

VOLTAGE SAG ANALYSIS BASED ON CLUSTER ANALYSIS AND CORRELATION ANALYSIS 128
Hao Liu, Qinghui Zeng, Shaohui Liu, Zhuojun Wu, Lei Li, Rongrong Ma

MODEL PREDICTIVE CURRENT CONTROL OF DUAL INVERTER FED OPEN-END WINDING PMSM BASED ON FAST VECTOR SELECTION 135
Di Wu, Rui Zhu, Jifeng Chen, Jiaquan Hu, Guowu Hua, Ying Wang

STUDY ON THE REACTIVE POWER COORDINATED CONTROL IN HYBRID PARALLEL HVDC SYSTEM 142
Yan Han, Song Rui, Wang Xuan, Zhao Shichang, Yang Wenli, Luo Min

A FIBER-OPTIC CURRENT SENSOR BASED ON FUZZY PI CONTROL 150
Zhaowei Diao, Zhongxiao Ji, Mei Liu, Yifei Wu, Yi Wu, Fei Yang

RESPIRATORY DETECTION USING NON-CONTACT SENSORS 156
Chen Xinqiang, Huang Zhicheng, Su Lumei

ELECTRICAL ENGINEERING AND AUTOMATION

PREDICTION OF SHORT-TERM PV OUTPUT POWER BASED ON PCA-STACKING UNDER DIFFERENT WEATHER CONDITIONS 161
Zhu Xiaofei

COMBINATION FORECASTING MODEL OF DAILY ELECTRICITY CONSUMPTION IN SUMMER BASED ON DAILY CHARACTERISTIC METEOROLOGICAL FACTORS 171
Ruimiao Wang, Xinyu Yao, Chunyan Li, Bo Hu, Kaigui Xie, Tao Niu, Mushui Li, Jian Fu, Qingsong Sun

ACTIVE MAGNETIC BEARING DESIGN AND BACKSTEPPING-ADAPTIVE CONTROL FOR HIGH-SPEED ROTORS 178
Umar Khayyam, He Boxia, Shaheer ul Hassan

STUDY ON CHARACTERISTICS OF NO-LOAD CONVERTER TRANSFORMER CLOSING ACCESSING TO HIGH-CABLE RATE NETWORK 187
Tongtong Wang, Shengyi Zhu, Haiqiang Jin, Jun Wen

SIMULATION ANALYSIS OF MAGNETIZING INRUSH CURRENT OF 500KV TRANSFORMER 193
Shengyi Zhu, Tongtong Wang, Haiqiang Jin, Jun Wen

STUDY ON THE TRANSFORMER PROTECTION SCHEME BASED ON DOUBLE THRESHOLDS AND WAVEFORM SIMILARITY 198
Yu Liang, Fan Yang, Dan Shi

RESEARCH ON DYNAMIC RELATIONSHIP BETWEEN URBAN RAIL TRANSIT AND POWER CONSUMPTION 204
Liu Zhe, Du Fengqing, Liu Xu

A NON-INTRUSIVE HOME LOAD IDENTIFICATION METHOD BASED ON ADAPTIVE REINFORCEMENT LEARNING ALGORITHM 210
Haoyang Li

RESEARCH ON THE CIVIL-MILITARY INTEGRATION LOGISTICS SUPPORT SYSTEM OF WARSHIPS 218
Xiaobai Liu, Fei Shao, Xingxing Zhang

A METHOD OF AUTOMATIC GENERATION OF DAILY WORK REPORT FOR POWER DISPATCHING BASED ON ROBERTA 223
Jie Li, Tao Zhang, Hao Li, Jixiang Lu, Jinjun Lu, Ziqi Li

CONTROL THEORY AND CONTROL SYSTEM

RECURSIVE-BIASED-GENERALIZED-INVERSE METHOD FOR THRUST ALLOCATION FOR DYNAMIC POSITIONING VESSELS WITH NON-ROTATABLE THRUSTERS AND FORCE CONSTRAINTS 228
Bo Zhao, Guoqing Xia

KALMAN FILTER BASED SENSOR FUSION FOR ALTITUDE ESTIMATION OF AERIAL VEHICLE 234
Muhammad Asad Bashir, Fahad Mumtaz Malik, Zeeshan Ali Akbar, Muhammad Uzair

A NON-LINEAR IMPROVED DOUBLE-INTEGRAL SLIDING MODE CONTROLLER (IDI-SMC) FOR MODELING AND SIMULATION OF 6 (DOF) QUADROTOR SYSTEM 242
Irfan Ahmad, Muwahida Liaquat, owais zarin khan, Fahad Mumtaz Malik

RESEARCH ON ADAPTIVE RECLOSING DISCRIMINANT METHOD BASED ON LMD SAMPLE ENTROPY 252
Liqun Shang, Yuan Zhang, Chenguang Yang

3-D LOCALIZATION OF UAV AND DETECTION BASED ON HARMONICS INDEX AND SPECTRAL ENTROPY CRITERIA 260
Muhammad Amjad Iqbal, Zhao Zhao, Xu ZhiYong, Saad Ur Rehman

DESIGN OF POWER WELL COVER WIRELESS MONITORING SYSTEM BASED ON FREERTOS AND NB-IOT TECHNOLOGY 267
Wei Wang, Tairong Xie, Xianhua Hu

THE OPTIMIZED LAYOUT STRATEGY OF RING OSCILLATOR NETWORK FOR TROJAN DETECTION 272
Lian Yang, Wenxiu Yang, Huan Li

ROBOT DEVELOPMENT AND MECHANICAL ENGINEERING

HUMANOID ROBOT TORSO MOTION PLANNING BASED ON MANIPULATOR POSE DEXTERITY INDEX 277
Baoshi Cao, Kui Sun, Yikun Gu, Minghe Jin, Hong Liu

GAIT PHASE CLASSIFICATION BASED ON SEMG SIGNALS USING LONG SHORT-TERM MEMORY FOR LOWER LIMB EXOSKELETON ROBOT 284
Ye Yuan, Ziming Guo, Can Wang, Shengcai Duan, Lufeng Zhang, Xinyu Wu

A MOBILE ROBOT RECOGNIZE BLURRED TARGETS VIA AN IMPROVED HISTOGRAM EQUALIZATION 291
Longzhi Zhang, Dongmei Wu

RESEARCH ON GAIT PLANNING OF HEXAPOD WHEEL LEGGED ROBOT BASED ON STM32 SINGLE CHIP MICROCOMPUTER 296
Chu Zenan, Wang Wei, Caisheng Yu

MINIMUM-TIME TRAJECTORY PLANNING FOR RESIDUAL VIBRATION SUPPRESSION OF FLEXIBLE MANIPULATOR CARRYING A PAYLOAD 302
Yong Liu, Zhe Du, Shaonan Zhang

DESIGN AND IMPLEMENTATION OF A DEVICE FOR REDUCING ROAD ACCIDENT AT FOGGY, RAINY AND DROWSY CONDITIONS 309
Shihab Uddin Ahmed, Sanzid Junaeid, Md. Monirul Islam, Bishwajit Banik Pathik

NEW DESIGN OF ALL-TERRAIN SMART CAR BASED ON ARDUINO 315
ZhengYang He, Ying Zhang, HongMing Zheng, Yi Zhang, HongTao Yu, MingNan Luo

COMPUTER AND INFORMATION ENGINEERING

ON COMPUTATION OF FRACTIONAL-ORDER DIFFERENTIATIONS 322
Soichiro Ebihara, Noriyuki Hori

COMPLEX NETWORK COMMUNITY DETECTION BASED ON GENETIC ALGORITHM USING K-CLIQUES 330
Jian Ma, Jianping Fan

R-MUCH: A CLUSTERING ROUTING ALGORITHM USING FUZZY LOGIC FOR WSNS 342
Alia Ghaddar, Mohammad Houssein Ghosn, Nathalie Mitton

UFSP-NET: A NEURAL NETWORK WITH SPATIO-TEMPORAL INFORMATION FUSION FOR URBAN FIRE SITUATION PREDICTION 354
Guangyin Jin, Cunchao Zhu, Xiaoxuan Chen, Hengyu Sha, Xingchen Hu, Jincai Huang

CATTLE IDENTIFICATION USING MUZZLE PRINT IMAGES BASED ON FEATURE FUSION 365
Cong Sian, Wang Jiye, Zhang Ru, Zhao Lizhi

AN ENHANCED STOCHASTIC ALGORITHM FOR OPTIMAL CONTRACT CAPACITY OF FURNITURE STORES 372
Chia-Sheng Tu, Ming-Tang Tsai, Chun-Chieh Huang, Kuei-Ching Su

COMPLEX MORPHOLOGICAL FILTERING ALGORITHM BASED ON MULTI-STRUCTURE ELEMENTS 379
Wang Yiyan, Luo Jiang

A DECISION-MAKING TECHNIQUE FOR SOLVING ORDER ALLOCATION PROBLEM USING A GENETIC ALGORITHM 388
Ijaz Ahmad, Yan Liu, Danish Javeed, Shahab Ahmad

A REVIEW OF ARTIFICIAL INTELLIGENCE TECHNIQUES FOR SELECTION & EVALUATION 397
Ijaz Ahmad, Yan Liu, Danish Javeed, Nadia Shamshad, Danish Sarwr, Shahab Ahmad

FATIGUE DETECTION BASED ON MULTI-FEATURE FUSION OF FATIGUE BEHAVIOR 403
Chen Xing, Su Lumei, Qin Meixin

RESEARCH ON TARGET DETECTION OF CARRIER-BASED AIRCRAFT BASED ON DEEP CONVOLUTIONAL NEURAL NETWORK 408
Zhu Xingdong, Tian Shaobing, Fan Jiali, Wang Zheng, Liu Tianqing

Author Index

IOP Conf. Series: Materials Science and Engineering **853** (2020) 011001 doi:10.1088/1757-899X/853/1/011001

PREFACE

The organizing committee of the 6th International Conference on Electrical Engineering, Control and Robotics (EECR 2020) and the Workshop the 3rd International Conference on Intelligent Control and Computing (ICICC 2020) aimed to facilitate interaction among participants for the latest progress and development in the fields of electrical engineering, intelligent control and robotics. Through the conference and workshop, the committee intends to enhance the sharing of individual experiences in their expertise and to encourage interdisciplinary and international collaborations.

The conference and workshop were held in Holiday Inn Express Xiamen Airport Zone, Xiamen, China, during January 10-12, 2020, co-organized by Sichuan Institute of Electronics and Huaqiao University, with the support of Concordia University, the South China University of Technology, the University of Electronic Science and Technology of China, Southwest Minzu University and others. Despite the high quality of most of the submissions, the final proceedings of EECR 2020 and ICICC 2020. includes 59 papers, which were selected after a thorough reviewing process and were resented at the conference and workshop.

The contributions of the technical program committee and the referees are deeply appreciated. Most of all, we would like to express our sincere thanks to the authors for submitting their most recent works and the organizing committee for their enormous efforts to turn this event into a smoothly running meeting.

We sincerely hope that this publication will prove to be an important resource for the scientific community.

With our warmest regards,

Chun-Yi Su

Concordia University, Canada

IOP Conf. Series: Materials Science and Engineering **853** (2020) 011002 doi:10.1088/1757-899X/853/1/011002

Conference Chairs

Chun-Yi Su, Huaqiao University, China
May Huang, International Technological University, USA

Conference Co-Chairs

Rencheng Zhang, Huaqiao University, China
Fan Yang, Huaqiao University, China

Program Chairs

Quang Ha, University of Technology Sydney, Australia
Xuelin Lei, East China University of Science and Technology, China

Local Organizing Chair

Feng Zhao, Huaqiao University, China

Local Organizing Committee

Chunqing Huang, Xiamen University, China
Xiangyu Luo, Huaqiao University, China
Yanzheng Zhu, Huaqiao University, China

Technical Committee

Subhash Rakheja, Concordia University, Canada
Xiuyu Zhang, Northeast Dianli University, China
Zhi Li, Northeastern university, China
Yonghua Liu, Guangdong University of Technology, China
Lianwei Ma, Zhejiang University, China
Ning Sun, Nankai University, China
Cheung-Chieh Ku, National Taiwan Ocean University, Taiwan
ANG Marcelo H. Jr, National University of Singapore
Chih-Jer Lin, National Taipei University of Technology, Taiwan
Jukkrit Kluabwang, Rajamangala University of Technology Lanna Tak,Thailand
Anthony Tzes, University of Patras, Greece
Muhammad Murtaza Yousaf, Punjab University, Pakistan
Ray-Hwa Wong, Hwa-Hsia University of Technology, Taiwan
Gheorghe-Daniel Andreescu, Politehnica University of Timisoara, Romania
Long Jin, Lanzhou University, China
Dong Huang, Chinese Academy of Sciences, China
Samuel P. Kozaitis, Florida Institute of Technology, USA

IOP Conf. Series: Materials Science and Engineering **853** (2020) 011003 doi:10.1088/1757-899X/853/1/011003

First name	Surname	Institution
Yanxin	Wang	Xi'an Jiaotong University, China
Lini	Li	Southwest Jiaotong University, China
Haiqiang	Jin	North China Electric Power University, China
Ping	Zhou	State Grid Chongqing Electric Economic Research institute,
Ping	Zhou	State Grid Chongqing Electric Economic Research institute,
Xiang	Shao	Nanjing university of posts and telecommunications, China
Shaotong	Bai	Nanjing University of Posts and Telecommunications, China
Le	Du	Nanjing University of Posts and Telecommunications, China
Qin	He	Nanjing University of Posts and Telecommunications, China
Husheng	Lin	Nanjing University of Posts and Telecommunications, China
Yang	Zhou	Nanjing university of posts and telecommunications, China
Tianyu	FANG	Shandong University, China
Yi	Li	Wuxi Power Supply Company, State Grid Jiangsu Electric
Qingsong	Sun	Chongqing University, China
Chuntao	Zhang	Beijing Institute of Technology, China
Qi	Gao	NARI Research Institute Beijing R&D Center of Nari
Haiqiang	Jin	North China Electric Power University, China
Hao	Liu	Foshan Power Supply Bureau of Guangdong Power Grid
Rui	Zhu	China University of Mining and technology, China
Han	Yan	Power Research Institute of QingHai power Grid company,
Zhaowei	Diao	Xi'an Jiaotong University, China
Xinqiang	Chen	Xiamen University of Technology, China
Xiaofei	Zhu	Nanjing University of Posts and Telecommunications, China
Li	Mushui	Chongqing University, China
Umar	Khayyam	Nanjing University of Science and Technology, China
Tongtong	Wang	North China Electric Power University, China
Tongtong	Wang	North China Electric Power University, China
Yu	Liang	Maintenance Branch Company of State Grid Fujian Electric
Liu	Zhe	State Grid Shanghai Electric Power Research Institute, China
Haoyang	Li	North China Electric Power University, China
Xiaobai	Liu	China Ship Development and Design Center, China
Tao	Zhang	NARI Group, China
Bo	Zhao	Harbin Engineering University, China
Muhammad Asad	Bashir	National University of Sciences and Technology Pakistan,
Irfan	Ahmad	National University of Sciences and Technology(NUST)
Yuan	Zhang	Xi'an University of Science And Technology, China

Muhammad	Iqbal	Nanjing University of Science and Technology, China
Tairong	Xie	Guangdong Yuanpeng Network Technology Co., Ltd, China
Wenxiu	Yang	Univ. of Elec. Sci. & Tech. of China
Baoshi	Cao	Harbin Institute of Technology, China
Yuan	Ye	Shenzhen Institutes of Advanced Technology, China
Longzhi	Zhang	Harbin Institute of Technology, China
Chu	Zcnan	Anyang Institute of Technology, China
Zhe	Du	Hefei University of Technology, China
Shihab Uddin	Ahmed	Hiperdyne Corporation, Japan
Zhengyang	He	Sichuan University Jinjiang College, China
Axu	Hu	Key Laboratory of China's Ethnic Languages and
Soichiro	Ebihara	University of Tsukuba,Japan
Jian	Ma	Beijing Jiaotong University, China
Alia	Ghaddar	Lebanese International University (LIU/BIU),Lebanon
Yaya	Sun	Xidian University, China
Guangyin	Jin	National University of Defense Technology, China
Sian	Cong	Minzu University of China
MIng-Tang	Tsai	Cheng-Shiu University,Taiwan, China
Yiyan	Wang	Sichuan University of Arts and Science, China
IJAZ	AHMAD	Changchun University of Science & Technology, China
IJAZ	AHMAD	Changchun University of Science & Technology, China
Xing	Chen	Xiamen University of Technology, China
Shaobing	Tian	Naval Aviation University Qingdao campus, China

IOP Conf. Series: Materials Science and Engineering **853** (2020) 011003 doi:10.1088/1757-899X/853/1/011003

E-mail address
yxwang199501@163.com
475857275@qq.com
18811313911@163.com
153326062@qq.com
153326062@qq.com
shaoxiangnjupt@163.com
690768034@qq.com
2817831123@qq.com
158984663@qq.com
1091814745@qq.com
yang_zhou0207@163.com
15953159053@163.com
liyi_ee@126.com
1025540819@qq.com
chuntao@bit.edu.cn
982739104@qq.com
18811313911@163.com
930125316@qq.com
749805964@qq.com
yh19940103@163.com
zhweidiao@163.com
mchenxinqiang@foxmail.com
1870559155@qq.com
934910580@qq.com
umarkhayyam@njust.edu.cn
pyyzwtt@163.com
pyyzwtt@163.com
935866302@qq.com
liuzheacyy@163.com
1586934450@qq.com
mail-fox@163.com
zhangtao11@sgepri.sgcc.com.cn
bo_zhao@hrbeu.edu.cn
asadbashir1993@gmail.com
irfankhandgk11@gmail.com
1224765156@qq.com

IOP Conf. Series: Materials Science and Engineering **853** (2020) 011003 doi:10.1088/1757-899X/853/1/011003

engramjad@njust.edu.cn

756148845@qq.com

201821010626@std.uestc.edu.cn

cbob.robot@gmail.com

13308348407@126.com

zlz.robot@gmail.com

165385778@qq.com

Zincdz@163.com

shihablitu09@gmail.com

cquxiaomage@qq.com

1415714973@qq.com

ebihara@digicon-

13112083@bjtu.edu.cn

alia.ghaddar@liu.edu.lb

yysun@stu.xidian.edu.cn

350020817@qq.com

1691806320@qq.com

k0217@gcloud.csu.edu.tw

wang_yiyan8207@163.com

ieec2017060@mails.cust.edu.cn

ieec2017060@mails.cust.edu.cn

362136278@qq.com

745157824@qq.com

IOP Conf. Series: Materials Science and Engineering **853** (2020) 011004 doi:10.1088/1757-899X/853/1/011004

Peer review statement

All papers published in this volume of *IOP Conference Series: Materials Science and Engineering* have been peer reviewed through processes administered by the proceedings Editors. Reviews were conducted by expert referees to the professional and scientific standards expected of a proceedings journal published by IOP Publishing.

IOP Conf. Series: Materials Science and Engineering **853** (2020) 012001 doi:10.1088/1757-899X/853/1/012001

GIS Partial Discharge Patterns Recognition with Spherical Convolutional Neural Network

Wei Yang[1], Guobao Zhang[1], Taiyun Zhu[1], Mengyi Cai[1], Hengyang Zhao[1], Jing Yan[2], Yanxin Wang[2]

1 State Grid Anhui Electric Power Company Limited Research Institute, Hefei, China.
2 State Key Lab of Electrical Insulation and Power Equipment,Xi'an Jiaoto University, Xi'an, China

yanjing@mail.xjtu.edu.cn

Abstract. The ubiquitous construction of the power Internet of Things provides a new idea for the real-time and accurate diagnosis of GIS partial discharge online monitoring fault diagnosis. However, the traditional partial discharge fault diagnosis method is difficult to solve the problem that the fault information of different online monitoring systems is different from the reference axis. In order to solve the problem that the fault information is difficult to identify in rotation and transformation, and improve the accuracy of fault diagnosis, this paper proposes a spherical convolutional neural network based on complex data sources. First, the PRPS picture transmitted to the ubiquitous power Internet of Things terminal is selected as the fault feature information. Secondly, a generalized Fourier algorithm (GFT) algorithm is used to construct a spherical convolution structure for PD pattern recognition. The algorithm can perform automatic feature extraction. Thirdly, the spherical convolutional neural network -based PD recognition method is applied to processing of the complex data sources with84.88% average accuracy rate. It shows that the PRPS 3D map is one of effective way to avoid the complexity of artificial feature extraction for spherical CNN and in the meantime, it can also improve the accuracy of fault diagnosis.

1. Introduction

Gas-insulated combination electrical appliances are widely used as protection and control devices in power system power systems duo to their high reliability, small footprint, and maintenance-free. However, there is a risk of failure in the actual running GIS, and the failure rate presents a 'potted' curve with the running time [1-2]. Once the GIS fails, it will cause an inevitable blackout and seriously threaten the safe operation of the power system. Therefore, online monitoring of power equipment to detect faults during the incubation period and effective maintenance is of great significance for ensuring the safety and reliability of the power grid.

GIS faults are mostly insulation faults, while insulation faults are manifested in the form of partial discharges. At the same time, the appearance of partial discharge will further aggravate insulation aging [3]. Fault detection is performed by detecting electromagnetic, acoustic, and optical phenomena excited by partial discharge. At present, the main detection methods are pulse current method, ultra-high frequency method, ultrasonic method, optical fiber detection method and chemical product decomposition method [4]. Among them, the UHF method is widely used due to its strong anti-interference and high detection sensitivity.

In view of GIS partial discharge pattern recognition, feature extraction is used to extract features of partial discharge UHF signals, and then machine learning methods are used for pattern recognition classification. At present, the main researches on the characteristic parameters of GIS partial discharge are mostly concentrated on the Weibull distribution parameter, time-resolved partial discharge (TRPD), Phase Resolved Pulse Sequence (PRPS) and phase-resolved partial discharge (TRPD) [5-7]. The principal component analysis, self-encoder and other methods are used to carry out feature dimension reduction and extract key feature parameters. The current main methods for pattern recognition classification are support vector machines, decision trees, neural networks, and their various improvements [8-10].

With the development of deep learning, Jiang Xiuchen et al. proposed a pattern recognition of partial discharge PRPS patterns using deep convolutional neural networks, which effectively weakened the interference of artificial feature engineering and accelerated the model's accuracy while improving the model's accuracy [11]. Identify the rate. Li Gaoyang et al. used multi-resolution convolutional neural networks combined with cyclic neural networks for pattern classification, which effectively improved the accuracy of the model [12]. However, convolutional neural networks cannot achieve recognition of rotated images. Under the background of ubiquitous power Internet of Things, because the fault data of different online monitoring systems is different, the device information represented by it is equivalent to the problem of multi-angle option.

In order to solve the problem that the image data representing the fault information under the ubiquitous power internet of things is invalid due to the flipping property, another deep convolution spherical neural network is proposed. While solving the data restaurant problem, it satisfies the real-time diagnosis and processing of big data under the ubiquitous power Internet of Things. In this paper, we use the spherical convolutional neural network as the pattern recognition classifier as well as use the PRPD three-dimensional model of the complex data source combined with the experiment and field as an input to train a general GIS partial discharge fault diagnostic tool. The rest of the paper is structured as follows. We shall first briefly introduce the spherical CNN in Section 2.The details of the spherical CNN model and the comprehensive framework used are discussed in Section 3. In the next section, the complex data of GIS is applied to test the classification accuracy. Finally, some conclusions are drawn in Section 5.

2. Spherical Convolutional Neural Network

Convolutional neural network is named for its unique convolutional calculations. With the sparse interactions, parameter sharing, and equivariant representations, it can to a large degree improve the efficiency of the system. Analogous to the translational invariance of traditional convolutional neural networks in 2D space, the spherical convolutional neural network has the characteristics of rotational symmetry in three-dimensional space to ensure the generalization performance of the system for 3D signals. A spherical CNN architecture is formed with convolutional layers, pooling layers, and full-connected layers [13]. The spherical convolution is calculated by taking inner product between a spherical signal and rotated spherical filter [14-15].

A detailed description of the characteristics of a spherical CNN is as follows: (1) Sparse interactions. The size of the convolution kernel is much smaller than the size of the input, which illustriously enhances the efficiency of the network. The cells in the deep layer of the network interact indirectly with most of the inputs. (2) Parameter sharing. In a spherical CNN, each element of the kernel acts on each location of the input. Therefore, convolution greatly outperforms the multiplication of dense matrices in terms of storage requirements and statistical efficiency. (3) Rotational symmetry. For spherical convolution, the special form of parameter sharing leads to rotational symmetry of the neural network layer. Thus the output signal changes with the same properties as the input signal.

IOP Conf. Series: Materials Science and Engineering **853** (2020) 012001 doi:10.1088/1757-899X/853/1/012001

3. Proposed method

3.1. Signal Preprocessing

Since the fault data set is composed of the field data and the experimental data, it is considered that the field data comes from different voltage levels and different manufacturers' equipment, and there are omissions, negligence and other problems in the data accumulation process, resulting in data missing, repeated and the format and size of unstructured data are not uniform. Therefore, it is necessary to use data cleaning to discover the most realistic law between fault signals and then identify fault types.

The field fault data is derived from the different voltage levels and equipment of different equipment manufacturers, resulting in different fault data sets, physical meanings, and so on. At the same time, different acquisition devices will also cause unstructured data to have different formats and sizes. Therefore, the fault set needs to be normalized to solve the same problem of the Subsequent spherical convolutional neural network input, and improve the generalization ability of the model to the field data. Using the moment-based image normalization technique [16], the basic working principle is as follows: Firstly, the parameters of the transformation function are determined by using the moments in the image which are invariant to the affine transformation, and then the original image is transformed by the transformation function determined by this parameter. Is a standard form of image (this image is independent of affine transformation). In general, the moment-based image normalization process consists of four steps: coordinate centering, x-shearing normalization, scaling normalization, and rotation normalization. The calculation formula is defined as:

$$y = \frac{x - MinValue}{MaxValue - MinValue} \tag{1}$$

Where x and y are the values before and after the conversion, and $MaxValue$ and $MinValue$ are the maximum and minimum values of the sample, respectively.

3.2. Fast Spherical Convolution with GFT

Analogous to planar convolution, spherical convolution is defined as the same as Taco S. Cohen's that the inner product between a spherical signal and rotated spherical filter. Square up the signal f and filter Ψ is vector-valued functions on the sphere. The S^2 convolution is defined as:

$$f * \Psi(R) = \int_{S^2} \sum_{k=1}^{K} f_k(x) \Psi_k(x)\left(R^{-1}x\right) dx \tag{2}$$

where the f: S^2 for K channels.

It is worth noting that the result of the convolution calculation in (1) is the function on the rotation group $SO(3)$ not on the sphere S^2. It is necessary to define the convolution calculation on $SO(3)$ as:

$$f * \Psi(R) = \int_{SO(3)} \sum_{k=1}^{K} f_k(R') \Psi_k(x)\left(R^{-1}R'\right) dR' \tag{3}$$

Analogy to planar convolution can be quickly calculated by Fast Fourier transform(FFT). For spherical convolution, there are corresponding generalized Fourier algorithm(GFT) and fast generalized Fourier algorithm(FGFT) on the rotating group. The GFT is defined as:

$$\hat{f}^l == \int_X f(x) \overline{U^l(x)} dx \tag{4}$$

Where X signify S^2 or $SO(3)$. For S^2, there is the spherical harmonics $Y_m^l(x)$ indexed by $l \geq 0$ and $-l \leq m \leq l$. For $SO(3)$,there are the Wigner D-functions $D_{mn}^l(R)$ indexed by $l \geq 0$ and $-l \leq m,\ n \leq l$.

The inverse $SO(3)$ is defined as:

$$f(R) = \sum_{l=0}^{b} (2l+1) \sum_{m,n=-l}^{l} \hat{f}_{mn}^l(R) \tag{5}$$

and similarly for S^2. The spherical convolution in the spectrum is shown in Fig. 1.

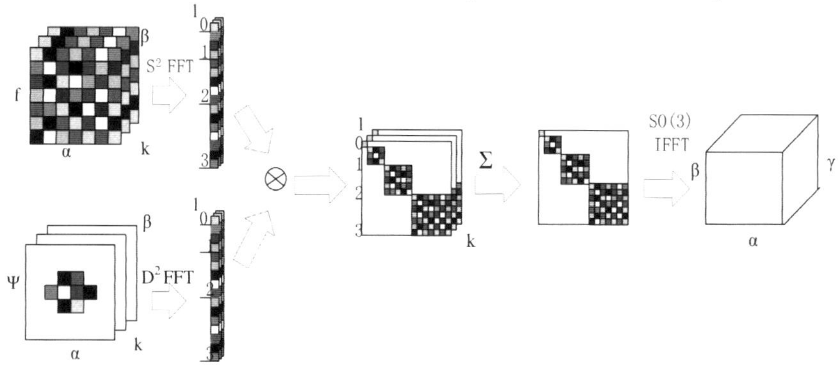

Figure 1. Spherical convolution in the spectrum.

The input signals f and the filter ψ are FFT, tensored, summed over input channels. α, β are the parameters of sphere. α, β, γ are the parameters of $SO(3)$.

4. Experiments

4.1. Data Acquisition

Taking into account the actual situation of the data under the ubiquitous power Internet of Things, the method of using laboratory data for flipping and transformation is used to construct complex partial discharge data to ubiquitous complex data sources under the power Internet of Things. Studies have shown that the partial discharge of the protruding defect under the lightning impulse voltage is much higher than the protruding defect under the power frequency. Therefore, the simulated GIS is subjected to lightning impulse and power frequency withstand voltage tests in the laboratory to obtain a rich discharge sample.

The GIS simulation model in the laboratory and the main insulation defects inside the GIS is shown in Figure 2. The UHF sensor detection frequency band is 300MHz-2000MHz, and is composed of an amplifier, a high-pass filter, a detector and a shielding case. The working bandwidth of the amplifier is 300MHz-1500Mhz, and the gain is 40dB.

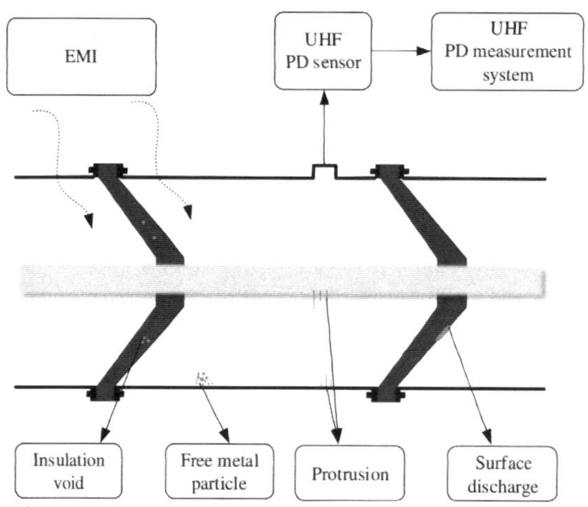

Figure 2. The GIS simulation model in the laboratory and the main insulation defects inside the GIS

During the experiment, 600 sets of GIS partial discharge experimental data were obtained, and the data was symmetrically transformed to simulate the complex data source under the actual ubiquitous power internet of things. For 1200 sets of data, 800 sets were selected as training data, and the remaining 400 sets constitute test data.

4.2. Result and Discussion

For this experiment testing set, BPMM model, SVM model, 2D DCNN model and spherical CNN model are used to train and identify PD sources. The recognition results are shown in Table 1.

Table 1. The recognition results

Category	Spherical CNN(%)	DCNN(%)	SVM(%)	BPNN(%)
Protrusion discharge	92.25	62.25	89.75	85.25
Particle discharge	91.75	60.50	90.25	85.5
Void discharge	83.75	58.75	77.75	68.25
Surface discharge	71.75	56.25	73.50	56.25
Overall	84.88	59.44	82.82	73.81

As shown in Table 1, the average accuracy of the spherical CNN is84.88%. The accuracy of the void discharge and the surface discharge are relatively low. The spherical CNN performs competitively with the state of the art results compared to other methods. The surface defects of the insulator are significantly more sensitive to lightning impulse voltage and overvoltage than the power frequency withstand voltage. However, studies have shown that there is almost no sign of the flash of the insulator. This defect recognition rate is relatively low due to the current detection method is difficult to detect the sign signal. Insulator internal defects mainly refer to small voids in the mold resin or voids in the layered region between the insulating material and the metal insert. With the accumulation of electric fields for a long time, the instability of partial discharge under such defects leads to low recognition accuracy.

The average recognition accuracy of SVM, DCNN and BPNN methods is 82.82% , 59.44% and 73.81%. Under the premise of the application of ubiquitous power Internet of Things, the spherical convolutional neural network significantly improves the accuracy of information fault recognition after the rotation and symmetric transformation of the partial discharge map. However, there is still a need to further improve the network layer parameters to further improve the accuracy of fault identification.

5. Conclusion

In the work reported in this paper, we have developed a novel deep learning architecture for UHF signal source recognition, a spherical CNN after GFT representation, which has the ability of feature extraction and diagnosis within the same architecture Spherical CNN based pattern recognition is more accurate for for the partial data of the partial discharge map under the ubiquitous power Internet of Things compared with the DCNN, BPNN and SVM methods. The conclusions can be drawn as follows:

• Spherical CNN has improved accuracy by 3.55%, 18.25% and 27% compared with DCNN, SVM and BPNN methods. As the data set expands and the number of network layers deepens, the recognition accuracy of the model will further increase..

• Under the application of the power Internet of Things platform, the convolutional neural network can solve the problem that the data collected by different online monitoring devices are not uniform in the coordinate axes. Therefore, it is more suitable for engineering applications under the power of the Internet of Things.

References

[1] Hikita, M.; Ohtsuka, S.; Teshima, T., et al.: ‘Examination of electromagnetic mode propagation

characteristic in straight and L-section GIS model using FD-TD analysis', IEEE Trans. Dielectr. Electr. Insul., 2007, 14, (6), pp. 3095– 3102
[2] Kaneko, S., Okabe, S., Muto, H., et al.: 'Electromagnetic wave radiated from an insulating spacer in gas insulated switchgear with partial discharge detection', IEEE Trans. Dielectr. Electr. Insul., 2009, 16, (1), pp. 60–68
[3] Gaoyang,L.,Mingzhe,R.,Xiaohua,W.,et al.,:'Partial diascharge patterns recognition with deep convolutional neural networks', International Conference on Condition Monitoring and Diagnosis,2016,Xi'an,pp324-327
[4] Kawada, M.: 'Fundamental study on locating partial discharge source using VHF-UHF radio interferometer system', Electr. Eng. Jpn., 2003, 144, (1), pp. 32–41
[5] Hamed,J.; Behzad,K.,:' Towards automated statistical partialdischarge source classification using pattern recognition techniques',IET Journals.High Voltage,2018,5048,(3),pp162-169
[6] Wu, M.; Cao, H.; Cao, J., et al.,:'An overview of state-of-the-art partial discharge analysis techniques for condition monitoring', IEEE Electr. Insul. Mag. 2015, 31, pp22–35.
[7] Li, L.; Tang, J.; Liu, Y,:' Partial discharge recognition in gas insulated switchgear based on multi-information fusion.', IEEE Trans. Dielectr. Electr. Insul. 2015, 22, pp1080–1087.
[8] Umamaheswari, R.; Sarathi, R,:' Identification of partial discharges in gas-insulated switchgear by ultra-high-frequency technique and classification by adopting multi-class support vector machines', Electr. Power Compon. Syst. 2011, 39, pp1577–1595.
[9] Wang, K.; Li, J.; Zhang, S., et al.,:' A hybrid algorithm based on s transform and affinity propagation clustering for separation of two simultaneously artificial partial discharge sources', IEEE Trans. Dielectr. Electr. Insul. 2015, 22, pp1042–1060.
[10] Mas'Ud, A.; Stewart, B.; McMeekin, S.,:' Application of an ensemble neural network for classifying partial discharge patterns.', Electr. Power Syst. Res. 2014, 110, pp154–162.
[11] Hui,S.,Jiejie,D.,Xiuchen,J.,et al.,:'GIS partial diascharge patterns recognition via deep convolutional neural network under complex data sources', IEEE Trans. Dielectr. Electr. Insul., 2018, 006930 pp. 678–685.
[12] Gaoyang,L.,Mingzhe,R.,Xiaohua,W.,et al.,:'Partial diascharge patterns recognition with deep convolutional neural networks', International Conference on Condition Monitoring and Diagnosis,2016,Xi'an,pp324-327.
[13] H. Shin, H.R. Roth, M. Gao, L. Lu, Z. Xu, I. Nogues, J. Yao, D. Mollura, R.M. Summers, "Deep convolutional neural networks for computer-aided detection: CNN architectures dataset characteristics and transfer learning", IEEE Trans. Med. Imaging, vol. 35, 2016, pp. 1285-1298.
[14] T.S. Cohen and M. Welling,:' Group equivariant convolutional networks. In Proceedings of The 33rd International Conference on Machine Learning (ICML)', volume 48, 2016, pp2990–2999.
[15] T.S. Cohen, M. Geiger, J. Koehler, and M. Welling,:'Convolutional networks for spherical signals. In ICML Workshop on Principled Approaches to Deep Learning', 2017.
[16] R. G. Casey.,:'Moment normalization of handprinted characters', IBM Journal of Research and Development, vol. 14, pp. 548 –557, Sep 1970.

Study on Insulation Characteristics of Scratch Defects in 35kV XLPE Cable Terminals

Wei Jiaxiang[1], Wang Chong[2], Gan JianWen[2], Lu Zhengxu[3], Liang Dawei[4], Li Lini[5]

[1] Guangxi Power Grid Power Dispatching Control Center, Guangxi, China
[2] Beijing Urban Construction Design and Development Group Co., Ltd., Beijing, China
[3] Guangxi Power Grid Materials Co., Ltd., China
[4] State Grid Zhejiang Electric Power Co., Ltd. Construction Branch, Zhejiang, China
[5] College of Electrical Engineering Southwest Jiaotong University, Sichuan China

475857275@qq.com

Abstract. In order to study the influence of different ring-shaped scratches on the insulation characteristics of 35kV Crosslinked polyethylene(XLPE) cable terminations, the model of air-gap defects caused by ring-shaped scratches was introduced at the end of the outer semi-conductor shielding layer, and the finite element simulation calculation was used to calculate the model. The electric field distribution characteristics inside the cable terminal were simultaneously made into cable end samples of different types of scratch defects, and the discharge characteristics were tested by a partial discharge test circuit. The results show that the air gap defect caused by the annular scratch defect increases the degree of electric field distortion. The maximum electric field strength is much larger than the breakdown field strength of the air, and it is prone to partial discharge activity. The Phase Resolved Partial Discharge(PRPD) of the cable terminal with ring scratch defects The spectrum shows a typical "rabbit ear" shape. As the scratch defects deepen, the PRPD spectrum begins to gradually transition to a short-feeling "cat ear" shape.

1. Introduction

The 35kV XLPE cable terminal is an important device for transmission lines, which is used to connect the cable body and other electrical equipment, and is responsible for the transmission of electrical energy [1]. Due to the particularity of the cable termination structure, the probability of failure is much greater than that of the main body. It is the weak link of the transmission line insulation, and its safe operation directly affects the effective operation of the transmission system. The XLPE cable terminal needs to be cut off part of the semiconductor layer during the installation of the field terminal, and the stress tube and the heat shrinkable tube are installed layer by layer, which causes the difference of the material parameters of the interlayer and the difference of the structure, resulting in uneven distribution of the internal electric field, and due to the immature manufacturing process, it is easy to cause scratch defects on the XLPE insulation surface of the cable terminal, and air gap defects are formed after the fabrication, so that the uneven distribution of the electric field is further deepened, and the insulation performance of the cable terminal is further weakened [2].

At present, domestic and foreign scholars have conducted some research on the defects of cable terminals. Yang Fengyuan of Shanghai Jiaotong University studied the cables of different defect types by partial discharge. The research shows that the partial discharge characteristics of cables with different defect types have significant characteristics, and the partial discharge information can be used as an important means of defect research [3]; HA Illias and Lee ZH use finite element analysis software to simulate the different defect sizes, locations and different insulation thicknesses of medium voltage cable joints, the main factors affecting the electric field strength distribution in cable joints are studied [4]; Sichuan University scholars use electric field simulation and aging experiments. In a combined way, the characteristics of the discharge at the air gap defect are explored. It is believed that the concentration of the electric field at the air gap defect is an important cause of the continuous ablation of the insulation and the final breakdown [5].
Based on previous research, due to the special materials and structures between the 35kV XLPE cable termination layers, when the scratch defects are introduced, the air gap is present, which has an important impact on the insulation performance of the cable termination. Through the finite element simulation analysis and partial discharge test of cable terminals with different air gap defects caused by different scratch depths, the electric field distortion and partial discharge characteristics are explored, which can provide an important basis for cable terminal fault diagnosis.

2. Simulation analysis

2.1. Model building.

According to the investigation, the 35kV XLPE cable terminal needs to cut off the water blocking layer and the outer semiconductor layer. Because the cable is a cylindrical structure and the thickness of the outer semiconductor layer is small, when the cutting is performed, the depth and strength of the feeding are difficult to control. The ring scratch defect is caused, and at the same time, due to the process reason, the treatment for the scratch defect is rough, and the scratch defect cannot be filled, thereby causing the existence of the air gap defect.
It is found through experiments that the annular scratch defect at the cut-off is most likely to occur. To study the insulation characteristics of the scratch defect of the 35kV XLPE cable terminal, according to the actual structure, the cable terminal model is built according to the 1:1 ratio. The simulation model is shown in Figure 1. The circular scratch defect is set with three parameters, which are circular scratch defects with a depth of 0.5mm, 1mm, 2mm and a thickness of 1mm at the cut. At the same time, considering the shape of the cut-off position, the material cannot be completely fitted, so that there is an air gap at the joint between the cut-off portion of the outer semi-conductor layer and the stress tube, and a triangular air gap with a height of 0.5 mm is provided at the joint of the cut-off portion and the stress tube.

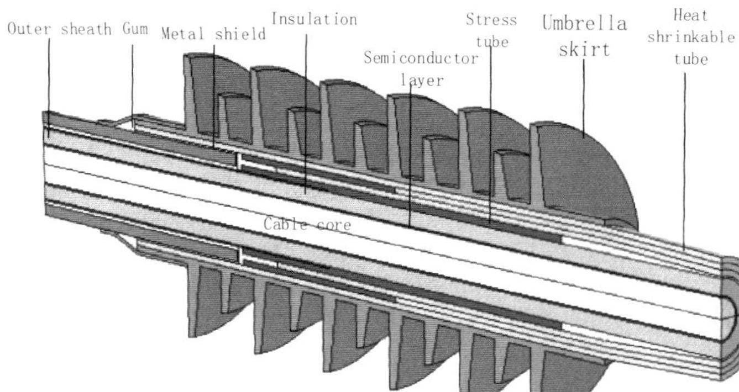

Figure 1. Simulation model diagram

2.2. Simulation results.

The 35kV XLPE cable terminal runs under power frequency. The electric field under the power frequency voltage changes slowly with time. The electric field can be solved according to the electrostatic field. According to the material test and literature, the material parameters are set. The air gap parameter is set to air and the metal shield is set. Layer grounding, electric field simulation for no scratch defects and three ring scratch defects respectively. Since the cable end is an axisymmetric structure and the defect is a circular scratch defect, the electric field distribution of the entire cable terminal can be characterized by studying a horizontal plane. The cross section of the maximum value of the electric field strength in the spatial region is studied, and the simulation results are shown in Figure 2.

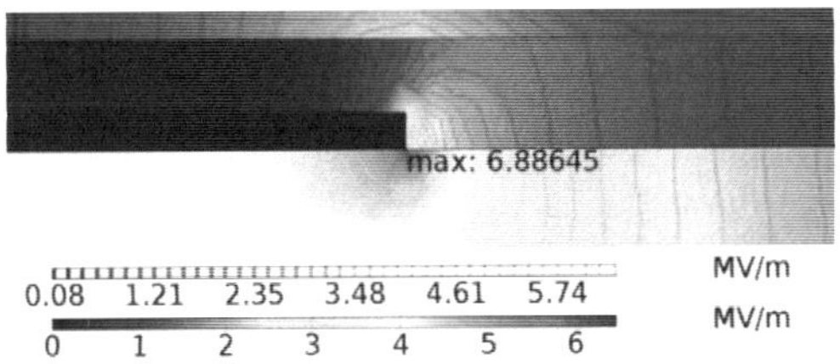

(a) Trouble-free cable terminal simulation diagram

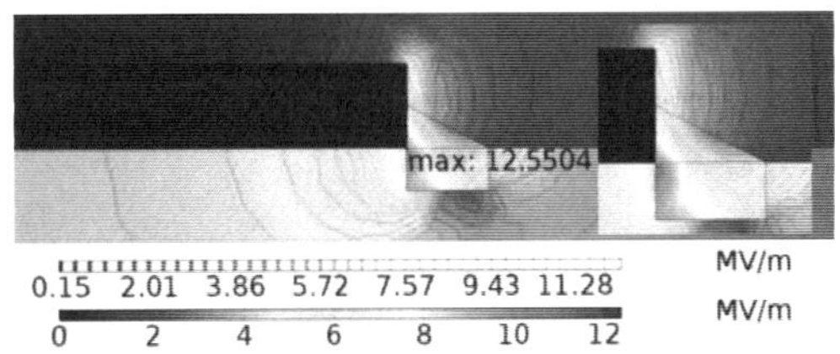

(b) 0.5mm scratch defect cable terminal simulation diagram

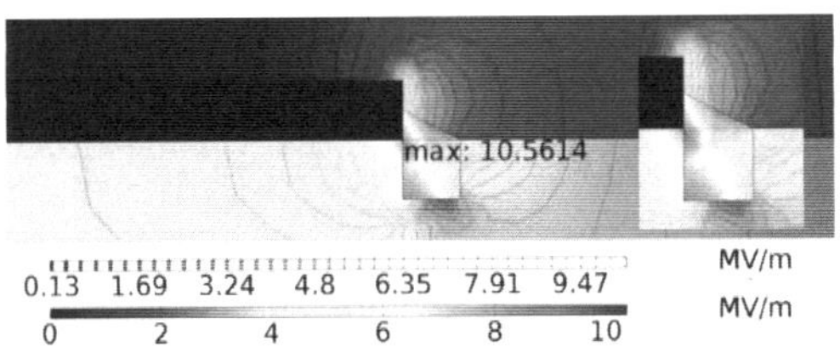

(c)1mm scratch defect cable terminal simulation diagram

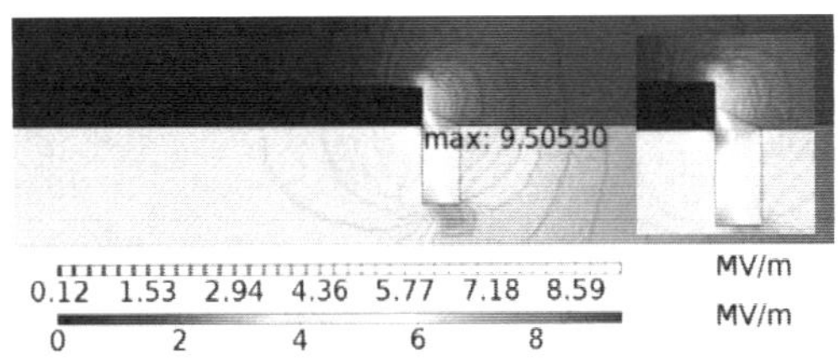

(d)2mm scratch defect cable terminal simulation diagram

Figure 2. Cable terminal simulation results

It can be seen from Figure 2 that when the XLPE cable terminal does not introduce a scratch defect, the electric field strength is at a maximum of 6.88 MV/m, which does not reach the breakdown field strength of the cross-linked polyethylene, and the maximum value is at the junction of the outer semiconductor layer cutoff and the insulating layer .The electric field strength at other locations is relatively evenly distributed ; When the annular scratch defect is introduced, the air gap defect is caused, as shown in Figure 2(b), the scratch depth is 0.5mm, and the air gap depth is 1mm. The maximum electric field strength is 12.55MV/ m, the breakdown field strength exceeding air [6] is 66.5% higher than the maximum terminal electric field strength when no scratch defect is introduced. It can be seen from the local electric field distribution magnified view that due to the abrupt change of the interlaminar material parameters and the junction structure The mutation, in the vicinity of the air gap defect, especially at the corresponding position of the outer semiconductor layer cutoff, the air gap electric field distortion is very serious, the air gap electric field strength is greater than the electric field of the surrounding insulating layer; As shown in Fig. 2(c), the scratch depth is 1mm at this time, and the formed air gap depth is 1.5mm. At this time, the maximum electric field strength is 10.56MV/m, which exceeds the breakdown field strength of air by 3 MV/m. It can be seen from the enlarged view of the local electric field distribution that although the scratch depth is deepened by 0.5 mm, the electric field intensity maximum is smaller than the scratch depth of 0.5 mm, but the electric field strength is small in the air gap region, especially in the insulating portion. The electric field distortion at both ends of the area is particularly serious; As shown in Fig. 2(d), the scratch depth is 2mm at this time, and the formed air gap depth is 2.5mm. At this time, the maximum electric field strength is 9.51MV/m, which is still higher than the breakdown field of air. It can be seen from the enlarged view of the electric field distribution that as the depth of the scratch is deepened, the maximum value of the electric field strength decreases at this time. The reason for the analysis is that the increase of the air gap defect region weakens the abrupt change of the junction structure, so that the electric field intensity is gradually distributed uniformly [7], but the electric field strength of the entire air gap region is still stronger than the electric field strength around the insulating layer.

It can be seen from Figure 2(b)-(d) that the introduction of scratch defects has a great influence on the electric field distribution inside the cable terminal, especially in the air gap portion, the maximum electric field strength is higher than the air breakdown field strength, which is extremely easy. Partial discharge occurs and the surrounding insulating material is burned. At the same time, the deepening of the scratch defect causes the thickness of the insulating layer to be drastically reduced, which reduces the insulation performance of the cable. Under the intense discharge activity and low insulation performance, the entire cable terminal' electrical performance will be greatly affected. At the same time, it can be seen from the comparison that the smaller the defect is, the more severe the electric field distortion is. Therefore, the extremely small air gap defect is also an extremely weak part of the cable terminal, and needs to be focused.

3. Study on the characteristics of partial discharge

The electric field simulation analysis of the 35kV XLPE cable terminal scratching defect can characterize its internal insulation characteristics, but there is no instrument to effectively detect the electric field distribution inside the cable terminal. Currently, the test for the insulation state of the cable terminal is generally through partial discharge. . The parameters related to the partial discharge test are closely related to the insulation state performance of the cable. The partial discharge quantity characteristic can be used to initially detect the insulation of the test cable, provide reference for the tester, and master the insulation state of different operating periods to reduce the operation. The sudden accident of the line provides great help [8,9].

As shown in Figure 3, it is a partial discharge test loop, in which the Partial Discharge (PD) uses JF2006. After the partial discharge signal is processed by the amplification module and the filter module in the signal conditioning module, the data is collected using DPO-5204B, finally transferred to the PC to achieve partial discharge data processing and display [10].

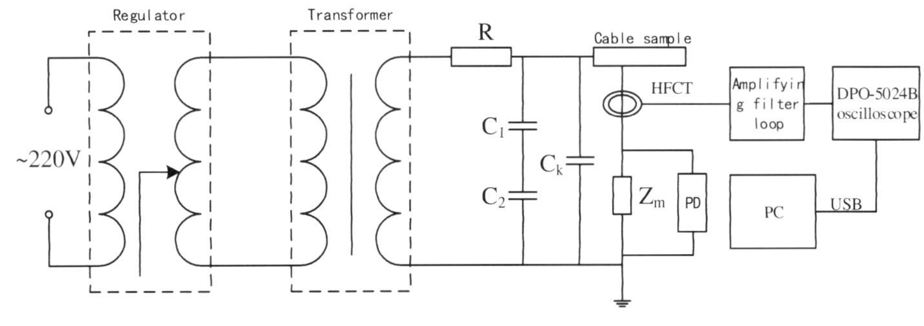

Figure 3. Partial discharge test circuit diagram

In order to fit the simulation experiment and the actual situation, this study carried out the production of four cable terminal samples. First, the water blocking layer and the outer semiconductor layer of the four cable bodies were cut off. No. 1 was no scratch defect, No. 2 is to make a circular scratch depth of 0.5 mm at the cutoff of the outer semiconductor layer, No. 3 is to make a circular scratch depth of 1mm at the cutoff of the outer semiconductor layer, No. 4 is to make a circular scratch depth of 2 mm at the cutoff of the outer semiconductor layer. Scratch defects are produced as shown in Figure 4.After the scratch is made, the stress tube, the heat shrinkable tube, and the shed are sequentially mounted in the specified order to form 4 cable termination samples.

Figure 4. Ring scratch defect production diagram

Firstly, the partial discharge system is calibrated, the background noise is controlled to be less than 10pC, and then the terminal sample of the No. 1 cable is connected to the partial discharge test circuit. The partial discharge signal is not detected at the rated working voltage of 35kV for 15 minutes to ensure the test.，that，the system is normal and noise interference is minimized[11].

Through the test, record the initial discharge voltage of each test and pressurize to 35kV, the time to form a stable PRPD spectrum, take the average of five tests of each cable terminal sample, and the terminal sample of 2-4 cable The initial discharge voltages were 7.8kV, 10.5kV, 11.2kV, respectively, and the stable discharge time were 75min, 64min, 60min respectively. After testing, the most representative PRPD of the five test samples of each cable terminal was selected. The spectrum is shown in Figure 5.

Fig. 5(a) is a PRPD spectrum of a circular scratch defect depth of 0.5 mm. It can be seen from the figure that the discharge phase is concentrated between 45°-100° and 225°-270°, and the positive and negative half-cycle discharge shapes are relatively thin and long, the whole discharge process presents a typical "rabbit ear" shape; the positive half cycle discharge is dense, the discharge amount is distributed at 10-1000pC; for the negative half cycle, the discharge amount is concentrated at 500-1000pC, and the discharge amount is less in the interval below 500pC, discharge The shape presents a floating state.

Fig. 5(b) is a PRPD spectrum of a circular scratch defect depth of 1 mm. It can be seen from the figure that the discharge interval is increased and the discharge phase is concentrated between 30°-100° and 225°-280°. The shape of the "rabbit ear" is relatively short and thick; the discharge trend is similar in the positive and negative half cycles, and the discharge area of each half cycle forms a "M" shape discharge shape, the discharge amount fault appears at about 252pC, and the high discharge amount is concentrated between 252-750pC. The discharge in the low discharge area is very dense.

Fig. 5(c) is a PRPD spectrum when the depth of the annular scratch defect is 2 mm. It can be seen from the figure that the discharge interval is increased, and each phase has a relatively obvious discharge activity, and the entire discharge process still exhibits "rabbit ear" shape, but has gradually transitioned to the "cat ear" shape; positive and negative half cycles exhibit different discharge characteristics, the discharge phase of the positive half cycle is concentrated at 75 ° -105 °, the discharge distribution is relatively uniform; the discharge phase of the negative half cycle is concentrated at 225 ° - 300°, the discharge intensity of the negative half cycle is greater than that of the positive half cycle, and the discharge amount is concentrated between 10-600pC, and the discharge in the low discharge region is relatively dense.

According to the partial discharge test, as the scratch defect deepens, the air gap region increases, so that the electron dispersion in the air gap region is relatively stable, and the partial discharge activity is relatively stable, however, the deepening of the scratch defect causes the thickness of the insulating layer to be lowered, the insulation performance is greatly impaired, and the breakdown failure of the cable terminal is extremely likely to occur.

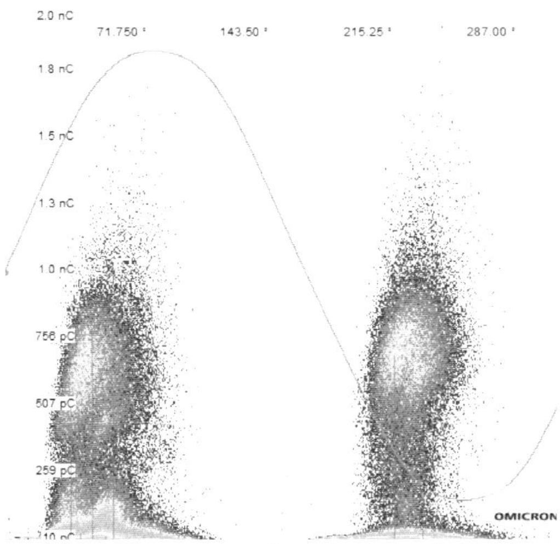

(a) PRPD spectrum of terminal sample No. 2

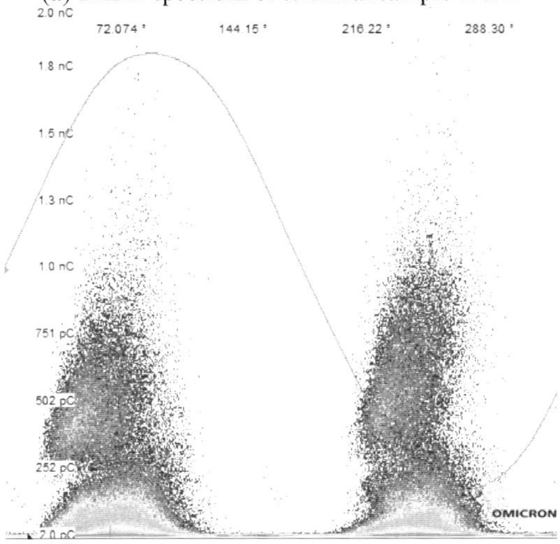

(b) PRPD spectrum of terminal sample No. 3

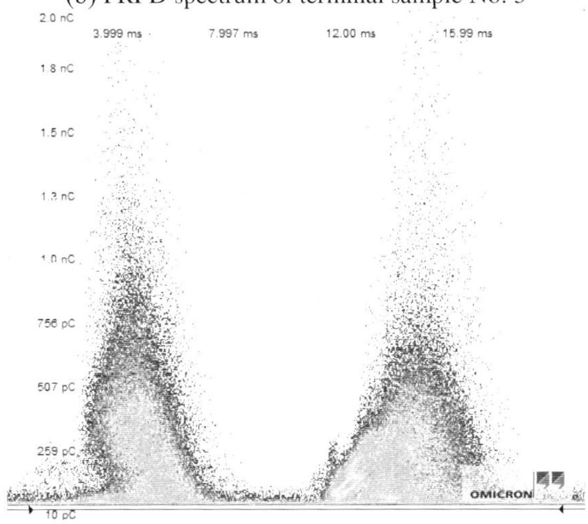

(c) PRPD spectrum of terminal sample No. 4

Figure 5. Partial discharge results

4. Conclusion

Based on the defects caused by the 35kV cable termination, the finite element simulation analysis and partial discharge test of the circular scratched cable terminal are carried out, and the following conclusions are obtained:

(1) When the cable terminal contains a circular scratch defect, it is easy to cause distortion of the internal electric field. The maximum electric field strength of the air gap defect area far exceeds the breakdown field strength of the air, which is easy to cause partial discharge activity;

(2) The simulation results of three types of cable terminations with ring-shaped scratch defects show that Due to the sudden change of the material parameters of the interlayer and the sudden change of the structure of the interface, the depth of the scratched air gap is shallow, and the degree of electric field distortion is more serious, so even small defects need to be focused;

(3) The partial discharge test results of the cable terminations with different types of scratch defects indicate that the shape of the partial discharge of the cable terminal with scratch defects is "rabbit ear" shape. As the depth of the scratch deepens, the shape of the discharge begins to transition from a "rabbit ear" shape to a short and fat "cat ear" shape, its partial discharge characteristics can be used as an important basis for insulation diagnosis.

(4) By studying the electric field and partial discharge characteristics of the cable termination with scratch defects, it can provide a good data foundation for the next cable fault insulation diagnosis.

Acknowledgments

Thanks to the support of Guangxi Power Grid Power Dispatching Control Center, Beijing Urban Construction Design and Development Group Co., Ltd. , Guangxi Power Grid Materials Co., Ltd. and State Grid Zhejiang Electric Power Co., Ltd. Construction Branch.

5. References

[1] Wan Hang, Zhou Kai, Zeng Qin, Zhang Chunshuo, Gong Jun, Huang Yonglu. Research on the location method of medium voltage cable terminal operating voltage based on phase location device[J]. Electrical Measurement & Instrumentation, 2019, 56(7): 30-34.

[2] Wu Ke, Ma Chunliang, Zhou Kai, et al.Partial discharge and surface ablation characteristics of air gap defects in 10 kV cable termination[J]. Insulating Material, 2015(7): 38–43. [3] Wu Ke, Ma Chunliang, Zhou Kai, et al.Partial discharge and surface ablation characteristics of air gap defects in 10 kV cable termination[J]. Insulating Material, 2015(7): 38–43.

[3] YANG Fengyuan, XU Yongpeng, ect. Tes Research on DC Partial Discharges of Cross Linked Polythylene Cable[J]. Proceedings of the CSEE, 2016, 36(24):6702-6710.

[4] Illias H A, Lee Z H, Bakar A H A, et al. Electric field distribution in 132 kV one piece premolded cable joint structures[C]. International Conference on Condition Monitoring and Diagnosis. IEEE, 2012: 643-646.

[5] LIU Gang,CHEN Zhijuan.Experiment of Main Insulation with Air Gap of 10kV Cross Linked Polyethylene Cable Terminal[J]. High Voltage Engineering, 2012, 38(3): 678-683.

[6] Wu Guangning. High Voltage Technology [M]. Beijing: Mechanical Industry Press, 2007: 5.

[7] WAN Li, ZHOU Kai 1,ect. Toward Understanding Development Mechanism of Partial Discharge in Air Gap Defects in Cable Termination by Analysis of Electric Field Characteristics[J].High Voltage Engineering, 2014,40(12):3709-3716.

[8] Tian Y, Lewin P L, et al. Partial discharge detection in cables using VHF capacitive couplers[J]. Dielectrics & Electrical Insulation IEEE Transactions on, 2003, 10(2): 343-353.

[9] Marc Qiang. Research and application of partial discharge pattern recognition technology for XLPE power cables [D]. North China Electric Power University, 2016.

[10] Zhang Ji Pei. The Study on Extraction and Processing Technology of High Frequency Signal of Partial Discharge in Cable[D]. Si Chuang : Southwest Jiaotong University, 2018.

[11] Ge Jingwei, Qiu Changrong et al. Partial discharge test [M]. Beijing: Mechanical Industry Press, 1984.

Analysis of The Influence of Flux Detector on the Subsynchronous Oscillation Characteristics of DFIG

Haiqiang Jin[1], Jun Wen[2], Tongtong Wang[2], Hanhua Zhang[2]

[1]School of Electrical and Electronic Engineering, North China Electric Power University, Changping District, Beijing 102206, China
[2]Electric Power Research Institute, State Grid Ningxia Electric Power Co.,Ltd., Yinchuan, China

18811313911@163.com

Abstract. There is a risk of subsynchronous oscillation after grid connection of doubly fed wind farm. The accurate detection of stator flux amplitude and phase will determine the accuracy of stator flux vector oriented control method, and then affect the subsynchronous oscillation characteristics of grid connected doubly fed wind turbine. Therefore, in PSCAD/EMTDC, this paper establishes the model of grid connected doubly fed wind turbine with u-i flux detector and u-ω flux detector respectively, and studies the influence of different flux detector on the subsynchronous oscillation. Simulation results show that different flux detector will participate in the subsynchronous oscillation to different extent.

1. Introduction

Doubly fed induction generator(DFIG) can run at variable speed and constant frequency, and can decouple the control of active power and reactive power. With the large-scale grid connection of wind power, the operation of power grid becomes more and more complex, facing the risk of subsynchronous oscillation.

At present, there are many examples of subsynchronous oscillations about doubly fed wind farms in the world. In October 2009, a grid fault occurred in Texas,, which caused a doubly fed wind farm to be connected to the grid through a 70% series-compensated line, resulting in a subsynchronous oscillation of about 20Hz[1]. A subsynchronous oscillation was also detected in the recording device of a wind power collection station in China. Analysis shows that the subsynchronous oscillation is probably caused by the internal reasons of the wind farm.

Based on the above background, it is of great significance to study the subsynchronous oscillation characteristics of doubly fed wind farms. In fact, the stator flux vector oriented control method widely used in DFIG depends on the accurate detection of the stator flux, which may affect the dynamic behavior of the system. Literature [2] pointed out that different phase-locked mechanisms and different phase-locked loop (PLL) parameters cause the wind turbine to participate in the electromechanical oscillation mode of power system in different degrees. In literature [3], the dynamic characteristics of the PLL are considered in the single machine infinite bus system. It is pointed out that the improper configuration of the PLL parameters will lead to system instability. Literature [4] show that the angle error measured by PLL will affect the power output of wind turbine. However, the influence of different flux detector on the subsynchronous oscillation characteristics of grid connected doubly fed wind turbine has not been reported.

In this paper, the mathematical models of u-i and u-ω stator flux detector are established firstly. Then in PSCAD/EMTDC, the models of grid connected doubly fed wind turbine using u-i and u-ω flux detectors are established respectively to verify the influence of different flux detector on the subsynchronous oscillation characteristics of grid connected doubly fed wind turbine.

2. Model of the system

Stator flux vector oriented control method

Stator flux vector oriented control method is the preferred scheme for rotor side control of doubly fed induction generator. If the stator flux is oriented on the d-axis of dq coordinate system, the d-axis and q-axis components of the stator flux are as follows

$$\begin{cases} \psi_{sd} = |\boldsymbol{\psi}_{sdq}| = \psi_s \\ \psi_{sq} = 0 \end{cases} \tag{1}$$

Where $\boldsymbol{\psi}_{sdq}$ is the stator flux vector and ψ_s is the amplitude of $\boldsymbol{\psi}_{sdq}$.

In practical application, especially for high-power induction generator, with the increase of power, the stator winding inductance is larger and the resistance is smaller. The stator resistance voltage drop is much smaller than the total voltage drops. Therefore, the angle between stator voltage vector and flux vector can be approximately π/2. Under this condition, the relationship between the stator flux amplitude and the voltage amplitude is

$$U_s = \omega_1 \psi_s \tag{2}$$

Flux detector

At present, there are two kinds of flux detector. One is the u-i flux detector based on stator voltage and current, the other is the u-ω flux detector based on stator voltage and grid angular frequency.

u-i flux detector. According to the relation that the induced electromotive force is equal to the rate of flux change, the flux can be obtained by integrating the electromotive force, thus the u-i flux detector can be constructed.

The stator voltage equation of doubly fed induction generator in two-phase static αβ coordinate system is

$$\begin{cases} u_{s\alpha} = R_s i_{s\alpha} + \dfrac{d\psi_{s\alpha}}{dt} \\ u_{s\beta} = R_s i_{s\beta} + \dfrac{d\psi_{s\beta}}{dt} \end{cases} \tag{3}$$

The αβ axis component of the flux can be obtained by integrating the two sides of the above equation

$$\begin{cases} \psi_{s\alpha} = \int (u_{s\alpha} - R_s i_{s\alpha}) \mathrm{d}t \\ \psi_{s\beta} = \int (u_{s\beta} - R_s i_{s\beta}) \mathrm{d}t \end{cases} \tag{4}$$

Then the amplitude and phase of the stator flux can be obtained by transforming the aβ axis component from the rectangular coordinate system to the polar coordinate system

If the stator resistance is ignored, the stator flux αβ components ψ_α and ψ_β can be obtained directly by integrating u_α and u_β, and then the stator flux amplitude and phase angle can be obtained by coordinate transformation. Both the two methods need to consider the problem of eliminating the DC bias introduced in the integration process. Therefore, in the above two methods, the integration link is replaced by the inertia link with a cut-off frequency of (0.5~1) Hz, so the value range of the inertia time constant of the inertia link is (0.159-0.318) s [5].

u-ω flux detector. According to the above, when the stator flux was oriented, if the stator resistance is ignored, the amplitude of the stator flux is constant

$$\psi_s = \frac{U_s}{\omega_1} \tag{5}$$

and the phase angle of stator flux lags behind the phase angle of stator voltage π/2. After Clarke transformation and transformation from rectangular coordinate system to polar coordinate system, the amplitude and phase of stator voltage can be obtained.The phase of stator flux can be obtained by subtracting π/2 from the phase of stator voltage,and the amplitude of stator flux can be obtained by dividing the amplitude of stator voltage by the grid angular frequency.The grid angular frequency can be obtained by PLL. The dynamic performance of PLL is determined by K_p(the proportional gain)and K_i(the integral gain)of PI link in PLL, so this parameter will have an impact on the dynamic performance of flux detector and even the grid connected doubly fed wind turbine.

3. Simulation analysis

Introduction to simulation model

In the model established in this paper, the grid side converter adopts the grid voltage vector oriented control method,the rotor side converter adopts flux vector oriented control method. The model of grid connected doubly fed wind turbine is shown in Figure 1. R_L and X_L represent the resistance and reactance of transmission line, R_S and X_S represent the equivalent resistance and reactance of the grid.

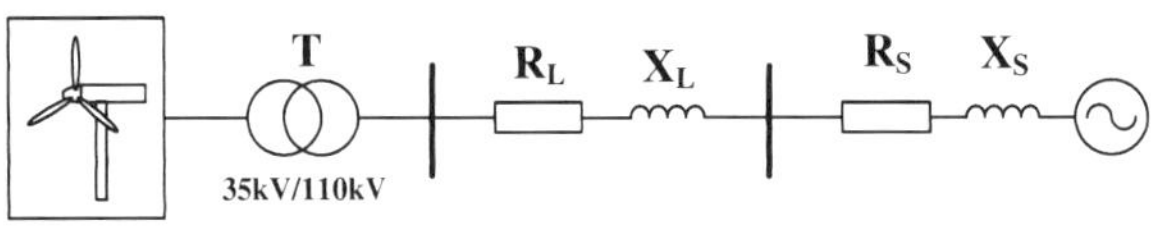

Figure 1. Model of grid connected doubly fed wind turbine.

3.1. The case of u-i flux detector

3.1.1. Ignore stator resistance. Under the condition of wind speed of 11.5m/s and rotating speed of 0.984pu, the frequency spectrum of the active power of the doubly fed wind turbine is shown in Figure 2(a). It can be seen from the figure that there is no obvious subsynchronous oscillation component in the active power.

In order to observe the influence of the time constant T of the internal inertial link of the flux detector, change the value of T within its range, and draw the curve of the frequency and amplitude of subsynchronous oscillation, as shown in Figure 2(b),(c). It can be seen from the figure that the change of inertia time constant of stator flux detector has no obvious influence on the frequency and amplitude of subsynchronous oscillation.

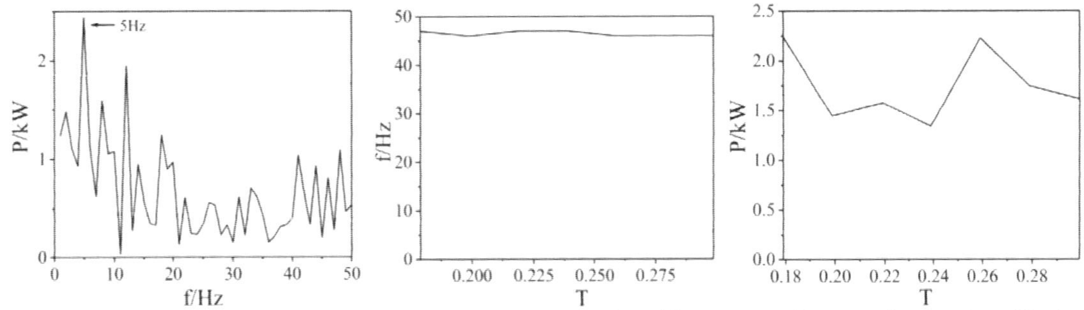

(a) Active power frequency spectrum (b) Frequency of oscillation (c) Amplitude of oscillation

Figure 2. Ignore stator resistance

3.1.2. Do not ignore stator resistance. Under the condition of wind speed of 11.5m/s and rotating speed of 0.984pu, the frequency spectrum of the active power of the doubly fed wind turbine is shown in Figure 3(a).It can be seen from the figure that the active power contains subsynchronous oscillation component near 47Hz, but the amplitude is very small, less than 2kW.

In this case, change the value of T within its range, and draw the curve of frequency and amplitude of subsynchronous oscillation, as shown in Figure 3(b),(c). It can be seen from the figure that the change of the inertia time constant has no obvious influence on the frequency and amplitude of the subsynchronous oscillation.

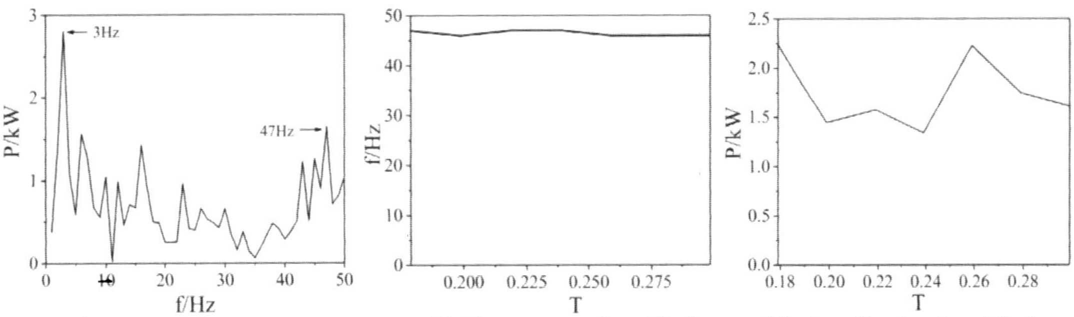

(a) Active power frequency spectrum (b) Frequency of oscillation (c) Amplitude of oscillation

Figure 3. Do not ignore stator resistance

After comparing the above two situations, we can see that, when the stator resistance is ignored in the flux detector, the sub synchronous oscillation component of the system is very small and can be ignored; when the stator resistance is not ignored, the sub synchronous oscillation component can be observed in the system.

3.2. The case of u-ω flux detector

The phase angle of the flux is obtained by directly calculating the measured signal, and the flux amplitude is obtained by dividing the grid voltage amplitude by the grid angular frequency measured by PLL in u-ω flux detector. Next, K_P and K_i of PI link in PLL are changed under the condition of ensuring the normal operation of PLL and the system, the frequency spectrum of active power is carried out under the condition of wind speed of 11.5m/s and rotating speed of 0.984pu.

Set the value of K_p to 0.5 and leave it unchanged,and increase K_i from 4 to 8 in steps of 0.5. Fig. 4 shows the curve of frequency and amplitude of subsynchronous oscillation when K_i changes. It can be seen from the figure that the frequency remains unchanged and the amplitude has no obvious tendency to change.

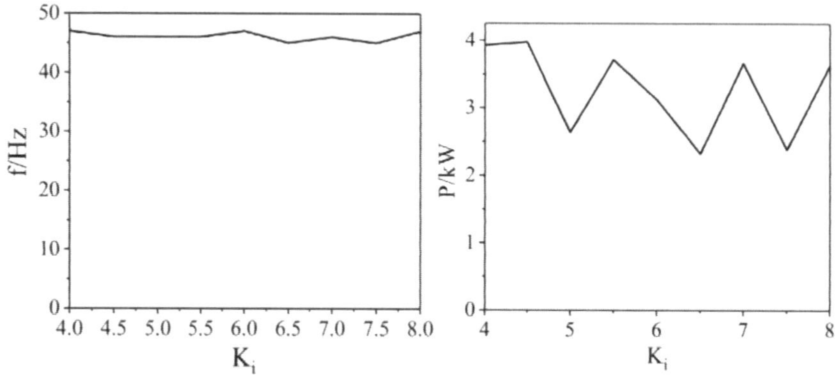

(a) Frequency of oscillation (b) Amplitude of oscillation

Figure 4. Influence of K_i on the subsynchronous oscillation.

Set the value of K_i to 333 and leave it unchanged,and increase K_p from 60 to 84 in steps of 4.Fig. 5 shows the curve of frequency and amplitude of subsynchronous oscillation when K_p changes. It can be seen from the figure that the frequency remains unchanged and the amplitude has no obvious tendency to change.

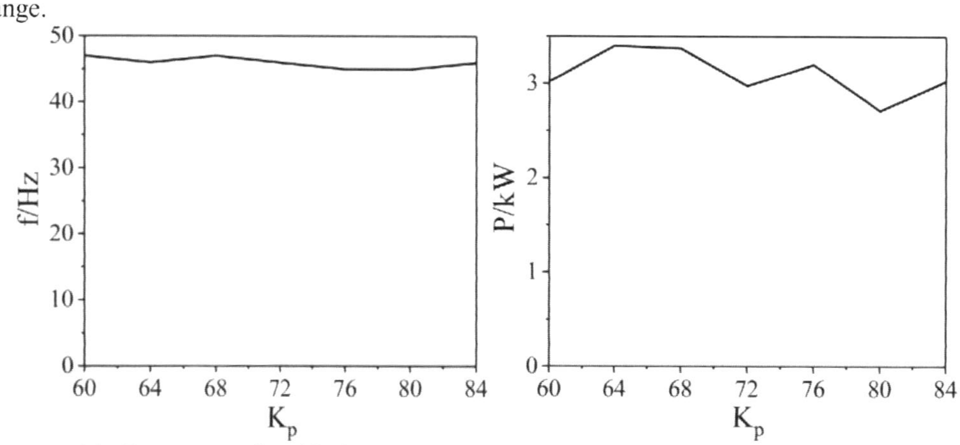

(a) Frequency of oscillation (b) Amplitude of oscillation

Figure 5. Influence of K_P on the subsynchronous oscillation.

In conclusion, under the existing system parameters and operation conditions, the change of PI parameters of PLL does not affect the sub synchronous oscillation characteristics of the system.

4. Conclusion

Through the above simulation, the following conclusions are obtained:

(1) For the doubly fed wind turbine with stator flux vector oriented control method on the rotor side, the subsynchronous oscillation characteristics of the system are different when different flux detecters are used.

(2) When the u-i flux detector ignoring the stator resistance is adopted, the subsynchronous oscillation component is very small, while the sub synchronous oscillation component can be observed when the stator resistance is not ignored. Changing the parameters of the u-i flux detector has no obvious effect on the subsynchronous oscillation.

(3) When the u-ω flux detector is used, the subsynchronous oscillation component of the system is larger than that of the u-i flux detector. Under the same operating condition, the u-ω flux detector has larger influence on the subsynchronous oscillation.

(4) The integral gain and the proportional gain of the PI link of the PLL in the u-ω flux detector do not have an obvious effect on the subsynchronous oscillation.

Acknowledgments
We are grateful to Dr Wen who put forward many valuable suggestions for revision, and to Miss Wang who provided a lot of help during the writing of the paper.

References
[1] Xie X.R, Liu H.K,He J.B, et al. Mechanism and Characteristics of Subsynchronous Oscillation Caused by the Interaction Between Full-converter Wind Turbines and AC Systems[J]. Journal of Chinese Electrical Engineering Science. 2016, 36(09): 2366-2372.
[2] Wang Z.W, Sheng C, Liu F, et al. Analysis on Impact of Doubly Fed Induction Generations with Different Phase Lock Mechanism on Power System Small Signal Stability[J]. Journal of Chinese Electrical Engineering Science. 2016, 36(09): 2366-2372.
[3] Zheng Y, Li Y. Stability analysis of doubly-fed wind power generation system based on phase-locked loop[Z]. 20082251-2254.
[4] Wang Z, Li G, Sun Y, et al. Effect of Erroneous Position Measurements in Vector-Controlled Doubly Fed Induction Generator[J]. IEEE Transactions on Energy Conversion. 2010, 25; 25(1; 1): 59-69.
[5] Ling Y, Modeling, Simulation and Control of Doubly Fed Wind Power Generation System[M]. China Machine Press. 2017.184-186.

Coordinated generation and transmission expansion planning approach considering probabilistic available transfer capability

Ping Zhou[1], Peng Kang[2], Ruiguo Chen[1], Hao Tian[1],

1. State Grid Chongqing Electric Economic Research institute, Chongqing 400000, China;
2. State Grid Chongqing, Tsinghua University, Chongqing 400000, China

153326062@qq.com; kp@cq.sgcc.com.cn; chenruiguo@cq.sgcc.com.cn; tianhao5@cq.sgcc.com.cn

Abstract. The available transmission capacity of the network has a great impact on the safety and reliability of the whole system. After the reform of electricity market, the grid is required to have the highest available transmission capacity (ATC) based on the existing load due to the uncertain factors. Therefore, it is necessary to consider the growing demand for available transmission capacity from the grid planning stage. Meanwhile, considering the uncertain factors have a great influence on the available transmission capacity (ATC), this paper first models the three uncertain factors and introduces the solving method of probabilistic available transfer capability, then proposes a grid expansion planning model considering the probabilistic available transmission capacity under the influence of uncertainties, which is applied to the six-node test system. Finally, the genetic algorithm is used to solve the final planning scheme. Compared with other planning schemes, this method can consider the comprehensive investment cost and the probability available transmission capacity. The case study shows that the method is effective and can achieve good results.

1. Introduction

The power market reform has become a trend in the development of the power industry in the world by breaking the traditional top-down, monopolistic power system management model, deregulating and introducing competition mechanisms to achieve optimal allocation of resources and improve the efficiency of power system operation. In this context, the calculation of existing transmission capacity becomes a necessary prerequisite for power market participants to complete power transactions [1]-[4]. The North American Electric Reliability Council (NERC) gives the definition of available transmission capacity (ATC) [5]: ATC refers to the remaining transmission capacity available for commercial use on the actual physical transmission network based on the existing transmission contract. In the traditional vertical control environment, the transmission capacity between regions is only a reference information. However, in the power market environment, the system operation uncertainty increases, the energy transaction changes rapidly, and the probability of failure will also increase. Therefore, it is necessary to study the available transfer capability (ATC) to improve the utilization rate of the power grid and optimize the grid structure. It can also provide important reference information for grid planning and operation.

At the same time, from the perspective of the power company itself, it is required that the power grid has the highest possible technical adaptability to the future external environment, and has the highest available transmission capacity on the basis of the existing load, that is, for all possible values of external factors that affect the available transmission capacity of the grid, the planning scheme should have sufficient capability to "disregard" their existence, that is, have sufficient ability to "eliminate" the uncertainty. This requires the available transmission capacity of the planning scheme to be as large as possible.

The calculation methods of ATC can be roughly divided into deterministic calculation methods and probabilistic calculation methods. The deterministic solution methods are mainly divided into DC power flow method (LPF) [6], Benders algorithm [7], optimal power flow method (OPF) [8]-[9] and continuous power flow method (CPF) [10]-[11]. The probabilistic calculation method combines the theory of probability theory and mathematical statistics analysis with the method of ATC solution, which is computationally intensive [12].

In reference [2], the conventional uncertainty factors such as generator output, load fluctuation and component failure are considered in the establishment of the ATC model, but the new energy factor is not considered. Reference [3] uses a hierarchical clustering algorithm to divide the sample samples, and adds wind power based on comprehensive consideration of conventional uncertainties, but does not take into account the correlation between wind power. In the latest research, the reference [4] and [13] used Latin hypercube sampling (LHS) sampling combined with Cholesky decomposition and Copula theory to model the correlation of input random variables in ATC calculation, respectively, which is convincing, but a little complicated.

Based on this, this paper firstly studies and establishes the mathematical model of uncertainty factors from load, generator and line components, then considers the influence of uncertainty factors, takes the probability of network available transfer capability (PATC) as the objective function, and takes the probability of network available transfer capability greater than the given threshold as the flexibility constraint of the model, establishes a grid flexible planning model that considers the network probability available transmission capacity (PATC). Finally obtain a grid planning scheme that meets the available transmission capacity level with a certain probability by genetic algorithm. The above method is used to expand the test system. The results of the example show that compared with other schemes, the model can minimize economic investment and increase the probability of available transmission capacity.

The structure of this paper is as follows. The second section introduces the uncertainty factor modeling method, the linear programming model for calculating the available transmission capacity (ATC), and the grid expansion planning model considering the probability available transmission capacity (PATC) and its solving algorithm. In the third section, the model system is validated by the model proposed in this paper. The fourth section is the conclusion of this paper.

2. The model formulation and its solving algorithm

2.1. Scenario generation method

The scenario generation in this paper considers the uncertainty of load, generator capacity and line fault. These three kinds of uncertainties cover three main basic components in the power network, which are highly representative.

1) Uncertainty model of loads

A probability model based on a normal distribution is used to represent the uncertainty of load growth. For the existing load node i, assuming that the original load is P_{Di0}, during the planning period, the change ΔP_{Di} of the load at that point is a random variable, obeying the normal distribution $\Delta P_{Di} \sim N(\mu_i, \sigma_i^2)$ where μ_i is the expected value, then the load at that point is $P_{Di} = P_{Di0} + \Delta P_{Di}$.

2) Uncertainty model of generation capacities

The discrete probability distribution is used to describe the uncertainty of generator capacity. For the power node i that may appear in the planning period, the probability of becoming a new power node is assumed to be p . For example, when the node i is determined to be a new power node, the installed capacity of the point is expected to be P_{Gik} , and the probability of each possible installed capacity is p_{ik} . In this way, the distribution of discrete random variables is:

$$\Pr(X = P_{Gik}) = p_{ik} \tag{1}$$

In the formula, $0 < p_{ik} < 1$, $\sum_{k=1}^{M} p_{ik} = 1$, $k = 1, 2, \ldots, M$, M is the possible installed capacity of each generating unit.

3) Uncertainty model of faults

The binomial distribution model is used to represent the uncertainty of line fault, in which 0 indicates that the line is in fault state and 1 indicates that the line is in normal operation state.

$$\Pr(X = k) = p^k (1 - p)^{1-k} \tag{2}$$

In the formula, $k = 0,1$, p is the forced outage rate of the line.
In the calculation, the corresponding load state, generator capacity state and line operation state are obtained by sampling according to the above three probability distributions.

2.2. Linear programming model for ATC calculation
Based on the DC power flow model, the linear programming model for calculating the available transmission capacity is as follows:

$$\max ATC = e^T d \tag{3}$$

$$s.t.\ \boldsymbol{Sf} + \boldsymbol{g} = \boldsymbol{l} + \boldsymbol{d} \tag{4}$$

$$f_{ij} - \gamma_{ij} n_{ij} (\theta_i - \theta_j) = 0 \tag{5}$$

$$\left| f_{ij} \right| \le n_{ij} f_{ij} \tag{6}$$

$$\boldsymbol{0} \le \boldsymbol{g} \le \overline{\boldsymbol{g}} \tag{7}$$

$$\boldsymbol{d} \ge \boldsymbol{0} \tag{8}$$

Where $\boldsymbol{l}, \boldsymbol{f}, \boldsymbol{g}, \boldsymbol{d}, \theta, \overline{\boldsymbol{g}}, \boldsymbol{e}$ are the load column vector, the line current column vector, the power generation output column vector, the load growth column vector, the node phase angle column vector, the power generation output upper limit column vector, and the unit 1 column vector, $\boldsymbol{S}$ is node branch association matrix, n_{ij}, γ_{ij}, $\overline{f_{ij}}$ are the number of lines, admittance values, and line limits between lines *ij*. By solving this model, the available transmission capacity of the network can be obtained.
Taking into account the random uncertainties in the power system, the probabilistic available transfer capability (PATC) is defined as follows: Firstly, based on the random characteristics of the power system, the possible operation modes of the system are determined by simulating the random breaking and load changes of the power transmission equipment, and then the appropriate ATC is used to solve the ATC of the system under these operating modes. Finally, a comprehensive analysis of the ATC values in each operating state gives the expected value of the system ATC value. In this paper, the

Monte Carlo simulation method and the linear programming method are combined to calculate the value of PATC. In the calculation process, each uncertain factor uses the mathematical model described in A. For the deterministic state obtained by each sampling, a deterministic ATC calculation method based on the linear programming model is used.

2.3. Flexible grid planning model considering PATC

The grid planning model considering the probabilistic available transmission capacity of the network under uncertain information is as follows:

$$Min\ v = \sum_{(i,j)\in\Omega} c_{ij}n_{ij} \tag{9}$$

$$Max\ PATC = E(e^T d) \tag{10}$$

$$\boldsymbol{Sf} + \boldsymbol{g} + \boldsymbol{r} = \boldsymbol{l} + \boldsymbol{d} \tag{11}$$

$$f_{ij} - \gamma_{ij}(n_{ij}^0 - n_{ij})(\theta_i - \theta_j) = 0 \tag{12}$$

$$\left|f_{ij}\right| \le (n_{ij}^0 + n_{ij})k_{ij}\overline{f_{ij}} \tag{13}$$

$$\boldsymbol{0} \le \boldsymbol{g} \le \boldsymbol{g}_{max} \tag{14}$$

$$\boldsymbol{0} \le \boldsymbol{r} \le \boldsymbol{l} + \boldsymbol{d} \tag{15}$$

$$\Pr(e^T d \ge h) \ge \beta \tag{16}$$

$$0 \le n_{ij} \le \overline{n_{ij}},\ d \ge 0 \tag{17}$$

Where $E, e, c_{ij}, n_{ij}^0, n_{ij}, \overline{n_{ij}}, \gamma_{ij}, \overline{f_{ij}}$ are respectively the expected value operator, 1 vector, the investment cost of adding a single line between the branches i-j, the number of original lines between the branches i-j, and the number of lines increased between the branches i-j, the maximum number of lines between the branches i-j, the admittance of a single line between the branches i-j, and the active transmission limit of a single line between the branches i-j, $\boldsymbol{l}$ is the predicted load active column vector. $\boldsymbol{g}_{max}$ is the generator active output upper limit column vector, $\boldsymbol{f}, f_{ij}, \boldsymbol{g}, \boldsymbol{r}, \theta$ are the active power column vector of the branch considering the available transmission capacity of the network, the active power between the branches i-j, the generator active output column vector, the minimum load shedding column vector, and the phase angle of the node i, $\boldsymbol{S}$ is node branch association matrix, $\boldsymbol{d}$ is an active power column vector that can be increased by each load node when considering the network probability available transmission capability, h is the minimum threshold of the available transmission capacity of the system considering the available transmission capacity, β is the confidence that the network available transmission capacity is greater than the minimum threshold h; Ω、Ω_s、v are the branch sets, load node sets and total investment costs of all lines that can be added.

Formula (9) and (10) are objective functions, which represent the minimum investment cost and the maximum available transmission capacity. In addition to the conventional power flow constraint, the constraint includes the probability constraint that the system PATC is greater than a given threshold (16) and the upper limit constraint of the number of available lines(17).

2.4. Solving algorithm

When solving the model (9)-(17), genetic algorithm is used to solve the model. Genetic algorithm is an adaptive heuristic global search algorithm based on the survival of the fittest rule and the mechanism of chromosome information exchange within the population in the process of biological evolution. It has been successfully applied in function optimization and other fields because of its simple, universal, robust, independent of problem model and suitable for parallel distributed processing. Genetic algorithm has been proved to have the ability to deal with nonconvex, nonlinear, mixed integer optimization problems [14], which is better than a series of mathematical methods in solving power grid expansion planning problems.

The general flow of using the genetic algorithm to solve the grid expansion planning model considering the probabilistic available transmission capacity is shown in Figure 1.

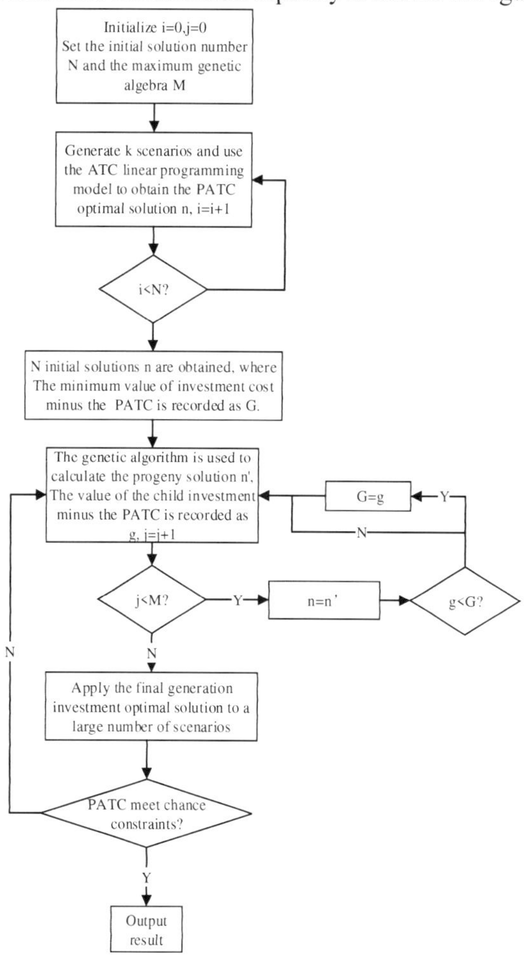

Figure 1. Flow chart for solving planning model considering PATC with genetic algorithm

First, using the ATC linear programming model to obtain N initial planning schemes as the initial solution of the genetic algorithm, the investment cost minus the PATC minimum is recorded as G, Then use the genetic algorithm to update the progeny solution and the investment cost minus the PATC minimum until the maximum genetic algebra is reached. Finally, the optimal solution is applied to a large number of scenarios. Finally, the optimal solution is applied to a large number of scenarios to verify whether the PATC of the scheme satisfies the chance constraint condition, and if so, the result is output, otherwise the calculation is recalculated.

3. Case studies

According to the above model of power grid expansion planning based on chance constraints, matlab program is compiled and applied to the test system for expansion planning. The system is a six node test system [15], the grid structure diagram of the original system is shown in Figure 2, the basic parameters of the original system such as load, generator and line are shown in Table 1, Table 2 and Table 3, and the expansion planning data is shown in Table 4.

All simulations have been run using GUROBI 6.5 under MATLAB, on a computer with two Intel Xeon E5-2680 (2.4 GHz) processors and 128 GB of random access memory.

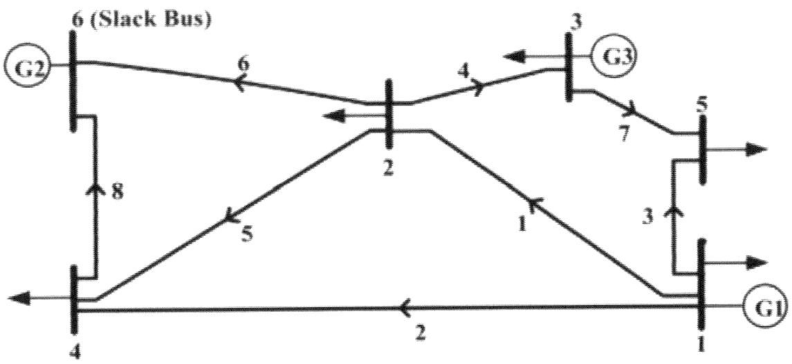

Figure 2. Six-node test system

Table 1. Bus load data

Bus	1	2	3	4	5	6
Load(MW)	95	284	47	190	284	0

Table 2. Generating unit data

Bus	Generating unit		
	Pmax(MW)	Pmin(MW)	Number of units
1	30	0	2
1	60	0	2
6	120	0	1
6	240	0	2
3	60	0	2
3	120	0	2

Table 3. Branch data

Branch	Branch			
	From bus	To bus	Reactance(P.U./100 MVA Base)	Capacity (MW)
1	1	2	0.4	100
2	1	4	0.6	80
3	1	5	0.2	100
4	2	3	0.2	100
5	2	4	0.4	100
6	2	6	0.3	100
7	3	5	0.2	100
8	4	6	0.3	100

Table 4. Extended planning data

Bus	Generating unit				
	Pmax(MW)		Pmin(MW)	Number of units	Price(¥)
1	180		0	1	8589200
3	180		0	1	8589200
5	180		0	1	8589200
Branch	Branch				
	From bus	To bus	Reactance(P.U./100 MVA Base)	Capacity (MW)	Price(¥)
9	2	3	0.2	200	872000
10	2	6	0.3	200	872000
11	3	5	0.2	200	872000
12	4	6	0.3	200	872000

The above model is solved by the process described in the second section, and the planning scheme 1 is obtained, which is compared with schemes 2 and 3, 1 means new construction, 0 means no construction. Considering the uncertain factors, 100 scenarios are extracted, and the total cost of investment and PATC in each scenario is calculated. The results are shown in Table 5.

Table 5. Results

Bus	Generating unit		Plan1	Plan2	Plan3
	Pmax(MW)	Pmin(MW)			
1	180	0	0	1	1
3	180	0	1	0	1
5	180	0	1	1	0
Branch	Branch				
	From bus	To bus			
9	2	3	1	1	1
10	2	6	1	1	1
11	3	5	1	1	1
12	4	6	1	1	1
Investment cost($\times 10^6$)			20.67	20.67	20.67
PATC(MW)			600.26	528.71	0

It can be found that using the planning model proposed in this paper, a power grid planning scheme that satisfies the available transmission capacity level of a certain probability can be obtained, and the probabilistic transmission capability of the network can be improved under a certain economic level, so that the comprehensive benefit of the planning scheme is optimal. Under the same investment cost, scheme 1 has the highest probability of transmission in 100 random scenarios, while the new route of scheme 2 is the same as scheme 1, and the change of generator position leads to a slight decrease in the probability of available transmission capacity. Scheme 3 also only changes the position of the newly built generator, and this change makes the probability available transmission capacity greatly reduced or even zero. It can be seen that the location of the new generator has a greater impact on the probability of transmission capacity.

In summary, the planning scheme obtained by the method proposed in this paper takes into account the investment cost and probabilistic available transfer capability (PATC), which is comprehensive and optimal, effective, and can achieve good results.

4. Conclusions

In this paper, three representative uncertain factors in power grid planning are studied and the specific mathematical models describing each uncertain factor is established, then a large number of random scenarios are generated. In addition, a grid expansion planning model considering the probabilistic available transmission capacity under the influence of uncertainties is proposed. Finally, the genetic algorithm is used to solve the model to obtain a power grid planning scheme that meets the transmission power level of a certain probability. The above method is used to expand the planning system. The results of the example show that compared with other schemes, the model can minimize economic investment under the condition of increasing the probabilistic available transmission capacity, and the overall benefit is better.

Acknowledgments
This work was supported in part by Key Technology Research and Application of Chongqing Power Grid Flexible Planning of Chongqing Electric Economic Research institute in 2019 (SGCQJY00GHJS1900074).

References

[1] Mu Yongzheng, Lu Zongxiang, Zhou Qinyong, et al. Optimized reliability balancing based transmission coordinating expansion planning for power grid with wind farms[J] . Power System Technology, 2015, 39(1): 16-22(in Chinese).
[2] Gao yajing. Research on available transfer capability considering uncertainties [D]. North China Electric Power University, 2007(in Chinese).
[3] Li Zhongcheng, Zhang Buhan, Duan Yao, et al. Fast calculation of probabilistic available transfer capability in power systems including large-scale wind farms[J]. Proceedings of the CSEE, 2014, 34(4): 505-513(in Chinese).
[4] Luo Gang, Chen Jinfu, Cai Defu. Probabilistic assessment of available transfer capability considering spatial correlation in wind power integrated system[J]. IET Gener Transm Distrib,2013, 7(12): 1527-1535.
[5] North American Electricity Reliability Council. Transmission transfer capability: a reference document for calculating and reporting the electric power transfer capability of interconnected electric systems. USA, 1995.
[6] Landgren G L,terhune H L,Angel R K. Transmission interchange capability analysis by computer IEEE[J]. Trans on Power Apparatus and Systems ,1992,91 (6): 748-754.
[7] Mohamed Shaaban, Liu H M, Li W X, et al. ATC calculation with static securityconstraints using Benders decomposition[J]. Proceedings of the CSEE, 2003,23(8): 7- 11.
[8] Pan X, Xu G Y. OPF based ATC calculation with static voltage stability constrains[J]. Proceedings of the CSEE, 2002, 22(11):86-91.
[9] Wang F, BaI X M. Opf based transfer capability calculation[J].Proceedings of the CSEE. 2002 ,22 (11) : 35-40.
[10] Breseti P, Lucarella D, Marannino P, et al. An OPF-based procedure for fast TTC analyses[C]. Power Engineering Society Summer Meeting, Chicago, 2002.
[11] Li G Q, Yao S W, Chen H H. Calculation of available transmission capability of AC / DC hybrid system based on interior point method[J].Automation of Electric Power Systems, 2009, 33(3):35- 39.
[12] Zhou ,M, Ran R J, LI G Y. Assessment of available transmission capacity for wind power grid - connected systems[J]. Proceedings of the CSEE, 2010(22): 14-21.
[13] Wang Jun, Cai Xingguo, Ji Feng. A simulation method of correlated random variables based on copula[J]. Proceedings of the CSEE, 2013, 33(22): 75-82(in Chinese).
[14] T.S. Chung, Y.Z. Li, Z.Y. Wang. Optimal generation expansion planning via improved genetic algorithm approach[J]. International Journal of Electrical Power and Energy Systems,2004,26(8).
[15] Yousefi G , S. Sayyadi S , Latify M A . Mid-term Vulnerability Analysis of Power Systems under Intentional Attacks[J]. IET Generation, Transmission & Distribution, 2016.

IOP Conf. Series: Materials Science and Engineering **853** (2020) 012005 doi:10.1088/1757-899X/853/1/012005

Coordinated generation and transmission expansion planning approach based on chance-constraints using genetic algorithm

Ping Zhou[1], **Peng Kang**[2], **Yuan Zhu**[1], **Ruiguo Chen**[1]

[1.] State Grid Chongqing Electric Economic Research institute, Chongqing 400000, China;
[2.] State Grid Chongqing, Chongqing 400000, China

153326062@qq.com; kp@cq.sgcc.com.cn; zhuyuan@cq.sgcc.com.cn; chenruiguo@cq.sgcc.com.cn

Abstract. The grid planning is a crucial and basic work in the power system, but there are various uncertain factors in the power system, which cause dramatic influences to the size of the grid power investment, as well as the power system security and stability for next few years. Traditional planning methods are gradually hard to adapt the grid planning under situations especially with large amounts of uncertainties. To reduce the impact of uncertain factors of the power system, it is necessary to consider increasingly significant uncertainties in system planning. However, if all extreme scenarios all taken into account, generally, the optimization results will be really conservative. Therefore, to prevent a relatively conservative optimization scheme, this paper proposes a grid-based expansion planning model based on chance constraints which take fully considerations of system uncertain factors using genetic algorithm. Compared with other planning schemes, this method comprehensively considers the investment cost and the level of load loss. Finally, a test system with six nodes is used to verify the effectiveness of the proposed method. The economics of the optimization scheme using the proposed approaches are observed using the test system.

1. Introduction

With the advancement of power market reforms, due to uncertainties such as future national policies, environmental changes, energy demand, and cyclical fluctuations in global economic development, the development of power grid planning schemes is beginning to face more and more complex uncertainties.

First of all, the traditional management mode of transmission, transmission and distribution has been broken. The site of the new power plant and its installed capacity have been decided by the power generation company. The grid planning workers cannot fully consider the impact of the power planning results in the power grid planning process, which makes the power supply become a major uncertainty in the power grid planning process. Secondly, with the continuous expansion of the breadth and depth of China's opening-up, the fluctuation of global economic development has more and more impact on China's economy [1]. The adaptability of the planned power grid based on the economy to this unpredictable load fluctuation obviously needs to be greatly reduced, therefore, the future load data is another unpredictable uncertainty factor in the process of power grid planning. Finally, an important factor affecting the reliability of power grid is line fault, which is also a kind of uncertainty factor.

IOP Conf. Series: Materials Science and Engineering **853** (2020) 012005 doi:10.1088/1757-899X/853/1/012005

In the method of dealing with uncertain factors [2]-[4], the paper [5] introduces the uncertainty of load and power supply into the mathematical model of transmission network planning, and obtains a more economic planning scheme which can adapt to the uncertainty of load and power supply at the same time. In reference [6], the uncertainty of load is treated by interval mathematics, and the load measurement value with error in interval number simulation is obtained, and the influence of the measurement error on the voltage value of power flow calculation is analyzed. In reference [7], the blind number BM model and its characteristics are introduced. The blind number is used to describe all kinds of uncertain information in power grid planning. A power grid planning model based on the blind number BM model is proposed, and a flexible power grid planning scheme with the best economy and reliability is obtained by using the cost-benefit analysis method. Based on the uncertainty of power supply and load, and taking into account the influence of environmental constraints, references [8]-[9] describe the adaptability of power grid planning scheme from multiple perspectives. In [11]-[11], from the aspects of stochastic theory, fuzzy set theory and credibility theory, the research results of domestic and foreign uncertain factor modeling methods and the research results of transmission network planning methods based on uncertainty theory are reviewed.
Based on this, this paper studies and establishes the specific mathematical model of the uncertain factors, including the uncertain factors from the load, generator and line components, and models them. Considering the influence of the uncertain factors, the chance constraint planning theory is introduced into the flexible planning of power grid, and the flexible expansion planning model of power grid based on the chance constraint planning is established. The above method is used to expand the planning system. The results of the example show that compared with other schemes, the model can minimize economic investment under the condition of satisfying the load demand as much as possible. The structure of this paper is as follows. The second section introduces the grid expansion planning model and solving algorithm based on the chance constraint. The third section uses the model proposed in this paper to verify the example system. The fourth section is the conclusion of this paper.

2. The model formulation and its solving algorithm

2.1. Scenario generation method

The scenario generation in this paper considers the uncertainty of load, generator capacity and line fault. These three kinds of uncertainties cover three main basic components in the power network, which are highly representative.

1) Uncertainty model of loads

A probability model based on a normal distribution is used to represent the uncertainty of load growth. For the existing load node i, assuming that the original load is P_{Di0}, during the planning period, the change ΔP_{Di} of the load at that point is a random variable, obeying the normal distribution $\Delta P_{Di} \sim N(\mu_i, \sigma_i^2)$ where μ_i is the expected value, then the load at that point is $P_{Di} = P_{Di0} + \Delta P_{Di}$.

2) Uncertainty model of generation capacities

The discrete probability distribution is used to describe the uncertainty of generator capacity. For the power node i that may appear in the planning period, the probability of becoming a new power node is assumed to be p. For example, when the node i is determined to be a new power node, the installed capacity of the point is expected to be P_{Gik}, and the probability of each possible installed capacity is p_{ik}. In this way, the distribution of discrete random variables is:

$$\Pr(X = P_{Gik}) = p_{ik} \tag{1}$$

In the formula, $0 < p_{ik} < 1, \sum_{k=1}^{M} p_{ik} = 1, k = 1, 2, \ldots, M$, M is the possible installed capacity of each generating unit.

3) Uncertainty model of faults

The binomial distribution model is used to represent the uncertainty of line fault, in which 0 indicates that the line is in fault state and 1 indicates that the line is in normal operation state.

$$\Pr(X = k) = p^k (1-p)^{1-k} \tag{2}$$

In the formula, $k = 0, 1$, p is the forced outage rate of the line.

In the calculation, the corresponding load state, generator capacity state and line operation state are obtained by sampling according to the above three probability distributions.

2.2. Planning model

In this paper, the objective is to minimize the investment cost of the transmission line, which allows the generated planning scheme to fail to meet the line overload constraint under some extreme ground conditions, but the probability of this situation must be less than a certain level of confidence. On this premise, a flexible planning mathematical model of power grid based on chance constraint planning is constructed.

$$Min\ invest = \sum_{(i,j)\in\Omega} c_{ij} n_{ij} \tag{3}$$

The objective function is to minimize the investment cost, where c_{ij} and n_{ij}^0 are the investment cost of adding a single line between branches i-j and the number of original lines between branches i-j .

$$s.t.\ \boldsymbol{Sf} + \boldsymbol{g} = \boldsymbol{l} \tag{4}$$

Equation (4) is constrained by Kirchhoff's first law, $\boldsymbol{S}$ is the node branch correlation matrix, $\boldsymbol{f}$ is the branch active power train vector, $\boldsymbol{g}$ is the generator active power train vector, $\boldsymbol{l}$ is the predicted load active random train vector.

$$f_{ij} - \gamma_{ij}(n_{ij}^0 + n_{ij})(\theta_i - \theta_j) = 0 \tag{5}$$

Equation (5) is constrained by Kirchhoff's second law, f_{ij} is the active power between branches i-j , γ_{ij} is the mobility of a single line between branches i-j , n_{ij}^0 is the number of original lines between branches i-j , n_{ij} is the number of lines actually added between branches i-j , and θ_i is the phase angle of node voltage.

$$\Pr(|f_{ij}| \le (n_{ij}^0 + n_{ij})k_{ij}\overline{f_{ij}}) \ge \alpha \tag{6}$$

Equation (6) is the limit crossing probability constraint of line power flow, which is based on the form of chance constraint, α is the set limit crossing probability value of line, $\overline{f_{ij}}$ is the upper limit of active power transmission of a single line between branches i-j , and k_{ij} is the active transmission load rate of a single line between branches i-j .

$$\mathbf{0} \le \boldsymbol{g}_1 \le \overline{\boldsymbol{g}} \tag{7}$$

Equation (7) is the upper limit constraint of generator output, and $\overline{\boldsymbol{g}}$ is the upper limit vector of generator active output.

$$\mathbf{0} \le \boldsymbol{r}_1 \le \boldsymbol{l} \tag{8}$$

Equation (8) is the node load shedding constraint, and $\boldsymbol{r}$ is the minimum load shedding column vector.

$$S_v^{(t)} < S_v^{\max},\ \boldsymbol{Sf} + \boldsymbol{g} = \boldsymbol{l}$$
$$t = 1, 2, \ldots m,\ v = 1,2, \ldots\ldots m,\ v \neq t \tag{9}$$

Equation (9) is the N-1 constraint of line fault, where $t = 1, 2, \ldots m$ represents the mode of m lines being disconnected respectively, and $v = 1, 2, \ldots m$ represents the running line. Inequality constraint $S_v^{(t)} < S_v^{\max}$ indicates that the apparent power of the line is less than its allowable transmission power $S_v^{\max}$ when line t is disconnected. This constraint indicates that when any line fails, there is no line out of limit or load shedding in the system.

$$0 \leq n_{ij} \leq \overline{n_{ij}} \tag{10}$$

Equation (10) is the upper limit of the number of circuits of the transmission corridor, and $\overline{n_{ij}}$ is the maximum number of lines that can be added between the branches i-j .

2.3. Solving algorithm

When solving the model equation reference goes here(3)-(10), genetic algorithm is used to solve the model. Genetic algorithm is an adaptive heuristic global search algorithm based on the survival of the fittest rule and the mechanism of chromosome information exchange within the population in the process of biological evolution. It has been successfully applied in function optimization and other fields because of its simple, universal, robust, independent of problem model and suitable for parallel distributed processing. Genetic algorithm has been proved to have the ability to deal with nonconvex, nonlinear, mixed integer optimization problems [11], which is better than a series of mathematical methods in solving power grid expansion planning problems.

The general process of using genetic algorithm to solve the flexible power grid planning model based on chance constrained planning is shown in Figure 1.

First of all, an initial planning schemes are obtained by using the minimum load shedding model as the initial solution of genetic algorithm. The optimal investment cost is recorded as g. the genetic algorithm is used to update the offspring solution and the optimal investment cost until the maximum genetic algebra is reached. The optimal solution is applied to a large number of scenarios to check whether it meets the chance constraints. If so, the result is output, otherwise the calculation is carried out again.

3. Case studies

According to the above model of power grid expansion planning based on chance constraints, matlab program is compiled and applied to the test system for expansion planning. The system is a six node test system [12], the grid structure diagram of the original system is shown in Figure 2, the basic parameters of the original system such as load, generator and line are shown in Table 1, Table 2 and Table 3, and the expansion planning data is shown in Table 4.

All simulations have been run using GUROBI 6.5 under MATLAB, on a computer with two Intel Xeon E5-2680 (2.4 GHz) processors and 128 GB of random access memory.

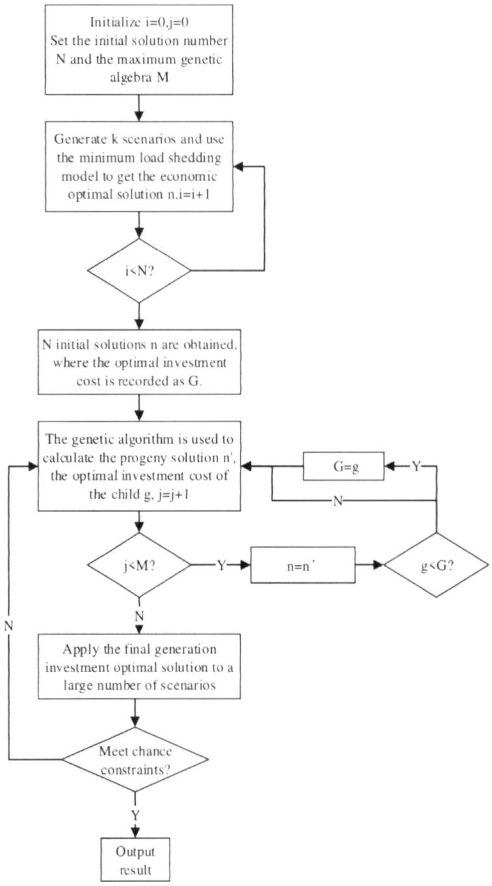

Figure 1. Flow chart of solving chance constrained power grid planning model with genetic algorithm

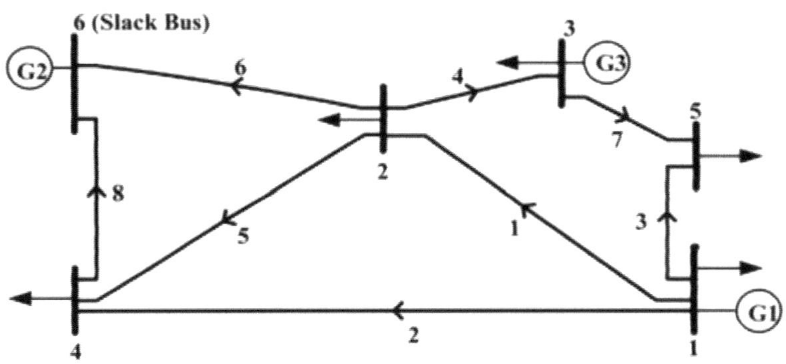

Figure 2. Six-node test system

Table 1. Bus load data

Bus	1	2	3	4	5	6
Load(MW)	95	284	47	190	284	0

Table 2. Generating unit data

Bus	Generating unit		
	Pmax(MW)	Pmin(MW)	Number of units
1	30	0	2
1	60	0	2
6	120	0	1
6	240	0	2
3	60	0	2
3	120	0	2

Table 3. Branch data

Branch	Branch			
	From bus	To bus	Reactance(P.U./100 MVA Base)	Capacity (MW)
1	1	2	0.4	100
2	1	4	0.6	80
3	1	5	0.2	100
4	2	3	0.2	100
5	2	4	0.4	100
6	2	6	0.3	100
7	3	5	0.2	100
8	4	6	0.3	100

Table 4. Extended planning data

Bus	Generating unit				
	Pmax(MW)		Pmin(MW)	Number of units	Price(¥)
1	120		0	1	4360000
3	180		0	1	8589200
5	180		0	1	8589200
Branch	Branch				
	From bus	To bus	Reactance(P.U./100 MVA Base)	Capacity (MW)	Price(¥)
9	2	3	0.2	200	872000
10	2	6	0.3	200	872000
11	3	5	0.2	200	872000
12	4	6	0.3	200	872000

The genetic algorithm process described in Section II is used to solve the problem and obtained that plan 1, 1 represents new construction and 0 represents no construction. Compared with plan 2 and plan 3, 100 scenarios are selected considering uncertainty factors. The total loss of load and total investment cost of each scenario are calculated. The results are shown in Table 5.

Table 5. Results

Bus	Generating unit		Plan1	Plan2	Plan3
	Pmax(MW)	Pmin(MW)			
1	120	0	1	1	1
3	180	0	0	1	0
5	180	0	1	0	1
Branch	Branch				
	From bus	To bus			
9	2	3	0	0	1
10	2	6	1	1	0
11	3	5	0	0	1
12	4	6	1	1	0
Investment			14.69	14.69	14.69

cost($\times 10^6$)				
Load loss(MW)		27.93	559.32	155.59

It can be found that the scheme obtained by using the chance constrained power grid expansion planning model allows load shedding in some extreme cases, so that the comprehensive benefit of the planning scheme is optimal. Under the same investment cost, the total load loss of scheme 1 is the smallest in 100 random scenarios. In scheme 2, the new line is the same as scheme 1, and the change of generator position leads to a great increase of load loss level. In scheme 3, the new generator position is the same as scheme 1, while the change of new line causes a slight increase of load loss level, but it is better than scheme 2.

To sum up, the planning scheme obtained by the method proposed in this paper considers both the investment cost and the level of loss of load, which is comprehensive optimal, effective and can achieve good results.

4. Conclusions

Aiming at the three typical uncertain factors in power grid planning, this paper studies and establishes a specific mathematical model to describe the uncertain factors, generates a large number of random scenes, considers the influence of uncertain factors, and establishes a flexible expansion planning model of power grid by using the chance constrained programming theory. The above method is used to expand the planning of the example system. The result of the example shows that compared with other schemes, the model proposed in this paper allows load shedding in extreme cases, and can minimize the economic investment and achieve the optimal comprehensive benefit of the planning scheme under the condition of satisfying the load demand as much as possible.

Acknowledgments

This work was supported in part by Key Technology Research and Application of Chongqing Power Grid Flexible Planning of Chongqing Electric Economic Research institute in 2019 (SGCQJY00GHJS1900074).

References

[1] Huifang Zhang. The role of load forecasting management in the new economic situation [J].Co-Operative Economy & Science, 2015(12):107.

[2] Haibao Zhai. Models and algorithms of electric power network flexible planning under multi uncertainty [D]. Doctoral Dissertation of ShanghaiJiaotong University，2007.

[3] Yan Zhang, Zhangchao Chen. An approach for transmission system planning incorporating uncertainties [J]. Power Syetem Technology, 1999, 23(3): 15-18.

[4] Haibao Zhai, Haozhong Chen, Chunlin Chen, et al. The Minimal Expectant Regret Concerned Electric Power Network Planning under Flexible Constraints [J]. Journal of Shanghaijiaotong University, 2005, 39(Sup.1):27-30.

[5] Escobar A H, Romero R A, Gallego R A. Transmission network expansion planning considering uncertainty in generation and demand[C]. Transmission and Distribution Conference and Exposition: Latin America, 2008 IEEE/PES. 2008 .1-6

[6] Wang Zian，Alvarado F L. Interval arithmetic in power flow analysis[J]. IEEE Trans. on Power Systems，1992，7(3)：1341-1349.

[7] Haozhong Chen, Haifeng Zhu, Jianmin Wang. Electric Power Networks Flexible Planning via the Blind Model (BM) of Unascertained Number [J]. Journal of Shanghaijiaotong University, 2003(09)：7-12.

[8] EI-Sheikhi, Farag Ali, Billinton, Roy. Load forecast uncertainty considerations in generating unit preventive maintenance scheduling for single system[C]. IEE Conference Publication, 1991, n 338: 241-245.

[9] Bresesti P, Capasso A, Falvo M C, et al. Power system planning under uncertainty conditions. criteria for transmission network flexibility evaluation[C]. Power Tech Conference Proceedings, 2003 IEEE Bologna .2003.2 :6.
[10] Alvarez J, Ponnambalam K, Quintana V H. Transmission Expansion under Risk using Stochastic Programming[C]. International Conference on Probabilistic Methods Applied to Power Systems, 2006. PMAPS 2006. 2006.1-7.
[11] Libo Zhang, Haozhong Chen, Pingliang Zeng, Qinyong Zhou, Lu Liu. Transmission Network Planning Approaches Based on Uncertainty Theories [J]. Automation of Electric Power Systems, 2016,40(16):159-167.
[12] T.S. Chung,Y.Z. Li,Z.Y. Wang. Optimal generation expansion planning via improved genetic algorithm approach[J]. International Journal of Electrical Power and Energy Systems,2004,26(8).
[13] Yousefi G , S. Sayyadi S , Latify M A . Mid-term Vulnerability Analysis of Power Systems under Intentional Attacks[J]. IET Generation, Transmission & Distribution, 2016.

A Novel Method for Implementing Harmonic Compensation with Multi-Inverters in a Micro-Grid

Xiang Shao

9 wenyuan road, Qixia district, Nanjing, Jiangsu province, China
Nanjing University of Posts and Telecommunications

shaoxiangnjupt@163.com

Abstract. This paper presents a new multi-inverter-based compensation scheme for the harmonic and reactive influences in a micro-grid, which not only realizes the integration of renewable energy, but also combines the functions of improving power quality at the point of common coupling (PCC) simultaneously. The proposed method uses the remaining power energy in the inverter to implement local harmonic compensation, which achieves the flexible customization of power quality and economics. Specifically, an accurate and stable modulated hopping Discrete Fourier Transform (mHDFT) algorithm with merits of computation of a sparse DFT result is applied to measure the harmonic current. In addition, a proportional resonant (PR) controller selected to track the command current due to its infinite gain for a specific point ensures there is no steady-state error (ess). Besides, delay compensation and discretization by pre-swapping bilinear transformation are performed to improve the stability of the current loop and enable the controller for digital implementation respectively. The simulation results based on Matlab/Simulink are given to verify the performance of the proposed method.

1. Introduction

Electrical Power has become the life-line of our civilization. However, the use of modern power electronic devices, such as switching power supplies and adjustable speed motor drives, have caused a serious power quality problem [1]. Among them, the potential harm of harmonics is quite serious, it will increase the cost of electricity production, transmission and usage, affect the service life of various components, generate electromagnetic interference, make the relay protection device think that the circuit is faulty and cut off the circuit, and reducing the accuracy of the measuring instrument [2]. In addition, harmonics will also generate resonance in the power grid, it could even cause spontaneous combustion of the line when the harmonic current is too much. Moreover, suitable equipment to implement the compensation is also needed. Therefore, it has been a major challenge to measure the harmonics accurately and compensate for it rapidly [3]. The methods to measure the harmonics are mainly based on the Discrete Fourier Transform (DFT) [4-6], instantaneous reactive power theory [7,8], Short Time Fourier Transform (STFT) and Wavelet Transforms[9-11]

At the beginning of the 21st century, the IEC electromagnetic compatibility standard requires the spectrum analysis only for the first 40 harmonics [12], so the calculation result of the FFT will be partially wasted. However, the Goertzel Algorithm [13] and the sliding DFT (SDFT) can be used.

These methods can compute sparse DFT results, such as only a single complex DFT spectral bin value for every N input time samples when the applications only cover a subset of all period points[14-16].
The most common method to improve the power quality in a micro-grid is passive governance, which uses additional power equipment (such as active power filter , unified power quality conditioner, dynamic voltage restorer) to compensate for harmonics and reactive power[17-20]. This paper present a new method to implement the harmonic compensation with multi inverters in micro-grid. The method combining the advantages of the two ways above and employs mHDFT algorithm to calculate the content of the required harmonic level, which not only reduces the amount of calculation, but also overcome instabilities and accumulated errors. Then a PR controller with delay compensation and discretization is designed to track the calculated command current, which make sure the accuracy of current loop. Finally, the method uses the remaining energy of the grid-connected inverter in the micro-grid (multiple inverters can be used in parallel) to perform fast and local compensation of harmonics and active current.
Compare to the traditional method, the strategy of the paper have the advantages that use the energy power which could be wasted to implement the compensation of harmonic and active current and the algorithm of mHDFT reduce much more calculation but not sacrifice the accuracy compare to the traditional FFT algorithm. The method of the paper improves the power quality of the micro-grid system and better adapts to the goal of flexible customization of power quality. The numerical experiments and simulation results demonstrate the precision and feasibility of the strategy.
The micro-grid act as an autonomous system with self-control and self-energy management can realize sending, distributing, and self-sufficient operation with distributed generation. It can also be flexibly switched between the two modes of grid-connected operation and isolated operation. However, harmonic and reactive power current in distributed generation systems and micro-grids seriously affect the power quality of the point of common coupling (PCC), even bring adverse effects into the control of the grid-connected inverters. To govern the quality of power at the PCC, the most common method is to install an active or passive filter, but this requires additional equipment and increases the size and cost of the system. Thus, this paper puts forward a new idea based on multi-converters.

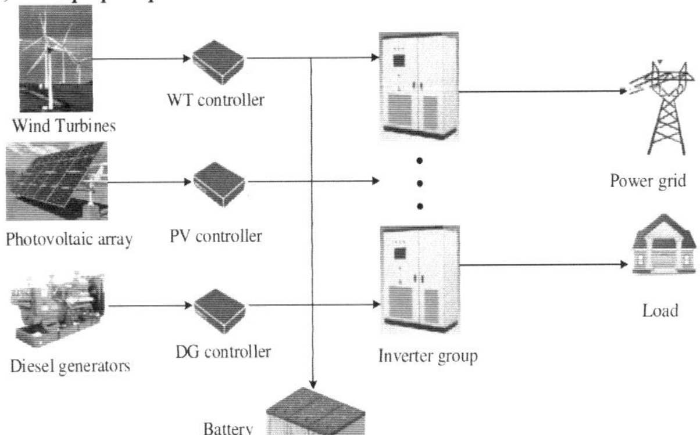

Figure 1. A campus Micro-grid with multi grid-tied inverters system
Figure 2 shows a campus micro-grid system diagram. The wind power generator and the photovoltaic array are connected as a distributed power source to the multi-grid inverter system, and the direct current (DC) is converted into alternating current (AC) by the inverter, and then connected to the large power grid and the load. It is not difficult to see that a part of the power will remain in the inverter when the inverter is working from the structure. According to the concept of this paper, the inverter also undertakes the function of improving the power quality of the micro-grid at the same time, that is, to use the remaining energy to compensate for harmonics and reactive current. At this time such a multi-inverter system is called a multi-function grid-tied inverter (MFGTI) by some scholars

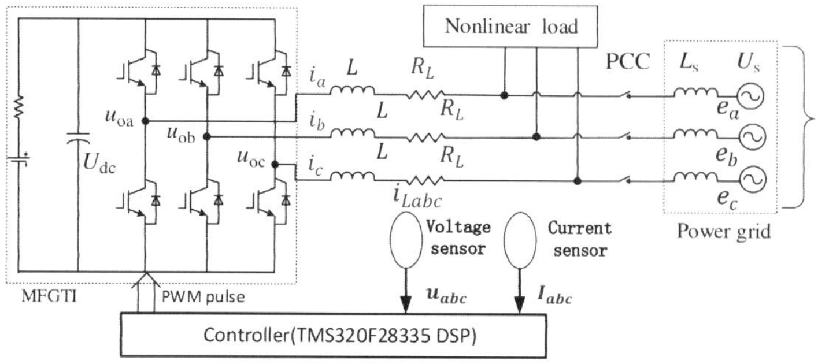

Figure 2. An MFGTI topology

Figure 2 show an MFGTI topology. The following equation can be obtained from this topology:

$$\begin{cases} L\dfrac{di_a}{dt} + R_L \cdot i_a = u_aN - e_a \\ L\dfrac{di_b}{dt} + R_L \cdot i_b = u_bN - e_b \\ L\dfrac{di_c}{dt} + R_L \cdot i_c = u_cN - e_c \end{cases} \tag{1}$$

Using the Clark transform and equation (1), the model of the controlled object in the αβ two-phase stationary coordinate system is obtained:

$$\begin{bmatrix} u_\alpha \\ u_\beta \end{bmatrix} = \begin{bmatrix} R_L & 0 \\ 0 & R_L \end{bmatrix} \begin{bmatrix} i_\alpha \\ i_\beta \end{bmatrix} + \begin{bmatrix} L & 0 \\ 0 & L \end{bmatrix} \frac{d}{dt} \begin{bmatrix} i_\alpha \\ i_\beta \end{bmatrix} + \begin{bmatrix} e_\alpha \\ e_\beta \end{bmatrix} \tag{2}$$

It can be expressed in vector form as follows:

$$u_{\alpha\beta} = R_L \cdot i_{\alpha\beta} + L \cdot \frac{di_{\alpha\beta}}{dt} + e_{\alpha\beta} \tag{3}$$

Assuming that the DC-side voltage remains constant for one switching period, and the converter modulating wave voltage exhibits zero-order retention characteristics in one switching cycle. Then a discrete model of a single-inductor grid-tied inverter in a stationary coordinate system can be obtained by combining the formula (3).

$$G_p^s(s) = \frac{1}{sL + R_L} \xrightarrow{zoh} G_p^s(z) = \frac{1}{R_L} \cdot \frac{1 - e^{-R_L \cdot T_s / L}}{z - e^{-R_L \cdot T_s / L}} \tag{4}$$

where T_s is the sampling period.

By analyzing the micro-grid system and topology of a single-inductor grid-tied inverter, some features can be found:

1. There will be residual energy when the multi-inverter system in the micro grid integrates renewable energy.
2. The discrete model of the single-inductance grid-tied inverter can contribute to the design of the current loop.
3. The load environment in the micro grid is ideal, most of them are air conditioners, computers, lights, etc., without large industrial equipment, which makes the power quality problem not very serious.

2. The Proposed Method

2.1. Harmonic Component Measurement

The DFT is the standard algorithm used for spectrum analysis. It is usually implemented by means of the fast Fourier transform, which reduces the amount of mathematical operations required and enhances the algorithm efficiency. Paper [14] also performed the DFT in an efficient way by using the SDFT, a recursive algorithm that calculates a unique spectral component of the signal over a sliding time window of length M.

the spectrum value of the kth frequency is:

$$X_n(\mathrm{k}) = W_M^k [X_{n-1}(\mathrm{k}) + \mathrm{d(n)}] \tag{5}$$

where $\mathrm{d(n)} = \mathrm{x(n)} - \mathrm{x(n - M)}$.

IOP Conf. Series: Materials Science and Engineering **853** (2020) 012006 doi:10.1088/1757-899X/853/1/012006

This is a single sliding DFT algorithm (the sliding step length L is 1), which adopts a sample-by-sample calculation. However, in some signal processing applications where real-time requirements are not strict, the single-step sliding DFT is computationally inefficient in such a low-time signal processing application.

Paper [16] promotes the sliding step size to any step length. In order to facilitate the derivation, consider the case of $L = 2^a, a \geq 0$. After L consecutive passes, the Hopping SDFT expression is:

$$X_n(\mathrm{k}) = W_M^{Lk}[X_{n-L}(\mathrm{k}) + D_n^L(k)] \quad (6)$$

The hopping SDFT is a recursive algorithm and computationally efficient, but the approach suffers from accumulated errors and potential instabilities caused by the finite precision representation of the twiddle factor.

Paper [15] finds that it is easy to observe that if frequency index k=0 when L=1, then the twiddle factor W_M^{Lk} will be null.

The X(k) DFT bin can be shifted to the index k=0 (zero Hz) by using the input signal x(n) to multiply a modulation sequence W_M^{km} in the time domain. The DFT output at time index n is calculated from the DFT result at the time $n - L$. Finally, the DFT result of the n-time frequency point k is obtained by phase correction.

Combining the advantages of papers [14-16], a high precision, permanently stable modulated HDFT algorithm is presented in paper [21]. The modulated HDFT after frequency domain shift process will be:

$$X_n(0) = X_{n-L}(0) + W_M^{km} G_n^L(k) \quad (7)$$

where $G_n^L(k)$ is the Revised updating vector transform (rUVT),

$$G_n^L(k) = \sum_{t=0}^{L-1} d(n-t)\, W_M^{-(t+L-1)k} \quad (8)$$

Therefore, the recursive method used in the mHDFT signal processing structure is shown in Figure 3.

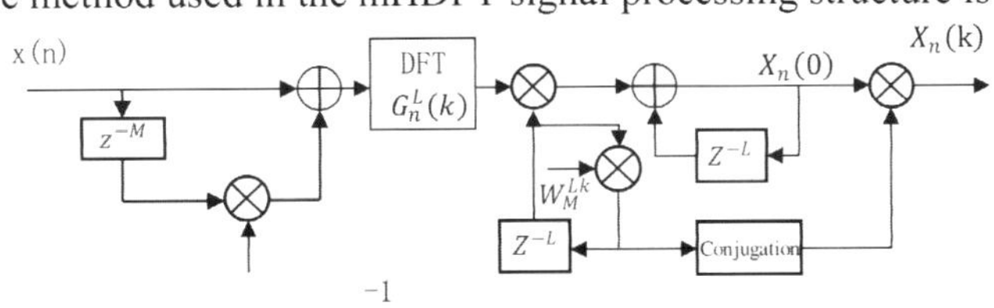

Figure 3. mHDFT structure with recursive computation of modulating sequence

Figure 4 shows the single-frequency-point mHDFT structure for L = 4 and M = 16.

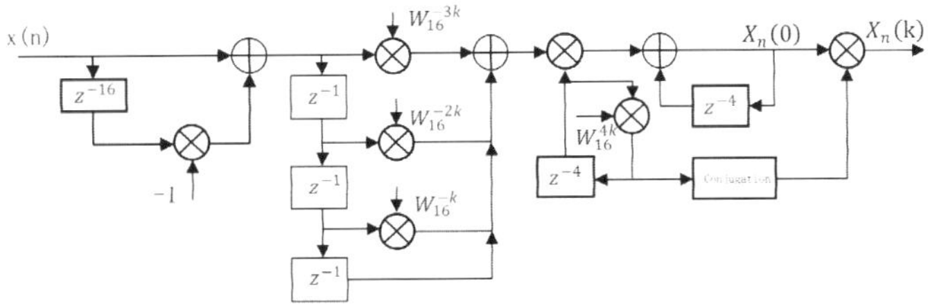

Figure 4. mHDFT structure for L = 4 and M = 16

2.2. Controller design

2.2.1 Controller choose. Figure 6 shows the whole system current loop structure, $G_C(z)$ represents the current controller (such as PR or VPI controller), and $G_{PL}(z)$ is the transfer function of the discrete domain of the controlled object.

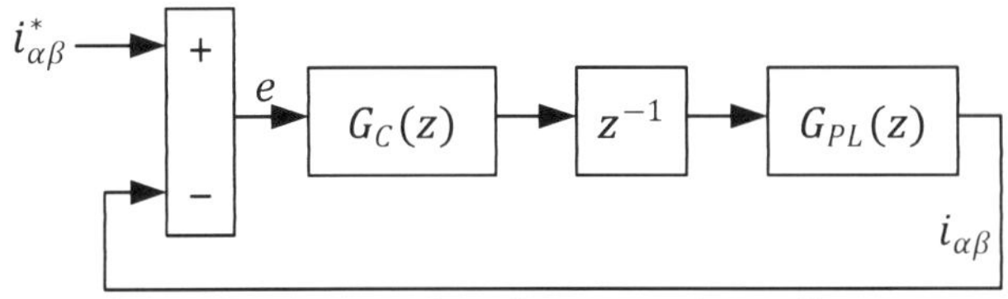

Figure 5. Discrete domain current loop diagram

$G_{PL}(z)$ is achieved by analyzing the topology structure of the MFGTI in section 2, the next step is to choose the current controller and the parameter design.

In this paper, considering that the proportional integral (PI) controller has infinitely gain for $\omega = 0$, thus ensuring there is no steady-state error. Nevertheless, the PR controller has infinite gain for the $\omega = \omega^*$ point to make sure there is no steady-state error, which is suitable to track command current. So the paper chooses the PR controller. The expression of the PR controller in the stationary coordinate system is:

$$G_{PR}(s) = K_P + K_i \frac{s}{s^2+\omega_e^2} \tag{9}$$

where K_P is the proportional integral gain, K_i is the resonant integral gain, ω_e is the resonant frequency.

2.2.2 The delay compensation of PR controller. Considering that the PR controller is superposed by Positive Sequence Synchronous Reference System PI (PS-SRF-PI) and Negative Sequence Synchronous Reference System PI (NS-SRF-PI). So the delay compensation principle of the PS-SRF-PI controller is extended to NS-SRF-PI. Thus, the block diagram of the delay compensation method of the PR controller is achieved. Where T_d is time constant of the delay term.

Further, using the SRF frequency transfer principle, combined with Figure 8, the PR controller transfer function after the delay compensation can be obtained, where φ is the compensation angle, which is defined as the phase difference between the after and before delay compensation at the resonance frequency ω_e.

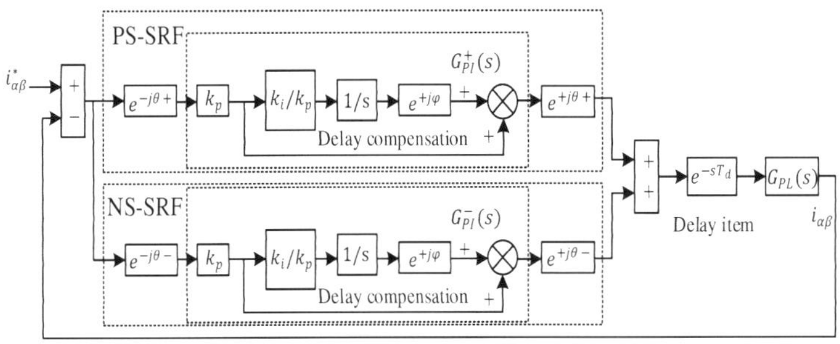

Figure 6. PR controller delay compensation

$$\begin{aligned} G^d_{PR}(s) &= 2k_p + 2k_i \cdot \frac{s \cdot \cos(\varphi) - \omega_e \cdot \sin(\varphi)}{s^2 + \omega_e^2} \\ &= K_p + K_i \cdot R^{1d}(s) \end{aligned} \tag{10}$$

For the delay compensation problem of the PR controller, paper [22] proposes that set delay compensation angle equal to the sum of the digital control one beat delay and the controlled object delay.

2.2.3 The discretization of PR controller. The analysis above is performed in a continuous domain, but the controller must be discretized in digital implementation. Paper [23] compared and analyzed the characteristics of eight discretization methods from the aspects of resonance frequency matching, resonance peak, system stability and delay compensation accuracy to research the discretization of the resonant controller. Finally, the paper uses a pre-swapping bilinear transformation to ensure the accuracy of the resonant frequency before and after the discretization.

$$s = \frac{\omega_e}{\tan(\frac{\omega_e T_S}{2})} \tag{11}$$

So:

$$R_{tp}^{1d}(z) = \frac{\frac{1}{2}(1 - z^{-2}) \cdot \cos(\varphi) \cdot \sin(\omega_e \cdot T_S) - (1 + 2z^{-1} + z^{-2}) \cdot \sin(\varphi) \cdot \left\{\sin(\frac{\omega_e \cdot T_S}{2})\right\}^2}{\omega_e \cdot (1 - 2z^{-1} \cdot \cos(\omega_e \cdot T_S) + z^{-2})} \tag{12}$$

The two formulas above can be combined to obtain the z-domain expression of the PR controller after the delay compensation.

$$G_{PR}^{d}(z) = K_p + K_i \cdot R_{tp}^{1d}(z) \tag{13}$$

As for the value of K_p and K_i, a rule called critical damping method in paper [24] can achieve nearly the minimum setting time in combination with negligible overshoot for reference change:

$$\begin{cases} K_p = k \cdot L \\ K_i = k \cdot R_L \end{cases} \tag{14}$$

where L is the value of the inductance, R_L is the value of the resistance and the coefficient k is calculated by:

$$k \approx 0.246 f_s \tag{15}$$

where f_s is the sample frequency. The PR controller based on the above design can better guarantee the tracking of the command current signal.

3. Simulation Results

In order to verify the correctness of the proposed algorithm on the MFGTI platform, this section is based on MATLAB/Simulink to build the corresponding grid-connected inverter model and the algorithm structure, to carry out experimental analysis and give experimental results to test the accuracy of proposed method in this paper. Figure 10 shows the calculation flow of the command current.

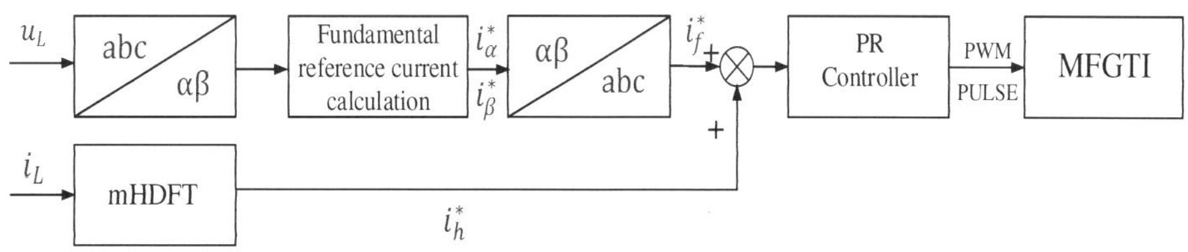

Figure 7. The calculation flow of the command current

The reference current consists of two parts: the active reference current and the harmonic reference current. The harmonic reference current is calculated by the mHDFT algorithm proposed above, and the active reference current is calculated as follows:

Firstly collecting a voltage signal between the grid and the load.

$$\begin{bmatrix} i_{fa}^* \\ i_{fb}^* \\ i_{fc}^* \end{bmatrix} = T_{2s-3s} * \begin{bmatrix} i_\alpha^* \\ i_\beta^* \end{bmatrix} \tag{16}$$

Thus, the command reference current is fully calculated:

$$i^* = i_f^* + i_h^* \tag{17}$$

A simulation model is built in Simulink according to the topology in Figure 2 and the current tracking method in Figure 5. The MFGTI is used to compensate for the reactive current and harmonic current at the PCC while implementing the inverter. Where the $R_L = 0.5\Omega$, L = 5mH, inverter switching frequency f = 10kHz, grid power command value is $P^* = 2kW, Q^* = 0var$. And the other system parameters are designed as section 2 and 3.

For convenience, only the phase A current is selected for the experimental results. The other two phases are the same as phase A, but with a phase difference. Figure 8 shows the current signal without any compensation, the 5th harmonic current is about 19.76% of the fundamental current, and the total harmonic distortion (THD) rate is 25.16%. The overall current distortion is more serious and the power quality is poor.

Figure 9 shows the grid-side current signal and its FFT analysis after the compensation of 5th harmonic. The 5th harmonic is almost zero after compensation, the current distortion and power quality have a significant improvement, and the THD is also reduced to 15.57%. However, this does not meet the requirement that THD must be less than 5%

For further governance, only a few MFGTIs are needed to be connected in parallel to compensate for the harmonics with large disturbances. As show in Figure 10, it is easily found that the current distortion is improved a lot after the compensation of several levels harmonics, the amount of each harmonic is significantly reduced, and the total THD is reduced to 2.48%.

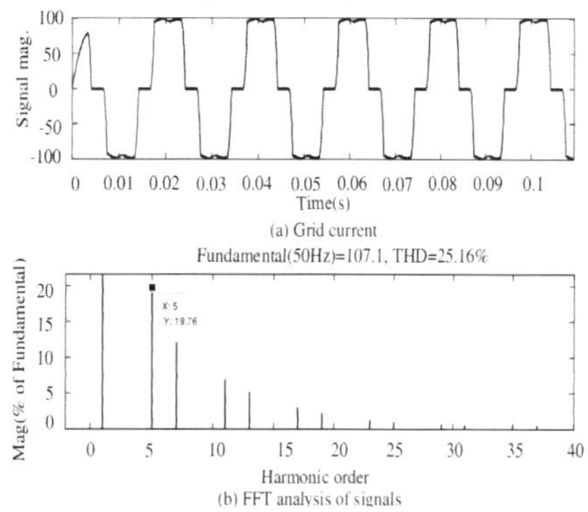

Figure 8. Uncompensated current signal and its FFT analysis

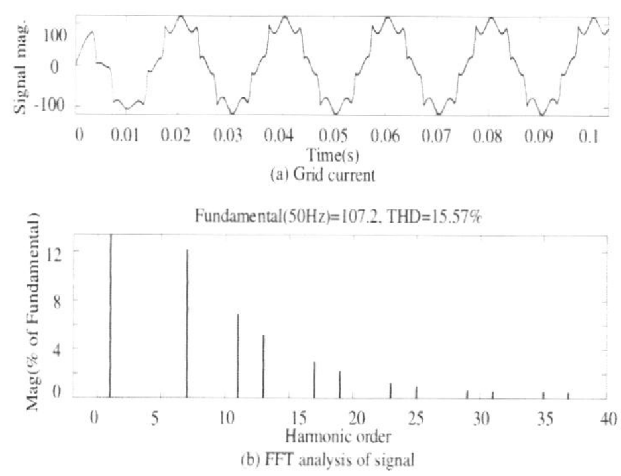

Figure 9. The current signal compensated 5th harmonic and its FFT analysis

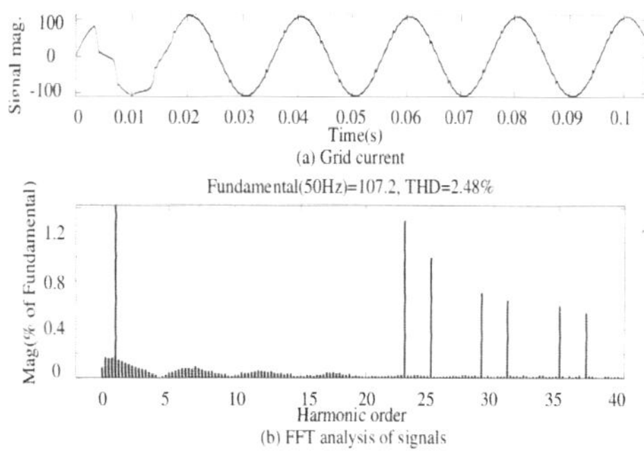

Figure 10. The current signal compensated multiple harmonic and its FFT analysis

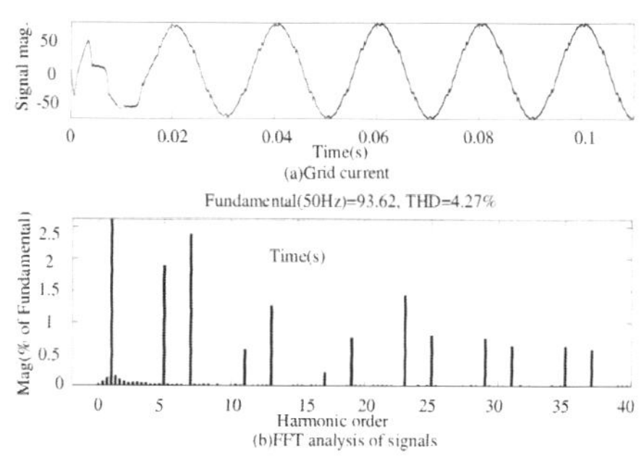

Figure 11. The current signal compensated multiple harmonic and active current and its FFT analysis

The compensation results of several levels harmonic and active current are shown in Figure 11. As the active injection, the fundamental component changed to 93.62 from 107.2, which cause the increase of total harmonic distortion, still below 5%.

4. Conclusion

A new strategy for harmonic measurement and compensation is presented in this paper. The method uses the remaining energy in grid-tied inverters to implement local harmonic compensation innovatively, with greatly reduced economic costs. The paper introduces the theory of recursive operation based on DFT to compute a single complex DFT spectral bin value for every M input time sample. Then use the DFT modulation property to exclude the accumulated errors and potential instabilities. Besides, this paper also extends the recursive interval from $L = 1$ to any step length to accommodate applications that do not require results output sample-by-sample. Moreover, this paper selects a PR controller according to the characteristics of a no steady-state error at the $\omega = \omega^*$ point, then a delay compensation and discretization are performed on the controller by pre-swapping bilinear transformation to improve the stability of the current loop and enable the controller to digital implementation. Most of all, several grid-tied inverters in parallel are used to achieve the goal of fast and local compensation.

The system is verified in MATLAB/Simulink with frequency steps (common in grid-connected transitions of micro-grid) and harmonics (produced by nonlinear loads) and the simulation results have demonstrated that the proposed method are robust and accurate in the case of highly distorted environments.

5. References

[1] A. Javadi and K. Al-Haddad, "A single-phase active device for power quality improvement of electrified transportation," IEEE Trans. Ind. Electron., vol. 62, no. 5, pp. 3033 – 3041, Feb.2015.

[2] F. G. Espin, E. Figueres, and G. Garcera, "An adaptive synchronous-reference-frame phase-locked loop for power quality improvement in a polluted utility grid," IEEE Trans. Ind. Electron., vol. 59, no. 6, pp. 2718 – 2731, Oct. 2012.

[3] H. Yoshida and K. Wada, "Third-Harmonic current suppression for power distribution systems under unbalanced installation of DG units," IEEE Trans. Ind. Electron., vol. 62, no. 9, pp. 5578 – 5585, Mar. 2015.

[4] B. N. Mohapatra and R. K. Mohapatra, "FFT and sparse FFT techniques and applications," 2017 Fourteenth International Conference on Wireless and Optical Communications Networks (WOCN). pp. 1 – 5, Feb. 2017.

[5] H. Wen, J. Zhang, Z. Meng, S. Guo, F. Li, and Y. Yang, "Harmonic estimation using symmetrical interpolation FFT based on triangular self-convolution window," IEEE Trans. Ind. Informat., vol.11, no.1, pp.16 – 26, Feb. 2015.

[6] Y. F. Wang and Y. W. Li, "Three-phase cascaded delayed signal cancellation PLL for fast selective harmonic detection," IEEE Trans. Ind. Electron., vol. 60, no. 4, pp. 1452 – 1463, Apr. 2013.

[7] T. Furuhashi, S. Okuma, and Y. Uchikawa, "A study on the theory of instantaneous reactive power," IEEE Trans. Ind. Electron., vol. 37, no. 1, pp. 86 – 90, Feb. 1990.

[8] M. Pradhan and M. K. Mishra, "Dual P- Q theory based energy-optimized dynamic voltage restorer for power quality improvement in a distribution system," IEEE Trans. Ind. Electron., vol. 66, no. 4, pp. 2946 – 2955, Apr. 2019.

[9] K. Satpathi, Y. M. Yeap, A. Ukil, and N. Geddada, "Short-time Fourier transform based transient analysis of vsc interfaced point-to-point dc system," IEEE Trans. Ind. Electron., vol. 65, no. 5, pp. 4080 – 4091, May. 2018.

[10] L. Coppola, Q. Liu, S. Buso, D. Boroyevich, and A. Bell, "Wavelet transform as an alternative to the short-time Fourier transform for the study of conducted noise in power electronics," IEEE Trans. Ind. Electron., vol. 55 , no. 2, pp. 880-887, Feb. 2008.

[11] X. Chen, "Real wavelet transform-based phase information extraction method: theory and demonstrations," IEEE Trans. Ind. Electron., vol. 56, no. 3, pp. 891 – 899, Mar. 2009.

[12] Electromagnetic Compatibility (EMC) Part 3-2, IEC Standard 61000-3-2, Aug. 2000.

[13] K. M. Singh, "Simultaneous estimation of moving-vibration parameters by sliding Goertzel algorithm in PLL technique," IEEE Trans. Instrum. Meas., vol. 68, no. 2, pp. 334 – 343, Feb. 2019.

[14] E. Jacobsen and R. Lyons, "The sliding DFT," IEEE Signal Processing Magazine, vol. 20, no. 2, Page s: 74 – 80, Mar. 2003.

[15] Krzysztof Duda, "Accurate, guaranteed stable, sliding discrete Fourier transform," IEEE Signal Process. Mag., vol. 27, no. 6, pp. 124 - 127, Nov. 2010.

[16] C. Park and S. Ko, "The Hopping Discrete Fourier Transform," IEEE Signal Process. Mag., vol. 31, no. 2, pp. 135-138, Mar. 2014.

[17] S. Rahmani, A. Hamadi, K. Al-Haddad, and L. A. Dessaint, "A combination of shunt hybrid power filter and thyristor-controlled reactor for power quality," IEEE Trans. Ind. Electron., vol. 61, no. 5, pp. 2152 – 2164, May. 2014.
[18] V. Cheung, H. Chung, and A. Lo, "A modular and scalable structure using multiparallel-connected series-voltage compensators for supply voltage regulation," IEEE Trans. Ind. Electron., vol. 31, no. 6, pp. 4096 – 4110, Jun. 2016.
[19] K. H. Kwan, P. L. So, an Y. C. Chu, "An output regulation-based unified power quality conditioner with Kalman filters," IEEE Trans. Ind. Electron., vol. 59, no. 11, pp. 4248 – 4262, Nov. 2012.
[20] A. M. Rauf and V. Khadkikar, "An enhanced voltage sag compensation scheme for dynamic voltage restorer," IEEE Trans. Ind. Electron., vol. 62, no. 5, pp. 2683 – 2692, May. 2015.
[21] Q. Wang, X. Yan and K. Y. Qin, "high-precision, permanently stable, modulated hopping Discrete Fourier Transform," IEEE Signal Processing Letters, vol. 65, no. 6, pp.748–751, Jun. 2015.
[22] Y. Alejandro G, L. Oscar, and F. Francisco D, "High-performance digital resonant controllers implemented with two integrators," IEEE Trans. Power Electron., vol. 26, no. 2, pp. 563-576, Feb. 2011.
[23] Y. Alejandro G, L. Oscar, and F. Francisco D, "Analysis and design of resonant current controller for voltage-source by means of Nyquist diagram and sensitivity function," IEEE Trans. Ind. Electron., vol. 58, no. 11, pp. 5006-5017, Nov. 2011.
[24] A. G. Yepes, A.Vidal, J. Malvar etc., "Tuning method aimed at optimized setting time and overshoot for synchronous proportional-integral current control in electric machines," IEEE Trans. Power Electron., vol. 29, no. 6, pp. 3041-3053, Jun. 2014.

IOP Conf. Series: Materials Science and Engineering **853** (2020) 012007 doi:10.1088/1757-899X/853/1/012007

Secondary control of island microgrid based on distributed multi-agent system under symmetric time delay

Bai Shaotong

No. 9 Wenyuan Road, Qixia District, Nanjing, Jiangsu, China

baixiaotong21920@163.com

Abstract. Aiming at the voltage and frequency instability caused by the primary control of the island mi-crogrid and the reactive power unevenness caused by the unbalanced circuit impedance, a quad-ratic control strategy based on the distributed multi-agent system is adopted. This paper is com-pleted to solve the voltage and frequency deviation due to droop control. The application of dis-tributed consistency algorithm under symmetric time delay is analyzed. The theoretical model is built in MATLAB/Simulink to verify the influence of delay on distributed consistency algorithm.

1. Introduction

At present, the microgrid technology is developing rapidly, and the microgrid can operate in both grid-connected and islanded states[1,2,3,4]. The three-layer microgrid control structure is the current mainstream architecture[5].In the traditional microgrid secondary control structure, a centralized communication structure is generally adopted, and a microgrid central controller (MGCC) is used to provide a reference value of the voltage frequency for the droop control. This approach has a number of disadvantages[6,7]:

(1) The requirements for the stability of the communication system are high. When a certain line fails, it will often lead to the overall flaw of the system;

(2) Strong dependence on the central controller (MGCC);

(3) “Plug and Play” capability is poor, and the use of microgrid is not used;

(4) Only the adjustment of a certain busbar can be completed.

These reasons lead to the fact that the distributed structure gradually replaces the centralized control structure and becomes a more mainstream microgrid control method[8]. In a distributed control structure, each distributed power source can be regarded as a node. Each node only needs to exchange information with its neighboring nodes that have communication links to complete its own stable operation and complete the microgrid It does not need to rely on the central controller like the centralized control structure. At present, the secondary control of microgrid based on distributed structure has already achieved some results[9,10,11]. In Document 1, The distributed control algorithm is used to control the voltage and frequency in the DC microgrid, and the distributed algorithm problem under the communication delay is considered, which solves the secondary control problem of the DC microgrid. In distributed secondary control, the time delay caused by communication connection is rarely taken into account, but in actual production life, the information transmission between distributed power sources will not be timely, so consider the communication delay. Distributed secondary control is very necessary. In the secondary control of distributed consistency algorithm, the influence of communication delay on secondary control is discussed. Finally, the relevant model was built in the simulation software and the simulation of the secondary control under the influence of delay was completed.

This paper is completed according to the following structure: The second part introduces the distributed structure of the primary control and secondary control of the microgrid. The third part introduces the construction of secondary control, the fourth part discusses the application feasibility of the distributed consistency algorithm in the case of time delay, and the fifth part is the communication structure of the model built on the experimental platform. The sixth part is The simulation results, the seventh part is the conclusion of the paper.

2. Microgrid distributed control structure

Figure 1 is a block diagram of the hierarchical control of the microgrid. It can be seen in the figure that the first-level control, that is, the local control is mainly responsible for stabilizing the inverter voltage; the secondary control mainly completes the correction of the voltage frequency deviation caused by the first-level control. This paper focuses on the study of primary control and secondary control and considered the impact of communication delay on the basis of traditional distributed secondary control. The microgrid model constructed by this is shown in Figure 2.

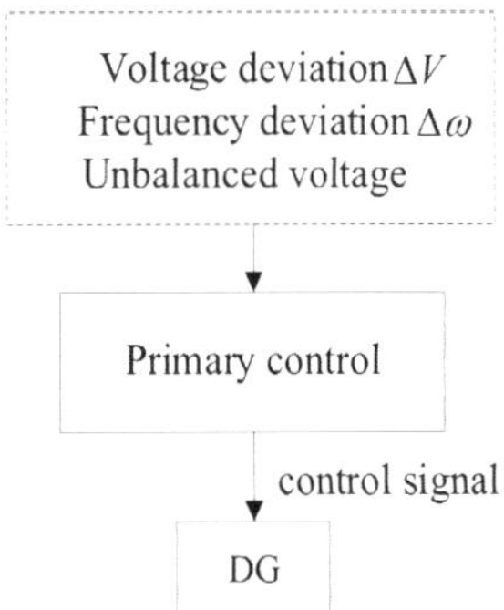

Figure 1. Distributed control structure hierarchy

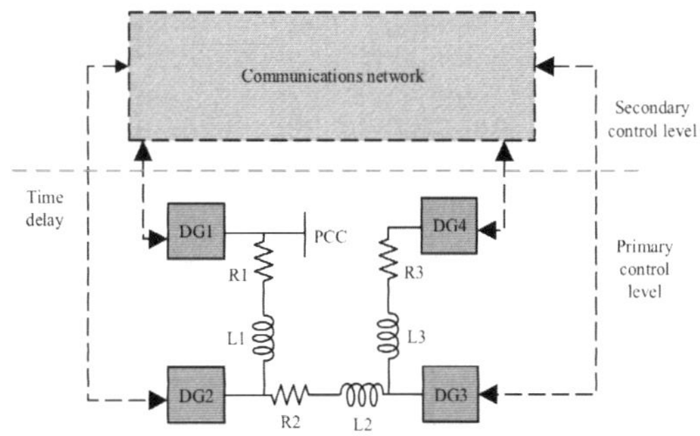

Figure 2. Microgrid secondary control structure with communication delay

The droop control shows the relationship between voltage and reactive power, frequency and active power. The voltage and frequency control is achieved as follows[12]:

$$\begin{cases} V_0 = V_n - nQ \\ \omega_0 = \omega_n - mP \end{cases} \tag{1}$$

In the equation, V_0 and ω_0 are the voltage amplitude and frequency generated by one control; V_n and ω_n are the reference values of the primary control of the voltage frequency, respectively, P and Q are the active and reactive power of the distributed power supply; m and n respectively The droop control coefficient corresponding to active and reactive power.

The active power and reactive power are calculated using the instantaneous power theory, and the DC component is used here. The DC component can be obtained by filtering the AC component of the first-order low-pass filter. The formula is as follows[13]:

$$\begin{cases} P = \frac{\omega_1}{s+\omega_1}(v_{od}i_{od} + v_{oq}i_{oq}) \\ Q = \frac{\omega_1}{s+\omega_1}(v_{oq}i_{od} - v_{od}i_{oq}) \end{cases} \quad (2)$$

In the equation, ω_1 is the cutoff frequency, and $v_{od}i_{od}$ and $v_{oq}i_{oq}$ are the d-axis and q-axis components of the distributed power supply output voltage and output current, respectively.
The resistance between the inverters has a certain influence on the droop control. If it is directly applied to the distributed control of the microgrid, it affects the stability of the system and the equalization of the power. The introduction of a virtual impedance loop in the line can attenuate the power coupling problem caused by the resistive load. The feedback of the voltage can solve the problem of reactive power unevenness caused by the impedance imbalance of the circuit, thereby improving system performance. The virtual impedance loop is designed as follows[14]:

$$\begin{cases} v_{Vd} = R_V i_{od}^+ - \omega L_V i_{oq}^+ \\ v_{Vq} = R_V i_{oq}^+ + \omega L_V i_{od}^+ \end{cases} \quad (3)$$

In the equation, $v_{Vd}v_{Vq}$ is the output voltage of the virtual impedance loop, R_V and L_V are the virtual resistor and the virtual inductor respectively. The "+" sign of the superscript represents the positive sequence component, and the negative sequence current passes through the virtual resistor, resulting in voltage imbalance, so do not join when calculating[15].In order to reduce the error caused by the uncertainty interference, the inner loop control uses robust control to enhance the stability of the system. The overall control block diagram of the system is shown in Figure 3.

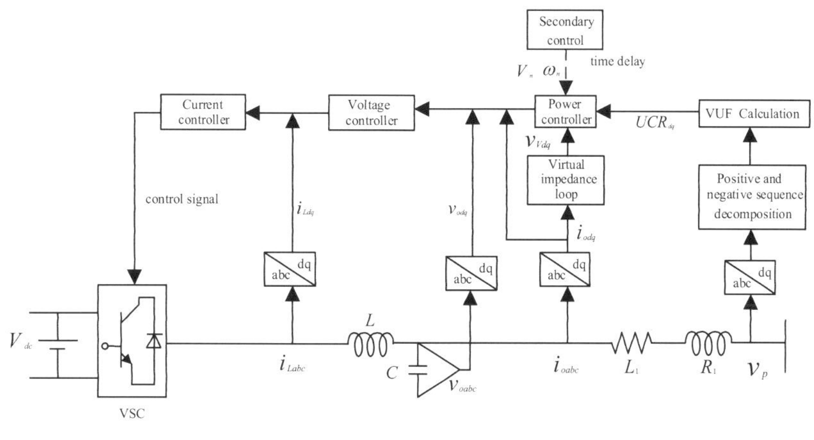

Figure 3. Microgrid primary/secondary control structure

3. Distributed cooperative secondary control

The traditional centralized control structure is shown in Figure 4(a). This control method has a high dependence on the central controller. Therefore, consider the distributed cooperative control structure,As shown in the distributed structure shown in Figure 4(b), distributed power supplies communicate with each other, simplifying the complexity of the organization and reducing communication requirements.

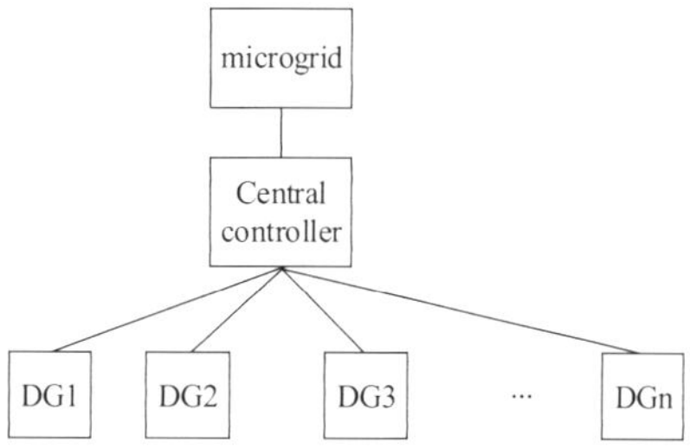

(a) centralized

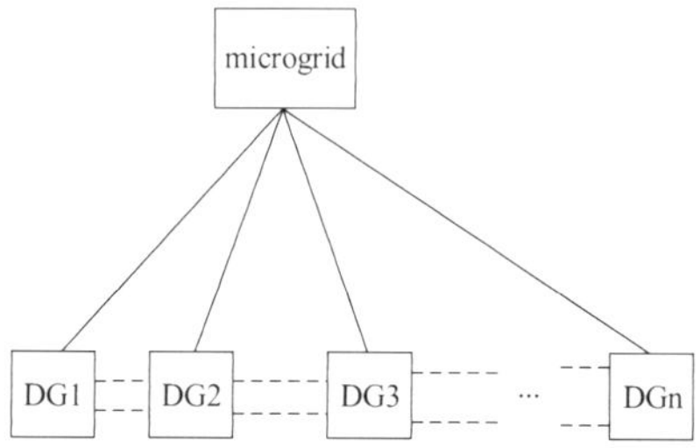

(b) distributed

Figure 4. Microgrid communication structure

3.1 Voltage controller design

The droop controller is constructed according to the aboveformula. In the dq coordinate system, the amplitude of the output voltage is:

$$v_{o,mag} = \sqrt{v_{odi}^2 + v_{oqi}^2} \qquad v_{oqi} = 0 \tag{4}$$

So the formula for first control can be modified to:

$$\begin{cases} v_{odi}^* = V_{ni} - n_i Q_i \\ v_{oqi}^* = 0 \end{cases} \tag{5}$$

In the formula, v_{odi}^*、v_{oqi}^* are the d-axis component and the q-axis component of the output voltage, and the subscript i represents the i-th distributed generation unit[16]. It can be seen from this that the control purpose of the secondary control is to make v_{odi} tend to v_{ref}, so that the frequency ω_i tends to ω_{ref}.
Since the voltage and current controller is in a state of dynamic balance, the above equation can be differentiated:

$$\dot{v}_{odi} = \dot{V}_{ni} - n_i \dot{Q}_i = u_{vi} \tag{6}$$

This process is called input-output feedback linearization, which can transform the voltage synchronization problem of N distributed power supplies into a first-order linear multi-agent synchronization tracking problem:

$$\dot{v}_{odi} = u_{vi},\ i = 1,2,3 \cdots N \tag{7}$$

The distributed power source DG in the microgrid is regarded as the target node. According to the knowledge of the graph theory, a directed graph can be used instead of the communication relationship between them. A leader node is set in the microgrid. This node provides the reference voltage and the reference frequency. Only a limited number of target nodes can receive the information of the reference node, and other nodes update the voltage and frequency through the communication connection. u_{vi}can be rewritten as:

$$u_{vi} = -c_v * e_{vi} = -c_v \left(\sum_{j \in M_j} a_{ij}(v_{odi} - v_{odj}) + g_i(v_{odi} - v_{ref})\right) \tag{8}$$

where in the formula, c_v is the control gain added to enhance the response speed of the system, M_j repre-

sents the set of adjacent nodes of the target node i, and the relationship between the nodes is constructed into a matrix, which is called adjacency The matrix A, in the adjacency matrix, the element a_{ij} represents the relationship between the nodes. If a_{ij}>0, it means that information is sent from node j to node i. When a_{ij}=0, it means that there is no relationship between them. The connection on the information. g_i is a gain coefficient. When g_i=1, it indicates that node i can receive the voltage reference signal sent by the leader node. If the signal cannot be received, g_i=0.
According to the proof in Document 5, the voltage reference value input to the droop controller is:

$$V_n = \int (u_{vi} + n_i \dot{Q}_i) dt \tag{9}$$

In the formula, $\dot{Q}$ is calculated as follows:

$$\dot{Q}_i = -\omega_c Q_i + \omega_c (v_{oqi} i_{odi} - v_{odi} i_{oqi}) \tag{10}$$

ω_c is the cutoff frequency of the low pass filter.
The specific structure diagram of voltage secondary control is shown in Figure 5.

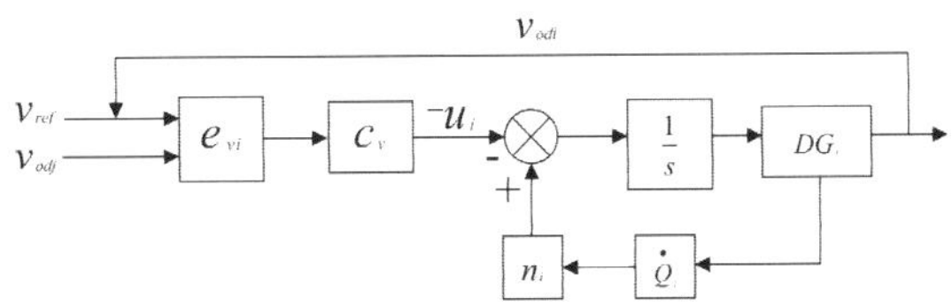

Figure 5. Voltage controller block diagram

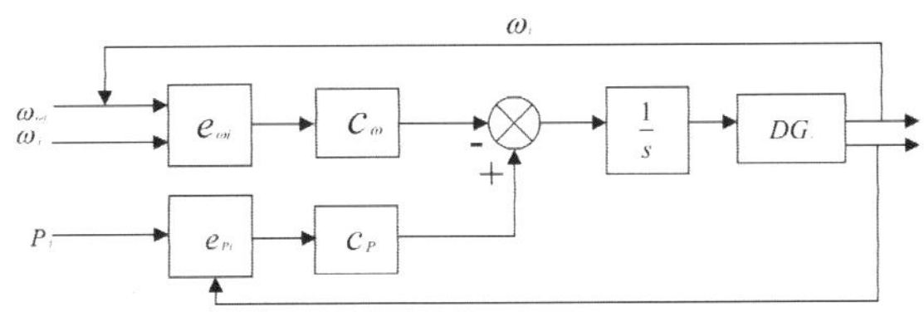

Figure 6. Frequence controller block diagram

3.2 Frequency quadratic controller design

The role of the frequency secondary controller is to cause the angular frequency ω_i of the distributed power source to tend to a given frequency reference value ω_ref. According to the definition of droop control:

$$\omega_i = \omega_{ni} - m_i P_i \tag{11}$$

Analogous to the design of the voltage quadratic controller, the equation can be differentiated to obtain the equation for frequency dynamic transformation:

$$\dot{\omega}_i = \dot{\omega}_{ni} - m_i \dot{P}_i = u_{\omega i} \; i = 1,2,3,\cdots,N \tag{12}$$

For $u_{\omega i}$,the frequency quadratic controller can be designed as follows:

$$u_{\omega i} = -c_\omega * e_{\omega i} = -c_\omega \left(\sum_{j \in M_j} a_{ij} (\omega_i - \omega_j) + g_i (\omega_i - \omega_{ref}) \right) \tag{13}$$

In the equation, c_ω is the control gain added to enhance the response speed of the system. Combining the above two modes, you can get the frequency input of the secondary control:

$$\omega_i = \int (u_{\omega i} + m_i \dot{P}_i) \, \mathrm{dt} \tag{14}$$

A slight difference from the voltage secondary control is that the frequency secondary control needs to achieve the equalization of the active power, that is, to achieve:

$$m_1 P_1 = m_2 P_2 = m_3 P_3 = \cdots = m_i P_i \; i = 1,2,3,\cdots,N \tag{15}$$

To maintain the balance of active power, the active power can be calculated as an input according to the idea of designing the voltage frequency controller:

$$m_i\dot{P}_i = u_{Pi} \tag{16}$$

$$u_{Pi} = -c_P * e_{Pi} - c_P\left(\sum_{j\in M_i} a_{ij}\left(m_i P_i - m_j P_j\right)\right) \tag{17}$$

Since the active power is not the input of the droop control, there is no intervention of the active power reference value in the calculation, and there is no gain term. Based on the above derivation, the design of the frequency secondary controller can be written as:

$$\omega_i = \int (u_{\omega i} + u_{Pi})\,\mathrm{dt} \tag{18}$$

The frequency control structure is shown in Figure 6.

4. Secondary control in case of delay

The traditional distributed consistency control algorithm is an ideal situation without considering the delay situation, but in the actual microgrid, communication delay always occurs inevitably. When considering the transmission delay, whether the existing distributed control algorithm can effectively complete the control of voltage and frequency is still a problem to be solved. Here, innovatively based on the traditional control algorithm, consider the transmission delay problem caused by problems such as data transmission and data processing. The structure of the continuity consistency algorithm is as follows:

$$\dot{x}_i(t) = -\sum_{j\in N_i} a_{ij}\left(x_j(t) - x_i(t)\right) \tag{19}$$

In the formula, $N_i$represents the adjacent agent, and a_{ij} represents the item in the adjacency matrix. When the two nodes are not connected, a_{ij} is 0, otherwise it is the weight of the corresponding connection edge. Here, the default is 1. The structural formula of the consistency algorithm can also be written as follows:

$$\dot{x} = -L \cdot x \tag{20}$$

In the formula, L is the Laplace matrix of the agent corresponding to the directed graph, $\mathrm{x} = [x_1, x_2, \cdots x_n]$. In actual calculations, for various reasons, the transmission between signals does not always arrive in real time, they often need to delay after a period of time to reach the destination, so consider the time delay in the secondary control of the distributed microgrid. It is very important. In this paper, considering the secondary control under transmission delay, a fixed time delay is added to simulate the transmission delay in the actual system based on the traditional distributed consistency algorithm. The algorithm structure at this time is as follows:

$$\dot{x}_i(t) = -\sum_{j\in N_i} a_{ij}\left(x_j\left(t - \tau_{ij}\right) - x_i(t - \tau_{ij})\right) \tag{21}$$

In the equation, τ_{ij} is the transmission delay of the simulation, where a fixed time is taken as the value of this item. For multi-agents, the system ultimately needs to achieve consistency, and it needs to meet certain conditions. When the system has time delay, the maximum time delay it can bear:

$$\tau_{max} = \frac{\pi}{2\lambda_n} \tag{22}$$

In the formula, λ_n is the maximum eigenvalue of the Laplace matrix corresponding to the multi-agent system.When the new transmission delay satisfies $\tau \le \tau_{max}$, the system can be operated in a steady state.

5. Distributed cooperative control communication structure

In the distributed topology, we choose each distributed unit to communicate only with its neighbors, specifying a specific node to receive reference information from the leading node. In this way, not only the communication connection is reduced, but also the communication burden caused by a large amount of complicated information exchange can be alleviated, and the information of each distributed node in the micro grid is included. In the process of simulation, we choose an island microgrid with four nodes. The topology of the microgrid is:

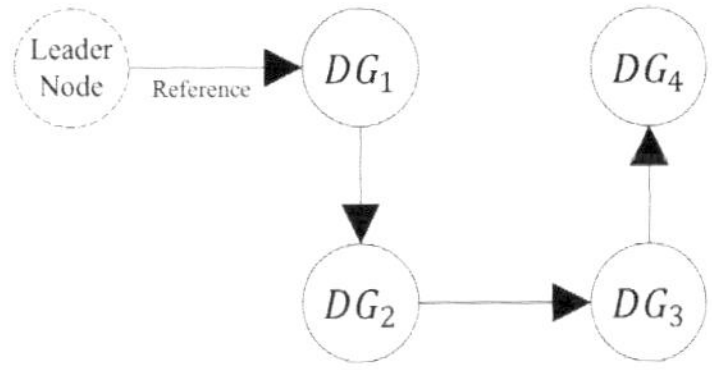

Figure 7. Frequence secondary controller block diagram

The adjacency matrix of the topology can be derived:

$$A = \begin{bmatrix} 0\,0\,0\,0 \\ 1\,0\,0\,0 \\ 0\,1\,0\,0 \\ 0\,0\,1\,0 \end{bmatrix} \tag{23}$$

The Laplace matrix of the multi-agent is:

$$L(G) = \Delta(G) - A(G) = \begin{bmatrix} 0\,0\,0\,0 \\ -1\,1\,0\,0 \\ 0-1\,1\,0 \\ 0\,0-1\,1 \end{bmatrix} \tag{24}$$

The maximum eigenvalue corresponding to this Laplace matrix is $\lambda_{max} = 1$. The maximum value of the delay obtained by calculation is[13]:

$$\tau_{max} = \frac{\pi}{2\lambda_{max}} = 1.57 \tag{25}$$

The simulation of the distributed multi-agent system is carried out with a delay of 0.5 seconds, and compared with the distributed multi-agent system without adding time delay, and the influence of the analysis delay on the secondary control of the distributed consistency algorithm is analyzed.

6. Simulation results

In order to verify the effect of delay on the distributed consistency algorithm, the model shown in Fig.2 is built in Matlab/Simulink. The grid operating voltage and frequency are set to 380V and 50Hz, and c_v、c_ω and c_p are set to 400. The simulation parameters are shown in Table 1.The microgrid operates in an island operation state. When there is no time delay in the transmission of voltage and frequency, the microgrid gradually enters a steady state after a period of time, and the voltage and frequency are basically stabilized at a preset value. When the delay link is added, there is some fluctuation in the voltage and frequency at this time, but eventually it can tend to set the value. As shown in the figure, after adding a time delay of 0.5 seconds, after applying the algorithm in this article, the voltage has stabilized before 0.5 seconds, and can still remain stable after a slight fluctuation at 0.5 seconds . As for the frequency, it was basically stable before 0.5 seconds, but there was still a certain degree of volatility. When the 0.5 second time delay appeared, it was obvious that the fluctuation was more severe and it could eventually approach a stable state. It shows that the method in this paper has a good regulation effect on island microgrids with time delay.

Table 1. System parameters used for testing

parameter	v_{dc}	m	n	V_n	ω_n	R_f	C_f
value	800V	9.4×10^{-5}	1.3×10^{-3}	311V	314rad/s	1Ω	600μf
parameter	L_f	L_c	k_{pv}	k_{pi}	k_{iv}	k_{ii}	
value	0.2mH	10mH	0.1	420	15	20000	

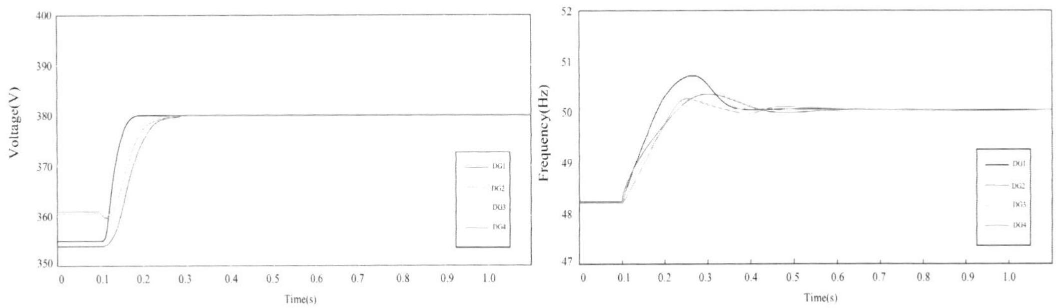

(a) Recovery of voltage deviation (b) Recovery of frequence deviation

Figure 8. Voltage/frequency with no time delay

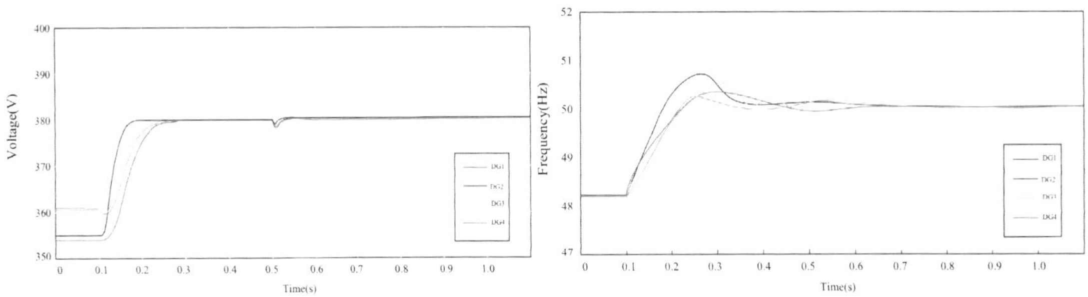

(a) Recovery of voltage deviation (b) Recovery of frequence deviation

Figure 9. Voltage/frequency with a delay of 0.5 s

7. Conclusion

In the quadratic control of the island microgrid based on the distributed consistency algorithm, the traditional consistency algorithm can solve the voltage and frequency deviation caused by the droop control, but it does not consider the impact due to the transmission time delay. Based on this, the paper considers the influence of delay and adds a fixed time delay to the control algorithm to simulate the delay phenomenon under actual conditions. The related model is built in Matlab/Simulink. The experimental results show that the fixed time delay does not affect the application of distributed consistency algorithm in quadratic control. When the fixed delay value is not greater than the calculated maximum At time lag, the system can still correct the deviation due to droop control.

Acknowledgments

This paper was completed under the careful guidance of my tutor Yue Dong. His serious research attitude and academic spirit, the work concept of excellence has deeply infected and inspired me. I would like to extend my most sincere gratitude and respect to my mentor. I would also like to thank my classmates and

friends who helped me during the writing process. It is because of your encouragement and help that I can overcome the difficulties and complete the writing of this paper.
Finally, I would like to express my sincere gratitude to the classmates and teachers who have cared and helped me.

References

[1] Dou C , Yue D , Zhang Z , et al. MAS-Based Distributed Cooperative Control for DC Microgrid Through Switching Topology Communication Network with Time-Varying Delays[J]. IEEE Systems Journal, 2017:1-10.
[2] Han H , Hou X , Yang J , et al. Review of Power Sharing Control Strategies for Islanding Operation of AC Microgrids[J]. IEEE Transactions on Smart Grid, 2015, 7(1):1-1.
[3] Olfati-Saber, R. Flocking for multi-agent dynamic systems: algorithms and theory[J]. IEEE Transac tions on Automatic Control, 2006, 51(3):401-420.
[4] Trends in Microgrid Control[J]. IEEE Transactions on Smart Grid, 2014, 5(4):1905-1919.
[5] Peng J , Xin A I , Yonggang W . Reactive Power Control Strategy of Microgird Using Potential Func tion Method[J]. Proceedings of the Csee, 2012, 32(25):44-51.
[6] Zhang Z , Chow M Y . Convergence Analysis of the Incremental Cost Consensus Algorithm Under Different Communication Network Topologies in a Smart Grid[J]. IEEE Transactions on Power Systems, 2012, 27(4):1761-1768.
[7] Nasirian V, Moayedi S, Davoudi A, et al. Distributed Cooperative Control of DC Microgrids[J]. IEEE Transactions on Power Electronics, 2014, 30(4):2288-2303.
[8] Weng S , Yue D , Dou C , et al. Distributed Event-Triggered Cooperative Control for Frequency and Voltage Stability and Power Sharing in Isolated Inverter-Based Microgrid[J]. IEEE Transactions on Cybernetics, 2018, PP(99):1-13.
[9] He J , Li Y W . An Enhanced Microgrid Load Demand Sharing Strategy[J]. IEEE Transactions on Power Electronics, 2012, 27(9):3984-3995.
[10] Singh B , Al-Haddad K , Chandra A . A review of active filters for power quality improvement[J]. IEEE Transactions on Industrial Electronics, 1999, 46(5):960-971.
[11] Shafiee Q , Guerrero J M , Vasquez J C . Distributed Secondary Control for Islanded Microgrids—A Novel Approach[J]. IEEE Transactions on Power Electronics, 2013, 29(2):1018-1031.
[12] Savaghebi M , Jalilian A , Vasquez J C , et al. Secondary Control Scheme for Voltage Unbalance Compensation in an Islanded Droop-Controlled Microgrid[J]. IEEE Transactions on Smart Grid, 2012, 3(2):797-807.
[13] Ovalle A, Ramos G, Bacha S, et al. Decentralized control of voltage source converters in microgrids based on the application of instantaneous power theory [J]. IEEE Transactions on Industrial Electronics,2015, 62(2): 1152-1162.
[14] Vasquez J C , Guerrero J M , Savaghebi M , et al. Modeling, Analysis, and Design of Stationary Reference Frame Droop Controlled Parallel Three-Phase Voltage Source Inverters[J]. IEEE Transactions on Industrial Electronics, 2013, 60(4):1271-1280.
[15] Graovac D , Katic V , Rufer A . Power Quality Problems Compensation With Universal Power Quality Conditioning System[J]. IEEE Transactions on Power Delivery, 2007, 22(2):968-976.
[16] Barrero F , Martinez S , Yeves F , et al. Universal and Reconfigurable to UPS Active Power Filter for Line Conditioning[J]. Power Engineering Review, IEEE, 2002, 22(8):72-72.

Double-layer Electricity Market Trading Strategy for Microgrids Based on Multi-agent System

Le Du, Xiaofei Zhu

Nanjing University of Posts and Telecommunications, Nanjing 210023,China

2817831123@qq.com; 1870559155@qq.com

Abstract. Aiming at the power generation and consumption of microgrid, this paper proposes a double-layer electricity market trading model for microgrids(MGs) based on multi-agent system(MAS). In the upper level, MG conducts power transactions with other MGs and grid companies in the electricity market. In the lower layer, the various components within the MG coordinate and cooperate to obtain the maximum profit. In order to the uncertainty problem in power forecasting, a rolling optimization model based on chance constrained programming is established by combining interruptible load and demand response in the real-time market. And the consequence of the case analyses by MATLAB simulation proves the rationality and validity of the model for optimal dispatch of virtual power plants given by this paper.

1. Introduction

Microgrids are small power systems that can more flexibly and reliably distribute energy over small geographic areas[1-3], typically using distributed energy resources (DERs), including distributed generator sets and energy storage facilities to meet local demand[4]. In addition to the environmental benefits of utilizing locally available renewable energy sources, MGs can also reduce transmission and distribution losses due to the physical proximity of DERs and loads. Therefore, microgrids play a dual role in the electricity market, depending on the amount of internal power generation and load demand[5]. When MGs generate more power than the load, the excess power is delivered to the large grid. And in the opposite case, they purchase electricity from the electricity market.

A number of novel models have been proposed methods and optimization techniques for improving electricity prices, regional bid strategies and RESs integration in the electricity trading market. Several game theory methods for microgrid energy trading are mentioned in [6-8]. These articles focus on solving the maximum profit problem of the microgrid using various game models, while ignoring internal optimization adjustments.

In addition, the uncertainty of power production and load forecasting in the microgrid also brings difficulties to the electricity market trading. Various forms of demand and supply issues have been studied in the literature to optimize the operation of MG in consideration of load demand forecasting, power generation forecasting and/or energy storage capacity[9,10]. However, due to the intermittent and large variability of renewable energy such as wind and solar energy, distributed generation using this unpredictable energy source is difficult to control. Especially with the large-scale deployment of renewable energy, it will be a huge challenge to balance the energy demand and supply while leveraging the capabilities of renewable energy.

To solve these proplem above, this paper proposes a double-layer electricity market trading strategy for microgrids based on MAS, taking microgrids consisting of photovoltaic cell(PV), wind turbines(WT), load and other distributed power generation resources(DR) as an example. The innovative model of the trading strategy is reflected in the following aspects:
1). Improved electricity market trading methods. Combined with multi-agent system, rationally distributes the power of the microgrid and eliminates fluctuations in electrical energy by means of grid companies. 2). Through the double-layer bidding model, the bidding and optimization process MG is realized, and the maximum profit is realized. 3). Reduce the impact of prediction errors through rolling optimization and improve the ability to absorb clean energy.

2. Double-layer electricity market trading model for microgrids based on MAS

2.1. Electricity market architecture for microgrids based on MAS

As a branch of artificial intelligence technology, MAS is a distributed autonomous system composed of multiple agents[11]. With the improvement of interaction, MAS provides a unified model and framework for various practical systems, which can well solve the communication problems of complex systems[12]. The MAS-based power market trading framework for MGs constructed in this paper is shown in Figure 1. It consists of the electricity market agent, microgrid agents and equipment agents. The layers are optimized and coordinated through information and energy exchange[13].

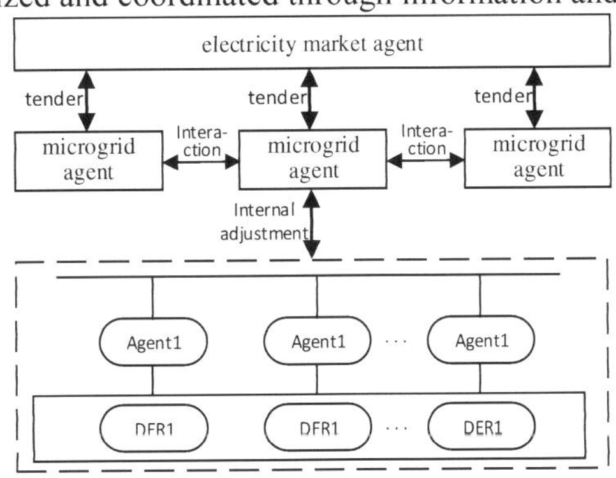

Figure 1. Microgrid structure diagram in the electricity market based on MAS.

1)Electricity Market Agent. Summarize the information of the purchase and sale of electricity and output range reported by the lower-level agent, consider the system security, calculate the electricity price decision, and send the electricity price and the transaction power result information to each MG agent, and monitor the MG agent's power status to judge Nash. Whether the condition of the equilibrium point is satisfied, and if it is satisfied, the process of the game is stopped.

2)Microgrid Agent. Combine the information of self-generation forecasting and internal load power forecasting to adjust the operation plan of the lower-level agent; electricity generation microgrid submits its bidding strategy, including bidding electricity price and power, while electricity purchase microgrid submits load demand; receive the clearing price and the bidding power issued by the power market agent and the trading strategy information of other MG agents; adjust the power generation/purchasing strategy with the goal of maximizing their own benefits, and finally determine the transaction amount of electricity to the micro grid and the sale/purchase amount to the grid company, and feedback the strategy to the superior agent.

3)Equipment Agent. PV Agent and WT Agent provide the prediction information of the clean energy generation to Microgrid Agent; DG Agent is responsible for providing MG Agent with information such as the operating status, output range, and power generation cost; Load Agent prioritizes according to the importance of the load, which will be different. And the type of load information will be sent to the upper agent. The equipment agent dynamically adjusts the power generation/load amount according to the control instructions of the upper Agent and its own constraint conditions, and feeds back its own power information to the upper-layer agent in time.

2.2. *Uncertainty characterization*

Wind and photovoltaic generation predictions can usually be described using Weibull distribution models[14] and Beta distribution models[15], and load prediction data is obtained through artificial neural networks[16]. Assuming that the prediction error of wind and solar power generation and load fluctuation are characterized by independent normal distribution with zero mean value, i.e.

$$\begin{cases} \delta_{WT}^{t} \sim N(0, \sigma_{WT}^{2}) \\ \delta_{PV}^{t} \sim N(0, \sigma_{PV}^{2}) \\ \delta_{L}^{t} \sim N(0, \sigma_{L}^{2}) \end{cases} \tag{1}$$

δ_{WT}^{t},δ_{PV}^{t}, and δ_{L}^{t} represent wind power output fluctuations and load fluctuations at time t, respectively. As the prediction accuracy increases with the decrease of the prediction time scale, this paper studies the optimal trading and real-time rolling optimization strategy to maximize the profit of microgrid power trading and balance the impact of prediction error.

2.3. *Double-layer bidding strategy for MGs*

The benefits of the MG W_i^t are mainly composed of two parts: the upper benefits W_{i-ext}^{t} from trading with other MGs or grid companies and the bottom revenue $W_{i-inner}^{t}$ determined by the internal benefits of the sale to the load and the cost of each power generation component.

$$\max W_i^t = \max(W_{i-ext}^{t} + W_{i-inner}^{t}) \tag{2}$$

2.3.1. Microgrid profit in the upper layer. The upper layer incom function of the MG in the electricity market W_{i-ext}^{t} can be expressed as:

$$W_{i-ext}^{t} = \begin{cases} \rho^t q_{i-M}^t + p_{i-PGs}^t (q_i^t - q_{i-M}^t) - e_{i-loss}^t (q_{i-M}^t)^2, & ,i \in n_s \\ \rho^t q_{i-M}^t + p_{i-PGb}^t (q_i^t - q_{i-M}^t) & ,i \in n_b \end{cases} \tag{3}$$

$$C_i^t(q_i^t) = 2a_i^t q_i^t + b_i^t \tag{4}$$

$$\rho_i^t = (1 + \mu_i) C_i^t(q_i^t), i \in n_s \tag{5}$$

$$\rho^t = \max(\rho_1^t, \rho_2^t, \dots, \rho_{n_s}^t) \tag{6}$$

$$\rho^t \in [\, p_{i-PGs}^t, p_{i-PGb}^t \,] \tag{7}$$

In the formula 3, q_{i-M}^t is the total amount of bid for the i-th MG, taking positive and negative respectively to represent the microgrid to sell and purchase electricity. q_i^t refers to the amount of electricity submitted to the power market. p_{i-PGs}^t and p_{i-PGb}^t represent the price of electricity absorbed and sold by the company. e_{i-loss}^t is the cost factor, which is converted from the network loss cost. n_s and n_b represent the number of power purchasers sellers in the microgrids. Equation 4 shows the marginal cost of MG $C_i^t(q_i^t)$, and a_i^t and b_i^t are their cost factors. ρ^t is the electricity clearing price.

To prevent the scalping behavior , MG only has one authority in the sale of electricity or electricity purchase at the same time and not allowed to purchase electricity with low electricity price and resell it with high electricity price. The following constraints(8-11) were proposed to achieve this condition.

$$q_i^t \in [\, q_{i,t}^{min}, q_{i,t}^{max} \,] \tag{8}$$

$$q_{i-M}^t \in [\, 0,\ q_i^t \,] \tag{9}$$

$$q_{i-M}^t \in [\, q_i^t,\ 0 \,] \tag{10}$$

$$\sum_{i=1}^{n_s} \rho_i^t = |\sum_{i=1}^{n_b} \rho_i^t| \tag{11}$$

2.3.2. Microgrid profit in the upper layer. Ignoring the cost of PV and WT generation, the objective function of the i-th MG $W_{i-inner}^t$ is.

$$W_{i-inner}^t = W_{i-L}^t - C_{i-DG}^t(q_{i-DG}^t) - C_{i-IL}^t(q_{i-IL}^t) = p_{i-L}^t(q_{i-L}^t - q_{i-IL}^t) - (a_{DG}(q_{i-DG}^t)^2 +$$

$$b_{DG}q_{i-DG}^{t} + d_{DG}) - (a_{IL}(q_{i-IL}^{t})^2 + b_{IL}q_{i-IL}^{t} + d_{IL}) \tag{12}$$

In the formula, q_{i-L}^{t}, q_{i-IL}^{t} and q_{i-DG}^{t} represent the power generation of L, IL, and DG respectively. The cost of IL includes the sales revenue and compensation cost of the cutting load. p_{i-L}^{t} is the internal selling electricity price. a_{DG}, b_{DG}, and d_{DG} are the power generation cost coefficients of DG. a_{IL}, b_{IL}, and d_{IL} are the interrupt cost coefficients of IL.
The reserve price p_{i-load}^{t} for selling its own load is defined and the actual load price p_{i-L}^{t} is considered a more favorable price. Besides, Equation 14 guarantees that the microgrid will not lose money. And Equation 15 is the power balance constraint. It is worth mentioning that q_{i-L}^{t}, q_{i-PV}^{t} and q_{i-WT}^{t} here refer to the predicted values of load power, PV and WT power generation.

$$p_{i-L}^{t} = \min(p_{i-load}^{t}, \rho^{t}) \tag{13}$$

$$p_{i-L}^{t} \geq C_i^t(q_i^t)/q_i^t \tag{14}$$

$$q_i^t + q_{i-L}^{t} - q_{i-IL}^{t} = q_{i-PV}^{t} + q_{i-WT}^{t} + q_{i-DG}^{t} \tag{15}$$

Considering the actual load and power generation, the electriciy market transactions for MGs should also meet the following constraints. $q_{i-DG,t}^{max}$, $q_{i-DG,t}^{min}$, R_{DG}^{U} and R_{DG}^{D} are the Upper and lower sections of power generation and climbing rate of DG. η is the maximum call rate of IL.

$$q_{i-DG,t}^{min} \leq q_{i-DG}^{t} \leq q_{i-DG,t}^{max} \tag{16}$$

$$-R_{DG}^{D}\Delta t \leq q_{i-DG}^{t} - q_{i-DG}^{t-1} \leq -R_{DG}^{U}\Delta t \tag{17}$$

$$0 \leq q_{i-IL}^{t} \leq \eta q_{i-L}^{t} \tag{18}$$

2.4. *Real-time rolling optimization strategy*

During the rolling optimization process, the MG performs direct load control by adjusting the output of the controllable unit DG and the price-based demand response DR to correct the error between the operational plan and the expected result. The objective function is

$$\min F = a_{DG}\,(q_{i-DG}^{t})^2 + b_{DG}q_{i-DG}^{t} + d_{DG} + a_{PV}q_{i-P}^{t} + a_{WT}q_{i-W}^{t}a_{IL}(q_{i-IL}^{t})^2 + b_{IL}q_{i-IL}^{t} \\ + d_{IL} + a_{DR}(q_{i-DR}^{t})^2 + b_{DR}q_{i-DR}^{t} - W_{i-ext}^{*} \tag{19}$$

$$W_{i-ext}^{*} = \begin{cases} \rho^t q_{i-M}^{t,*} + p_{i-PGs}^{t}(q_i^t - q_{i-M}^{t,*}) - e_{i-loss}^{t}(q_{i-M}^{t,*})^2, & i \in n_s \\ \rho^t q_{i-M}^{t,*} + p_{i-PGb}^{t}(q_i^t - q_{i-M}^{t,*}), & i \in n_b \end{cases} \tag{20}$$

q_{i-P}^{t}, q_{i-W}^{t} is the amount of abandoned wind and abandoned light generated by PV and WT. a_{PV} and a_{WT} are the cost coefficients of abandoning light and wind. W_{i-ext}^{*} is the revenue of microgrids in the current power market transactions, including revenue from participating in the electricity market and internal load sales. $q_{i-M}^{t,*}$ is the actual winning power of MG in the market in the past few days. The constraints(8-11,13-18) are also required.
In addition, in order to achieve the dual goals of reliability and economy, it is also necessary to meet the opportunity constraint planning($set\ \delta$ 0.95 $in\ this\ paper$).

$$\mathrm{P}\{\,(q_{i-WT}^{t} + \delta_{WT}^{t} - q_{i-W}^{t}) + (q_{i-PV}^{t} + \delta_{PV}^{t} - q_{i-P}^{t}) + q_{i-DG}^{t} + q_{i-IL}^{t} + q_{i-DR}^{t} \geq q_i^t + \\ (q_{i-L}^{t} + \delta_L^t)\,\} \geq \delta \tag{21}$$

In some cases, adjusting the output of components such as DG and DR through real-time scrolling optimization still cannot meet the day-ahead interactive power limit. At this point, it is necessary to update the day-to-day trading model and dynamically correct the trading power with other microgrids and grid company according to the forecasted data of wind and photovoltaic power output and load.
The tasks of the real-time rolling phase include: adjusting the MT output, determining the amount of abandoned wind and IL removal, and determining the total trading power and DR.

3. Opitimization process and solution

The Nash equilibrium game strategy of the upper and lower layers of the microgrid can be expressed as

$$W_i^t(S_i^{t,*}, S_{-i}^{t,*}) \geq W_i^t(S_i^t, S_{-i}^{t,*}), i \in U(n_s, n_b) \tag{22}$$

$$\boldsymbol{q}_i^{*} = \arg\max W_{i-inner}^{t}(\boldsymbol{q}_i) \tag{23}$$

In the electricity market, the bidding strategy for each microgrid is $S_i^t(q_i^t, \rho_i^t)$ (sufficient power) or $S_i^t(q_i^t)$ (lack of power). Use $S_i^{t,*}$ to represent the optimal strategy of the microgrid i, and $-i$ denotes other MGs apart from microgrid i. q_i represents the power generation strategy of the internal controllable components, including the output of PV, WT and DG and the power consumption of L and IL. n is the total number of internal controllable components inside the microgrid.

Since the above model involves the random variable prediction and the opportunity constraints of the rolling optimization stage, it is a solution to convert it into an equivalent determination form and reuse the linear/nonlinear programming problem. However, in practical applications, especially when there are many random variables, this conversion is more difficult. Therefore, this paper uses a genetic algorithm to solve.

For the microgrid in the electricity market and the internal game model, the equilibrium can be achieved through a finite number of games, so a nested genetic algorithm is used to solve the problem and the game flow chart is shown in Figure 2.

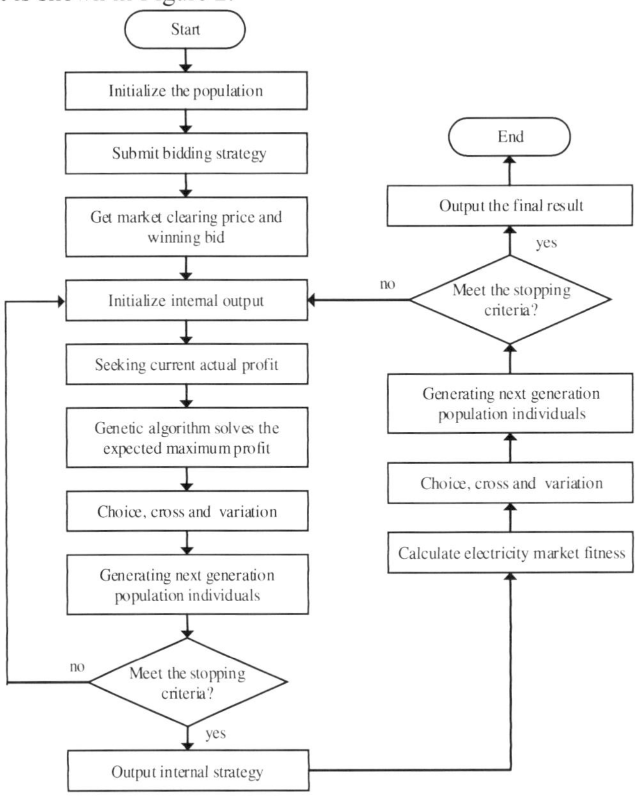

Figure 2. Flowchart of the algorithm.

4. Case Analysis

This part simulates the trading strategy of the electricity market. Assume that there are five microgrids in the area. Each MG consists of PV, WT, DG, L and IL, as is shown in Figure 3.

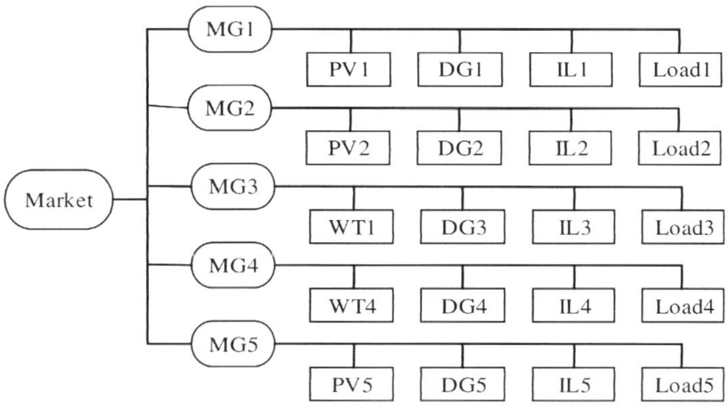

Figure 3. structure chart of of the power market

The cost parameters of each DG are shown in Table 1. The maximum call rate of IL is 10%, $a_{IL} = 0.005$, $b_{IL} = 0.06$, and $d_{IL} = 1.32$. DDR accounts for 5% of the total load, $a_{DR} = 0.006$, $b_{DR} = 0.06$, and $d_{DR} = 1.38$. The electricity sale and purchase price of the electricity market are 0.4 yuan and 0.7 yuan, and the load standby electricity price p_{i-load}^{t} is 0.55 yuan. The daily and real-time predicted fluctuation variances of WT, PV output and load demand are 0.3, 0.3 and 0.05.

Table 1. Related parameters of DG

	a	b	d	Pmin	Pmax	RmtU	RmtD
DG1	0.63	0.103	4.02	100	300	80	80
DG2	0.724	0.176	3.03	60	180	60	60
DG3	1.11	0.215	4.03	0	120	40	40
DG4	0.984	0.167	2.27	80	200	60	60
DG5	1.304	0.329	4.03	0	90	30	30

4.1. Day-ahead double-layer trading strategy

The bidding power and real winning power of MGs result are shown in the figure4.

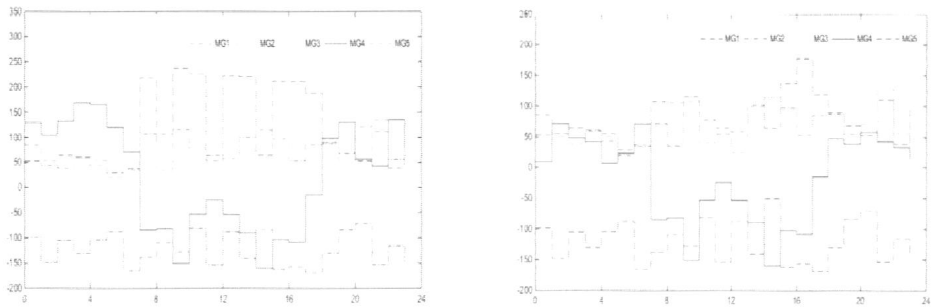

Figure 4. Comparison chart of bidding power and real winning power of MGs

MG1 and MG2 always sell electric energy while MG5 always purchases electrical energy in the electricity market. MG3 and MG4 sell or purchase electric energy at different times according to the internal load power consumption. And the bidding power is slightly different from the actual winning bid for MGs.

For convenience of comparison, this paper considers two trading strategies: microgrids operate independently and directly cooperate with the grid company to purchase/sell electricity for maximum profit or microgrids trade with others and grid companies. The cost and benefits of each microgrid are shown in Table 2.

Table 2. Cost and income statement for each MG

Index	Cost		Income	
	Strategy 1	Strategy 2	Strategy 1	Strategy 3
MG1	1246.76	1308.01	2002.58	2246.97
MG2	1054.55	973.07	1491.78	1477.69
MG3	736.73	689.61	952.21	956.79
MG4	1401.56	1259.03	2065.3	1983.41
MG5	789.26	709.75	869.13	996.83
Total	5228.86	4939.47	7381	7661.69

It can be seen from the table that the cost coefficients of MG1, MG2 and MG4 are small, the profit is relatively high, and the profit of MG3 and MG5 is small. Besides, the income of the three microgrids using the microgrid trading model proposed in this paper is higher than the respective revenues under the individual operating conditions, which is an effective way of income distribution and can improve the economic benefits of VPP operation.

4.2. Real-time scrolling optimization results

This section is the result of verifying real-time scrolling optimization. Considering the MG scheduling results directly determined by daily optimization and or by rolling optimization proposed in this paper two modes. The results of the two modes of scheduling and the required spare capacity for rotation are shown in Figure 9, using MG3 as an example.

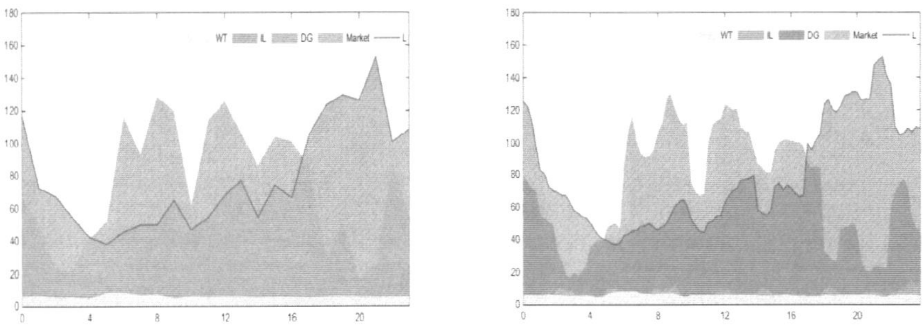

Figure 5. Optimization results by strategy1 and strategy2

As can be seen from Figure 6, when using the forecast data of the day before, the running result is smoother than Strategy 2 because there are fewer sampling points (24 data points in 24 hours). In real-time rolling optimization, the microgrid can dynamically correct the amount of distributed power and removable load based on current forecast data while satisfying operational constraints. The specific power purchase and power sales strategies also change.

5. Conclusion

The double-layer electricity market trading strategy for microgrids proposed in this paper shows that,

1) The two-layer optimization model can make the microgrid more profitable.

2) The appropriate economic load reduction during peak hours or insufficient power generation can improve the overall economic benefits of the system and adding certain economic compensation to users can increase the enthusiasm of users.

3) Consider rolling optimization strategy and opportunity constrained planning, which can more accurately coordinate the distributed power output and improve system reliability.

Acknowledgments

I would like to acknowledgement the valuable cooperation of Dr.Chen Jianbo and my classmate Zhu XiaoFei in the preparation of this application note.

References

[1] Hatziargyriou N , Asano H , Iravani R , et al. Microgrids[J]. IEEE Power and Energy Magazine, 2007, 5(4):78-94.
[2] Rana M , Li L . An Overview of Distributed Microgrid State Estimation and Control for Smart Grids[J]. Sensors, 2015, 15(2):4302-4325.
[3] Shayeghi H, Shahryari E. Integration and management technique of renewable energy resources in microgrid. Springer International Publishing; 2017.
[4] Ajao A , Luo J , Liang Z , et al. Intelligent Home Energy Management System for Distributed Renewable Generators, Dispatchable Residential Loads and Distributed Energy Storage Devices[J]. 2017.
[5] Park S , Lee J , Hwang G , et al. Event-Driven Energy Trading System in Microgrids: Aperiodic Market Model Analysis With a Game Theoretic Approach[J]. IEEE Access, 2017, 5:26291-26302.
[6] Zhao M, Shen C, Liu F, et al. A game-theoretic method for prediction of microgrid growing trends in China[J]. Science China, 2015, 58(10):1760-1766.
[7] Nian L , Jing Z , Jie W , et al. A Trading Model of PV Microgrid Cluster Based on Cooperative Game Theory[J]. Transactions of China Electrotechnical Society, 2018.
[8] G. E. Asimakopoulou, A. L. Dimeas, and N. D. Hatziargyriou, "Leaderfollower strategies for energy management of multi-microgrids," IEEE Trans. Smart Grid, vol. 4, no. 4, pp. 1909–1916, Dec. 2013.
[9] Hytowitz R B , Hedman K W . Managing solar uncertainty in microgrid systems with stochastic unit commitment[J]. Electric Power Systems Research, 2015, 119:111-118.
[10] A. Chaouachi, R. M. Kamel, R. Andoulsi, and K. Nagasaka, "Multiobjective intelligent energy management for a microgrid," IEEE Trans. Ind. Electron., vol. 60, no. 4, pp. 1688–1699, Apr. 2013.
[11] Azevedo R D, Cintuglu M H, Ma T, et al. Multi-Agent Based Optimal Microgrid Control Using Fully Distributed Diffusion Strategy[J]. IEEE Transactions on Smart Grid, 2016, PP(99):1-1.
[12] Sun Qiuye, Han Renke, Zhang Huaguang, et al. A multiagent-based consensus algorithm for distributed coordinated control of distributed generators in the energy internet[J]. IEEE Transactions on Smart Grid, 2015, 6(6): 3006-3019.
[13] Wei Chao, Fadlullah Z M, Kato N, et al. GT-CFS: A game theoretic coalition formulation strategy for reducing power loss in micro grids[J]. IEEE Transactions on Parallel and Distributed Systems, 2014, 25(9):2307-2317.
[14] Ozay C, Celiktas M S. Statistical analysis of wind speed using two-parameter Weibull distribution in Alaçatı region[J]. Energy Conversion & Management, 2016, 121:49-54.
[15] Y. Ying, Y. Wu, Y. Su, R. Fu, X. Liang and H. Xu, "Dispatching approach for active distribution network considering PV generation reliability and load predicting interval," in The Journal of Engineering, vol. 2017, no. 13, pp. 2433-2437, 2017.
[16] S. F. Abdelsamad, W. G. Morsi and T. S. Sidhu, "Impact of Wind-Based Distributed Generation on Electric Energy in Distribution Systems Embedded With Electric Vehicles," in IEEE Transactions on Sustainable Energy, vol. 6, no. 1, pp. 79-87, Jan. 2015.

Safety Evaluation of Microgrid Using Chaotic Time Series and RBF Neural Network

He Qin, Wenbo Tang

Nanjing University of Posts and Telecommunications, No. 9 Wenyuan Road, Qixia District, Nanjing, Jiangsu, China

158984663@qq.com

Abstract. In order to evaluate safety of the Microgrid(MG) after distributed energy resources and different types of loads accessed in or disconnected, chaotic time series and RBF neural network are essential and beneficial tools. In this paper, Voltage Security Assessment Index(VSAI) is established to assess whether the MG is able to achieve equilibrium state for voltage. The application of Solar Photo Voltaic(SPV) or Wind Turbine(WT) generation alone is tested in the process of the phase space reconstruction and compared to the consorted hybrid example. Chaos algorithm can be used to define and optimal the RBF center and the connection weights of output layer, which can further improve the convergence speed of RBF neural network. Comparing the result of output between actual and prediction in three different conditions, if the difference of their values is above or below a certain threshold, the safety of MG will be judged unsafe. Test is conducted on an autonomous MG bus feeder to verify the usefulness of the posed system. The research conclusion shows that chaotic time series and RBF neural network is effective and feasible for the safety evaluation of MG.

1. Introduction

A microgrid is a small-scale power grid that can be operated independently or in combination with the main electrical grid. Distributed renewable energy sources (such as wind turbines, photo-voltaic), energy storage systems and loads are composition of a MG [1,2]. When a disturbance happening, MG will disconnect from main grid and switch to islanded mode, otherwise it may operate in grid-connected mode. Considering randomness of renewable energy sources, the safety of MG with island operation capability faces many challenges such as voltage flicker, fluctuation and dip which should be solved [3].

Many studies and papers have been published to discuss the impact of microgrid on safety of distribution network, however, few publications handle the safety of interior microgrid, especially in aspect of voltage. Paper [4] proposes an approach based on fault tree analysis(FAT) to evaluate the safety of islanded microgrid with the specific operation time, but the approach may be not applicable in common situation. Combine the advantages of rough set with the advantages of artificial neural network, [5] proposed a way that pretreat the input data of neural network in the way of rough set, use the key part of input data as input to the network and improves the level of evaluation. Combination of safety engineering, factor analysis and neural network in Paper [6] to evaluate whether the system is safety, however this new security assessment way only for transmission grid system and it can not handle the safety of the MG with DGs and different loads.

This paper is organized as follows. Section 2 focuses on dynamic voltage security assessment at point of common coupling(PCC) of microgrid. Section 3 describes the proposed a evaluation method of the microgrid safety which uses chaotic time series and RBF neural network. In Section 4, simulation studies are offered to prove the usefulness of the posed method. Finally, Section 5 concludes the paper.

2. Dynamic voltage security assessment of microgrid

In this paper, a dynamic voltage security criterion originated from the VSAI [7] is used to evaluate whether the system can achieve an equilibrium state for voltage after disturbance.

Voltage Security Assessment Index [7]: VSAI is calculated from a time series data of the rms values of the voltage at point of common coupling(PCC) in the microgrid. Assuming the ith node voltage sequence as $V_i = [V_i^1, V_i^2, ..., V_i^m]^T$, and the following steps about computation of the VSAI.

a) Calculate the moving average value of the ith node voltage at jth moment from N available measurements as follows:

$$\begin{aligned} &\text{if}(j \le N): v_j = \sum_{k=1}^{j} V_i^k / j, \\ &j \in 1,2,...,N \\ &\text{if}(j > N): v_j = \sum_{k=j-N+1}^{j} V_i^k / N, \\ &j \in N+1, N+2,...,m \end{aligned} \tag{1}$$

b) Calculate the percentage discrepancy C_i^j between the node voltage measured at jth moment V_i^j and the moving average value $V_i^{(j)}$.

$$C_i^j = \left|V_i^j - V_i^{(j)}\right|, (j \in 1,2,...,m) \tag{2}$$

c) Calculating the value by eliminating the area under the percentage discrepancy curve by N at jth moment as follows:

$$\begin{aligned} &\text{if } j \le N: U_i^{(j)} = \sum_{k=1}^{j} (C_i^k + C_i^{k-1}) / 2j, \\ &j \in 1,2,...,N, \\ &\text{if } j > N: U_i^{(j)} = \sum_{k=j-N+1}^{j} (C_i^k + C_i^{k-1}) / 2N, \\ &j \in N+1, N+2,...m. \end{aligned} \tag{3}$$

d) At jth moment the VSAI is define as:

$$VSAI_i^{(j)} = U_{th} - U_i^{(j)} \tag{4}$$

Where U_{th} is security threshold value.

3. An evaluation method based on chaos theory and RBF neural network

3.1. Judgement of the chaotic time series

Lyapunov exponent can judge whether the dynamic voltage time sequence with chaotic characteristic [8]. The computing steps as follows:

Suppose the dynamic voltage time series is $V\{v_1, v_2, ..., v_N\}$.

1) Count the delay time τ and the embedding dimension m of the dynamic voltage time series, and the average period p.

2) Reconstruct the phase space with the help of τ and m.

$$V(t_i) = (v_i, v_{i+\tau}, \cdots, x_{i+(m-1)\tau}) \in R^n \tag{5}$$

where $i = 1,2,\cdots,M; N = M + (m-1)\tau$ and define the reconstructed phase space as $\{V_j, j = 1,2,\cdots,M\}$.

3) Find the nearest point $V_{\hat{j}}$, and calculate the shortest distance to the particular reference point V_j, shown as follows:

$$d_j(0) = \min_X \left\| V_j - V_{\hat{j}} \right\| \tag{6}$$

4) Count the distance $d_j(i)$ between every dot of the reconstructed phase space after i discrete-time paces.

$$d_j(i) = \left| V_{j+i} - V_{\hat{j}+i} \right| \tag{7}$$

Where, $i = 1,2,\cdots,\min(M-j, M-\hat{j})$.

5) Compute the mean line $v(i)$.

$$v(i) = \frac{1}{q\Delta t} \sum_{j=1}^{q} \ln d_j(i) \tag{8}$$

in which, q means the number of non-zero $d_j(i)$ and Δt is the sampling time of the dynamic voltage time series.

6) Obtain the largest Lyapunov exponent λ by computing the slope of regression line.

3.2. Reconstruction of phase space

Assuming the ath node voltage sequence of length b as $\{V_a^b, a = 1,\cdots,13; b = 1,2,...,n\}$,and according the theory of Takens [9],the new reconstructed phase space as follows:

$$V = [V_a^1, V_a^2, ..., V_a^c]^T \tag{9}$$

where c is the number of phase points in reconstructed phase space and expressed as:

$$c = b - (m-1)\tau \tag{10}$$

where m is embedding dimension and τ is delay time.

And according (10), it is necessary to find appropriate the embedding dimension m and the delay time τ in the process of reconstruction.

3.2.1. The selection of the delay time τ. Considering the non-liner characteristics between dynamic voltage time series in microgrid, this paper utilizes the mutual information function to determine the delay time, and comparing impacts of DGs in different situations on the delay time, and the calculation formula as follows:

$$I(\tau) = \sum_{v_a^b, v_a^{b+\tau}} P(v_a^b, v_a^{b+\tau}) \log_2 [\frac{P(v_a^b, v_a^{b+\tau})}{P(v_a^b) P(v_a^{b+\tau})}] \tag{11}$$

Where $P(v_a^b)$ and $P(v_a^{b+\tau})$ are the normalized distributions of v_a^b and $v_a^{b+\tau}$,and $P(v_a^b, v_a^{b+\tau})$ is their joint distribution.

3.2.2. The selection of the embedding dimension m. In this paper, using the GP method proposed by Grassberger and Procaccia, and the main steps are as follows:

1) Give a smaller value m_0 first and calculate the delay time for the dynamic voltage time series $\{V_a^b, a = 1,2,\cdots 30, b = 1,2,...,n\}$, which corresponds to a reconstructed phase space.

2) Calculate the correlation function :

$$C(r) = \lim_{N\to\infty} \frac{1}{N} \sum_{i,j=1}^{N} \theta(r - |Y(t_i) - Y(t_j)|) \tag{12}$$

where $|Y(t_i) - Y(t_j)|$ is the distance between the phase point $Y(t_i)$ and $Y(t_j)$, $\theta(z)$ is the Heaviside function, $C(r)$ represents the odds that the distance between two dots on the phase space attractor is less than r.

3) In an suitable scope of r, the relationship between the dimension d of attractor and the cumulative distribution function $C(r)$ should fulfill the logarithmic linear, $d(m) = \ln C(r)/\ln r$,Thus fit out the correlation dimension $d(m_0)$ corresponding to m_0 .

4) Increase the value of the embedding dimension $m_1 > m_0$, and repeat the process of the steps 2 and 3 until the estimated value of the correlation dimension $d(m)$ is no more change with the increase of m in a fixed scope of deviation. The d obtained at this moment is the correlation dimension of the attractor.

3.2.3. The forecast of the RBF neural network. In this paper, the steps of prediction as follows:

1) Use the given dynamic voltage data of microgrid as the predicted time series, and according to the G-P algorithm and the mutual information method in chaos theory, calculating the optimal embedding dimension m and the delay time τ . Then getting the reconstruction of the phase space:

$$V = [V_a^1, V_a^2, ..., V_a^c]^T \tag{13}$$

2) Use the dynamic voltage data of microgrid after reconstructed as the input data, half of the data as the training sample input and the other half as the test sample input.

3)Train the neural network by using dynamic voltage data at PCC in MG system as training samples, and build network construction by debugging reasonable parameter, finally obtain the forecast data by the test sample.

4. Simulation

The method introduced in the article is tested on a MG bus feeder which includes an actual 115KV/4.16KV 50-HZ distribution line. The model shown in Figure.1 is taken from the IEEE 13 feeder test system, and the distributed generation are imploded into the MG feeder at bus 9 for simulation. According to a design standard in [10], bus 9 is chosen as a connected bus.

Bus 14 represents a concentrated load between distributed load lines 1 and 4.. It is invisible and situated at 2/3 the distance from bus 4.

This paper uses dynamic voltage data at bus 14 of the MG, under the situation of the solar PV/wind turbo integrated into the MG independently or hybrid of them connected with the MG. The voltage magnitude may change with the access of solar PV and turbo energy to the MG, and further threaten the safety of MG.

The base voltage size is 0.985 pu at 14 point without any distributed generations through power flower calculation. Three different colour lines represent voltage changes at three different condition in Figure 2.

4.1. Demonstrate the effectiveness of the VSAI.

There are many factors that affect the intensity of solar radiation at a place in one day, such as rainfall, cloud cover and temperature etc, and most of them occur in the transient [10]. The curve of voltages may change at node 14 in short time, while the entire MG transfers from the state of parallel operation of the solar PV and wind turbine to the state in which only the wind turbine is in operation at this time. And when solar PV stop working, as shown in Fig. 3, the voltage starts to fall at t = 12s, the corresponding curve of VSAIs at 14th node of MG system also begins to fall from 0. When the fault at the 14th node stops and the voltage at 14th node is in another safe situation, the VSAI also returns to 0.

Therefore, whether the node voltage is in a safe state can be determined according to the value of VSAI.

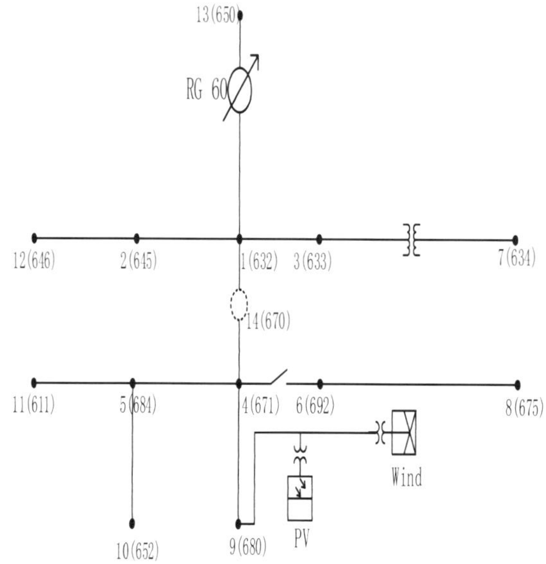

Figure 1. The structure of MG bus feeder.

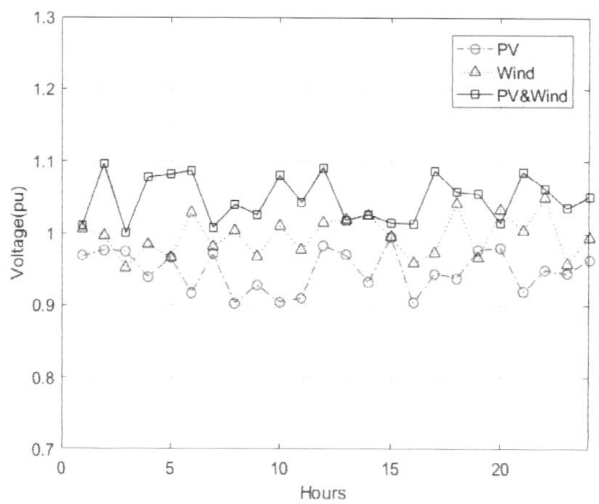

Figure 2. Voltages graph at bus 14 because of PV, wind and hybrid PV/wind.

4.2. Determine the existence of chaos.

According to the theory of the small amount of data method which is proposed in the front part of the article.The largest Lyapunov exponent of the dynamic voltage data is $\lambda = 1.3972$ by the calculation of the Matlab programming. Because of $\lambda > 0$,the chaotic feature of dynamic voltage sequence can be determined.

4.3. Reconstruction of the phase space.

Reconstruct the new phase space of dynamic voltage sequence (9):

$$V = [V_a^1, V_a^2, ..., V_a^c]^T \tag{14}$$

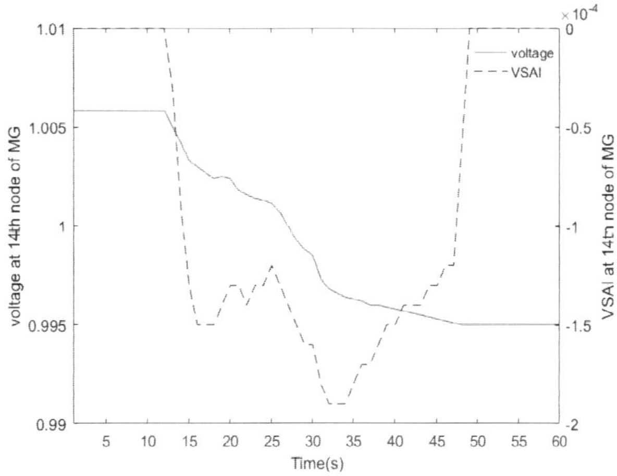

Figure 3. Safety case---voltage and VSAI at critical MG.

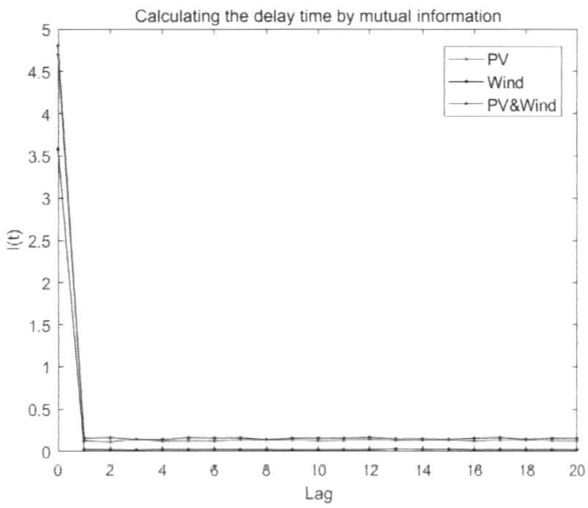

Figure 4. Delay time of the new phase space.

The delay time $\tau_{PV} = \tau_{\mathrm{Wind}} = \tau_{PV\&Wind} = 1$ at three different conditions is shown in Figure.4 which means the delay time has no relationship with the ways of the distributed generation assess to the MG system.

The following step is computing the embedding dimension through the way of GP algorithm and which is shown in Fig. 5. The embedding dimension is $\mathrm{m} = 6$ when the solar PV generation provide energy only for the MG system, the $\mathrm{m} = 9$ when the Wind turbine operates independently in the MG system, and the $\mathrm{m} = 10$ when both of them work together. And the bigger of the embedding dimension, the more evolutionary information of dynamic voltage time series will be revealed when it is in appropriate interval. So hybrid of SPV and WT generation integrated into the MG system can provide a more suitable embedding dimension and a more stable voltage output.

4.4. Chaos theory and neural network prediction.

The output of distributed generation is always affected by environmental factors, and changes in environmental factors occur in an instant, and due to this reason, the evaluation system must be able to assess changes in the system over a short period of time.Evaluate the safety of the MG system within half an hour, take seconds as the sampling time unit, and the evaluation system can assess the safety of the MG system sensitively according to the environmental changes.

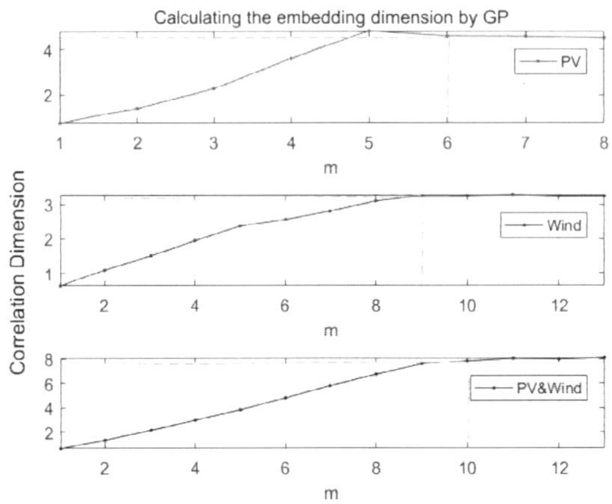

Figure 5. Embedding dimension of the new phase space.

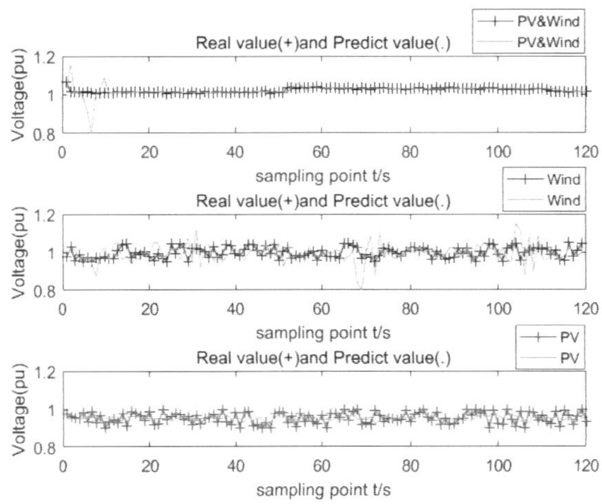

Figure 6. The contradistinction of real value and predict value.

According to Fig.6, the voltage magnitude fluctuates around 0.9 pu or 1.0 pu when the SPV or WT integrated into the Microgrid system independently,and the voltage magnitude is more stable around 1.05 pu when hybridization of SPV and WT connected with the Microgrid system.

The absolute error of the hybridization of SPV and WT is the smallest and the smoothest in Fig.7, and the maximum absolute predict error of it is 0.18,and the other is between $0 \sim 0.2$.

5. Conclusion

A new method has been suggested for evaluating the safety of MG, following an accident, promoting self-healing of the MG.The proposed scheme can judge whether the dynamic voltage at PCC of the MG is in safety status through a dynamic voltage safety criterion drafted exploiting a Voltage Security Assessment Index(VSAI). This scheme uses chaotic time series method to analysis the data of dynamic

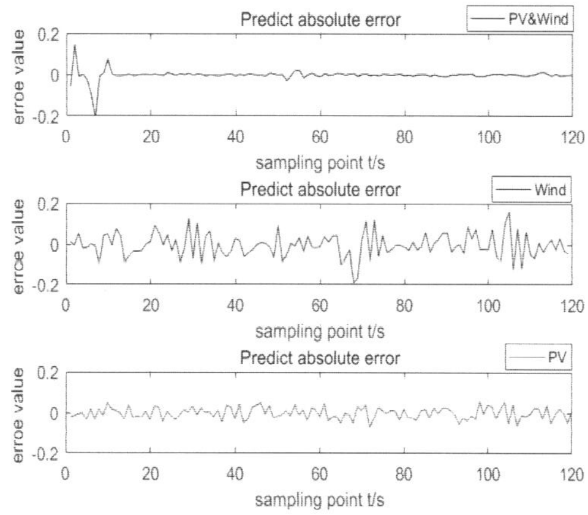

Figure 7. The prediction absolute error curve.

voltage at PCC of the MG, and calculated that, which means the series of dynamic voltage with chaotic characteristics. The hybridi-zation of SPV and WT generation can provide a more stable voltage output and it also acts as a remedy to voltage sup-pression problem in the MG system by comparin three different situations in which distributed generation is integrated into the MG. The result of the maximum absolute error is 0.18 proves the advantage of combining chaotic time series theory with RBF neural network.

References

[1] S. Anand, B.G. Fernandes, and J.M. Guerrero, "Distributed control to ensure proportional load sharing and improve voltage regulation in low-voltage DC microgrid," IEEE Trans. Power Electron., vol. 28, pp. 1900–1913, Apr. 2013

[2] G. A. Pagani and M. Aiello, "Towards decentralization: A trpological investigation of the medium and low voltage grid," *IEEE Trans. Smart Grid.*, vol. 2, pp. 538–547, Sep. 2011.

[3] S.Chowdhury, S.P.Chowdhury , and P.Crossley, "Microgrids and Active Distribution Networks," IET RENEWABLE ENERGY SERIES 6, 2009.

[4] Zhenjie Li, Yue Yuan and Furong Li, "Evaluating the reliability of islanded microgrid in an emergency mode," 45th International Universities Power Engineering Conference UPEC2010, Cardiff, Wales, 2010, pp. 1-5.

[5] J. Li, Y. Zhao and J. Li, "Power Grid Safety Evaluation Based on Rough Set Neural Network," 2008 International Conference on Risk Management & Engineering Management, Beijing, 2008, pp. 245-249.

[6] Z. Yang, Z. Mao and G. Ma, "A safety assessment approach using factor analysis and neural network for transmission grid system," 2009 IEEE International Conference on Intelligent Computing and Intelligent Systems, Shanghai, 2009, pp. 149-153.

[7] Kim, "System and method for calculating voltage stability risk-index in power system using time series data," U.S. 7 236 898 B2, Jun.2007

[8] Chen Min, Ye Xiaozhou. Research on the chaotic time series decision method[J]. Information Technology. 2008(6),22~26

[9] G. A. Leonov and S. M. Seledzhi, "Lyapunov Dimension of Attractors of the Lorenz-Like Differential Equations," *2016 Third International Conference on Mathematics and Computers in Sciences and in Industry (MCSI)*, Chania, 2016, pp. 119-121.

[10] M. Shaaban, and J. O. Petinrin, "Sizing and siting of distributed generation in distribution systems for voltage improvement and loss reduction," International Journal of Smart Grid and Clean Energy, vol. 2, pp. 350-356, 2013.

IOP Conf. Series: Materials Science and Engineering **853** (2020) 012010 doi:10.1088/1757-899X/853/1/012010

Research on Optimizing the Recovery Strategy of a Distribution Network Communication System Under Extreme Disasters

Husheng Lin, Xia Zhou, Jianqiang Lu

9 wenyuan road, Qixia district, Nanjing, Jiangsu province, China
Nanjing university of posts and telecommunications

1091814745@qq.com

Abstract. Extreme disasters, such as typhoons, cause great damage to distribution network communication systems, affecting the perception of disaster situations and the repair speed of the distribution networks. To realize the rapid recovery of a distribution network, an optimal recovery strategy of the distribution network communication system is proposed in this paper from the perspectives of predisaster prediction and postdisaster rush repair. First, based on multivariate data, the space-time mesh fusion method is used to predict the fault position of the distribution network communication system, guiding the accurate allocation of emergency repair resources, and the fault position is corrected using man-made statistics after a disaster. Then, combining fiber optics, mobile nets and emergency communication vehicles, rush repair rules for the distribution network communication system after extreme disasters are put forward, and the shortest route search algorithm is used to solve the minimum rush repair time under the rules, which is the optimal rush repair scheme for the distribution network communication system after an extreme disaster. Finally, the proposed strategy is applied to a practical topology, and the results demonstrate that the strategy is reasonable and effective.

1. Introduction

In recent years, due to extreme disasters, accidents in distribution network communication systems have occurred frequently, which has seriously hindered postdisaster rush repair work in disaster-influenced areas. In July 2016, typhoon 'Nibote' caused a total of 2,784 communication base stations to be retired in Fujian, and many regional communications were interrupted. The command center could not obtain the fault information in time, which seriously delayed the recovery of the distribution network [1]. To improve the recovery ability of distribution networks and reduce the losses caused by disasters, it is of great significance to accelerate the restoration of distribution network communication systems.

At present, existing research on rush repair work in distribution networks, including predisaster failure predictions [2], allocation of repair resources [3], trouble information acquisition after the disaster [4], and the formulation of a repair strategy [5], has formed a mature system. However, such research is based on the complete availability of information and has not yet taken into account the impact of communication system failures. Once a distribution network communication system fails due to extreme disasters, the commander may not be able to obtain the status information in time and thus cannot make the optimal decision. Therefore, it is necessary to repair the distribution network communication system in a timely manner.

Compared with existing studies, the advantages and contributions of the proposed strategy are as follows: An improved predisaster fault prediction method based on space-time meshes and a postdisaster optimal rush repair strategy coordinated by using fiber optic repair, mobile base stations and emergency communication vehicles are proposed. In terms of predisaster fault prediction, domestic and foreign scholars have used statistical analysis [6] and smart assessment [7] to predict the failure of physical and communication equipment in a distribution network under a single disaster. These methods have received a large amount of research and have matured, but the accuracy of the position of the fault equipment is not fine enough. Moreover, the application of geographic grids lacks time dimensions and has basically been performed for a single disaster type [8], with a lack of judgment of communication equipment failure under multidimensional disaster types; in the postdisaster repair of distribution network communication systems, the applications of fiber optics [9], emergency communication vehicles [10] and mobile nets [11] in electricity emergency communication have been researched, and there is a lack of a combination of wired and wireless communication. Using only fiber optic repair, the recovery speed of the communication system is very slow. Wireless repair has improved the speed of communication restoration, but the particularity of the lifeline load nodes in the distribution network is neglected; such nodes need to be connected to the control center through a cable network. In the process of repairing communication systems, there is a lack of corresponding rush repair principles, mainly relying on human experience to give priority to the restoration of important nodes, which is highly subjective. Under extreme disasters, the operating state of the system changes rapidly, and human judgment alone is bound to lead to problems.

2. Structure of the Power Communication System

As shown in Figure.1, the distribution network communication system layer is similar to the power layer and has a single- or multiple-loop structure. The distribution network communication system includes a control center, communication intermediate nodes, mobile base stations, and normal communication nodes. The 110 kV power node corresponds to a communication intermediate node and is directly connected to the control center. The state data are transmitted in both directions, and the control signal is transmitted from top to bottom in one direction. This signal can only be transmitted from the control center to a communication intermediate node and then transferred to a normal communication node. A mobile base station is a special communication intermediate node. To ensure the security of the business, it can only transmit state data, not control signals [12].

Due to the increasing degree of coupling between the communication system and the power system, after an extreme disaster, the communication system can not only monitor the fault information of the distribution network point-to-point but also facilitate the command center to guide the optimal scheduling of emergency personnel. Once the communication system fails on a large scale, the fault information cannot be obtained in a timely manner, the repair personnel are disorganized, and the repair work cannot be smoothly carried out. Therefore, it is particularly important to restore the communication system. From the perspectives of predisaster prediction and postdisaster emergency repair, the optimal recovery strategy for a distribution network communication system is studied in this paper.

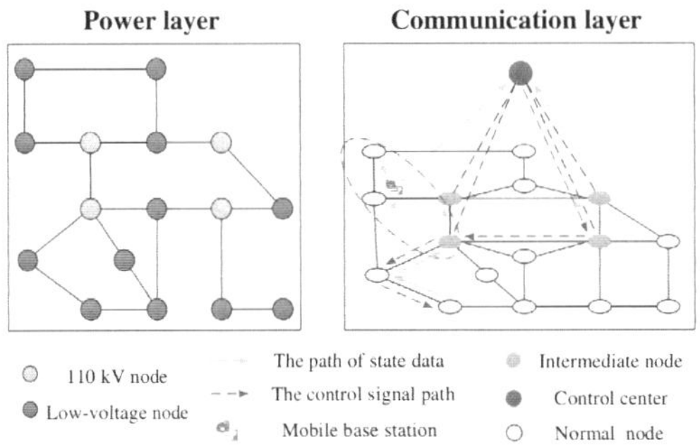

Figure 1. Structure diagram of the power communication composite system

3. A Fast Method for Determining the State of the Communication Network

To speed up the rush repair of the communication system after an extreme disaster, real-time judgment of the fault is required. Due to communication failures, the efficiency of fault information acquisition is very low. Therefore, according to the predisaster prediction, the fault position is initially completed based on the forecast information of extreme disasters, and the allocation of emergency repair resources is guided. Then, after the disaster, the fault position is corrected to guide personnel scheduling.

In terms of predisaster fault prediction, multivariate information, such as geographic information, meteorological information, and communication network structure information, is used to predict the state of the communication network. The amount of multivariate information is large, and the range is wide; it is difficult to process data on a large scale. A method of space-time meshes is adopted in this paper and has the following four steps:

Step 1: Divide the disaster-influenced area into small grids, encode the grids, and quickly obtain the location of each device;

Step 2: Add the time dimension on the basis of the geographic grid and superimpose the forecast disaster and the actual disaster, including the typhoon path, rainstorm information and other multidimensional disaster types, on the geographic grid;

Step 3: Model and analyze the disaster strength and invulnerability of the equipment within the small grids, and predict the probability of equipment failure. Equipment with a failure probability greater than a certain threshold is considered a faulty device. According to the overall communication system failure situation, emergency repair resources are allocated reasonably;

Step 4: According to man-made statistics, the final failure state of the communication system is corrected, which is conducive to accurate repair.

4. Repair Strategy of the Distribution Network Communication System

4.1. The Repair Methods

At present, repair methods of distribution network communications systems mainly use fiber optic repair. With the advantage of safety, this approach can safely transmit control signals, but the speed is slow and it is not conducive to the repair of electricity; in addition, wireless communication of distribution network nodes can be achieved through mobile base stations and emergency communication vehicles. A single mobile base station can replace multiple distribution network communication nodes; this method shortens the recovery time and is conducive to disaster perception. The disadvantages are that the communication nodes restored in this way cannot transmit control signals, and the nodes still need fiber optic repair, which increases the workload.

4.2. The Repair Rules

Priority is given to repairing the communication nodes corresponding to lifeline loads, including hospitals, government buildings, and the water supply, and ensuring that such nodes can be automatically controlled by the command center. To realize the rapid recovery of the communication system after an extreme disaster, the advantages of comprehensive wireless communication and wired communication are considered. Based on giving priority to repairing the lifeline load communication nodes, the following repair rules for the distribution network communication system are made in this paper: (1) Priority is given to repairing the communication nodes corresponding to lifeline loads, and the method used is fiber optic repair; (2) Mobile base station repair is better than single-communication-node repair; and (3) Emergency communication vehicles give priority to repairing communication intermediate nodes.

4.3. The Repair Targets

After an extreme disaster, the large-scale damage to the communication system affects the disaster perception of the distribution network, which in turn affects the repair speed. Due to the different locations of communication fault nodes, mobile base stations, emergency communication vehicles and repair teams, the task arrangement of repair teams and the repair order of communication nodes will influence the overall optimal recovery goal.

In multifault rush repair of distribution network communication systems, the most important concerns in engineering are the economics and timeliness, which can be equivalent to a two-layer optimization model. The upper target is the lowest cost of the communication system failure, and the lower target is the shortest repair time needed for the distribution network communication system to achieve emergency communication.

$$Val_{total} = \sum_{i=1}^{m} Val_i \cdot t_i \tag{1}$$

Val_{total} is the total failure cost, Val_i is the failure cost in unit time, m is the total number of failure nodes, and t_i is the failure time. Under the same type of communication nodes, the longer the failure time is, the greater the failure cost. Therefore, the upper target can be equivalent to shortening the repair time as much as possible under the premise of priority restoration of important value nodes. Then, the solution of the two-layer optimization problem can be converted into solving the shortest repair time under the rules. The optimal solution can be acquired quickly through the shortest path search algorithm [13], avoiding problems such as a local optimal solution [14] of the two-layer optimization model.

The shortest repair time in this paper is determined by the team that spends the most time in the repair tasks assigned by the J repair teams:

$$f_1 = \max\{(T_1, T_x, \cdots, T_J)\} \tag{2}$$

where T_x is the time that it takes repair team x to complete the allocated task and J is the number of repair teams.

With mobile base stations and emergency communication vehicles, the objective function is updated to:

$$f_2 = \max\{\max(T_1, T_2 \cdots T_J), \max(T_1, T_2 \cdots T_K), \max(T_1, T_2 \cdots T_L)\} \tag{3}$$

where K is the number of mobile base station repair teams and L is the number of emergency communication vehicles.

The constraints are as follows:

(1) Each fault communication node can only be assigned and must be assigned to a repair team, and x_{jk}, has a value of 1 when the node is assigned to any repair team.

$$\sum_{k=1}^{J} x_{jk} = 1 \tag{4}$$

(2) At least one of the fibers connected to each distribution network communication node is normal, where l_{ij} is fiber j of communication node 'i'. If the fiber is good, it takes the value 1; otherwise, it takes the value 0. P is the total number of fibers connected to the communication node.

$$\sum_{i=1}^{P} l_{ij} \geq 1 \tag{5}$$

IOP Conf. Series: Materials Science and Engineering **853** (2020) 012010 doi:10.1088/1757-899X/853/1/012010

(3) Any repaired communication node can be connected to a communication intermediate node via fiber optic. x_Z is a node that has resumed communication, and l_{Zi} is the fiber between any failed node and the recovery node, where the fiber has a normal value of 1 and a break value of 0.

$$\exists x_z, l_{zi} * \sum_{k=1}^{J} x_{jk} = 1 \tag{6}$$

5. Experimental results

Take a practical topology in Zhangzhou as an example, the distribution network communication nodes correspond to power nodes one by one. The topological damage and resource distribution information of the communication system are obtained by the space-time mesh fusion method. Nodes 9, 12, 13, 15, 50, 51 are lifeline load. It is assumed that the repair times of a mobile base station, communication intermediate node and normal communication node are 40 min, 30 min and 20 min, respectively. For a unit distance, the fiber optic repair, emergency communication vehicles and repair teams travel 15 min, 10 min and 5 min, respectively; the value of a communication node corresponding to a lifeline load is 10; the value of a communication intermediate node is 5; and the value of a normal communication node is 1. There is one repair team for mobile base stations and two repair teams for fiber optics.

The prediction of the fault position guides the allocation of emergency repair resources. In the absence of wireless communication, the teams repair nearby communication fault nodes. Figure.2 shows that the positions of repair teams affect the repair time and failure cost of the communication system after an extreme disaster. The closer a repair team is to a lifeline load node, the lower the cost caused by failure of the communication system. The closer a repair team is to a communication intermediate node, the shorter the repair time, and it is unfavorable to place a repair team at a nonfaulty node. Therefore, it is very important to predict the fault position of the distribution network communication system, and the accuracy of the prediction directly affects the allocation of emergency repair resources and the repair speed.

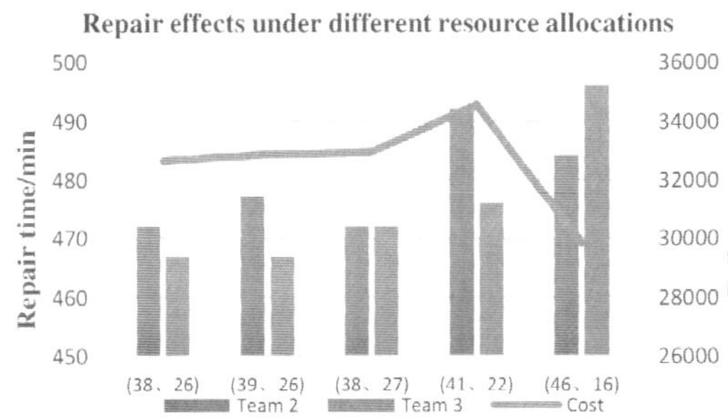

Figure 2. Repair effects under different resource allocations

If the existing repair resources and resource allocations are determined, whether the rules mentioned in this paper are applied, the corresponding repair time and failure cost of each method are shown in table 1.

Table1. Repair effects corresponding to whether the rules are followed.

		Only fiber optic	With mobile base station	With vehicle
Not following the rules	Time/min	472	352	274
	Cost/Val_i*t_i	32664	28643	20292
Following the rules	Time/min	499	471	402
	Cost/Val_i*t_i	14513	13911	8236

Even if the repair of a mobile base station restores some communication functions to some communication nodes corresponding to lifeline loads, the station can only transmit state information and cannot accept the control center signal. Therefore, this approach does not meet the requirements of emergency communication because these types of nodes should be repaired by fiber optic repair.

Table 1 shows that, although the repair time increases with the rules, the overall failure cost decreases significantly by close to 60%, which is in accordance with the repair requirements of the communication system after extreme disasters.

To show the comprehensiveness and rationality of the repair rules proposed in this paper, with the aim of guiding the repair work of the distribution network communication system after extreme disasters, quantitative discussion on the repair effects of only following some of the rules is conducted, and the results are shown in Table 2.

Table2. Repair effects corresponding to different rules.

Rules	Time/min	Cost/$Val_i * t_i$
None	272	20292
1	442	10224
1&2	436	10021
1&3	414	8432
1&2&3	402	8236

Table 2 shows that the rules can not only reduce the repair time but also reduce the failure cost. Combining the results of Tables 1 and 2, it can be seen that the first rule is strongly related to the failure cost, so the first rule is most important in terms of the failure cost.

In summary, the repair rules in this paper are effective, comprehensive and reasonable. By following the rules, the optimal repair plan of the case study is shown in Figure. 3

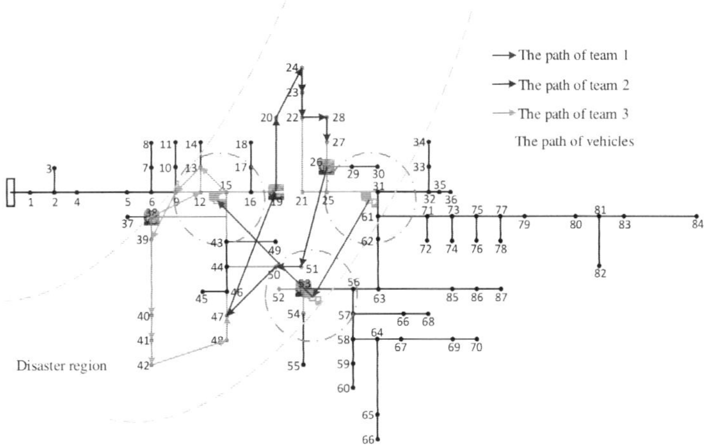

Figure 3. Optimal repair scheme for the communication system after an extreme disaster

6. Conclusion

This paper mainly studies the optimization of the recovery strategy of a distribution network communication system under extreme disasters and considers the combination of fiber optics, mobile base stations and emergency communication vehicles. The following conclusions can be drawn:

(1) The repair effects of the distribution network communication system after an extreme disaster are not only related to the repair strategy after the disaster but also affected by the allocation of emergency resources before the disaster and the speed of fault correction after the disaster. The coordination of wireless communication and fiber optic repair can optimize the repair effects of the distribution network communication system.

(2) After an extreme disaster, once the communication system fails on a large scale, priority is given to fiber optic repair of the communication nodes corresponding to lifeline loads, the remaining nodes are given priority to restore communication through a mobile base station, and an emergency communication vehicle is given priority to restore the communication intermediate nodes.

The communication system restoration of the distribution network and power recovery complement each other. This paper focuses only on the communication nodes corresponding to lifeline loads and does not fully consider the coupling relationship between the distribution network communication system and power system. This is the research work that needs to be carried out in the future.

Acknowledgments
This work is supported by National key research and development plan projects(Security analysis and assessment method of cyber physical system (CPS) for distribution network attack ,2017YFB090300); State Key Laboratory of Smart Grid Protection and Operation Control(SGNR0000GZJS1808074); National Natural Science Foundation of China(Research on Cyber Physical Security Theory and Active Defense Technology for Cyber Attacks, 61833008)

References
[1] L. X. Wang, "Let the signal pass through the wind and rain," Fujian Mobile Anti-Typhoon "Nibote" documentary, China Telecom, vol.07, pp.16-18, 2016.
[2] S.L. Zhang, Y.X. Wang, M.Q. Liu, et al., "Data-Based Line Trip Fault Prediction in Power Systems Using LSTM Networks and SVM," IEEE Access, vol.6, pp.7675-7686, December 2017.
[3] W.Q. Zhao, Q.G. Meng, W.H. Paul, "A Heuristic Distributed Task Allocation Method for Multivehicle Multitask Problems and Its Application to Search and Rescue Scenario," IEEE Transactions on Cybernetics, vol.46, no.4, pp.902-915, April 2016.
[4] R. Krishnathevar, E.E. Ngu, "Generalized Impedance-Based Fault Location for Distribution Systems," IEEE Transactions on Power Delivery, vol.27, no.1, pp.449-451, January 2012.
[5] Z.G. Lu, D. Li, X.J. Lu, et al., "The multi-fault emergency repair strategy of the distribution network under the ice disaster with distributed power supply," Journal of Electrical Technology, vol.33, no.2, pp423-432, 2018.
[6] R. Billinton, G. Singh, "Application of adverse and extreme adverse weather: modelling in transmission and distribution system reliability evaluation," IEE Proceedings-Generation, Transmission and Distribution, vol.153, no.1, pp.115-120, January 2006.
[7] Y. Liu, C. Singh, "A Methodology for Evaluation of Hurricane Impact on Composite Power System Reliability," IEEE Transactions on Power Systems, vol.26, no.1, pp.145-152, May 2010.
[8] H. Liu, R. A. Davidson, D. V. Rosowsky, et al., "Negative binomial regression of electric power outages in hurricanes," J. Infrastruct. Syst., vol. 11, no. 4, pp. 258-267, 2005.
[9] C. Ma, J. Zhang, Y.L. Zhao, et al., "Traveling repairman problem for optical network recovery to restore virtual networks after a disaster," IEEE Journal of Optical Communications and Networking, vol.7, no.11, pp. B81 - B92, 2015.
[10] R. Oorni, A. Goulart, "In-Vehicle Emergency Call Services: eCall and Beyond," IEEE Communications Magazine, vol.55, no.1, pp.159-165, 2017.
[11] N. Zhao, W.D. Lu, M. Sheng, et al., "UAV-Assisted Emergency Networks in Disasters," IEEE Wireless Communications, vol.26, no.1, pp.45-51, February 2019.
[12] R. Fu, X.J. Huang, Y.S. Xue, et al., "Security Assessment for Cyber Physical Distribution Power System under Intrusion Attacks," IEEE Access, vol.7, pp.75615-75628, July 2018.
[13] Y.M. Wang, X.P. Li, R.B. Ruiz, "An Exact Algorithm for the Shortest Path Problem With Position-Based Learning Effects," IEEE Transactions on Systems, Man, and Cybernetics: Systems, vol.47, no.11, pp.3037-3049, May 2016.
[14] W.A. Bukhsh, A. Grothey, K. I. M. McKinnon, et al., "Local Solutions of the Optimal Power Flow Problem," IEEE Transactions on Power Systems, vol.28, no.4, pp.4780-4788, August 2013.

IOP Conf. Series: Materials Science and Engineering **853** (2020) 012011 doi:10.1088/1757-899X/853/1/012011

A security Assessment of Distribution Network CPS Based on Association Matrix Modeling Analysis

Zhou Yang[1], Xia Zhou[1], Ming Ni[2], Manli Li[2], Husheng Lin[1]

[1]Nanjing University of Posts and Telecommunications, Nanjing 210023, China
[2]NARI Group Corporation (State Grid Electric Power Research Institute) , Nanjing 211106, China

yang_zhou0207@163.com

Abstract. The addition of the information system has improved the operation of the distribution network CPS, but it also brings certain risks and impacts to the safety of the distribution network CPS operation. Therefore, the distribution network CPS security assessment needs to consider the impact of information systems. This paper proposes a security Assessment of Distribution Network CPS Based on Association Matrix Modeling Analysis. The research framework of information physical fault impact is proposed, and the correlation and impact of information physical failures are analyzed. The CPS model of distribution network based on association matrix is established, and the safety evaluation index and evaluation method are proposed. Case study shows the rationality and effectiveness of the proposed method.

1. Introduction

With the advancement of smart grid and energy Internet strategies, the high coupling of information and physics has become a key feature of smart distribution networks [1]. The addition of information systems brings convenience to information collection, information transmission and optimization control. At the same time, it also increases the security risk of distribution network CPS, which leads to an increased possibility of attack on CPS of distribution network [2].The physical coupling mechanism leads to cascading failures, causing the entire distribution network CPS to even threaten the safe operation of the large power grid. In the distribution network CPS, the degree of influence of information on the physical system depends mainly on the role of information functions in power grid fault handling and recovery [3]. When the physical system fails, the simultaneous failure of the information will affect the fault handling process and deteriorate the system state. If the system fails, the feeder circuit breaker control function failure will cause the cascade fault to occur [4].
At present, the research on distribution network CPS has been improved from the initial architecture, modeling [5-7], to reliability[8-10], to risk assessment [11-12], to the impact analysis of information attacks [13].However, there is less research on the impact of research information-physical prediction failure on the safety of distribution network CPS. In the actual operation of the distribution network CPS, most of the safety problems are caused by the failure of internal functions, such as the failure of the information acquisition equipment, the rejection or misoperation of the control operation of the secondary equipment, and the interruption or extension of the transmission of the information channel. Time. The failure of the functions of these information systems will induce the formation of combined

faults in the failure of the CPS physical system of the distribution network, which will affect the recovery of the CPS faults in the distribution network and bring secondary losses to the distribution network CPS.

This paper focuses on the impact of information system failures on the security of power systems. Firstly, the basic framework of fault research on distribution network CPS is proposed. According to the physical coupling characteristics of the information, a model based on the correlation matrix is established. A method for evaluating the safety of the distribution network CPS considering the influence of information is proposed. Finally, it was verified in the example system.

2. Analysis of the impact of information physical failure

2.1. Distribution network CPS hierarchy

Modern power distribution network is a typical information physical fusion system, which is divided into three hierarchical structures: distribution network backbone layer, access layer and terminal layer. As shown in Figure 1. The main layer includes the power distribution master station system, SCADA server, etc., and uses SDH technology to connect the power distribution main station and each of the distribution stations. The access layer uses EPON technology to connect the distribution station and each distribution terminal to realize real-time communication between the distribution terminal and the distribution station. The terminal layer contains feeders, circuit breakers, segment switches, and FRTUs, etc. The framework is divided into hierarchical layers, including the information center layer, the information physical coupling layer, and the physical entity layer. As shown in Figure 2.

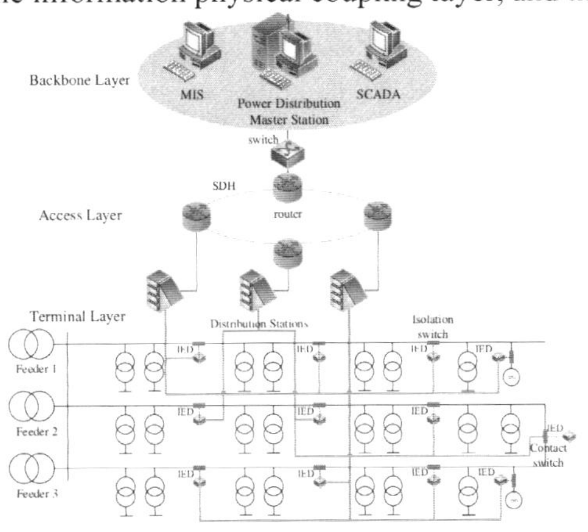

Figure 1. Structure of Distribution Network CPS

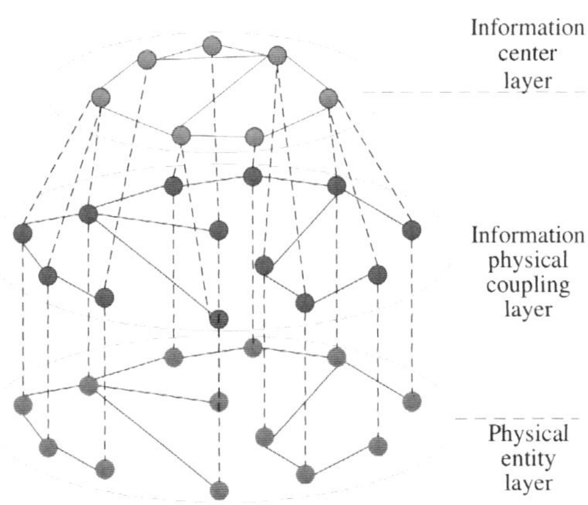

Figure 2. Framework of Hierarchical Layers

2.2. Information System Failure Impact Research Framework

2.2.1. Failure correlation analysis. The correlation between information failure and physical failure is divided into direct correlation and indirect correlation. Direct correlation: The functions completed by the information system directly affect the primary equipment of the power. This type of information failure will directly cause the primary equipment to fail to work properly and cause physical failure. Indirect correlation: The functions performed by the information system do not directly affect the primary equipment of the power, but will affect its performance. At the same time as physical faults occur, the simultaneous occurrence of information faults will affect the process of fault recovery and deteriorate the system state. The impact of physical faults on information faults is primarily a direct impact. The primary device of the physical system provides power for the communication device in the information system. However, in practice, most of the CPS information systems of the distribution network have an uninterruptible power supply, so the physical failure has less impact on the information failure.

2.2.2. Analysis of Impact of Information Failure on Physical System. The impact of the communication network on the physical system is mainly reflected in the delay, error and interruption generated during the information transmission. The function of the secondary equipment network is to realize the collection and transmission of information and transmission, and on the other hand, the real-time analysis and processing of the corresponding data [14].Its influence on the physical system mainly manifests as the effect of information processing (such as the accuracy of fault identification, the accuracy of signal acquisition) and the delays and errors generated during the information processing. Therefore, the influence of the information system on the physical system can be expressed by equation (1).

$$F = f[R(c_{ij}), R(s_{ij}), R(l_{ij})] \tag{1}$$

In the formula, f is evaluation function for the state of the distribution network power business after its execution; $R(c_{ij})$, $R(s_{ij})$, $R(l_{ij})$ is execution results of related power services after considering the status of the communication node or link, secondary device node or channel, information node.

3. A security assessment Method for Distribution network CPS

This paper proposes a correlation method based on the correlation matrix model for the assessment of the safety of the distribution network CPS, in accordance with the "fault analysis - model establishment - indicator evaluation" approach.

3.1. Distribution Network CPS Association Matrix Modeling

Firstly, the fusion information and physical distribution network CPS system should be modeled and analyzed. The correlation matrix model can accurately describe the topological relationship (structure) and logical association (control logic) between layers [15]. The modeling process is as follows.

3.1.1. Network modelling. Take power network modeling as an example. Other networks will adopt the same method. Physical layer model P: According to the distribution network topology, the logical relationship between the physical layer is represented by two logical elements of "0" and "1". For a distribution network with m power nodes, establish an m-square matrix.

$$P = \begin{bmatrix} p_{11} & \cdots & p_{1j} & \cdots & p_{1m} \\ \vdots & & & & \vdots \\ p_{i1} & \cdots & p_{ij} & \cdots & p_{im} \\ \vdots & & & & \vdots \\ p_{m1} & \cdots & p_{mj} & \cdots & p_{mm} \end{bmatrix} \tag{2}$$

In the formula, p_{ij} represents the nodes and lines of the physical entity layer; When p_{ij} is equal to 0, it means no connection; When p_{ij} is equal to 1, it means that there is a connection.

3.1.2. Association modelling. The association matrix describes the relationship between the layers of the distribution network CPS. The following is an example. Secondary equipment-communication association model: For the distribution network CPS network including k secondary device nodes and m communication nodes, the secondary device node-communication node association characteristic matrix is used to describe the information uploading and delivery process.

In the next formula, $s\text{-}c_{ij} = [s\text{-}c_{T.ij}, s\text{-}c_{PB.ij}]$ is used to describe the relationship between the communication node and the secondary device node. $s\text{-}c_{T.ij}, s\text{-}c_{B.ij}$ represents the delay and interruption of the communication node i and the secondary device j node respectively.

$$S\text{-}C=\begin{bmatrix} s\text{-}c_{11} & \cdots & s\text{-}c_{1j} & \cdots & s\text{-}c_{1m} \\ \vdots & & \vdots & & \vdots \\ s\text{-}c_{i1} & & s\text{-}c_{ij} & & s\text{-}c_{im} \\ \vdots & & \vdots & & \vdots \\ s\text{-}c_{k1} & \cdots & s\text{-}c_{kj} & \cdots & s\text{-}c_{km} \end{bmatrix} \tag{3}$$

3.2. Security assessment index of distribution network CPS

In the distribution network CPS, the load loss caused after the physical failure occurs may include two parts: load loss of the fault section and load loss of the non-fault section. The former is fixed, which is caused by the isolation of the fault after the fault occurs; the latter is not fixed, due to the radiation network structure of the distribution network, which causes the downlink power supply area to lose power after the fault isolation. Load loss can be recovered by grid-connected and off-grid: grid-connected recovery refers to the use of alternate tie lines for adjacent feeder load transfer; off-grid recovery refers to the use of non-faulty segments for distributed The power supply is supplied. In the CPS fault recovery process of the distribution network, the information system plays a pivotal role. If information failure occurs again, it will once again affect the safety of the distribution network CPS. Based on this, this paper proposes the post-fault loss level indicator and the resilience index as the CPS safety assessment index of the distribution network.

3.2.1. System loss degree index after failure. The post-fault loss indicator includes two parts: the degree of load loss and the degree of user loss.

The degree of load loss is the ratio between the sum of the load reduction of the fault section after the failure of the distribution network and the load reduction of the non-fault section and the total load of the distribution network.

$$\rho_l = \frac{\sum_{i\in M}\mu_i P_i + \sum_{i\in N}\mu_i P_i}{\sum_{i\in S}\mu_i P_i} \tag{4}$$

In the formula, μ_i is level weight of the node i; P_i *is* active power load of node i; M, N, S is load node sets of the fault zone, the non-fault zone and all。

The degree of user loss is reflected in the degree of impact on the user side after the failure of the distribution network CPS, including the proportion of power outages and the proportion of user power outage losses.

$$\rho_u = (\varpi_1 \frac{X}{Y} + \varpi_2 \frac{\sum_{j\in X} w_j}{\sum_{j\in Y} w_j}) \tag{5}$$

In the formula, ϖ_1, ϖ_2 are the weighting factors for the proportion of power outages and the proportion of user power outage losses. w_j is the rating factor of user j. X, Y are the number of all lost users and the total number of users of the system after the failure.

3.2.2. System recovery capability index after failure. The timely recovery of the distribution network after the failure of the distribution network CPS is an important basis for maintaining the safe and stable operation of the distribution network. It is mainly evaluated from the two aspects of failure recovery rate and failure recovery speed.

Failure recovery rate: The power restoration process after the CPS failure of the distribution network generally starts from the failure, and continues until the fault is completely cleared and the normal power supply of the entire network is restored. Therefore, during this period of time, the recovery of the load should be considered in consideration of the load importance level.

$$\rho_r = \frac{\sum_{t=1}^{T} \sum_{i \in M+N} \sigma_i P_{i,t} \Delta t}{\sum_{t=1}^{T} \sum_{i \in M+N} \sigma_i L_{i,t} \Delta t} \tag{6}$$

In the formula, $P_{i,t}$ is the actual recovery power of the load i; $L_{i,t}$ is the original power demand of the load i at the moment; σ_i is the rank weight of the node load i; Δt is the duration of the fault; M, N are load node sets of the fault zone and the non-fault zone.

Fault recovery speed: In order to reduce the load loss and shorten the power outage time of the user, the distribution network requires timely power restoration for the non-faulty section after the fault occurs, and quick repair of the fault section.

$$T_f = T_1 + T_2 \tag{7}$$

In the formula, T_1, T_2 are fault zone positioning time and fault isolation and non-fault zone recovery time.

3.3. Security assessment method based on expected fault set

The safety of the distribution network CPS refers to any time section in the continuous operation of the system. For a set of physical or communication expected faults, it can keep the distribution network CPS system running continuously to maintain the power supply capacity to the load. This paper proposes a CPS security assessment method based on the expected fault set. The flow is shown in Figure 3.

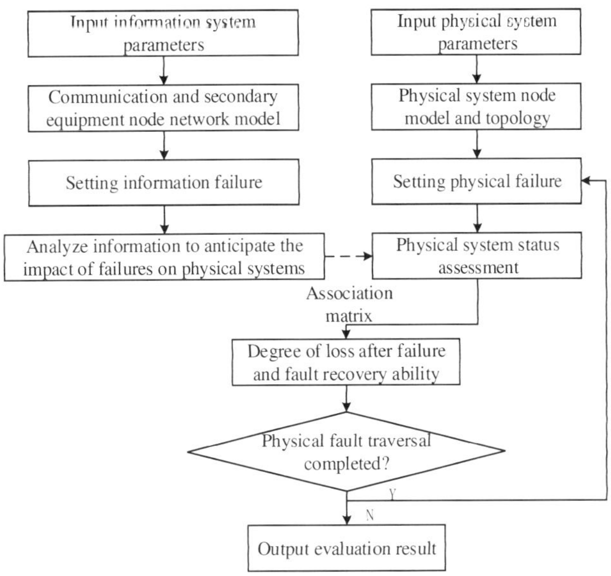

Figure 3. Security Assessment Process for Distribution Network Cyber Physical System

4. Case analysis

Figure 4 is a structural diagram of the distribution network 3 feeder system and its corresponding information system. In the figure, the distribution physical system has a total of 8 distributed power supplies to support the island operation, and a communication switch between the feeder 2 and the feeder 3 is provided for load exclusive supply. In order to make the calculation convenient, the 33 load nodes are divided into 11 sections, corresponding to 11 kinds of physical expected faults, and the sections are provided with isolating switches to ensure the isolation of faulty sections and non-faulty sections.

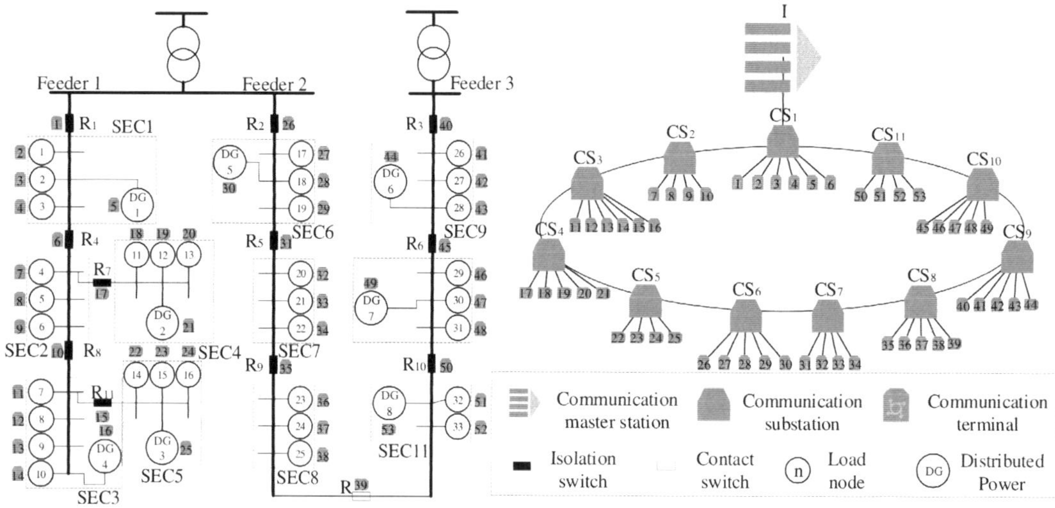

Figure 4. Structure of Case

4.1. Result analysis

In this paper, the security of the distribution network CPS is studied under the normal and fault scenarios of the information system.

4.1.1. The information system is normal. When the information system is normal, the communication and control functions between the communication device and the secondary device and the physical node are normal, and the load, switch, and the like in the physical system of the distribution network can be monitored and controlled. On this basis, the fault prediction is carried out, and the self-regulating ability of the distribution network CPS is used for recovery, and the calculation safety index is shown in Table 1.In Table 1, after the occurrence of the failure *SEC11*, the load and user loss are the lowest and the failure recovery rate is the highest. This is because the load level of the section is low and the number of users is small, and the load transfer can be quickly realized through the feeder transfer. It is expected that the load and user loss will be higher after the occurrence of the fault *SEC5*. This is because the load of the section is more important and the fault recovery rate is lower because it can only be restored by the power supply of the island.

Table1. Calculation Result (Normal Information System).

Predictive fault set	Safety indicator calculation result			
	ρ_l / %	ρ_u / %	ρ_r / %	T / s
SEC1	2.34	2.40	96.14	120
SEC2	7.17	4.79	88.16	120
SEC3	15.51	16.94	74.38	120
SEC4	9.24	11.98	94.92	40
SEC5	35.87	19.55	80.26	40

SEC6	2.34	2.40	96.79	100
SEC7	3.02	7.00	95.84	100
SEC8	7.55	11.98	95.43	40
SEC9	2.34	2.40	96.79	100
SEC10	16.35	18.98	81.55	100
SEC11	0.47	1.60	99.74	40

4.1.2. Fault information upload failure. The uploading of CPS fault information of the distribution network is mainly the two parts of the monitoring equipment collecting fault information and the communication network transmission fault information. This article addresses the expected failures occurring at *SEC1*, setting information to monitor for error faults. The physical expectation fault occurs between the load nodes P_2-P_3, but because the information monitoring equipment of the load node P_3 fails, the fault information cannot be transmitted, and the monitoring information of the load node P_4 is successfully transmitted, the fault section is located at *SEC1* and *SEC2*. Therefore, the security indicators for the expected fault *SEC1* in different scenarios are shown in Table 2 below.

Table 2. Calculation Result (Information Transmission Failure).

Fault information upload status	Safety indicator calculation result			
	ρ_{load} / %	ρ_{user} / %	$\rho_{recovery}$ / %	T
Normal	2.34	2.40	96.14	120s
Malfunction	10.51	4.79	88.16	120s
Manual positioning	2.34	2.40	90.51	30min

It can be seen from Table 2 that the fault location range becomes larger due to the failure of the fault information uploading. If the decision control center does not find this information fault, the non-faulty section *SEC2* is also isolated, resulting in load loss and user loss increase; If this information failure is discovered in time, manual fault location will be taken, which increases the time for failure recovery.

4.1.3. Control command issued failure. The fault information is uploaded to the information center, and after the decision is made, a control command is sent to the fault isolation switch and the load transfer switch. If the interrupt is generated by the control command, the above recovery operation cannot be completed in time. When the expected fault occurs in the *feeder 2* and the *feeder 3*, the fault occurs during the issuance of the control command, the contact switch *R* cannot be closed, and the load transfer failure between the feeders fails, and the non-faulty section cannot be restored in time. The safety indicators are shown in Table 3 below.

Table 3. Calculation Result (Command Transmission Failure).

Predictive fault set	Safety indicator calculation result			
	ρ_l / %	ρ_u / %	ρ_r / %	T / s
SEC1	2.34	2.40	96.14	120s
SEC2	7.17	4.79	88.16	120s
SEC3	15.51	16.94	74.38	120s
SEC4	9.24	11.98	94.92	40s
SEC5	35.87	19.55	80.26	40s
SEC6	12.91	21.38	88.06	60min
SEC7	10.57	18.98	91.27	60min
SEC8	7.55	11.98	95.43	40s
SEC9	2.34	2.40	96.59	120s
SEC10	16.35	18.98	81.43	120s
SEC11	0.47	1.60	99.74	40s

Comparing the results of Table 1 and Table 3 can be seen: Since there is no load transfer between the *feeder 1* and the *feeder 2/3* through the tie switch *R*, each expected fault on the feeder has no effect on various indicators. By comparing the expected fault *SEC6*, due to the failure of the node control device, after the expected fault occurs, the fault isolation is performed. For the downlink *SEC7* and *SEC8* segments, the load transfer cannot be completed in time, thus causing the non-faulty segment to be powered off, so the load and The degree of user loss is greatly affected, and the failure recovery rate is also reduced. Compared with the expected *SEC9-11*, the contact switch is faulty, but since the feeder can also be powered by the distributed power supply, the impact is small, but it is turned into the island power supply through the contact switch. And the speed of failure recovery has an impact.

Acknowledgments
This work is supported by National key research and development plan projects(Security analysis and assessment method of cyber physical system (CPS) for distribution network attack ,2017YFB090300); State Key Laboratory of Smart Grid Protection and Operation Control(SGNR0000GZJS1808074); National Natural Science Foundation of China(Research on Cyber Physical Security Theory and Active Defense Technology for Cyber Attacks, 61833008).

References

[1] MA Zhao, ZHOU Xiaoxin, SHANG Yuwei, et al. Form and development trend of future distribution system[J]. Pro-ceedings of the CSEE, 2015, 35(6): 1289-1298.
[2] WANG Yufei，GAO Kunlun，ZHAO Ting，et al. Assessing the harmfulness of cascading failures across space in electric cyber-physical system based on improved attack graph[J].Procee- dings of the CSEE，2016，36（6）：1490-1499.
[3] LIU Wenxia, GONG Qi, GUO Jing, et al. Reliability Simulation of ADN Cyber-Physical System Based on Hybrid Communication Network[J]. Proceedings of the CSEE, 2018, 38(06): 1706-1718+1907.
[4] Zhao Ting, Wang Dong, Lu Dongxu, et al. A risk assessment method for cascading failure caused by electric cyber-physical system(ECPS)[C]//Proceedings of the 2015 5th International Conference on Electric Utility Deregulation and Restructuring and Power Technologies (DRPT). Changsha, China: IEEE, 2015: 787-791.
[5] LI Peikai, CAO Yong, XIN Huanhai, et al. Discussion on Cooperative Control Architecture of Cyber-physical Distribution Network System[J]. Electric Power Automation Equipment, 2017, 37(12):2-7+15.
[6] GONG Gangjun, LUO Anqin, CHEN Zhimin, et al. Cyber Physical System of Active Distribution Network Based on Edge Computing[J]. Power System Technology, 2018,42(10):3128-3135.
[7] CHEN Bin, HUANG Jianye, ZHANG Gonglin, et al. Typical Characteristics and Joint Test Verification of Cyber Physical System in Distribution Network[J]. Electric Power Construction, 2017, 38 (05):46-51.
[8] JIANG Zhuozhen, LIU Junyong, XIANG Yue. Key Technologies for Reliability Assessment of Distribution Network Cyber Physical System[J]. Electric Power Automation Equipment, 2017, 37(12): 30-42.
[9] LIU Wenxia, GONG Qi, GUO Jing, SHAN Xiaodong, WANG Zhiqiang. Reliability Simulation of ADN Cyber-Physical System Based on Hybrid Communication Network[J]. Proceedings of the CSEE, 2018, 38(06): 1706-1718+1907.
[10] WANG Bingjia, DAN Shuheng, XIA Xin. Reliability Evaluation of Distribution Network CPS Considering EPON[J/OL]. Proceedings of the CAU-EPSA:1-7.
[11] WANG Zibo, MU Yunfei, WANG Yufei, et al. Cyber Physical System Risk Assessment of Distribution Network Considering Communication Failure of Multi-agent System for Power Restoration[J]. Automation of Electric Power Systems, 2016,40(17):51-58.
[12] AN Yu, LIU Dong, CHEN Fei, XU Weiwei.Risk Analysis of Physical Distribution Network

Operation Considering Cyber Attack[J/OL]. Power System Technology: 1-8 [2019-05-07].
[13] ZHANG Li, LIU Junyong, LIU Youbo.Static security risk assessment of power system considering wind speed correlation[J].Electric Power Automation Equipment, 2015, 35(04):84-89.
[14] TANG Yi, LI Feng, WANG Qi, NI Ming. Quantitative eval-uation of communication system fault effect on real-time load control of power system[J]. Electric Power Automation Equipment, 2017, 37(02): 90-96.
[15] XUE Yusheng, LI Manli, LUO Jianbo, et al. Modeling Method for Coupling Relations in Cyber Physical Power Systems Based on Correlation Characteristic Matrix[J]. Automation of Electric Power Systems, 2018, 42(02): 11-19.

The Modification of Line Parameters in Distribution Network Considering the Electro-thermal Coordination

FANG Tianyu, ZHANG Hengxu, SHI Fang, HOU Yanruo

The Key Laboratory of Power System Intelligent Dispatch and Control of the Ministry of Education (Shandong University), 17923 Jingshi Road, Jinan, PR China

15953159053@163.com

Abstract. Compared with the transmission network, the distribution network resistance parameters cannot be ignored. The resistance is closely related to the temperature, and the influence of temperature should be considered in the process of parameter estimation. In this paper, the line resistance is corrected in real time by combining the line temperature, resistance and power flow closely through the heat balance equation. Based on the theory of electro-thermal coordination, a method of temperature resistance correction for distribution network is proposed. This method integrates the heat balance equation and the least square method of state estimation, and proposes an extended Jacobian matrix circuit parameter correction algorithm considering the state vector. Then, this paper derives the deviation of the SCADA measurements, PMU measurements and the line temperature in detail. And the Jacobian matrix is extended to make the line temperature and node state can be calculated at the same time, which effectively corrects the overall line resistance of the distribution network. Finally, based on PSCAD / EMTDC IEEE 33 node distribution network model, this paper obtains the simulation data considering load fluctuation and measurement error and analysis the consistency deviation, which verifies the effectiveness and convergence of the algorithm. The results show that the algorithm significantly reduces the estimation error.

1. Introduction

The progress of condition monitoring and data analysis technology greatly promotes the improvement of power system automation level and plays an increasingly important role in the safe and economic operation of power grid. Accurate grid parameters and network topology are the prerequisite for reliable operation of advanced applications such as state estimation. The improvement of SCADA and the application of PMU in the distribution network make it possible to modify the line parameters based on the measurement data, which provides a way to improve the line parameter management level [1-3]. Especially, the parameter identification method using PMU synchronous phasor at both ends of the line has the advantages of strong adaptive ability, high identification accuracy and small calculation.

Compared with the transmission network, the R/X ratio of the distribution network is large. The resistance is closely related to the temperature, while the line temperature is related to the current carrying capacity, environmental temperature and other conditions. In the current research, the parameters of transmission lines are often regarded as constant values, it results the errors in grid parameters.

IOP Conf. Series: Materials Science and Engineering **853** (2020) 012012 doi:10.1088/1757-899X/853/1/012012

Electric -thermal coordination (ETC) is a method that uses electricity to coordinate the line temperature. It can reduce the temperature related errors in power flow calculation, state estimation and correlation analysis. Its basic principle is to establish the analytical relationship among the temperature, line current and environmental weather conditions of the transmission line through the heat balance equation (HBE). The HBE method is used to calculate the temperature of transmission line, and the resistance of transmission line is modified to improve the accuracy of power flow calculation and state estimation. The research of ETC in power system analysis mainly focuses on the following aspects:

(1) In the field of state estimation, reference [4] proposes a tracking estimation of transmission line temperature based on PMU measurement. Using the monotonic relationship between line temperature and resistance, the continuous estimation of line temperature is realized through the estimation of line resistance. However, most of the existing literature deal with the estimation of line temperature and state separately, which leads to a large error in the estimation results[5].

(2) Line parameters correction of distribution network. The ratio of R/X is large in distribution network, so it is more significant to transplant ETC into distribution network to improve the accuracy of power flow calculation. Shu Jun et al [6] introduced ETC into power flow calculation of distribution network for the first time, proposed an extended push back method, and studied the influence of external environmental factors and load levels on power flow distribution.

On the basis of the above research, this paper proposes a modified model of distribution network line parameters, which takes the line temperature into account, in the traditional weighted least square state estimation model. The heat balance equation of the line is taken as pseudo quantity measurement, and the line temperature is introduced into the state quantity. The change of the augmented ratio matrix H after the temperature state quantity increasing is analysed. The proposed algorithm can improve the accuracy of state estimation of the whole network while modifying the parameters of distribution network. Finally, the IEEE 33 node distribution network model based on PSCAD is used to obtain the simulation data considering load fluctuation and measurement error.

2. State estimation of Power Grid Considering electric heat correlation

2.1. Heat balance principle of overhead lines

It is assumed that the overhead transmission line in the distribution network is an ideal conductor with uniform material. According to IEEE-738 standard, when the heat exchange process of the transmission line reaches dynamic balance, the line temperature will be constant, and the heat balance equation can be expressed as follows:

$$f[T(t)] = I^2 R[T(t)] + q_s(t) - q_c[T(t)] - q_r[T(t)] = 0 \tag{1}$$

where T is the temperature of transmission line(°C); $R(T)$ is the AC resistance value of transmission line per unit length (Ω); I is the current value of transmission line (A); q_s、 q_c and q_r are the solar radiation heat absorption of transmission line per unit length quantity (W/m), convective heat transfer (W/m) and radiation heat dissipation (W/m).

2.2. Mixed Measurement and State Vector

In this section, symmetrical lumped parameter line model has been selected. The circuit model is shown in the figure. Wherein, $\dot{U}_{1,2}$ is the voltage phasor at both ends of the line to be identified, $\dot{I}_{1,2}$ is the current phasor at both ends, $P_{1,2}$ is the active power at both ends and the reactive power at both ends.

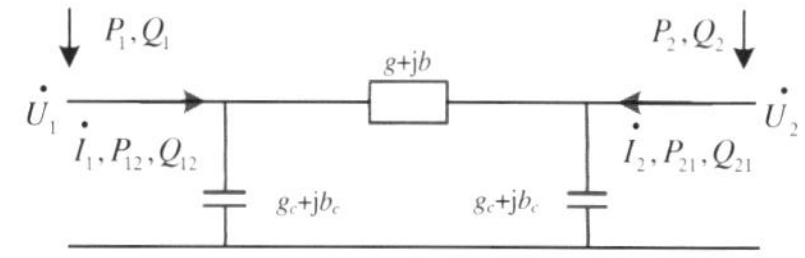

Figure 1. Line model of distribution network

In this paper, by modifying the influence of line temperature on line resistance, the measurement and state vectors of the model are extended, in which the measurement vectors are:

$$z_{\text{aug}} = [z_{\text{SCADA}} \quad z_{\text{PMU}} \quad z_{\text{wea}}]^T \tag{2}$$

$$\boldsymbol{z}_{\text{ele}} = \big[\underbrace{U_1 \ P_1 \ Q_1 \ P_{12} \ Q_{12}}_{\boldsymbol{z}_{\text{SCADA}}} \ \underbrace{U_1 \ \theta_1 \ I_{1,\text{r}} \ I_{1,\text{i}}}_{\boldsymbol{z}_{\text{PMU}}}\big]^T \tag{3}$$

$$z_{\text{wea}} = \left[Q_{\text{solar}} \ \ \vartheta \ \ V_{\text{w}} \ \ T_{\text{a}} \ \ \phi\right]^T \tag{4}$$

The measurement equation is:

$$\boldsymbol{h}_{\text{aug}} = \left[\boldsymbol{h}_{\text{SCADA}} \ \ \boldsymbol{h}_{\text{PMU}} \ \ \boldsymbol{h}_{\text{HBE}}\right]^T \tag{5}$$

The state quantity is:

$$\boldsymbol{x}_{\text{aug}} = [\boldsymbol{U} \quad \boldsymbol{\theta} \quad \boldsymbol{T}_{1\times p}]^T \tag{6}$$

In the formula, z_{SCADA}, $\mathbf{z}_{\text{PMU}}$, $\boldsymbol{h}_{\text{SCADA}}$, $\boldsymbol{h}_{PMU}$ respectively represent the measurement of SCADA and PMU and the corresponding measurement equation; U_l and θ_l are the amplitude and phase angle of the phasor $\overset{g}{U}_1$; Q_{solar}, ϑ, V_w; $\boldsymbol{h}_{\text{HBE}}$ is the steady-state heat balance equation of the line.

2.3. The mixed measurement equation

After considering the electric heat correlation, the line admittance becomes the function of the line temperature, as follows:

$$G_{12}(T) = -g_{12}(T) = -\frac{R_{12}(T)}{R_{12}(T)^2 + X_{12}{}^2} \tag{7}$$

$$B_{12}(T) = -b_{12}(T) = \frac{X_{12}}{R_{12}(T)^2 + X_{12}{}^2} \tag{8}$$

At this time, the non-linear model measured by SCADA and PMU is as follows:

$$\begin{cases} I_{1,\text{r}} = U_1 g\cos\theta_1 - U_2 g\cos\theta_2 - U_1 b\sin\theta_1 \\ \qquad + U_2 b\sin\theta_2 + U_1(g_c\cos\theta_1 - b_c\sin\theta_1) \\ I_{1,\text{i}} = U_1 b\cos\theta_1 - U_2 b\cos\theta_2 + U_1 g\sin\theta_1 \\ \qquad - U_2 g\sin\theta_2 + U_1(b_c\cos\theta_1 + g_c\sin\theta_1) \\ P_1 = U_1 \sum\limits_{1\in 2} U_2[G_{12}(T)\cos\theta_{12} + B_{12}(T)\sin\theta_{12}] = 0 \\ Q_1 = U_1 \sum\limits_{1\in 2} U_2[G_{12}(T)\sin\theta_{12} - B_{12}(T)\cos\theta_{12}] = 0 \\ P_{12} = U_i^2(g+g_c) - U_1U_2 g\cos\theta_{12} - U_1U_2 b\sin\theta_{12} \\ Q_{12} = -U_1^2(b+b_c) - U_1U_2 g\sin\theta_{12} + U_1U_2 b\cos\theta_{12} \\ U_1 = U_1 \\ \theta_1 = \theta_1 \end{cases} \tag{9}$$

According to IEEE-738 standard and meteorological data, the expression of heat loss is obtained as follows:

$$h_{\mathrm{HBE}} = I^2 R(T) + q_{\mathrm{s}} - q_{\mathrm{c}}(T) - q_{\mathrm{r}}(T) = 0 \tag{10}$$

$$R(T) = R_{\mathrm{r}}[1 + \alpha_l (T - T_{\mathrm{d}})] \tag{11}$$

where q_S, q_c and q_r are the solar heat gain, convective heat loss and radiation heat loss rate respectively; α_1 is the temperature coefficient of resistance; T_d is the reference temperature, and RR is the rated resistance under the temperature T_d.

2.4. The derivation of the Jacobian matrix based on the least square method

The circuit temperature and node voltage of this algorithm are calculated at the same time. The state vector is consistent with x_{aug} in the previous section. The measurement vector z_{aug} is:

$$\boldsymbol{z}_{aug} = [\underbrace{u_1\ P_1\ Q_1\ P_{12}\ Q_{12}}_{\boldsymbol{z}_{\mathrm{SCADA}}}\ \underbrace{u_1\ \theta_1\ I_{1,r}}_{\boldsymbol{z}_{\mathrm{SCADA}}}\ \underbrace{I_{1,i}\ \boldsymbol{0}_{1\times p}}_{\boldsymbol{z}_{\mathrm{HBE}}}]^T \tag{12}$$

Compared with the traditional least square method, the measurement dimension and state vector of the model are increased. However, in addition to the change of matrix dimension, the model is still a set of overdetermined equations, so it can still be solved by the weighted least square method. The measurement equation can be expressed as:

$$\boldsymbol{z}_{\mathrm{aug}} = \boldsymbol{h}(\boldsymbol{x}_{\mathrm{aug}}) + \boldsymbol{v}_z \tag{13}$$

where $\boldsymbol{v}_z$ is the vector of measurement errors.
In order to minimize the residual, the objective function formula correction vector in *k*th step is:

$$J(\boldsymbol{x}_{\mathrm{aug}}) = [\boldsymbol{z}_{\mathrm{aug}} - \boldsymbol{h}(\boldsymbol{x}_{\mathrm{aug}})]^T \boldsymbol{R}^{-1}[\boldsymbol{z}_{\mathrm{aug}} - \boldsymbol{h}(\boldsymbol{x}_{\mathrm{aug}})]. \tag{14}$$

$$\begin{cases} \Delta \boldsymbol{x}_{\mathrm{aug},k} = [\boldsymbol{H}^T(\boldsymbol{x}_{\mathrm{aug},k})\boldsymbol{R}^{-1}\boldsymbol{H}(\boldsymbol{x}_{\mathrm{aug},k})]^{-1}\boldsymbol{H}^T(\boldsymbol{x}_{\mathrm{aug},k})\boldsymbol{R}^{-1}\Delta \boldsymbol{z}_{\mathrm{aug},k} \\ \boldsymbol{x}_{\mathrm{aug},k+1} = \boldsymbol{x}_{\mathrm{aug},k} + \Delta \boldsymbol{x}_{\mathrm{aug},k} \end{cases} \tag{15}$$

The extended Jacobian matrix is shown as follows. In order to simplify, block representations of *H*-matrix (the expressions of sub matrix H_{11}, H_{12}, H_{21} and H_{22} are the same as those of general state estimation) are

$$\boldsymbol{H}_{aug} = \frac{\partial \boldsymbol{h}(\boldsymbol{x})}{\partial \boldsymbol{x}} = \begin{bmatrix} \boldsymbol{H}_{11} & \boldsymbol{H}_{12} & \boldsymbol{H}_{13} \\ \boldsymbol{H}_{21} & \boldsymbol{H}_{22} & \boldsymbol{H}_{23} \\ \boldsymbol{H}_{31} & \boldsymbol{H}_{32} & \boldsymbol{H}_{33} \end{bmatrix} = \begin{bmatrix} \partial \boldsymbol{h}_{SCADA} / \partial \boldsymbol{U} & \partial \boldsymbol{h}_{SCADA} / \partial \boldsymbol{\theta} & \partial \boldsymbol{h}_{SCADA} / \partial \boldsymbol{T} \\ \partial \boldsymbol{h}_{PMU} / \partial \boldsymbol{U} & \partial \boldsymbol{h}_{PMU} / \partial \boldsymbol{\theta} & \partial \boldsymbol{h}_{PMU} / \partial \boldsymbol{T} \\ \partial \boldsymbol{h}_{HBE} / \partial \boldsymbol{U} & \partial \boldsymbol{h}_{HBE} / \partial \boldsymbol{\theta} & \partial \boldsymbol{h}_{HBE} / \partial \boldsymbol{T} \end{bmatrix} \tag{16}$$

Take $H_{23\cdot 1}$ as an example, which represents the derivative between the measurement and the corresponding line temperature when PMU is configured on bus 1.

$$\boldsymbol{H}_{23.1} = \begin{bmatrix} & & & 0 & 0 & \\ \mathrm{L} & \mathrm{M} & \mathrm{M} & \mathrm{M} & \mathrm{M} & \mathrm{L} \\ & \frac{\partial U_1}{\partial T_l} & \frac{\partial \theta_1}{\partial T_l} & \frac{\partial I_{1,r}}{\partial T_l} & \frac{\partial I_{1,i}}{\partial T_l} & \\ \mathrm{L} & \mathrm{M} & \mathrm{M} & \mathrm{M} & \mathrm{M} & \mathrm{L} \\ & & & 0 & 0 & \end{bmatrix}^T \tag{17}$$

It should be noted that the partial derivative is not equal to 0 only when the temperature T_L corresponds to the transmission line 12. That is to say, each line has only one element and is not equal to 0. In the submatrix H_{31}, at most two elements in each row are not equal to 0, as follows:

$$H_{31} = \begin{bmatrix} & \cdots & & \cdots & & \cdots & \cdots & \\ 0 & \cdots & \dfrac{\partial h_{HBE.l}}{\partial U_i} & \cdots & 0 & \cdots & \dfrac{\partial h_{HBE.l}}{\partial U_j} & \cdots & 0 \\ & \cdots & & \cdots & & \cdots & \cdots & \end{bmatrix} \begin{matrix} \\ \rightarrow line\ l \\ \\ \end{matrix} \quad \begin{matrix} bus\ i & & bus\ j \end{matrix} \tag{18}$$

where,

$$\frac{\partial h_{HBE.l}}{\partial U_i} = (G_{ij}^2 + B_{ij}^2)\cdot(2U_i - U_j\cos\theta_{ij})\cdot R$$

Similarly, the expression of H_{32} can be derived. Finally, the sub matrix H_{33} is a diagonal matrix, because the *HBE* of transmission line *L* is only related to the corresponding line temperature T_l, namely:

$$H_{33} = \begin{bmatrix} \ddots & & & 0 \\ 0 & \dfrac{\partial h_{HBE.l}}{\partial T_l} & \cdots & 0 \\ 0 & \cdots & \dfrac{\partial h_{HBE.m}}{\partial T_m} & 0 \\ 0 & & & \ddots \end{bmatrix} \tag{19}$$

The relevant elements in the augmented Jacobian matrix *H* have been completely derived. As shown in the figure above, the ETC – WLS state estimation model is very similar to the WLS state estimation model, which only extends the measurement and state vector, and it can correct the parameters of the line in real time during the iteration. Besides, Newton method can still be used to solve the problem.

3. Case study

3.1. The influence of power flow on resistance in distribution network

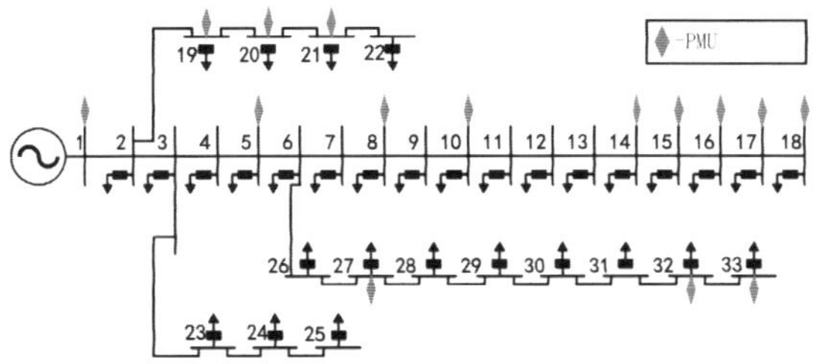

Figure 2. IEEE 33 node distribution network model

The IEEE 33 node distribution network system is used to explain the influence of power flow on line resistance. The system network diagram is shown below (all nodes are equipped with SCADA measurement), and all branches are considered electrothermal correlation. The following table shows the calculation results of line current and line resistance of some branches by adopting the Newton-Raphson method and ETC power flow algorithm. It can be seen that after taking ETC into account, the line resistance increases substantially. At the same time, the change of line resistance affects the node voltage and power flow distribution. Compared with the traditional Newton-Raphson method, the node voltage of the system after taking etc into account decreases by 0.1822% on average. Furthermore, when the climate conditions and transmission current are different, the maximum resistance of the line increases by 18.4537% compared with its rated value. The minimum resistance of the line decreases by 9.1%.

Table1. Comparison of voltage & resistance value of N-R method and ETC

Line Number	Node Voltage			Line Resistance		
	N-R method (p.u.)	ETC (p.u.)	Rate of change %	N-R method (p.u.)	ETC (p.u.)	Rate of change%
1-2	1.0004/0.9973	0.9995/0.9965	0.0914/0.0792	0.0058	0.0063	8.7701
15-16	0.9163/0.9149	0.9160/0.9146	0.0349/0.0357	0.0465	0.0500	7.3612
19-20	0.9967/0.9931	0.9960/0.9925	0.0732/0.0645	0.0938	0.1017	8.4185
24-25	0.9741/0.9707	0.9697/0.9657	0.4515/0.5179	0.0559	0.0662	18.4537
32-33	0.9168/0.9165	0.9165/0.9162	0.0274/0.0294	0.0213	0.0227	6.7617

To sum up, the line power flow and environmental conditions have a significant impact on the line resistance. In power flow calculation and state estimation, if the influence of temperature on the resistance is not considered, the high-level application error will be larger.

3.2. Correction of line parameters considering electrothermal combination

The above example is used to verify the effectiveness of the proposed parameter correction method and to prove that it can effectively improve the state estimation accuracy. The PMU installation location is shown in the above figure. The electrothermal power flow solution is used as the real value of measurement, power measurement and voltage amplitude measurement add 3% and 2% Gaussian white noise to the real value respectively. Monte Carlo simulation is used to randomly generate 1000 sets of measurement data. Traditional WLS estimation and electrothermal joint estimation models are used to calculate the error distribution, assuming that the environmental parameters remain constant during the study period. RMS is defined as follows:

$$H_{33} = \begin{bmatrix} \mathrm{O} & & & 0 \\ 0 & \frac{\partial h_{HBE,l}}{\partial T_l} & \mathrm{L} & 0 \\ 0 & \mathrm{L} & \frac{\partial h_{HBE,m}}{\partial T_m} & 0 \\ 0 & & & \mathrm{O} \end{bmatrix} \tag{20}$$

Table.2 shows the resistance and temperature estimation results of some circuits. It can be seen that, compared with the rated resistance of the line, the real value of the line resistance increases by more than 6% after considering the electrothermal correlation. It has been verified for many times that the error of all line resistances estimated by traditional WLS in this example is more than 5%, and most of them are more than 7%. The average error of the line temperature obtained by the two-step electrothermal estimation method is 0.20%, which shows that the proposed model can effectively correct the error of resistance parameters and improve the estimation accuracy.

Table2. Resistance and temperature estimation results of several circuits

Line	R/p.u.	T/°C	e_{RMS}/%	Line	R/p.u	T/°C	e_{RM}/%
1-2	0.0060	34.67	0.21	19-20	0.1015	35.63	0.30
9-10	0.0078	34.63	0.20	24-25	0.0670	34.97	0.20
15-16	0.0499	34.59	0.23	32-33	0.0008	34.58	0.17

Figure 3 shows the probability density diagram of voltage amplitude error of two estimation methods. It can be found that the estimation errors of both methods are normal distribution, but the traditional results on the left are more scattered and larger. After considering the correction of the line temperature to the resistance parameters, the error of state estimation decreases obviously, and the average error of amplitude decreases from 0.32% to 0.17%. It is further proved that the proposed model can effectively correct the error of resistance parameters and improve the accuracy of state estimation.

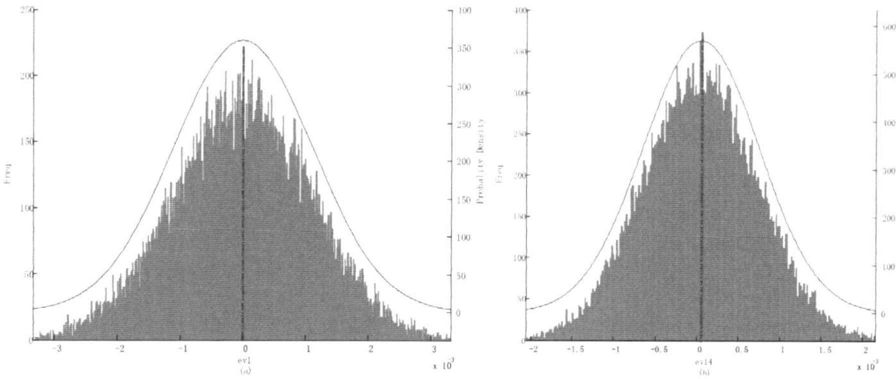

Figure 3. Voltage amplitude error probability density diagram of WLS & ETC-WLS

4. Conclusion

The line resistance parameters of distribution network are constantly changing under the influence of meteorological parameters (such as air temperature, wind speed, solar radiation intensity, etc.) and line current. It is necessary to use the line temperature to modify the line resistance parameters in real time. Based on the traditional WLS, this paper proposes a new method to modify the line parameters of the distribution network, which takes account of the combination of electric heating and electric heating. It can process the temperature estimation and state estimation together, and use the real-time estimation of the line temperature to modify the resistance parameters, improving the accuracy of the line parameters of the distribution network and the precision of state estimation.

Acknowledgments

Supported by National Natural Science Foundation of China (No. 51627811).

References

[1] Karnouskos S , Colombo A W . Architecting the next generation of service-based SCADA/DCS system of systems[C]// IECON 2011 - 37th Annual Conference on IEEE Industrial Electronics Society. IEEE, 2011.

[2] LIHongkui, ZENG Wenting , LI Fujian, et al. A method of line dynamic parameter identification based on SCADA measure-ment[J]. Electrical Measurement & In-strumentation, 2019, 56(04):127-134+147

[3] QI Wenbin, XIAO Xiaochun, LI Dongping, et al. The Difference Between SCADA and WAMS Real-Time Data in Dispatching Cen-ter[J].Southern Power System Technology, 2013, 7(5):87-91

[4] CHEN Fang,HAN Xueshan,LI Huadong,WANG Mengxia,YANG Ming,Tracking Estimation of Transmission Line Temperature Based on PMU Measurement[J], Automation of Electric Power Systems, 2009,33(19):25-29.

[5] CHEN Fang,Zha Hao,HAN Xueshan, et al. Transmission Line Temperature Estimation Based on Semi-parametric Adjustment Model [J]. Automation of Electric Power Systems, 2015, v.39;No.571(21):86-91+117.

[6] SHU Juan, GUAN Rui, et al. Calculation of Distribution Network Power Flow Considering Electrothermal Coupling Effect [J]. Automation of Electric Power Systems, 2015(19):54-59.

Low frequency oscillation online identification based on synchronous phasor of distribution network

Li Yi, Yan Fen, Wang Kun, Sun Chaohui, Zhang Wenwei

Wuxi Power Supply Company, State Grid Jiangsu Electric Power Company, Wuxi 214000, China

liyi_ee@126.com

Abstract. With the advancement of the interconnection process of power grid, the low-frequency oscillation (LFO) problems in the interconnected power grid greatly affect the economic and stable operation of the system. In recent years, the research of synchronous measurement system of distribution network has made great progress, which makes it possible to construct a new generation of low-cost, flexible and reliable online monitoring system of low-frequency oscillation based on the synchronous phasor of distribution network (LFO-DPMU). Based on this, the physical structure and processing flow of LFO-DPMU are constructed in this paper. The LFO-DPMU can extracts dominant mode of oscillation from the distribution network information in real time. For the identification algorithm, this paper selects Prony and HHT algorithm, which two kinds of commonly used oscillation identification algorithms in high-voltage power grid, and compare the characteristics of algorithm in anti-noise, stability and analysis efficiency. Then, combined with the characteristics of distribution network signal, this paper analyses the applicability of the two algorithms. Finally, two algorithms are used to extract the oscillation mode of the measured synchronous phasor signal of distribution network. The analysis result shows the feasibility and accuracy of the whole identification system.

1. Introduction

With the development of large power grid interconnection, UHV transmission lines with heavy load and long distance become the power exchange channels of regional power grid. But at the same time, the low-frequency oscillation (LFO) in the interconnected power grid has a great impact on the economic and stable operation of the system [1]. Among them, the continuous oscillation caused by the weak damping mode seriously restricts the transmission power of the interconnection line. The amplification oscillation caused by the negative damping mode is very easy to affect the stability of the power grid, and even lead to large-scale blackouts. Therefore, it is of great practical significance to identify the dominant mode of system oscillation and take effective measures to suppress it.

At present, two main methods for low-frequency oscillation analysis are model analysis and signal processing. The method based on the grid model analyzes the stability of the system according to Lyapunov's first theorem after establishing the algebraic and differential equations [2]. This method is easy to form dimension disaster with the increase of power grid scale, and it is difficult to meet the needs of real-time analysis. The other method based on signal processing can identify the oscillation characteristics directly according to the measured data in real-time, but it also requires the high precision of the measured data and the identification algorithm [3].

IOP Conf. Series: Materials Science and Engineering **853** (2020) 012013 doi:10.1088/1757-899X/853/1/012013

Wide area monitoring system (WAMS) can provides high-precision data support for signal analysis of low-frequency oscillation [4]. However, due to the high cost of PMU under WAMS and the strict requirements of installation location and transmission line, the coverage of PMU is limited. In recent years, great progress has been made in the research of synchronous measurement system for distribution network. The system places the measurement device DPMU in the low-voltage network, which has the characteristics of low cost and flexible installation, making the dynamic monitoring of distribution network possible. Depending on the high-precision and high-density synchronous phasor signal provided by DPMU, a new generation of on-line monitoring system for low-frequency oscillation (LFO-DPMU) can be established in the distribution network.

In the identification of low-frequency oscillation, common algorithms include Fourier transform, wavelet analysis, Prony algorithm, HHT algorithm. [5]. When the distribution network is connected with a large number of distributed generation and flexible load equipment, its dynamic characteristics are characterized by high disturbance and nonlinearity. When the above identification algorithm is applied to the distribution network, there are problems in the identification effect due to the mechanism of the algorithm. Therefore, it is necessary to study the applicability of the existing identification algorithm according to the characteristics of the measured signal.

On the basis of the above, this paper presents the physical structure and algorithm flow of LFO-DPMU, and compares the identification performance of Prony and HHT algorithm in oscillation mode analysis under synchronous phasor of distribution network. Finally, according to the measured synchronous phasor signals of distribution network in central and north China, the applicability of the two algorithms is compared, and the on-line identification system is further verified accuracy.

2. System architecture of LFO-DPMU

2.1. System physical architecture

The physical structure and processing flow of LFO-DPMU are shown in figure 1. The whole system consists of four layers: synchronous monitoring layer, data transmission layer, data reception layer and LFO online identification layer. The first two layers are distributed at the key nodes of wide area power grid, and the second two are set at the master station of the system.

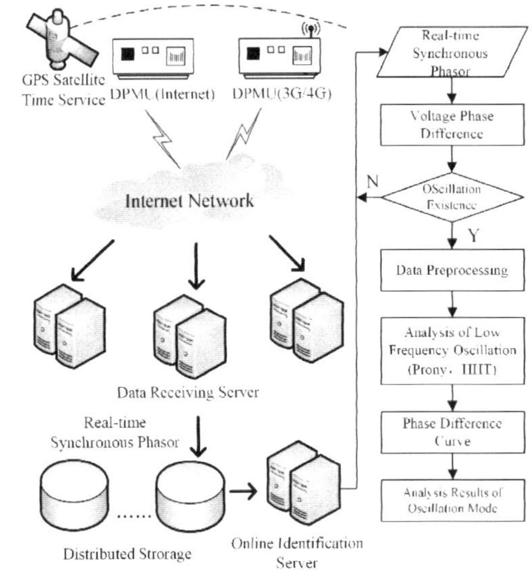

Figure 1. The architecture of LFO-DPMU

DPMU is installed on the key node or bus on the 220V or 10kV side of the low-voltage power grid to record the dynamic behavior of the power grid, and the synchronous time scale is sealed under GPS satellite timing. Because there is no optical fiber laying in the power line of the distribution network, the data transmission network of the device adopts Internet wired network and 3G/4G wireless network. The firewall set in the master station can manage the access rights of the internal and external network of the dispatching center, so as to prevent the illegal network intrusion from threatening the data security. After decoding and extracting the real-time data, the receiving server stores data in the distributed database. Based on this, the identification application server extracts the oscillation modes in real time and presents the analysis results in the front end.

2.2. Identification processing flow

The on-line identification algorithm of LFO is built in the application server. After extracting the phase difference signal according to the voltage phasor in the two regional power grids, the phase difference waveform is detected if there is oscillation feature. Then the oscillation signal is sent to the modal identification module for real-time analysis after preprocessing. The data preprocessing module mainly includes the following three aspects:

(1) Phase mutation detection. When calculating the phase difference according to the measured phase data, as the phase change period is 360 °, the continuous decline from 0 ° will step to 360 °. And the continuous growth from 360 ° will suddenly change to 0 °. This results in the existence of pulse signal in the phase difference waveform. It is necessary to ± 360 ° in the signal step to remove the pulse caused by the angle mutation.

(2) DC and low-pass filtering. The frequency of typical low frequency oscillation signal is generally 0.1~2.5 Hz. Among them, oscillation of 0.1~0.7 Hz usually occurs in the long-distance transmission line with heavy load. Considering the influence of multiple signal interference sources, the high frequency and DC signals should be filtered.

(3) Sliding mean filtering. In the process of measurement and transmission, it is easy to introduce random noise disturbance. With the help of sliding mean filtering, the random noise pollution in the measurement data can be eliminated, and the calculation amount is small, which is suitable for the application in the real-time data analysis process.

3. Adaptability analysis of low frequency oscillation algorithm

Prony and HHT algorithm are two common methods of low-frequency oscillation analysis in WAMS system of high-voltage power grid at present. Therefore, these two algorithms are selected to analyse the adaptability of oscillation mode extraction from synchronous phasor of distribution network.

3.1. The Prony algorithm

Prony algorithm will instead the discrete sampling data $x(n)$ (i=0...N-1) as p exponential functions $\hat{x}(n)$ with any amplitude, phase, frequency and attenuation factor. According to the expression of the exponential function, the characteristic parameters of the measured data can be approximately obtained. In low frequency oscillation analysis, each approximate exponential function represents an oscillation mode, then the original signal is composed of p modes, which can be expressed as:

$$\hat{x}(n) = \sum_{i=1}^{p} b_i z_i^n \qquad b_i = A_i \exp(j\theta_i) \qquad z_i = \exp((\alpha_i + j2\pi f_i)\Delta t) \tag{1}$$

Among them, A_i, θ_i, α_i, f_i (i=1...p) are the amplitude, phase, attenuation factor and oscillation frequency of the signal, Δt is the sampling interval. In order to make the approximate function close to the actual sampling data, the error ε between them should be minimized

$$\varepsilon = \sum_{n=1}^{N} (x(n) - \hat{x}(n))^2 \tag{2}$$

The above equation is solved by the least square method, and eventually can be reduced to the following formula

$$Zb = \hat{x} \tag{3}$$

$$Z = \begin{bmatrix} 1 & 1 & \dots & 1 \\ z_1 & z_2 & \dots & z_p \\ \dots & \dots & \dots & \dots \\ z_1^{N-1} & z_2^{N-1} & \dots & z_p^{N-1} \end{bmatrix} \qquad b = [b_1, b_2, \dots b_p]^T \qquad \hat{x} = [\hat{x}(0), \hat{x}(1), \dots \hat{x}(N-1)]^T \tag{4}$$

Then the amplitude A_i, phase θ_i, attenuation factor α_i and oscillation frequency f_i (i=1...p) of each oscillation mode are calculated directly:

$$\begin{aligned} A_i &= |b_i| \\ \theta_i &= \arctan[\mathrm{Im}(b_i)/\mathrm{Re}(b_i)]/2\pi\Delta t \\ \alpha_i &= In|z_i|/\Delta t \\ f_i &= \arctan[\mathrm{Im}(z_i)/\mathrm{Re}(z_i)]/2\pi\Delta t \end{aligned} \tag{5}$$

3.2. The HHT algorithm
The HHT algorithm consists of empirical mode decomposition (EMD) and Hillbert transform. EMD can decompose the sampled data into multiple intrinsic mode function (IMF) components and residual components. Each IMF component shall meet the following two conditions:

(1) The envelope of the maximum point and the minimum point is symmetric with respect to the time axis;

(2) The number of maximum and minimum points is equal to or at most different from the number of zero crossing points by 1.

Furthermore, Hillbert transform is used to calculate the instantaneous amplitude, frequency and phase of each IMF component. Considering the large margin error of HHT algorithm, the middle analysis result is selected as the initial value, and the amplitude, damping coefficient, frequency and phase parameters of each mode are fitted according to the least square method.

3.3. Applicability analysis of two algorithms
According to the above algorithm mechanism, the performance of Prony and HHT algorithm in noise interference, signal stability and solution speed are compared as follows:

(1) Noise interference. Prony algorithm is sensitive to the influence of noise. The algorithm can get better identification results only when the signal-to-noise ratio is not less than 50~60 dB[6]. Therefore, it is necessary to build a prefilter before using Prony algorithm. HHT algorithm has the ability to filter out the high-frequency disturbance, so it has strong anti-noise interference ability.

(2) Stationary signal identification. The two algorithms can achieve good identification results for stationary signals. Prony algorithm has better identification results for small amplitude components and stronger signal resolution.

(3) Nonstationary signal identification. Prony algorithm is a global oriented method, which can only match a set of oscillation modes. Therefore, the identification results cannot show the transient characteristics of non-stationary signals. HHT algorithm introduces the concept of instantaneous frequency and amplitude, which can extract the oscillation mode of non-stationary signal.

(4) Identification speed. With the increase of the scale of the measurement data set, the order of the equation needed to be solved by Prony algorithm is increasing. And the solution of the high dimension equation limits the speed of identification. HHT algorithm mostly involves elementary operation, and the analysis speed is linear with the data scale.

According to the above four points, the applicability of the two identification algorithms is considered in combination with the characteristics of distribution network. Because the distribution network has many harmonic sources, the noise content of the measured signal is high. And the measurement

IOP Conf. Series: Materials Science and Engineering **853** (2020) 012013 doi:10.1088/1757-899X/853/1/012013

signals of power system are non-stationary and non-linear. Therefore, HHT algorithm is more suitable to extract the oscillation mode from the synchronous phasor of distribution network. When the measurement data is stationary, the oscillation information can be extracted by Prony algorithm after the measurement signal is filtered, so as to obtain higher accuracy identification results.

4. Case study

In this paper, the measured synchronous phasor signal of distribution network in China's central and north interconnection power grid is taken by LFO-DPMU as the research object. The signal data is got from WAMS Light, a wide area measurement system of distribution network exploited by Shandong University [7]. PMU light, a measurement device under the WAMS Light, has been widely used in six major power grids of China. The device is arranged at the low-voltage user side and collects high-precision synchronous phasor of distribution network to the main station all day long, which has become a representative application of distribution network dynamic monitoring in China.

Figure 2 shows the measured phase waveforms of Jinan and Wuhan distribution networks under WAMS light. It shows that the phase difference between the two places is about 40 °.

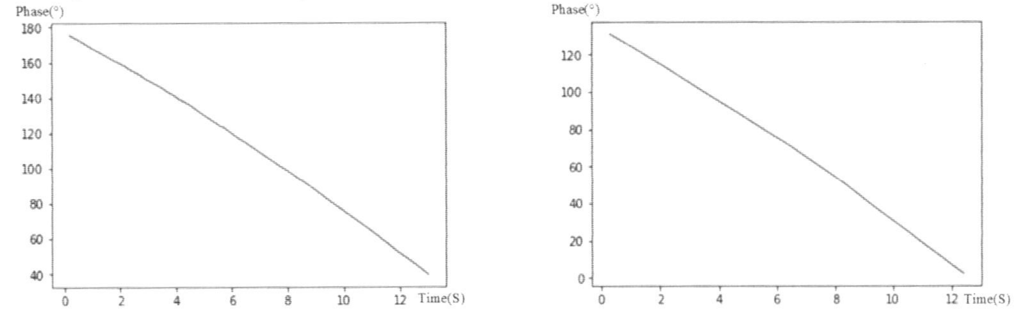

(a) Rgional distribution network in Jinan (b)Rgional distribution network in Wuhan

Figure 2. Measured phase waveform

The phase difference is calculated according to the measured phase data. After the mutation, low-pass filtering and DC elimination, the phase difference waveform is shown in figure 3 (a).

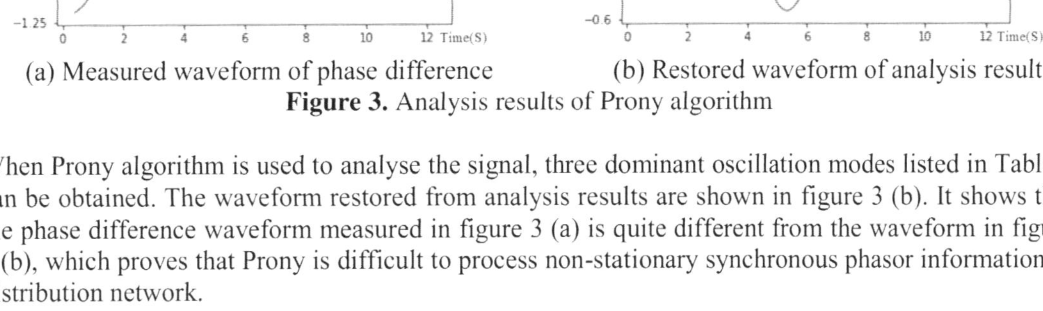

(a) Measured waveform of phase difference (b) Restored waveform of analysis result

Figure 3. Analysis results of Prony algorithm

When Prony algorithm is used to analyse the signal, three dominant oscillation modes listed in Table 1 can be obtained. The waveform restored from analysis results are shown in figure 3 (b). It shows that the phase difference waveform measured in figure 3 (a) is quite different from the waveform in figure 3 (b), which proves that Prony is difficult to process non-stationary synchronous phasor information of distribution network.

Table1. Prony analysis results of measured phasor in distribution network

Oscillation mode i	f_i	α_i	A_i	θ_i
1	0.178	-0.154	1.426	116.405
2	0.184	-0.566	1.278	-113.228
3	1.280	-5.993	0.114	112.316

The HHT algorithm is used to analyse the measured phase difference data in figure 3 (a). The amplitude A_i , phase θ_i , attenuation factor α_i and oscillation frequency f_i of the dominant oscillation mode are 1.197, 2.339, 0.116 and 0.157 respectively. The waveform restored from HHT analysis results are shown in figure 4 (a), which is very close to the original data waveform. The error between the original waveform and restored waveform is shown in figure 4 (b), which shows that the error is large at the boundary. This is because the boundary effect of Hillbert transform, and in the middle of the waveform, HHT algorithm has accurate identification results for the oscillation mode.

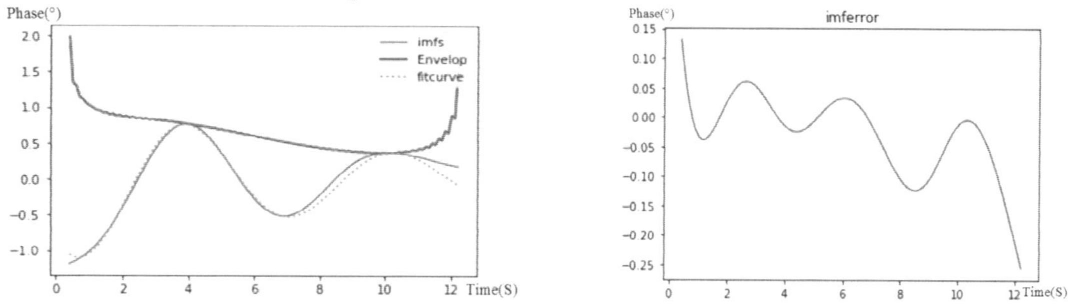

(a) Restored waveform of HHT analysis result (b) Error waveform

Figure 4. Analysis results of HHT algorithm

5. Conclusion

Based on the synchronous phasor of distribution network, this paper constructs an online monitoring system of low-frequency oscillation (LFO-DPMU), which makes full use of the low-cost, flexible and reliable characteristics of measuring equipment, and realizes the real-time extraction of low-frequency oscillation in low-voltage distribution network. The whole process of measurement data acquisition from distribution network to modal extraction is completed with four-layer physical architecture. The mechanism and performance of Prony and HHT are compared, and the applicability of the two algorithms is analysed based on the characteristics of distribution network signals. Finally, the feasibility and accuracy of the LFO-DPMU are proved by the measured synchronous phasor of distribution network, which makes it possible to monitor the low frequency oscillation of the interconnected power grid at the distribution network side.

References

[1] Dunwen S , Xuetao Y et al. A Survey on Analysis on Low Frequency Oscillation in Large-Scale Interconnected Power Grid and Its Control Measures[J]. Power System Technology, 2011.

[2] Li H , Chen H , Yang C , et al. Modal analysis of the low-frequency oscillation of power systems with dfig-based wind farms[J]. Proceedings of the CSEE, 2013(28).

[3] Yang D , Tang W, et al. Extraction of the oscillatory parameters of low frequency oscillation based on improved local mean decomposition[J]. Proceedings of the Csee, 2013, 33(4):135-140.

[4] Yang D, Ding J. WAMS based characteristic discrimination of negative damping low-frequency oscillation and forced power oscillation[J]. Automation of Electric Power Systems, 2013(37).

[5] Anna LI. Research on identifying low frequency oscillation modes based on morphological filtering theory and Prony algorithm[J]. Power System Protection and Control, 2015.

[6] Tang X , Sun Y , Qi Z . Analysis of wind power fluctuation characteristics and its impact on power system low frequency oscillation based on HHT[J]. Power System Technology, 2015.
[7] Zhang HX, Jing ZS, et al. Wide-area measurement system light and its application in China[J]. Automation of Electric Power Systems, 2014(38).

Calculating the winter base load and AC load in Chongqing considering the correlation of temperature and power system load

Ruimiao Wang[1], Zheng Liao[1], Chunyan Li[2], Bo Hu[2], Kaigui Xie[2], Tao Niu[2], Qingsong Sun[2], Gang Wang[2]

[1]State Grid Chongqing Electric Power Research Institute, Chongqing, China
[2]State Key Laboratory of Power Transmission Equipment & System Security, College of Electrical Engineering, Chongqing University, Chongqing, China

1025540819@qq.com

Abstract. The widely use of air conditioning (AC) equipment have brought a large amount of AC load to the power grid, which has an increasingly impact on the safe and economic operation of the power system. In order to correctly estimate the scale of AC load, this paper proposes a new winter AC load calculation method considering the correlation of temperature and load. Firstly, abnormal temperature days (ATD) are removed from the set of representative days (RD) by calculating the correlation coefficient of temperature and load in RD screening process. Secondly, taking into account the daily growth of the base load, the regression analysis is used to fit the base load curve for each day in winter. Then the spring festival (SF) coefficient is proposed to modify the system base load in SF because power system load decreases dramatically during SF. Finally, the winter AC load is equal to the difference between winter power system load and winter base load. The winter AC load in Chongqing 2018 is calculated using the pro-posed method and the result shows the effectiveness of this method.

1. Introduction

In recent years, air conditioning (AC) load has become a significant part in power system load, especially in the south part of China, where the AC equipment is the main device providing heat in winter. With the increase of AC load, the power system load factor decreases and the peak-valley difference increases, making it harder to stabilize the system voltage. At the same time, it poses a severe challenge to the power grid peaking shaving and frequency regulation [1]-[2]. In the voltage collapse accident in Tokyo 1987, the AC load increased sharply at noon and the system suffered from the lack of re-active load. Then the system voltage lost stability and finally the grid voltage collapsed [3].

System operators can't get statistic data of all the AC equipment directly, limited to the present technical and management ability. Thus, the AC load can only be evaluated based on the power system load data. Peak-load comparison method (PLCM) and basic-load comparing method (BLCM) are the widely-adopted methods by system operators to calculate AC load. In PLCM, the daily peak AC load is equal to the difference between daily peak load in winter and daily peak load of the day without AC load [4]-[5]. In BLCM, the winter base load is obtained by calculating the average value of all the representative days (RD) in spring and autumn. Then the AC load is equal to the difference between power system load and base load. However, there are two problems in these two conventional methods. Firstly, power system load has a natural growth trend, which is neglected in both PLCM and BLCM. Secondly, a fixed period is used as RD set in BLCM, during which ab-normal temperature day (ATD) may occur. It means that the power load of the selected RD set may be affected by temperature and the AC load can be miscalculated.

In this paper, a new method to evaluate the AC load is proposed considering the correlation between temperature and power system load. At first, ATD is identified and removed by calculating the correlation coefficient of temperature and load of RD set. Then a set of new RD is get, whose power system load is hardly influence by temperature. To account for the natural growth of base load, the regression analysis is used to fit the base load curves for each day in winter. Power system load decreases dramatically during spring festival (SF). Thus, the SF coefficient is proposed to modify the system base load in SF. Finally, the winter AC load is the difference between winter power system load and winter base load.

The rest of the paper is organized as follows. A RD choosing method is proposed considering the correlation of temperature and power system load in section 2. A base load calculation method is proposed in section 3 based on regression analysis. The influence of SF to power system load is discussed and SF coefficient calculation method is proposed in section 4. We show the calculation process of the paper in section 5. Section 6 is the case study and section 7 is the conclusion.

2. A new RD choosing method considering the correlation of temperature and power system load

AC device is mainly used for heating during winter and cooling in summer. The ab-normal temperature has direct relationship with AC load. The temperature is suitable for people in spring and autumn, generally no AC load appears. Thus, spring days and autumn days are suitable as the RD for base load calculating [6]-[7]. However, the temperature in spring and autumn may be abnormal, like late spring coldness phenomenon. In other words, not all the days in spring and autumn are suitable as the RD. No AC load is allowed to appear in an ideal RD set and the power system load of which shows no relationship with temperature. When abnormal weather occurs in spring and autumn, the daily power system load will increase, which results in a certain correlation between load and temperature. In this paper, a RD choosing method is proposed based on the calculation of correlation coefficient of temperature and power system load. In order to calculate the AC load of this winter, we need to select RD set from this fall and next spring.

If we choose a RD subset from autumn, the choosing process is as follows:

1) Choose typical months and select all days in these months to form the initial RD set $\Omega_{B,aut0}$.
2) The daily electrical energy consumption is obtained from the hourly load of RD in the RD set.

$$E_{aut,j} = \sum_{i=1}^{24} P_{aut,j,i} \Delta t \tag{1}$$

where, $E_{aut,j}$ is the j_{th} daily electrical energy consumption in the RD set. $P_{aut,j,i}$ is the i_{th} hourly load in the j_{th} day. Δt is duration. To be simple, the subscript 'aut' is omitted in the follow part of the paper.
3) Calculate the correlation coefficient ρ between daily base load and daily average temperature. The formulation is shown in (2). The smaller ρ indicates the weaker correlation. To evaluate the strength of correlation, ρ=0.4 is set as the threshold. If ρ<0.4 , the load and temperature is in weak correlation [8]. Then the present set is the ideal RD set and the process is put into an end. If not, go step 4.

$$\rho = \left| \frac{\sum_{j=1}^{D_N}(E_j - \bar{E})(T_{mean,j} - \bar{T}_{mean})}{\sqrt{\sum_{j=1}^{D_N}(E_j - \bar{E})^2 \sum_{j=1}^{D_N}(T_{mean,j} - \bar{T}_{mean})^2}} \right| \tag{2}$$

where, D_N is the number of elements in present set. $\bar{E}$ is the average daily electrical energy consumption and the formulation is shown in (3). $T_{mean,j}$ is the average temperature of j_{th} day in RD set. $\bar{T}_{mean}$ is the average temperature of all day in RD set and the formulation is shown in (4). $|\cdot|$ is the symbol of absolute value.

$$\bar{E} = \frac{1}{D_N}\sum_{j=1}^{D_N} E_j \tag{3}$$

$$\bar{T}_{mean} = \frac{1}{D_N}\sum_{j=1}^{D_N} T_{mean,j} \tag{4}$$

4) Find the day with the biggest $\Delta T_{mean,j}$, where $\Delta T_{mean,j}$ is the difference between daily average temperature and average temperature of the present RD set. Then remove that day from the set to obtain the new set of RD. Go back to step 2.

The RD subset of next spring can be obtained in the same way. The RD subset of this autumn and next spring make up the complete RD set for winter AC load calculating.

3. Winter base load calculation based on regression analysis

In winter, power system load is not only influenced by temperature, but also by the economy growth and other factors. Power system load has the trend of increasing steadily. To evaluate the winter AC load precisely considering the increasing trend, regression analysis is used in this section.

Regression analysis is a classical data analysis method in statistics. The mathematical relationship between dependent and independent variables is determined by establishing regression equation in regression analysis. In this paper, the independent variable is the date of each day in winter, and the dependent variable is the base load of h_{th} hour in each day of winter. Due to the different load characteristic of weekdays and weekends, weekdays subset of RD is used to calculate the base load of winter weekdays, weekends subset of RD is used to calculate the base load of winter weekends.

A quantitative fitting method is deployed to get the hourly load in winter weekdays and weekends considering the growth trend. Taking the calculation of base load value at 1 o'clock in winter weekdays as an example, the choosing process is as follows:

1) Firstly, four models are used to fit the load.

a) Linear model

$$P_{B,w,1} = a_1 t + b_1 \tag{5}$$

b) Parabolic model

$$P_{B,w,1} = a_2 t^2 + b_2 t + c_2 \tag{6}$$

c) Exponential model

$$P_{B,w,1} = a_3 e^{b_3 t} \tag{7}$$

d) Logarithmic model

$$P_{B,w,1} = a_4 + b_4 \ln t \tag{8}$$

where, $a_1, a_2, a_3, a_4, b_1, b_2, b_3, b_4, c_2$ are unknown parameters to be estimated. In this paper, the least square method is used to solve the parameters.

2) Calculate the determination coefficient R^2 of each model, and the indicator is used to test the fitting effect of different regression models. Larger determination coefficient shows a better

effectiveness of regression model[9]. The model with the largest coefficient is selected as the optimal fit equation.

3) Calculate the base load at 1 o'clock in winter weekdays using the selected model.

The base load value from 2 o'clock to 24 o'clock in winter weekdays can be obtained easily in the same way. Similarly, the base load in weekends is calculated by using the weekends data. The winter base load curve is then obtained by splicing all the hourly load value.

4. Modifying the winter base load considering the SF coefficient

The base load curve calculated in section 3 does not take into account the impact of Spring Festival (SF) on social electrical energy consumption. The SF generally in January or February and the power system load during the SF will go through a significant decline due to the shutdown of the factory and public holidays [10].The load characteristic during SF is different from that of ordinary days. A modification should be made towards winter base load considering the SF factor.

In order to take into account the influence of the SF, five days before New Year's Eve to fifteen days after New Year's Eve are set as the SF impact period. Two weeks before the SF impact period is set as the reference period for calculating the SF coefficient. SF coefficient calculation formula is in (9).

$$\lambda_n = \begin{cases} \dfrac{E_n}{\overline{E}_{BF,w}}, n \in \Omega_{SF,w} \\ \dfrac{E_n}{\overline{E}_{BF,o}}, n \in \Omega_{SF,o} \\ 1, n \notin \Omega_{SF} \end{cases} \tag{9}$$

where λ_n is the SF coefficient for n_{th} day. Ω_{SF} is the set of SF impact period. $\Omega_{BF,w}$ and $\Omega_{BF,o}$ are the subset of weekdays and weekends of Ω_{SF}. E_n is the daily electrical energy consumption for n_{th} day, which is calculated by formula (1). $\overline{E}_{BF,w}$ is the average electrical energy consumption in weekdays in SF reference period, while $\overline{E}_{BF,o}$ is the average electrical energy consumption in weekends in SF reference period.

The base load curve considering SF influence is calculated in (10):

$$P'_{B,n,h} = \lambda_n P_{B,n,h} \tag{10}$$

where $P_{B,n,h}$ is the h_{th} hourly load in n_{th} day before modification and $P'_{B,n,h}$ represent the modified hourly load.

5. Winter AC load calculation process

Firstly, select all the days in typical spring month and autumn month to form the initial RD set. Execute the screening process to initial RD by considering the correlation between temperature and power system load. The subset of weekdays and weekends are obtained accordingly. Secondly, regression analysis is deployed to get the winter base load and SF coefficient is calculated to modify the winter base load. Finally, the winter AC load is the difference between winter power system load and winter base load.

$$P_{AC,n,h} = P_{L,n,h} - P'_{B,n,h} \tag{11}$$

In this formula, $P_{AC,n,h}$ is the h_{th} hourly AC load in n_{th} day. $P_{L,n,h}$ is the h_{th} hourly power system load in n_{th} day.

The AC load calculation process is shown in Fig. 1.

6. Case study

6.1. Case Introduction

In this section, the winter AC load in Chongqing 2018 is calculated base on the hourly load data and daily average temperature data from September 1st, 2018 to July 30th, 2019.

Chongqing has a typical continental monsoon climate. The heating demand in Chongqing winter is high and air-conditioning equipment is the main equipment for providing heat. In this paper, September 1st, 2018 to October 31st, 2018 is set as the autumn of 2018. November 1st, 2018 to February 28th, 2019 is the winter of 2018. March 1st, 2019 to April 30th, 2019 is the spring of 2019. The hourly load curve of Chongqing winter 2018 is shown in Fig. 2. We can see that the daily load curves in winter have the similar shape. There are two peaks in the curve occurring in the morning and night in every day. In addition, it is obvious that the load level decreases during SF.

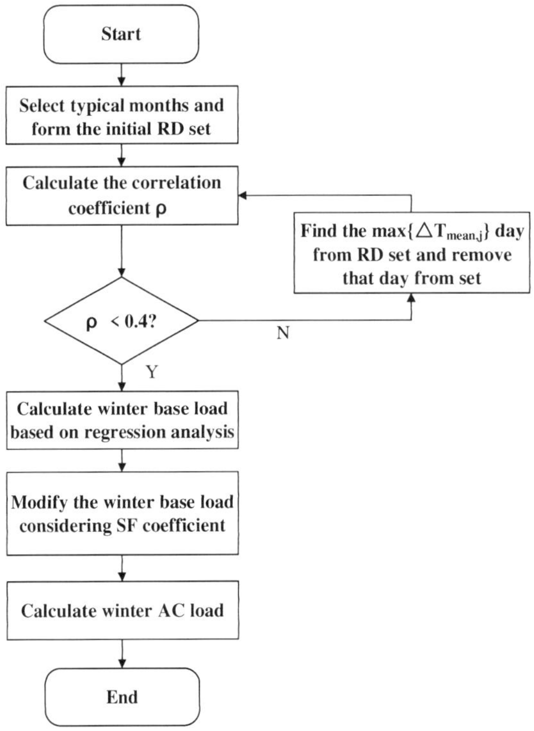

Figure 1. AC load calculation process

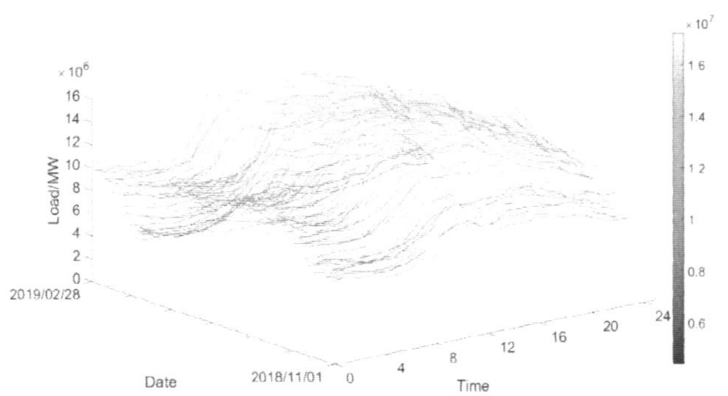

Figure 2. Winter daily load curve of Chongqing 2018

6.2. Analysis to the winter AC load in Chongqing 2018

Base on the RD choosing method in section 2, all the ATD in fall 2018 and spring 2019 is removed from the original RD set. Then we get the spring RD set and autumn RD set. Take autumn 2018 as an example, Fig.3 shows the daily electrical energy consumption and average temperature in autumn 2018 in Chongqing. It shows that the average temperature went through an increase in both September 1st, 2018 and September 19th, 2018. At the same time, the AC load increased too, which caused two load peaks in the two periods. As a result, the correlation between power system load and temperature in autumn 2018 is relatively high.

Table 1 shows the average temperature of ATD and the correlation coefficient ρ in the screening process of autumn 2018. In the process, abnormal temperature data is detected and removed from initial the RD set. It can be seen that the two high-temperature weather processes in September are included in removed ATD set. As the screening process progresses, and the correlation between power system load and temperature decreases.

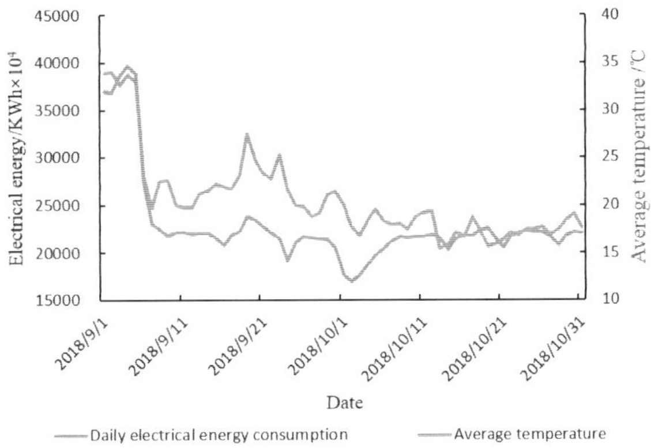

Figure 3. Daily electrical energy consumption and average temperature in autumn Chongqing 2018

Table 1. The RD set screening process in autumn 2018

Screening Time	Removed ATD	Average Temperature of ATD	Correlation Coefficient ρ after Screening
1	2018/09/02	34	0.579
2	2018/09/01	33.9	0.552
3	2018/09/04	33.7	0.515
4	2018/09/05	33.1	0.496
5	2018/09/03	32.7	0.506
6	2018/09/19	27.5	0.494
7	2018/09/23	25.3	0.472
8	2018/09/20	24.7	0.471
9	2018/09/21	23.4	0.431
10	2018/09/06	23.2	0.449
11	2018/09/18	23.2	0.436
12	2018/09/22	22.8	0.407
13	2018/09/09	22.6	0.374

Regression analysis is used to fit the hourly base load in winter based on the daily load curve of autumn RD and spring RD. January 30th, 2019 to February 19th is set as the SF influence period and January 16th, 2019 to January 29th is set as SF reference period. Based on the daily load data, the SF coefficient during SF is calculated. Table 2 shows SF coefficient of four days around SF It can be seen that the power system load in Chongqing decreased before SF and increased after SF but the impact of SF towards power system load didn't completely disappear after fifteen days of the Chinese New Year's Eve. The original base load is multiplied by SF coefficient and the adjusted base load curve in

Chongqing 2018 is obtained accordingly. The winter AC load in Chongqing 2018 is equal to the difference between power system load and base load, which is shown in Fig. 4.

Table 2. SF coefficient of four days in winter of 2018

Date	SF Coefficient
2019/02/01	0.740
2019/02/04	0.537
2019/02/05	0.487
2019/02/19	0.884

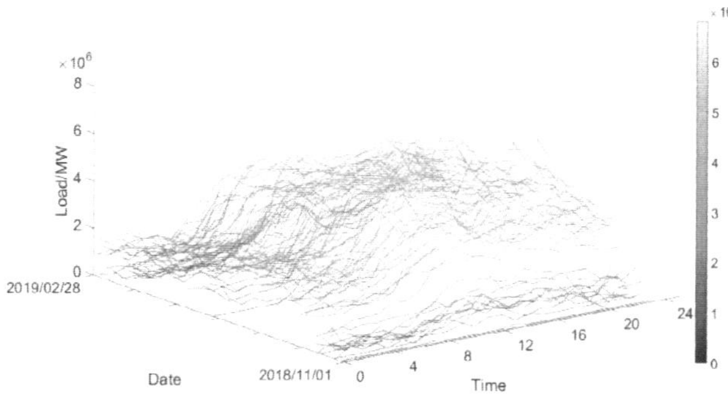

Figure 4. Winter AC load curve in Chongqing 2018

6.3. Analysis of calculation results compared to BLCM

To verify the effectiveness of the proposed AC load calculation method, BLCM is used to calculate the winter AC load in Chongqing 2018 and Fig.5 shows the winter AC peak load curves in Chongqing 2018 calculated by the proposed method and BLCM. It can be seen that the two curves in Fig.5 have the same trend. However, the AC peak load in early winter calculated by BLCM is lower than that by the method proposed in this paper. It is because that BLCM using the same base load curve for the entire winter. The base load curve is obtained by calculating the average value of RD in autumn 2018 and spring 2019, which leads to the calculated based load in early winter is higher and the AC load is lower.

We also calculated correlation coefficient between winter peak AC load and different types of temperature, including daily peak temperature, daily valley temperature and daily average temperature. The results are shown in Table 3. We can see that the peak AC load has the strongest relationship with daily average temperature, compared to that with daily peak temperature and daily valley temperature, calculated in both methods. It is also easy to find that AC load calculated by proposed method has stringer correlation with different types of temperature, which illustrates the correctness and effectiveness of the proposed AC load calculation method.

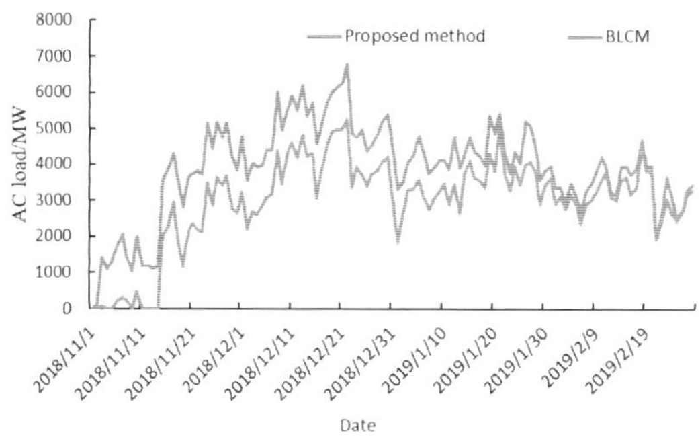

Figure 5. Winter AC peak load in Chongqing 2018 using two methods

Table 3. Correlation coefficient of winter peak AC load and different types of temperature in two methods

Type of Temperature	Proposed Method	BLCM
Daily Peak Temperature	-0.6772	-0.5846
Daily Valley Temperature	-0.6057	-0.5725
Daily Average Temperature	-0.6863	-0.6214

7. Conclusions

A new winter AC load calculation method is proposed in this paper. A screening process is deployed to remove the ATD from the initial RD set. After the screening process, a set of RD hardly influenced by temperature is obtained. Regression analysis is carried out to consider the growth trend of base load and SF coefficient is calculated to consider the impact of SF towards power system load. Finally, the proposed method is used in calculating the winter AC load in Chongqing 2018. The results are compared to the results using BLCM. It is concluded that the load calculated by the proposed method has higher correlation with temperature, which illustrate the correctness and effectiveness of the proposed method.

Acknowledgments

The work is supported by the Science and Technology Project of State Grid Chongqing Power Company under the project of "Key technologies and methods of Chongqing social electricity consumption prediction considering the influence of multiple factors".

References

[1] B. R. Williams, W. R. Schmus and D. C. Dawson, "Transmission voltage recovery delayed by stalled air conditioner compressors," *IEEE Trans. Power Syst.*, vol. 7, no. 3, pp. 1173-1181, Aug. 1992.

[2] S. M. R. Sanhueza, F. L. Tofoli, F. L. de Albuquerque, J. C. de Oliveira and G. C. Guimaraes, "Analysis and evaluation of residential air condtioners for power system studies," *IEEE Trans. Power Syst.*, vol. 22, no. 2, pp. 706-716, May 2007.

[3] A. Kurita and T. Sakurai, "The power system failure on July 23, 1987 in Tokyo," *Proceedings of the 27th IEEE Conference on Decision and Control*, Austin, Texas, pp. 2093–2097, 1988.

[4] J. Zheng, S. Zhu, G. Wang and X. Wang, "Key characteristics and modeling of air conditioner load," *Proceedings of the CSEE*, vol. 29, no. 10, pp. 67-73, Apr. 2009.

[5] Q. Wen, J. Li and J. Zhao, "Air-conditioner load calculation methods and its application," *Power Demand Side Management*, vol. 7, no.4, pp. 16-18, Jul. 2005.

[6] H. Song, G. Tang, Y. Lu and Y. Li, "Analysis on summer air-conditioning loads composition in Jiangsu Province and estimation of peak load shifting effect by DSM measures," *Power system technology*, vol. 30, no. 17, pp. 88-96, Sep. 2006.

[7] M. Xie, J. Deng, M. Liu, J. Li, Q. Sun and L. Tan, "Temperature-lowering load estimation method based on meteorological data and entropy weight theory," *Automation of electric power systems*, vol. 40, no. 3, pp. 135-139, Feb. 2016.

[8] J. Rodgers, W. Nicewander. "Thirteen ways to look at the correlation coefficient," *American statistician*, vol. 42, no. 1, pp. 59-66, 1988.

[9] P. Kamphuisen *et al*. "Familial clustering of fator VIII and von Willebrand factor levels," *Throbosis & Haemostasis*, vol. 79, no. 2, pp. 323-327, 1998.

[10] L. Chen, J. Mai, J. Yin. " Parameters Estimated Subsection Predicting Model for Electrical energy consumption in the Spring Festival," *Internatioal*

IOP Conf. Series: Materials Science and Engineering **853** (2020) 012015 doi:10.1088/1757-899X/853/1/012015

Deadbeat Predictive Current Control with Disturbance Observer for Permanent Magnet Synchronous Motor

Zhang Chuntao[1]; Dang Ruina[2]; Wang Yang[2]; Zhang Chengning[1]; Zhang Shuo[1]

[1]School of Mechanical Engineering, Beijing Institute of Technology, Beijing, China
[2]China North Vehicle Research Institute, Beijing, China

chuntao@bit.edu.cn; ruinadang@163.com;

Abstract. Deadbeat predictive current control (DPCC) is a kind of motor control method with excellent dynamic response performance. To increase the parameter robustness of DPCC, a disturbance observer (DO) is designed based on the extended Kalman filter (EKF). Firstly, this paper derives the DPCC method based on the mathematical model of permanent magnet synchronous motor (PMSM). Then, the DO is designed based on the EKF. The voltage disturbances caused by parameters mismatch are selected as the state vector, and the voltage disturbances at each time are calculated by the EKF iteratively. Thus, a novel control algorithm combining DPCC with DO is proposed. The simulation results of DPCC method are compared with the results of DPCC+DO method to verify the effectiveness of the proposed method.

1. Introduction

For permanent magnet synchronous motor (PMSM), the robustness and response speed of internal current loop plays a decisive role in the performance of the whole control system. Deadbeat predictive current control (DPCC) as a non-linear control method has better dynamic response performance than traditional PI control [1]. However, DPCC is based on motor model with accurate parameters. When the parameters of motor changed due to temperature or other factors, the calculated reference voltage will deviate from the expected value. Therefore, how to improve the parameter robustness of DPCC is of great significance to the stable operation of the motor [2].

Some researchers have studied methods of parameter identification to mitigate the performance degradation caused by model uncertainty [3-5]. However, the methods are susceptible to interference and have the shortcomings of large amount of calculation. In addition, approaches that combining the DPCC and disturbance observer (DO) can improve control performance as well [6]. Zhang [7] proposed a DO based on sliding mode exponential reaching law, but the inherent high-frequency chattering of sliding mode control will affect the performance of the motor. It is a relatively effective method to correct the output voltage of DPCC by compensation. However, how to design a simple and accurate DO remains to be studied.

In order to overcome the shortcomings of the existing DO, a DO based on extended Kalman filter (EKF) is designed in this paper. The DO can observe the system voltage disturbance caused by parameter mismatch accurately and improve the parameter robustness of DPCC.

2. DPCC and Disturbance Observer

2.1. Mathematical Model of PMSM and DPCC

In this paper, the surface-mounted PMSM is taken as the research object. The voltage equation of the motor in the α-β coordinate system is as follows:

$$\begin{cases} u_\alpha = R_s i_\alpha + L_s p i_\alpha - \omega_e \psi_r \sin\theta \\ u_\beta = R_s i_\beta + L_s p i_\beta + \omega_e \psi_r \cos\theta \end{cases} \tag{1}$$

In the equation, u_α and u_β are the stator voltage in the α-β coordinate system; i_α and i_β are the stator current in the α-β coordinate system; ψ_r is the rotor flux; R_s is the stator resistance; L_s is the stator inductance; ω_e is the rotor electrical palstance; θ is the rotor position angle; p is the differential operator.

The model described in equation (1) is ideal, that is, assuming that the parameters of motor remain unchanged. Considering the variation of motor parameters, equation (1) is rewritten as follows:

$$\begin{cases} u_\alpha = R_{se} i_\alpha + L_{se} p i_\alpha - \omega_e \psi_{re} \sin\theta + f_\alpha \\ u_\beta = R_{se} i_\beta + L_{se} p i_\beta + \omega_e \psi_{re} \cos\theta + f_\beta \end{cases} \tag{2}$$

Where, R_{se} 、 L_{se} 、 ψ_{re} are the estimated value of motor parameters, and f_α、 f_β are the voltage disturbance caused by the parameter mismatch. The voltage disturbance(f_α、 f_β) are defined as follows:

$$\begin{cases} f_\alpha = \Delta R_s i_\alpha + \Delta L_s p i_\alpha - \Delta\psi_r \omega_e \sin\theta \\ f_\beta = \Delta R_s i_\beta + \Delta L_s p i_\beta + \Delta\psi_r \omega_e \cos\theta \end{cases} \tag{3}$$

If the sampling time T_{sc} is short enough, the voltage equation (1) can be discretized by using the first-order Euler equation, and the voltage equation of PMSM in the discrete state can be obtained. Based on the voltage equation in the discrete state, the DPCC can be expressed as follow:

$$\begin{bmatrix} i_\alpha(k+1) \\ i_\beta(k+1) \end{bmatrix} = (1 - \frac{T_{sc}R_s}{L_s}) \cdot \begin{bmatrix} i_\alpha(k) \\ i_\beta(k) \end{bmatrix} + \frac{T_{sc}}{L_s} \cdot \begin{bmatrix} u_\alpha(k) \\ u_\beta(k) \end{bmatrix} + M(k) \tag{4}$$

$$\begin{bmatrix} u_\alpha(k+1) \\ u_\beta(k+1) \end{bmatrix} = \frac{L_s}{T_{sc}} \cdot \left\{ \begin{bmatrix} i_\alpha^*(k+2) \\ i_\beta^*(k+2) \end{bmatrix} - (1 - \frac{T_{sc}R_s}{L_s}) \cdot \begin{bmatrix} i_\alpha(k+1) \\ i_\beta(k+1) \end{bmatrix} - M(k+1) \right\} \tag{5}$$

Where, $i_\alpha^*(k+2)$ and $i_\beta^*(k+2)$ represent the reference current values at k+2 time; M is defined as follows:

$$M(k) = \begin{bmatrix} \frac{T_{sc}}{L_s} \omega_e(k) \psi_r \sin\theta & -\frac{T_{sc}}{L_s} \omega_e(k) \psi_r \cos\theta \end{bmatrix}^T \tag{6}$$

2.2. Disturbance observer

The calculation process of EKF can be divided into two parts: prior prediction and posterior correction.Based on the EKF algorithm, the state vector x and the control vector u can be selected. It is simpler to use the output voltage of DPCC as the control vector than to collect the output voltage of the inverters as the control vector.

$$x = \begin{bmatrix} i_\alpha & i_\beta & \omega_e & \theta & f_\alpha & f_\beta \end{bmatrix}^T \tag{7}$$

$$u = \begin{bmatrix} u_\alpha & u_\beta \end{bmatrix}^T \tag{8}$$

Then, by rewriting equation (2) in the form of current state equation, the corresponding non-linear equation of PMSM can be written out as equation (9). The change of motor parameters are slow relative to the change of current, so it can be considered that the change of voltage disturbances caused by parameter mismatch are slow relative to the change of current, and they remain unchanged during the interval of a sampling period, i.e., the derivative of disturbances are zero.

$$\dot{x} = f(x) + Cu + w \tag{9}$$

Where, w is the system noise, C is the gain matrix of control variables, and f (x) is the state transition function.

$$f(x)=\begin{bmatrix} f_1 \\ f_2 \\ f_3 \\ f_4 \\ f_5 \\ f_6 \end{bmatrix}=\begin{bmatrix} (-R_{se}i_\alpha+\psi_{re}\omega_e\sin\theta-f_\alpha)/L_{se} \\ (-R_{se}i_\beta-\psi_{re}\omega_e\cos\theta-f_\beta)/L_{se} \\ 0 \\ \omega_e \\ 0 \\ 0 \end{bmatrix} \tag{10}$$

$$C=\begin{bmatrix} L_{se}^{-1} & 0 & 0 & 0 & 0 & 0 \\ 0 & L_{se}^{-1} & 0 & 0 & 0 & 0 \end{bmatrix} \tag{11}$$

The stator current, rotational speed and rotor position angle are selected as the observation vector. Then the corresponding measurement equation of motor control system can be obtained.

$$y=\begin{bmatrix} i_\alpha & i_\beta & \omega_e & \theta \end{bmatrix}^T \tag{12}$$

$$y=h(x)+v \tag{13}$$

Where, v is the measurement noise and h (x) is the measurement function.

$$h(x)=\begin{bmatrix} i_\alpha & i_\beta & \omega_e & \theta \end{bmatrix}^T \tag{14}$$

The DO in this paper has the function of observing the voltage disturbance caused by parameter mismatch in real-time. Using EKF algorithm to linearize the non-linear motor model can minimize the amount of on-line calculation and ensure the good dynamic current tracking characteristics in the process of motor operation. Linearization of f(x) and h(x) is carried out as follows:

$$F(x(t))=\frac{\partial f}{\partial x}\Big|_{x=x(t)} \tag{15}$$

$$H(x(t))=\frac{\partial h}{\partial x}\Big|_{x=x(t)} \tag{16}$$

2.3. DPCC with DO

Combining the designed DO with DPCC by feed-forward compensation, the output voltage of DPCC can be corrected by the DO, which can effectively improve the parameter robustness of DPCC. The block diagram of DPCC with DO is shown in Fig. 1.

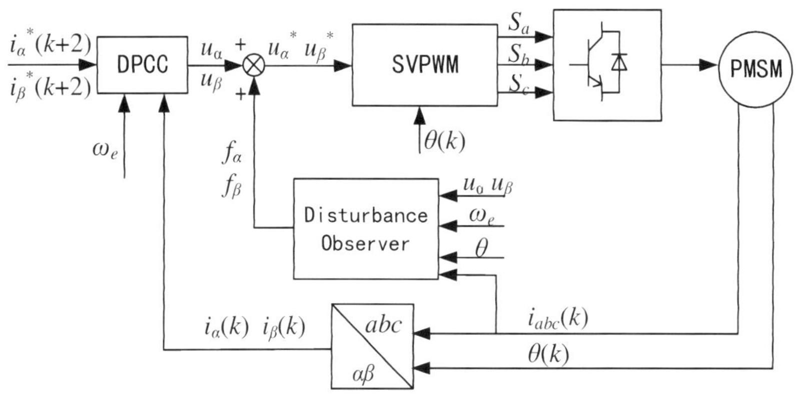

Figure 1. Block diagram of DPCC with DO.

3. Results and Discussions

In this part, the verification test was carried out on Matlab/Simulink. The test was conducted in a comparative way. Firstly, the traditional DPCC method was simulated for parameter mismatch, and then the DPCC+DO method was simulated for the same parameter mismatch. The effect of DO on improving the parameter robustness is analysed through the different performances of the motor. The parameters of the PMSM used in the simulation are shown in Table 1. As for the initialization of the DO, the initial value of the covariance matrix of the state vector as well as the initial values of the covariance matrix of system noise and measurement noise were set as follows:

$$P_0 = [0.1 \quad 0.1 \quad 500 \quad 0.5 \quad 10 \quad 10]^T \tag{17}$$

$$Q_d = diag(0.1, 0.1, 300, 0.2, 2, 2) \tag{18}$$

$$R = diag(0.001, 0.001, 0.001, 0.001) \tag{19}$$

Table1. Parameters of PMSM for Test.

Motor parameters	Value
d-axis and *q*-axis inductances	0.001225 mH
Stator phase resistance	0.365 Ω
Flux linkage of permanent magnet	0.1667 Wb
Rotational inertia	0.00194 kg·m^2
Number of pole pairs	4

In the simulation, the initial target speed was set to 800 r/min and the motor started without load; the load torque of 10N·m was applied to the motor when the test had been carried out to 0.04s; and the target speed of the motor was raised to 1200 r/min when the test had been carried out to 0.1s.

Fig. 2 is the current variation diagram of DPCC under different parameters mismatch. Based on the feed-forward compensation method shown in Fig. 1, a DPCC+DO simulation model was built. The results are shown in Fig. 3. By comparing Fig. 2 with Fig. 3, it is concluded that: the DPCC+DO method can reduce the ripple caused by inductance mismatch effectively and compensate for the deterioration of dynamic characteristics caused by resistance mismatch and flux mismatch to a certain extent.

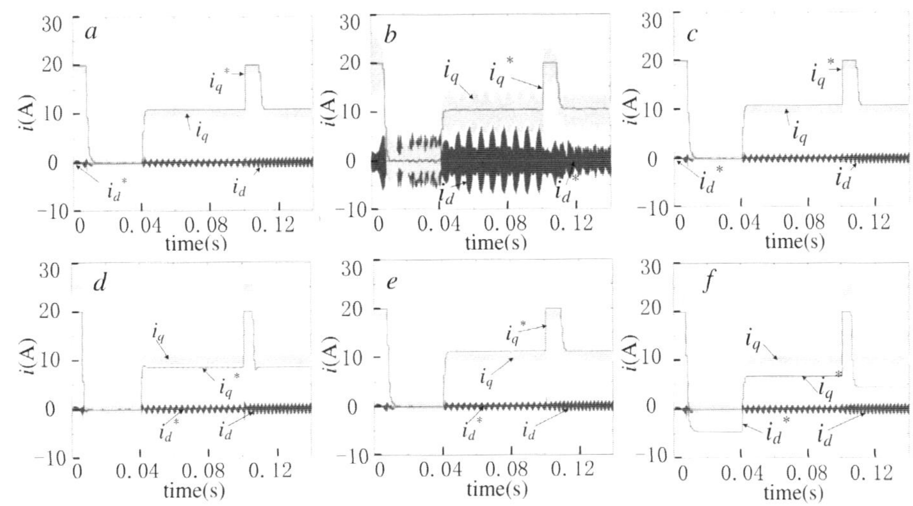

Figure 2. Current variation diagram of DPCC under parameter mismatch.
(*a*) No mismatch of parameters (*b*) $L_{se}=2L_s$ (*c*) $L_{se}=0.5\ L_s$ (*d*) $R_{se}=10R_s$ (*e*) $R_{se}=0.1R_s$ (*f*) $\psi_{re}=2\psi_r$

Considering that only the inductance mismatch has a significant impact on the steady-state of the motor, the amplitude of current ripple in the case of inductance mismatch (L=2L_0) is recorded in Table 2. It can be concluded that when the inductance mismatch occurs, the amplitude of current ripple in q-axis can be reduced by 80.27% (0.04s-0.1s) and 70.9% (0.1s-0.14s) respectively, and the amplitude of current ripple in d- axis can be reduced by 89.14% (0.04s-0.1s) and 70.96% (0.1s-0.14s), respectively. The results reveal that the DO can effectively reduce the ripple caused by the mismatch of inductance.

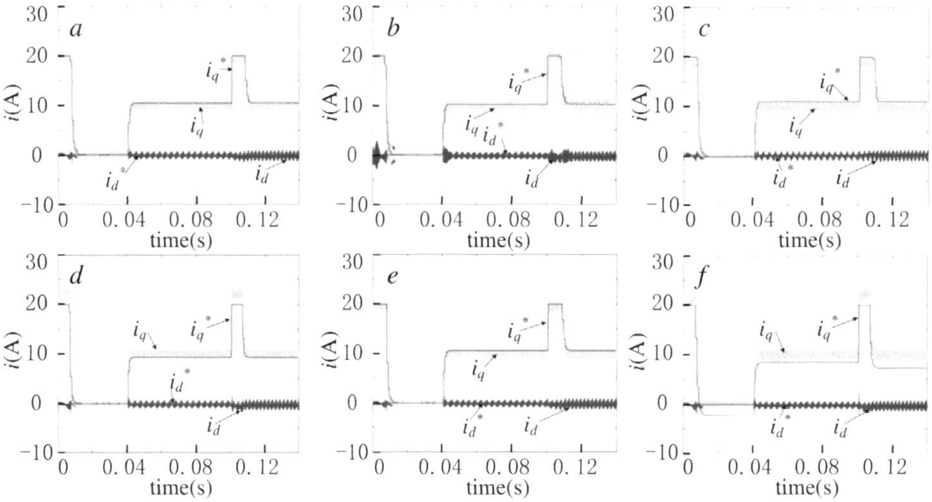

Figure 3. Current variation diagram of DPCC+DO under parameter mismatch.
(a) No mismatch of parameters (b) L_{se}=2L_s (c) L_{se} =0.5 L_s (d) R_{se}=10R_s (e) R_{se} =0.1R_s (f) ψ_{re}=2ψ_r

Table 2. Quantitative comparison of DPCC and DPCC+DO under inductance mismatch (L_{se}=2L_s).

	Method	DPCC	DPCC+DO
i_q	0.04s-0.1s-0.14s	9.82 A & 7.63 A	1.94 A & 2.22 A
i_d	0.04s-0.1s-0.14s	14.54 A & 7.82 A	1.58 A & 2.27 A

4. Conclusion

In this paper, the DO based on EKF is combined with DPCC by feed-forward compensation, and the novel method improves the parameter robustness of DPCC. Using the outputs of the DO, the state of the motor can be monitored in real time. Also we admit it that if the convergence speed of EKF can be faster, e.g. by reducing its dimension, the dynamic characteristics of the observer will be better.

Acknowledgments
This research was funded by R&D plan in key areas of Guangdong province (NO.2019B090910001).

References

[1] Zhang Y, Xia B, Yang H. Performance evaluation of an improved model predictive control with field oriented control as a benchmark[J], IET Electric Power Application, 2017, 11: 677-687.
[2] Xia C, et al. Robust model predictive current control of three-phase voltage source PWM rectifier with online disturbance observation[J], IEEE Trans. on Industrial Inform., 2012, 8: 459-471.
[3] Morimoto S, Sanada M, Takeda Y. Mechanical sensorless drives of IPMSM with online parameter identification[J]. IEEE Trans. on Industrial Application, 2006, 42: 1241-1248.
[4] Mobarakeh B, Meibody F, Sargos F. On-line identification of PMSM electrical parameters

based on decoupling control[C]. IEEE Industry Applications Conference, 2001: 266-273.
[5] Hamida M, Leon J, Glumineau A, et al. An adaptive interconnected observer for sensorless control of PM synchronous motors with online parameter identification[J]. IEEE Trans. on Industrial Electronics, 2013, 60(2): 739–748.
[6] Kim K. A Current Control for a Permanent Magnet Synchronous Motor with a Simple Disturbance Estimation Scheme[J]. IEEE Trans. on Control Systems Tech., 1999, 7: 630-633.
[7] Zhang X, Hou B, Mei Y, et al. Deadbeat predictive current control of permanent-magnet synchronous motors with stator current and disturbance observer[J], IEEE Trans. on Power Electronic, 2017, 32(5): 3818-3814.

The Realization of Intelligent Knowledge Adaptive Learning Method in the Field of Substation Operation and Maintenance

Jianlin Yin[1], Yali Liang[1], Qi Gao[2], Lu Liu[2], Yabin Lei[1], Xilan Zhao[1]

[1.] State Grid Gansu Electric Power Company, Lanzhou 730030, Gansu Province, China;
[2.] NARI Research Institute Beijing R&D Center of Nari Group, Haidian District, Beijing 100094, China

982739104@qq.com

Abstract: In view of the lack of targeted and personalized knowledge learning in existing substation operators, this paper proposes an intelligent adaptive knowledge learning method in the field of substation operation and maintenance. First, an adaptive knowledge learning system is established including domain model, user model, teaching model and adaptive engine. In the user model, the item response theory (IRT) is used to estimate the learner's master degree of knowledge points contained in the domain model, and then their learning needs are obtained. Meanwhile, the Bayesian network method is used to speculate the cognitive style of learners. On the basis of a comprehensive consideration of learning needs and cognitive style, the particle swarm optimization algorithm (PSO) in the teaching model is used to recommend the optimal learning resource for learners to realize the high-efficiency personalized knowledge learning in substation operation and maintenance.
Key words: substation operation and maintenance, adaptive learning, cognitive style, item response theory, particle swarm optimization algorithm

1. Introduction

The safe and stable operation of substation is of great significance to ensure the safety of the power grid. With the development of ubiquitous power internet of things and smart grid, the application of abundant new technologies, new equipment and new systems in substations has accelerated the updating of the knowledge on substation operation and maintenance, which demands higher abilities of substation personnel. However, the traditional training methods for the personnel featured with large amounts and different ability level are mostly short of pertinence and personalization. Taking the online university platform owned by the State Grid Corporation as an example, it simply moves offline learning resources and examination questions to the web page but cannot generate dynamically the learning content special for each user, descending the training outcomes and quality.

In recent years, the rapid development of artificial intelligence provides the possibility to effectively and further enhance the knowledge and skill level of substation personnel. Among them, the adaptive learning technology [1, 2] is able to select, assemble, and present personalized learning content for users through a deep and real-time analysis of their learning behaviors and requests. This technology has been successfully applied in systems like Knewton, Squirrel AI Learning, etc. on the area of K-12 education. Adaptive learning technology involves a variety of computer science and artificial intelligence methods, including ontology [3], data mining [4, 5], item response theory [6], fuzzy logic [7], Bayesian network [8], evolution algorithms [9, 10], etc. As for the operation and maintenance training with the test results as one of the main assessment indexes, the item response theory (IRT) establishes the correlation between the ability level and the accuracy of answering questions, based on which each trainee's ability on different knowledge points can be scientifically tested and estimated. With the aid of genetic algorithm (GA) [9], particle swarm optimization (PSO) algorithm [10] or other evolution algorithms, personalized resource recommendations as well as learning navigations can be offered to trainees to promote themselves effectively. In addition, considering the learning habit and cognitive style when selecting the matching type of learning resources can further improve the training efficiency, continuing to provide intellectual support and talent guarantee for power grid development. In this paper, an adaptive learning method applied to the field of substation operation and maintenance is established. The knowledge level and cognitive style of the trainee selected in this paper are evaluated using IRT and Bayesian network, respectively. Integrating the above two learning characteristics, the PSO algorithm is introduced to recommend the optimal learning resource for the trainee to realize the adaptive learning process.

2. Adaptive learning system

Figure 1 illustrates the fundamental construction of the adaptive learning system designed in this work. The structure refers to the adaptive educational hypermedia system (AEHS), a general reference model proposed by Brusilovsky et al. [11] from the University of Pittsburgh, USA. The system is mainly composed of four parts:

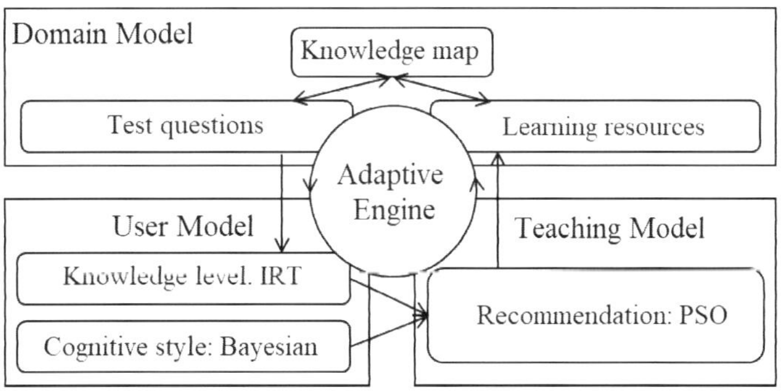

Figure 1. The fundamental construction of the adaptive learning system

(1) Domain Model: it describes the knowledge structure of substation operation and maintenance and stores the materials including test questions, courseware, etc. Through ontology technology, domain knowledge is summarized and stored in the form of knowledge map with the content covering operating principle, equipment testing, fault recovery and so on. The design and revision of the knowledge map is done by a dozen of experts in this field. Moreover, the experts connect all types of learning resources (video, text, etc.) as well as test questions with different knowledge points.

(2) User Model: it builds the trainee's "portrait", namely learning characteristics. Typical characteristics include multiple dimensions like basic information, knowledge level, cognitive style, user emotion, society relationship, etc. In the training of substation personnel, the knowledge level is always viewed as one of the most important characteristics in evaluation. Therefore, we use IRT in this work to estimate the trainee's knowledge level of different knowledge points. Based on the results, the current weakness of the trainee is obtained and his learning needs can be further speculated. Besides, our user model also lays emphasis on the dimension of cognitive style. Different from traditional unified teaching materials, the recommended learning resource not only matches the trainee in the aspect of resource contents, but also in resource types, as to improve the efficiency in the personalized learning process.

(3) Teaching Model: it declares the rules of recommending the optimal learning resource(s) to the trainee. In this work, the knowledge level and cognitive style are chosen as the two main criteria in making the rule and the PSO algorithm is adopted to search the optimal recommendation.
(4) Adaptive Engine: it drives the domain model, user model, and teaching model to realize the adaptive learning process. According to the knowledge map, the engine examines the trainee's knowledge level of selected knowledge points, and recommends the optimal learning resource with consideration of both the trainee's knowledge level and cognitive style.

3. Model development

3.1. Domain model

As mentioned above, the knowledge map in the domain model is designed and revised by the expert in the substation operation and maintenance. Taking a 500 kV substation as an example, the knowledge mapping refers to a professional teaching material [12] and starts from chapters to units, which covers fundamental knowledge, basic skills (monitoring, inspection, maintenance, and switching operation), accident handling, and equipment repair. Knowledge points are summarized from the unit content in each chapter and connected with each other forming the network-like knowledge map that follows the principle of accuracy, cleanness, and uniqueness.
The links between knowledge points and test questions as well as learning resources are further built. Each test question is linked with the most relevant knowledge point and labeled with its difficulty value d, which are done artificially by the expert to guarantee the accuracy. For K knowledge points with each one linked with Q test questions, a difficulty matrix $D=[d_{kq}]_{K\times Q}$ can be established. Based on IRT, d_{kq} ranges from -4.0 to 4.0. The higher d_{kq} is, the more difficult the test question is.
Different from test questions, one learning resource on substation operation and maintenance often involves many knowledge points, and the resource type involves video, animation, PPT, text, etc. We establish the learning resource matrix $R=[r_{mk}]_{M\times(K+1)}$ to describe the relationship between M learning resources and K knowledge points. r_{mk} is defined as:

$$\begin{aligned}&\text{when } k \le K,\ r_{mk} = \begin{cases}1 & \text{No. } m \text{ learning resource involves No. } k \text{ knowledge point}\\ 0 & \text{otherwise}\end{cases}\\ &\text{when } k = K+1,\ r_{mk} = \begin{cases}1 & \text{video-type learning resources}\\ 0 & \text{text-type learning resources}\end{cases}\end{aligned} \quad (1)$$

Here, the video-type learning resources include video, animation, and figure-based PPT, whereas the text-type learning resources include text and word-based PPT.

3.2. User model

3.2.1. Knowledge level. For K knowledge points, randomly select I ($I \le Q$) test questions from every knowledge point to conduct the test, guaranteeing the difficulty value d uniformly distributed in the range [-4, 4]. After the test, we can obtain the test result matrix $A=[a_{ki}]_{K\times I}$ as:

$$a_{ki} = \begin{cases}1 & \text{correctly answer No. } i \text{ question related to No. } k \text{ knowledge point}\\ 0 & \text{otherwise}\end{cases} \quad (2)$$

Item response theory (IRT) is a modern psychometric theory that establishes the association between test results and potential characteristics. In this paper, the single-parameter Logistic model (termed as Rasch model) based on this theory is used to estimate the knowledge level c_k [13]. Rasch model describes the probability of a person with knowledge level c_k correctly answering question (a_{ki}=1) as:

$$p(a_{ki}=1)=\frac{1}{1+e^{-(c_k-d_{ki})}} \tag{3}$$

The probability of not correctly answering equals to 1-p. Based on the test result matrix A, the maximum likelihood estimation (MLE) method is adopted to estimate the knowledge level c_k, using the following likelihood function L:

$$L\left(a_k|c_k,d_k\right)=\prod_{i=1}^{I}p\left(a_{ki}|c_k,d_{ki}\right) \tag{4}$$

The knowledge level on the No. k knowledge point equals to the c_k that makes ln(L) reaches its maximum. We use the Newton-Raphson iterative algorithm to solve this mathematical problem and finally obtain the knowledge level vector C=$[c_k]_K$ related to the K knowledge points. c_k ranges from -4.0 to 4.0.

The learning need is further deduced from the knowledge level. Basically, if one owns a low knowledge level at a certain knowledge point, he/she has a high level of learning need at this point. Therefore, we use the following function to quantify the relationship between the knowledge level and the learning need:

$$n_k=1-\frac{1}{1+e^{-c_k}} \tag{5}$$

n_k stands for the learning need on the No. k knowledge point and ranges from -1.0 to 1.0. Based on the knowledge level vector C, the learning need vector N=$[n_k]_K$ is calculated.

3.2.2. Cognitive style. The cognitive style considered in this work mainly refers to the preference on the type of learning resources, and is quantified with the cognitive style vector S=[s_1, s_2, s_3]. Specifically, s_1 stands for the percentage of video-type (including animation) learning resources in all learning resources that are ordinarily learned. In similar, s_2 and s_3 stand for the percentage of PPT-type and text-type learning resources, respectively. Before the adaptive learning, the incipient cognitive style can be obtained by a questionnaire investigation. The update of the cognitive style afterwards can be done on the basis of the online learning record.

The Bayesian method is used to ascertain the cognitive style. Bayesian network is a probability-based directed acyclic graph (DAG). Referring to literature [14], we design a video-style/text-style probability scale, as shown in Table 1. √ stands for selecting this type of learning resources while × stands for not selecting. Based on the cognitive style vector S, the cognitive style Sty can be calculated as:

$$Sty=0.85s_1s_2s_3+1.0s_1s_2\tilde{s}_3+0.6s_1\tilde{s}_2s_3+0.7s_1\tilde{s}_2\tilde{s}_3+0.4\tilde{s}_1s_2s_3+0.7\tilde{s}_1s_2\tilde{s}_3+0\tilde{s}_1\tilde{s}_2s_3+0.5\tilde{s}_1\tilde{s}_2\tilde{s}_3 \tag{6}$$

Here, $\tilde{s}$=1.0-s. Sty ranges from 0 to 1.0. If Sty approaches to 1, the cognitive style is more like video-style. Otherwise, it is more like text-style.

Table 1. Video-style/text-style probability scale used in this work

Video (including animation)	√				×			
PPT	√		×		√		×	
Text	√	×	√	×	√	×	√	×
Video-style	0.85	1.0	0.6	0.7	0.4	0.7	0	0.5
Text-style	0.15	0	0.4	0.3	0.6	0.3	1.0	0.5

3.3. Teaching model

Based on the learning need and cognitive style, the optimal learning resource will be selected in this part. First, we define a decision vector X=$[x_m]_M$ in which:

$$x_m = \begin{cases} 1 & \text{select No. } m \text{ learning resource and recommend} \\ 0 & \text{otherwise} \end{cases} \tag{7}$$

The decision vector X determines which one in the M learning resources is ultimately recommended. Corresponding to the decision vector, the target functions are formulated as follows:

$$\begin{aligned} F_1 &= \frac{\sum_{m=1}^{M}\sum_{k=1}^{K} x_m \left| r_{mk} - n_k \right|}{\sum_{m=1}^{M} x_m} \\ F_2 &= \frac{\sum_{m=1}^{M} x_m \left| r_{m(K+1)} - Sty \right|}{\sum_{m=1}^{M} x_m} \\ F &= \min\left(\xi_1 F_1 + \xi_2 F_2\right) \end{aligned} \tag{8}$$

F_1 function presents the proximity between the knowledge points involved by the learning resource and those related to the learning need. F_2 presents the proximity between the type of the learning resource and the cognitive style. The lower the value of F_1 and F_2 are, the more suitable the selected learning resource is. F is the sum of F_1 and F_2, with the weighting factor of ξ_1 and ξ_2. Thus, the decision vector that makes F reaches its minimum will be regarded as the best one.

As above, the recommendation process has been simplified into an optimization problem. The particle swarm optimization (PSO) algorithm is one of the typical optimization algorithms that simulates the process of bird flock foraging and have unique advantages in solving multi-object optimization problems. In this algorithm, a group of particles are first generated and each of them carries a possible solution to the problem. In every generation, each particle will adjust its velocity (value and direction) according to the particle best position and the group best position. During the competition and cooperation, the best position in the solution space will be ultimately found. In this work, we adopt this algorithm to find the best decision vector X. Each particle is given a decision vector X_{id} when first generated. *id* is the identity of the particle and every initial x_{id} in X_{id} is randomly set as 0 or 1. The searching velocity, v_{id}, is calculated with the following equation:

$$v_{id}(t) = \omega v_{id}(t-1) + l_1 \text{rand}(1)\left(pbest_{id} - x_{id}(t-1)\right) + l_2 \text{rand}(1)\left(gbest - x_{id}(t-1)\right) \tag{9}$$

Here, t stands for the generation number. $pbest_{id}$ is the best position of the No. *id* particle and *gbest* is the group best position, which are chosen in the t-1 generation. rand(1) generates a random number in (0, 1). l_1 and l_2 are the model coefficients. ω is the inertia weight. In the early stage, big ω is preferred to enhance the searching ability of the algorithm in the entire solution space. When it comes to the late stage, small ω is beneficial to the local precise searching. Referring to literature [15], we assume a linearly changing ω:

$$\omega = \omega_{\max} - \frac{\left(\omega_{\max} - \omega_{\min}\right)}{T} t \tag{10}$$

T is the total generation number. $\omega_{\max}$ equals to 0.9 and $\omega_{\min}$ equals to 0.4. Considering the discreteness of x_{id} (only 0 and 1), we use the following equation to update the position of particles [16]:

$$x_{id}(t)=\begin{cases}1 & \dfrac{1}{1+e^{-v_{id}(t)}}>\text{rand}(1)\\ 0 & \text{otherwise}\end{cases} \tag{11}$$

4. Results and discussion

4.1. Case setting

We select 30 knowledge points on switching from the knowledge map of substation operation and maintenance (K=30), together with 15 related learning resources (M=15). The knowledge points cover the switching operation of high-voltage line, bus, voltage transformer, compensation device, secondary equipment, etc. and the types of learning resources include video, animation, PPT, and text. All selected knowledge points and learning resources are re-numbered with Kn. 1-Kn. 30 and Re. 1-Re. 15, respectively. Based on eq. (1), the learning resource matrix R is established. For each knowledge point, 10 relevant test questions are selected (I=10) with the marked difficulty ranging from -4.0 to +4.0. We use all these questions to test the trainee and then record his test result to establish the test result matrix A. Afterwards, the knowledge level vector C as well as the learning need vector N are derived based on eq. (2)-eq. (5). By means of a questionnaire investigation, we obtain the trainee's current cognitive style Sty=0.708.

The parameters in the PSO algorithm are set as: total number of particles=200, T=100, ξ_1=0.6, ξ_2=0.4, l_1= l_2=1.49618.

4.2. Result analysis

Figure 2 illustrates the value of target function F as a function of the generation number. As present, the PSO algorithm finds the best solution of the our case after 18th generation and the best decision vector is X_{best}=[0, 0, 0, 0, 0, 0, 0, 0, 0, 1, 0, 0, 0, 0, 0]. The result indicates that Re. 10 is most suitable learning resource to recommend to the trainee.

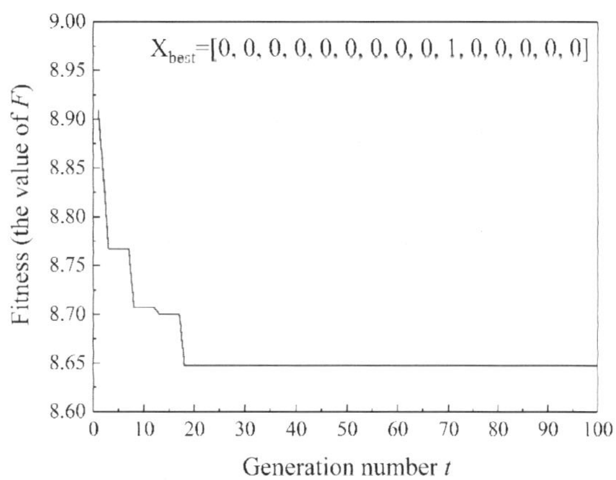

Figure 2. The value of F as a function of the generation number t

In order to verify whether the recommended learning resource matches the learning need of the trainee, we compare the knowledge points involved in the learning resource with those related to the learning need, as shown in Fig. 3. It can be seen that the recommended learning resource involves 16 knowledge points in the aspect of switching operation. Among them, all the three knowledge points of which the knowledge level is less than -1.5 (learning need$\geqslant$0.818) are covered. For the knowledge points of which the knowledge level is less than -1.0 (learning need$\geqslant$0.731), the learning resource

covers 66.7% (four of six) of them. The recommended learning resource meets the trainee's learning need pretty well.
To further clarify the fitness of the recommended learning resource in all the selected ones, we define the deviation σ between the learning resources and the learning need as:

$$\sigma=\sqrt{\frac{\sum_{k=1}^{K}\left(r_{mk}-n_k\right)^2}{K}} \tag{12}$$

From the view point of learning need, the lower σ is, the more suitable the resource is. Fig. 4 illustrates the σ of all the fifteen selected learning resources. Among all, Re. 14, Re. 10, and Re. 15 are the top 3 resources that are closest to the learning need, with σ equalling to 0.552, 0.555, and 0.561.

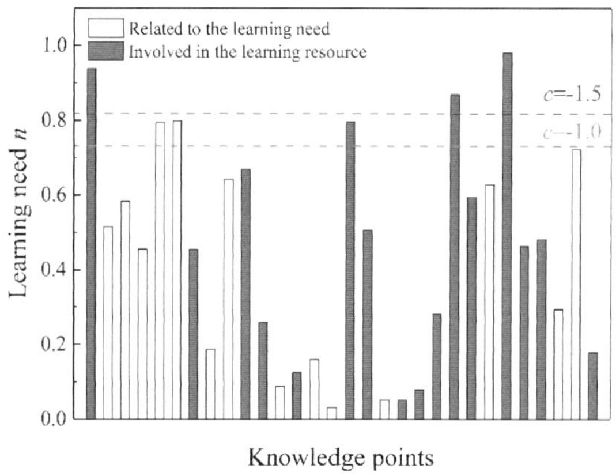

Figure 3. The knowledge points corresponding to the recommended resource and the learning need

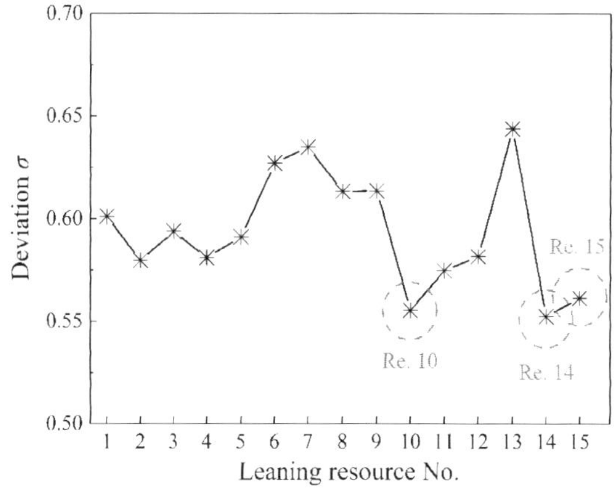

Figure 4. The deviation between the learning resources and the learning need

Specifically, the details of Re. 14, Re. 10, and Re. 15 are extracted from the resource matrix and listed in Table 2. It is demonstrated that Re. 10 and Re. 15 are video-type, while Re. 14 is text-type. For a trainee with a cognitive style of Sty=0.708, video-type Re. 10 is the optimal recommendation.

Table 2. The top 3 learning resources close to the learning need

	Kn. 1	Kn. 2	…	Kn. 30	Type
Re. 10	1	0	…	1	1
Re. 14	0	0	…	0	0

Re. 15	1	0	...	1	1

5. Conclusion

In this paper, an intelligent knowledge adaptive learning method is proposed and realized in the training of substation operation and maintenance. Item response theory (IRT) and Bayesian network are adopted to obtain the knowledge level and cognitive style of the trainee, respectively. Based on these two key characteristics, the particle swarm optimization (PSO) algorithm is used to successfully find the optimal learning resource. For a case with 30 knowledge points and 15 relevant learning resources, the optimal solution is obtained at ~20th generation in the PSO algorithm. The recommended learning resource using our adaptive learning method covers 100% of the knowledge points with the knowledge level of the trainee less than -1.5, and 66.7% of those with the knowledge level less than -1.0. Under the premise of meeting the learning need, the recommended resource type well fits the video-preferred cognitive style of the trainee. The adaptive learning method proposed in this work integrates knowledge level and cognitive style to realize the personalized learning process. Further studies can focus on more learning characteristic dimensions or more detailed recommendation rules, which, along with the method in the work, has broad application prospects in the intelligent training of substation operation and maintenance.

Acknowledgments
Special thanks are due to the project “Key technology research and development of the "Internet+" intelligent power knowledge service platform”.

References

[1] Lan G S, Guo Q, Lv C J, et al. Constructing Intelligent Education with Intelligent Technology in the Era of“Intelligence +”: Key Points and Considerations of Horizon Report (2019 Higher Education Edition)[J]. Open Education Research, 2019, 25(3): 22-35.
[2] Brusilovsky P. Adaptive and intelligent technologies for web-based eduction[J]. Ki, 1999, 13(4): 19-25.
[3] Vesin B, Ivanović M, KlašNja-MilićEvić A, et al. Protus 2.0: Ontology-based semantic recommendation in programming tutoring system[J]. Expert Systems with Applications, 2012, 39(15): 12229-12246.
[4] Salehi M, Kamalabadi I N, Ghoushchi M B G. Personalized recommendation of learning material using sequential pattern mining and attribute based collaborative filtering[J]. Education and Information Technologies, 2014, 19(4): 713-735.
[5] Wang X Y. Research on Sequential Pattern Mining in Self-Adaptive Learning System[D]. Northeast Normal University, 2008.
[6] Wang L P. Research on Learner Model and Pedagogical Model in Adaptative Learning System[D]. Northeast Normal University, 2017.
[7] Chrysafiadi K, Virvou M. Evaluating the integration of fuzzy logic into the student model of a web-based learning environment[J]. Expert Systems with Applications, 2012, 39(18): 13127-13134.
[8] Song T. Research on Learning Analysis of Bayes Network Hidden Markov Chain Model[D]. Capital University of Economics and Business, 2016.
[9] Huang M J, Huang H S, Chen M Y. Constructing a personalized e-learning system based on genetic algorithm and case-based reasoning approach[J]. Expert Systems with Applications, 2007, 33(3): 551-564.
[10] Yin C, Guo G Q, Li W B, et al. Multi-objective particle swarm optimization algorithm based on self-adaptive learning[J]. Application Research of Computers, 2012, 29(9).
[11] Brusilovsky P. Methods and techniques of adaptive hypermedia[J]. User modeling and user-adapted interaction, 1996, 6(2-3): 87-129.
[12] Li B X. Special Teaching Material for Vocational Ability Training of Production Skilled

Personnel of State Grid Corporation of China, Substation operation (500kV)[M], 2010.
[13] Hambleton R K, Swaminathan H. Item response theory: Principles and applications[M]. Springer Science & Business Media, 2013.
[14] Jiang Q. Research on Supported Model and Implementing Mechanism for Adaptive Learning System[D]. Northeast Normal University, 2012.
[15] Eberhart R C, Shi Y. Comparing inertia weights and constriction factors in particle swarm optimization[C]// Proceedings of the 2000 congress on evolutionary computation. CEC00 (Cat. No. 00TH8512). IEEE, 2000, 1: 84-88.
[16] Yang C. Learning resource recommendation method based on particle swarm optimization algorithm[J]. Journal of Computer Applications, 2014, 34(5): 1350-1353.

DFIG Model Suitable for SSR Research

Hanhua Zhang[1], Haiqiang Jin[2], Tongtong Wang[2], Jun Wen[2]

[1] Electric Power Research Institute, State Grid Ningxia Electric Power Co., Ltd., Yinchuan, China

[2] School of Electrical and Electronic Engineering, North China Electric Power University, Changping District, Beijing 102206, China

18811313911@163.com

Abstract. The simulation model of doubly fed induction generator (DFIG) is the basis of the research on subsynchronous resonance (SSR) of DFIG. In this paper, the conditions that the model of DFIG is suitable for the research of subsynchronous oscillation (SSO) are analyzed firstly. Then, the drive system, doubly fed induction motor, AC / DC / AC power converter and control strategy of DFIG are modeled, and these models are unified as electromechanical model. At last, the validity of the model is verified by the example of SSR when the DFIG is sent out through series compensated transmission line.

1. Introduction

DFIG is widely used in power industry because of its mature technology and high power generation efficiency. With the advantages of long-distance and large transmission capacity, series compensated transmission line is often used to transfer large capacity wind energy to remote load center. However, there are many examples of SSR when wind farms are sent out through series compensated transmission line.

In the process of analyzing the mechanism of SSR of DFIG, an accurate model of DFIG is the basis of research. In reference [1], the equivalent model of wind farm and risk assessment of wind farm considering random wind speed are studied. In reference [2], a mathematical model is established for the SSR caused by the series compensated transmission line of doubly fed wind farms. In reference [3], the existing model of DFIG is improved, the unified electromechanical model of DFIG is established, and the causes of SSO are analyzed.

Although the oscillation mechanism of DFIG can be accurately analyzed by using a detailed mathematical model, the detailed mathematical model is a multivariable differential equation, and the numerical solution process is very complex, which affects the simulation speed and efficiency. Therefore, it is necessary to establish the equivalent electromagnetic transient model to meet the needs of simplified analysis in practical projects.

Based on the above background, this paper firstly analyzed the conditions that the model of DFIG suitable for SSO research needs to meet, then established the mathematical models of each component of DFIG and its control strategy, and finally simulated the SSR caused by DFIG and series compensated transmission line in PSCAD / EMTDC. Simulation results show that the model can meet the needs of SSR research.

2. The model of DFIG

The DFIG model suitable for SSO research needs to meet the following conditions: It can reasonably reflect the external dynamic electrical characteristics of DFIG; It is suitable for the simulation of subsynchronous time scale and can reflect the characteristics of DFIG in the frequency range of 5-100Hz; It can reduce the complexity of the model as much as possible to reduce the amount of calculation and improve the simulation speed.

DFIG consists of wind turbine and drive system, doubly fed induction motor, AC / DC / AC power converter and its control system.

2.1. Wind turbine and drive system

Wind turbine and drive system are important parts of DFIG, which determine the output power of the wind turbine. The drive system is mainly composed of low-speed shaft, high-speed shaft and gearbox. The commonly used wind turbine shaft models are lumped mass model, two mass model and three mass model. The two mass model can accurately reflect the dynamic performance of shaft and simplify the calculation [4]. The two mass model of the shaft is shown in Figure 1.

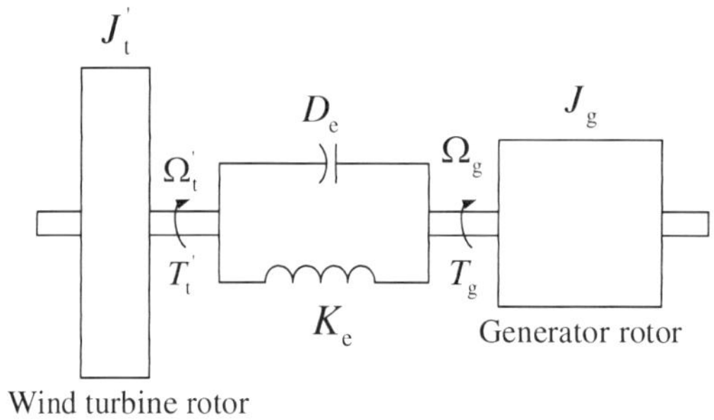

Figure 1. Two mass model of the shaft

2.2. Doubly fed induction motor

Doubly fed induction motor is a complex nonlinear time-varying electromechanical system. In order to achieve high-performance control, vector control method must be used, and vector control is realized by coordinate transformation [5].According to the motor convention, the voltage equation and flux equation of doubly fed induction motor in the dq rotating coordinate system are as follows

$$\begin{cases} u_{\mathrm{sd}} = R_{\mathrm{s}} i_{sd} + \dfrac{\mathrm{d}\psi_{\mathrm{sd}}}{\mathrm{d}t} - \omega_1 \psi_{\mathrm{sq}} \\ u_{\mathrm{sq}} = R_{\mathrm{s}} i_{\mathrm{sq}} + \dfrac{\mathrm{d}\psi_{\mathrm{sq}}}{\mathrm{d}t} + \omega_1 \psi_{\mathrm{sd}} \\ u_{\mathrm{rd}} = R_{\mathrm{r}} i_{\mathrm{sd}} + \dfrac{\mathrm{d}\psi_{\mathrm{rd}}}{\mathrm{d}t} - (\omega_1 - \omega_{\mathrm{r}})\psi_{\mathrm{rq}} \\ u_{\mathrm{rq}} = R_{\mathrm{r}} i_{\mathrm{sq}} + \dfrac{\mathrm{d}\psi_{\mathrm{rq}}}{\mathrm{d}t} + (\omega_1 - \omega_{\mathrm{r}})\psi_{\mathrm{rd}} \end{cases} \tag{1}$$

$$\begin{cases} \psi_{\mathrm{sd}} = L_{\mathrm{s}} i_{\mathrm{sd}} + L_{\mathrm{m}} i_{\mathrm{rd}} \\ \psi_{\mathrm{sq}} = L_{\mathrm{s}} i_{\mathrm{sq}} + L_{\mathrm{m}} i_{\mathrm{rq}} \\ \psi_{\mathrm{rd}} = L_{\mathrm{m}} i_{\mathrm{sd}} + L_{\mathrm{r}} i_{\mathrm{rd}} \\ \psi_{rq} = L_{\mathrm{m}} i_{\mathrm{sq}} + L_{\mathrm{r}} i_{\mathrm{rq}} \end{cases} \tag{2}$$

where $u$sd、$u$sq、urd、$u$rq are respectively the d and q axis components of stator and rotor voltage; $i$sd、$i$sq、ird、$i$rq are respectively the d and q axis components of stator and rotor current; ψsd、ψsq、ψrd、ψrq are respectively the d and q axis components of stator and rotor flux; Ls、Lr、Lm are respectively the self inductance of stator winding, self inductance of rotor winding and mutual inductance between stator and rotor windings;Rs、Rr are respectively the resistance of stator and rotor winding;ω1、ωr are the angular frequency of grid and angular speed of rotor.

2.3. AC / DC / AC power converter

AC / DC / AC power converter consists of two PWM converters with the same structure, which are respectively called grid side converter (GSC) and rotor side converter (RSC).

2.3.1. GSC. The main function of GSC is to keep the voltage of DC bus stable, input current sinusoidal and control power factor. The decoupling control of power transmission between rotor and GSC can be realized by grid voltage vector oriented control.In the dq rotating coordinate system, when the grid voltage vector $\boldsymbol{Ug}$ is oriented to the d-axis ,the voltage equation of the grid side filter is as follows

$$\begin{cases} L_{\mathrm{g}}\dfrac{\mathrm{d}i_{\mathrm{gd}}}{\mathrm{d}t} = -R_{\mathrm{g}}i_{\mathrm{gd}} + u_{\mathrm{fd}} + \omega_1 L_{\mathrm{g}} i_{\mathrm{gq}} - u_{\mathrm{gd}} \\ L_{\mathrm{g}}\dfrac{\mathrm{d}i_{\mathrm{gq}}}{\mathrm{d}t} = -R_{\mathrm{g}}i_{\mathrm{gq}} + u_{\mathrm{fq}} - \omega_1 L_{\mathrm{g}} i_{\mathrm{gd}} \end{cases} \tag{3}$$

where, $u$gd is the d axis components of grid voltage; $u$fd and $u$fq are the d and q axis components of GSC voltage with grid neutral point as reference; $i$gd and $i$gq are the d and q axis components of grid side filter current; Rg and Lg are the resistance and inductance of grid side filter.

It can be seen from the above formula that in order to realize the decoupling control of d-axis and q-axis components of the grid side filter current and eliminate the influence of grid voltage, it is necessary to introduce current feedback to decouple, and introduce grid voltage to carry out feedforward compensation [6].Based on the above analysis, the control system of GSC is shown in Figure 2(a).

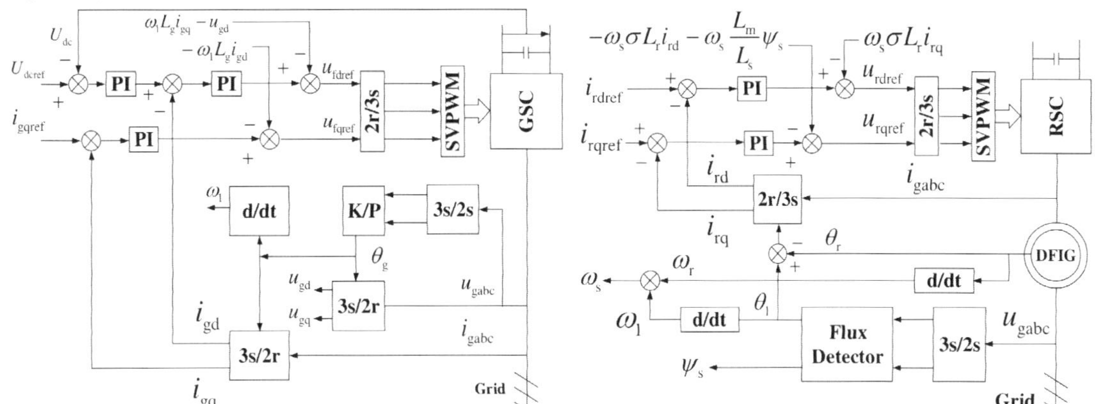

(a) Control block diagram of GSC (b) Control block diagram of RSC

Figure 2. Control system of power converter.

2.3.2. RSC. The maximum power tracking and reactive power control can be realized by controlling the RSC. Stator flux vector oriented control is a commonly used method.The closed-loop controller of rotor current is designed according to the voltage equation of rotor in dq rotating coordinate system

$$\begin{cases} u_{rd} = R_r i_{rd} + \sigma L_r \dfrac{di_{rd}}{dt} - \omega_s \sigma L_r i_{rq} \\ u_{rq} = R_r i_{rq} + \sigma L_r \dfrac{di_{rq}}{dt} + \omega_s (\dfrac{L_m}{L_s} \psi_s + \sigma L_r i_{rd}) \end{cases} \tag{4}$$

where ψs is the amplitude of $\boldsymbol{\psi s}$, which can be regarded as a constant when the grid voltage is ideal;σ is the leakage coefficient of the generator; ωs is the slip angular speed and $\omega s=\omega 1-\omega r$.
Similarly, in order to realize the decoupling control of d axis and q axis components of rotor current, it is necessary to introduce current feedback for decoupling and excitation current for feedforward compensation. According to the above analysis, the control system of RSC based on stator flux vector orientation is shown as Figure 2(b).

2.3.3. DC side. Capacitor is the key to realize the decoupling control of RSC and GSC. Due to the existence of Capacitor, the functions of *GSC* and *RSC* are independent of each other.The equation of DC side in dq rotating coordinate system is

$$C\frac{dU_{dc}}{dt} = \frac{3}{2}(S_d i_{gd} + S_q i_{gq}) - i_{load} \tag{5}$$

where *U*dc is the voltage of DC side; *i*load is the load current; *S*d and *S*q are the d and q axis components of switch function.

3. Simulation for SSR

When DFIG is sent out through series compensated transmission line, the SSR component of line current may have self-excited oscillation in the case of small disturbance. Based on the established model, this section further establishes the model of DFIG sent out by series compensated transmission line in PSCAD / EMTDC, and simulates the SSR phenomenon.The model of DFIG sent by series compensated transmission line is shown in Figure 3.

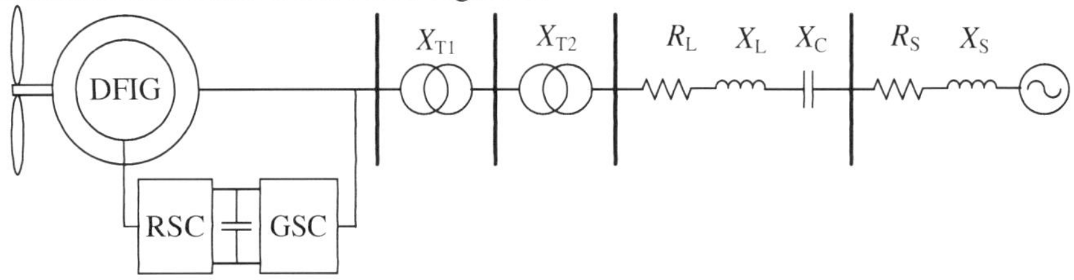

Figure 3. The model of DFIG sent by series compensated transmission line

The system shown in Figure 3 has a natural resonance frequency *f*er

$$f_{er} = f_1\sqrt{\frac{X_C}{X_{L\Sigma}}} \tag{6}$$

where *X*LΣ=*X*DFIG+*X*T1+*X*T2+*X*L; *f*1 is the grid synchronous frequency and *f*1=50Hz;*X*T1 is the leakage reactance of the internal transformer of DFIG; *X*T2 is the leakage reactance of the step-up transformer; *R*L and *X*L are the equivalent resistance and reactance of the transmission line; *R*S and *X*S are the equivalent resistance and reactance of the grid;*X*C is the capacitance reactance under synchronous frequency.If *k* is the series compensation degree ($k < 1$), then *X*C=*kX*L.
The DFIG is set to operate under the condition of wind speed of 12 m/s and rotating speed of 1.1pu. When the system runs to a stable state, a small disturbance is applied to the system to cause the oscillation of the line current. When the series compensation degree of the line is 30% and 50%, the current oscillation waveform and spectrum of current of A-phase of the line is shown in Fig. 4(a),(b).It can be seen that under the two series compensation degrees, the current contains not only the

fundamental frequency component of 50Hz, but also the resonance frequency component of 15Hz and 20Hz.

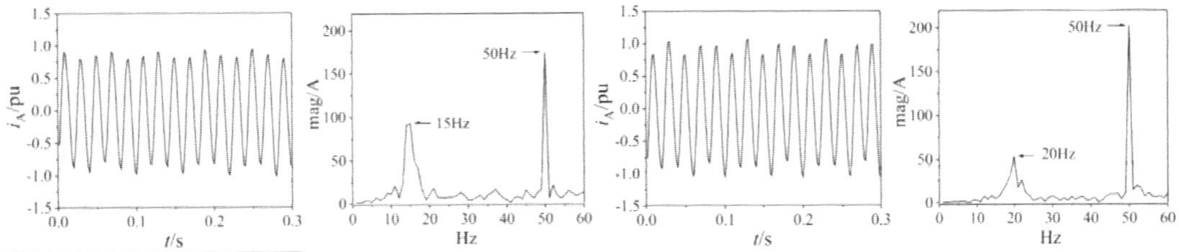

(a) k=30% (b) k=50%

Figure 4. Current oscillation waveform and spectrum of current of A-phase in different series compensation degrees.

4. Conclusion

On the basis of the traditional model of DFIG, considering the requirements of the model suitable for the study of SSO, this paper establishes the electromechanical simulation model of DFIG in PSCAD / EMTDC, and verifies the accuracy of the model with the example of the DFIG sent by series compensated transmission line. It can be seen from the simulation that the model established in this paper can accurately reappear the SSR, and can be used for the research of SSR of DFIG.

Acknowledgments
We are grateful to Dr Wen who put forward many valuable suggestions for revision, and to Miss Wang who provided a lot of help during the writing of the paper.

References

[1] Wang D.H,Equivalent Modeling and Risk Assessment Method for Sub-synchronous Oscillation in a Wind Farm[D]. Jiangsu University, 2018
[2] Tian J,Zhang W.Y,Wang Y.P, et al. Modeling and Analysis of Sub-synchronous Oscillation in Wind Power Grid Connection System[J]. Distributed Energy, 2018, 3(05): 22-27.
[3] Ding G.D, The Study on Subsynchronous Oscillations of Doubly Fed Wind Turbine[D]. Shanghai Dianji University, 2015
[4] Yang S.Y, Converter and Control for Doubly Fed Induction Generator -Based Wind Power Generation[D]. Hefei University of Technology, 2007
[5] He Y.K,Hu J.B,Xu L, Operation and Control of Grid Connected Doubly Fed Asynchronous Wind Turbine[M]. China Electric Power Press, 2011
[6] Ling Y, Modeling, Simulation and Control of Doubly Fed Wind Power Generation System[M]. China Machine Press. 2017.184-186.

Voltage sag analysis based on cluster analysis and correlation analysis

Hao Liu, Qinghui Zeng, Shaohui Liu, Zhuojun Wu, Lei Li and Rongrong Ma

Fenjiang South Road, Chancheng District, Foshan City, Guangdong Province

77163134@qq.com

Abstract. Based on data mining algorithm, a voltage sag analysis method is proposed, which first performs clustering analysis and then correlation analysis.Firstly, the climatic factors are used to cluster the data, and then the correlation analysis of each cluster is carried out according to the selected voltage sag feature dimensions, finally, the strong association rules are obtained. Through the analysis of the example, it is found that there are some certain correlations between the climatic factors, the voltage sag dimensions and the causes of the voltage sag. These correlations can provide a theoretical basis for the prevention and control of voltage sag.

1. Introduction

Voltage sag is a typical power quality problem[1]. According to statistics, over 80% of power quality problems are caused by voltage sags[2]. Voltage sag refers to "short-term voltage drop", which is caused by short circuit, lightning stroke, start-up of large motors, equipment failure, etc[3]. Once a voltage sag accident occurs, it will cause the motor to stop, data loss, etc., and the equipment will be damaged directly, which will seriously affect the normal life and operation of the users. Some surveys have shown that a voltage sag accident can cause a loss of 200-300 million yuan[4]. Therefore, the prevention and control of voltage sag has great practical significance.

At present, many scholars try to use data mining methods to study voltage sag. Luo Yan et al.[5] studied the historical events of voltage sag, selected seven dimensions related to voltage sag to analyze the association rules, and obtained strong association rules between voltage sag type and other dimensions. Xu Zhong et al.[6] used two-step clustering to study the voltage sag monitoring data in a certain area. Finally, the voltage sag was divided into four categories, and the characteristics of each cluster were given. Shen Xiang et al.[7] adopted the adaptive Gaussian cloud transform algorithm and gray target theory to establish a matching model between the actual scene and the association rules, and screened out the knowledge of the influence of the voltage sag on the nodes in the actual scene. The above research uses the data mining method to analyze the voltage sag historical data, but all of them are excavated from the perspective of some voltage sag data dimensions, which have certain one-sidedness.

This paper proposes a voltage sag analysis method that clusters first and then associates. Through this process the historical monitoring data and historical climate data accumulated by the power quality monitoring system will be analyzed, which proves that there is a certain relationship among the climate, load nature, voltage level, occurrence area, week, and time. That can provide a theoretical basis for the prevention and control of voltage sag.

2. Algorithm principle

2.1. *Cluster analysis*

As an important algorithm of data mining, cluster analysis mainly refers to the process of dividing a set into multiple classes which are composed of similar elements.

First, select one point as the initial cluster center, and calculate the distance between the remaining points and the initial cluster center, and select the farthest point as the second cluster center. The distance is calculated as:

$$d_{f,i} = \sqrt{(x_1^f - x_1^i)^2 + \ldots + (x_n^f - x_n^i)^2} \tag{1}$$

Where $d_{f,i}$ represents the distance between the measurement day and clustering center i, and $x_1, \ldots, x_n$ represents each cluster indicator.

Then, according to the same method, select the cluster center i until there are no more cluster centers.

Finally, calculate the distance from the remaining points to each cluster center, and assign each point to each class according to the principle of minimum distance.

2.2. *Correlation Analysis*

The purpose of correlation analysis is to mine the potential connections between data items which from the data. Correlation analysis can connect things that seem completely unrelated to data.

According to the association rule, each sample data is called a "transaction", transaction database D consists of n transactions and each transaction is specified by multiple attributes, which are recorded as "items", and multiple items constitute an "item set". The probability P(C) of item set C is called support, and the item set exceeding the minimum support is called frequent item set.

If item set $X \cap Y \neq \varnothing$ and both of them are all included in D, then X→Y is called the association rule. The probability that D contains the union of X and Y is called support, and the degree of support represents the importance and number of occurrences of association rules. The support degree of the association rule X→Y is:

$$S = P\{Y \mid X\} = \frac{\|\{t \in D \mid X \cup Y \in t\}\|}{\|D\|} \tag{2}$$

When the confidence of the rule X→Y is not less than the confidence of the set of the minimum support set by the artificial, the rule is a strong association rule. The confidence of the rule X→Y is:

$$P\{X \to Y\} = P\{Y \mid X\} = \frac{P\{X \cap Y\}}{P\{X\}} \tag{3}$$

3. Voltage sag analysis algorithm based on cluster analysis and correlation analysis

3.1. *Data preprocessing*

3.1.1. Historical climate data collection. According to statistics, most of the voltage sag events are caused by lightning strikes. It can be seen that the climate has a certain impact on the voltage sag. This paper selects temperature, humidity, and wind speed as climatic factors, and obtains historical climate data from the Internet.

3.1.2. The value of voltage sag characteristic dimensions. The voltage sag data in this paper is mainly derived from the power quality detection system. In the power quality detection system, there are more than ten dimensions related to voltage sag. In order to simplify the calculation, this paper chooses the six dimensions of load nature, voltage level, occurrence area, week, time and sag cause as the characteristic dimension of voltage sag for correlation analysis. The results are shown in the table 1 below.

Table1. The value of voltage sag characteristic dimensions

Characteristic dimensions	Data value
Load natures	Ordinary load、heavy load、new energy、sensitive user
Voltage level	500kv and abovevoltage levels、330 kv、220 kv、110 kv、35 kv、10kv and below voltage levels
Voltage sag area	Geographical area of the voltage sag event
Week	The nature of the voltage sag event occurring in week
Time	The time when the voltage sag event occurred in 1 day
Cause of voltage sag	The main cause of voltage sags

Link historical climate data to time and voltage sag data records and proceed to the next step.

3.1.3. Discretization of characteristic dimensions of voltage sag. Discrete the characteristic dimensions of voltage sag. Based on the classification in the power quality detection system, we divide the load properties into common load, heavy load, new energy, and sensitive users. According to China's voltage levels (220V, 380V, 10kV, 24kV, 35kV, 110kV, 220kV, 500kV, 800kV and 1000kV, respectively), the voltage levels are reclassified into five categories: 500 kV and above, 330 kV, 220 kV, 110 kV, 35 kV, 10 kV and below. In order to facilitate the processing, the voltage will be temporarily numbered according to the area. According to the date attributes and the household electricity time rules, the week of voltage sag will be divided into working days and holidays, and the time of voltage sag is divided into 00:00- 8:00, 8:00-18:00, 18:00-24:00. According to the classification of the main cause of voltage dip, the reasons will be temporarily divided into short circuit, heavy load, new energy and others. The discretization data for each dimension is shown in the following table 2.

Table 2. Voltage sag characteristic dimensions discretized data items

Characteristic dimensions	Identifier	Data values
Load nature	U	Ordinary load、heavy load、new energy、sensitive user
Voltage level	L	500kV and above voltage levels、330 kV、220 kV、110 kV、35 kV、10kV and below voltage levels
Voltage sag area	R	01、02、03、…、10
Week	W	Working days(0)、holidays(1)
Time	T	00:00-8:00、8:00-18:00、18:00-24:00
Cause of voltage sag	O	Short circuit, heavy load, new energy and others

3.2. Cluster analysis

This article classifies voltage pause records based on temperature, humidity, and wind. The steps are as follows:

1) Arbitrarily select a point as the first cluster center Z1.

2) Calculate the distance d_{X_i,Z_1} between the remaining points {X1, X2,...Xn}(i=1,2,...,n) to Z1, select the point farthest from Z1, which is $\max\{d_{X_1,Z_1},...,d_{X_n,Z_1}\}$, as the new cluster center Z2.

3) Calculate the distance d_{X_i,Z_1}, d_{X_i,Z_2} from the remaining points to Z1, Z2, and take the minimum values min{ d_{X_i,Z_1}, d_{X_i,Z_2} } of the two distances, if the value is greater than the distance between the two cluster centers, i.e.:

$$\min\{ d_{X_i,Z_1}, d_{X_i,Z_2} \} > \sqrt{(Z_1 - Z_2)^2} \tag{4}$$

Then use this point as the cluster center Z3.

4) And so on, find all cluster centers until the number of cluster centers does not increase.

5) Calculate the distance between the remaining points and each cluster center, and assign them to each class according to the principle of minimum distance.

6) Repeat the previous step until all points have been divided.

The algorithm flow chart of the cluster analysis provided in this paper is shown in the figure 1 below.

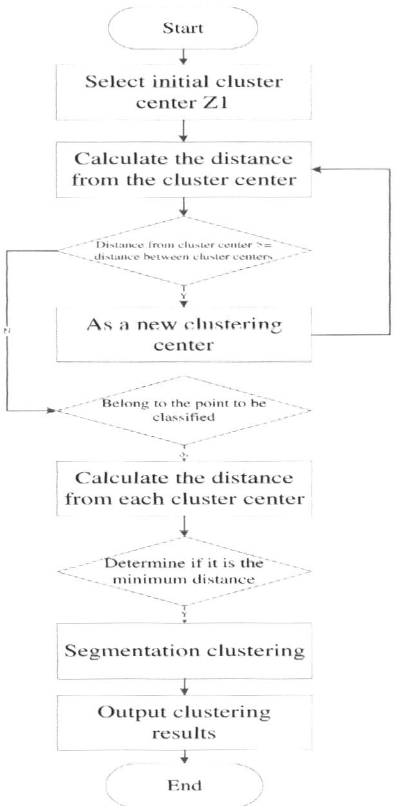

Figure 1. Clustering algorithm flow chart

3.3. Correlation Analysis

According to the clustering result of 3.2, the correlation analysis is carried out in each cluster, and it is proposed to find out the association rules of the causes of voltage sag and the other factors in each cluster.

3.3.1. Building a dimension matrix. Traditional correlation analysis requires frequent scanning of the database. In order to reduce the scanning time and speed up the algorithm, this paper generates an array of all the *information* after a database scan, and then scans the database to scan the array.
The dimension matrix A is established according to discrete data items of each feature dimension.

$$A = \begin{bmatrix} 1\, U_1\ L_1\ R_1\ Q_1\ W_1\ T_1\ O_1 \\ 2\, U_2\ L_2\ R_2\ Q_2\ W_2\ T_2\ O_2 \\ \cdots\ \cdots\ \cdots\ \cdots\ \cdots\ \cdots\ \cdots\ \cdots \\ n\, U_n\ L_n\ R_n\ Q_n\ W_3\ T_n\ O_n \end{bmatrix} \tag{5}$$

3.3.2. Looking for frequent item sets. Next, using the dimension matrix A for correlation analysis, then a frequent item set is obtained.
First, each value in the dimension matrix A is used as a candidate 1-item set; then, by scanning the matrix A, the statistical values in the row of each value are accumulated to obtain the minimum support degree; the candidate 1-item concentration support does not match the item set required for minimum support is culled, resulting in a frequent 1-item set.
Next, the rows in the matrix A that do not contain the frequent 1-item set are deleted, and the dimension matrix A_2 is obtained. At the same time, the item sets in the frequent 1-item set are connected to obtain the candidate 2-item set, and the candidate 2-item centralized support is less than the minimum support degree, and the frequent 2-item set is obtained.
By analogy, the dimension matrix A_{k+1} is obtained by deleting the rows of the dimension matrix A_k that do not contain the frequent k-item set, and the candidate (k+1)-item set is obtained according to the frequent k-item set self-joining, and then delete the items in candidate (k+1)-items set that do not meet the minimum support, the frequent (k+1)-item set is obtained. When a new frequent item set cannot be generated, the operation ends.

3.2.3. Output strong association rules. According to frequent item sets, the non-empty subset is used to get multiple association rules $X \rightarrow Y$. Among them, item set Y is the cause of voltage sag, and item set X is one or more dimensions of the other six dimensions, namely load nature, voltage level, occurrence area, occurrence quarter, week and time. Calculate the confidence of the association rule and compare it with the set minimum confidence. If it is not less than the minimum confidence, it is a strong association rule.

4. Analysis of results

Taking the historical power quality detection record as the data source, and 6060 voltage sag records of 70 detection points in 10 areas were selected for mining. Clustering by climatic factors yielded three clusters, and the characteristics are shown in the table 3.

Table3. Clustering result feature distribution

Cluster	I	II	III
Proportion	39%	44%	17%
Temperature	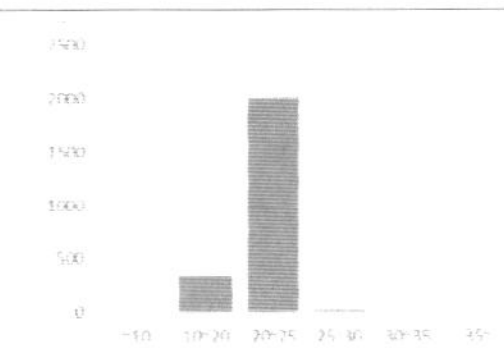	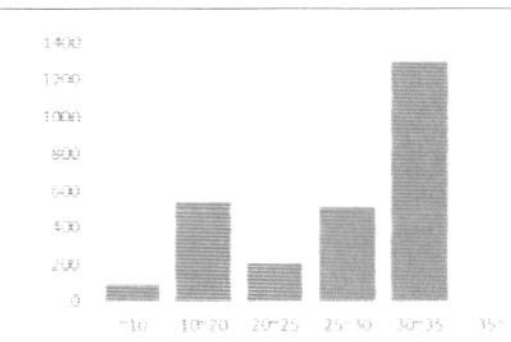	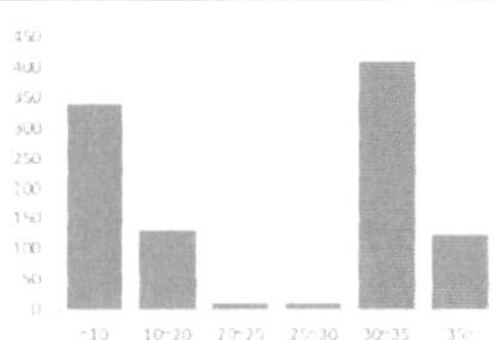

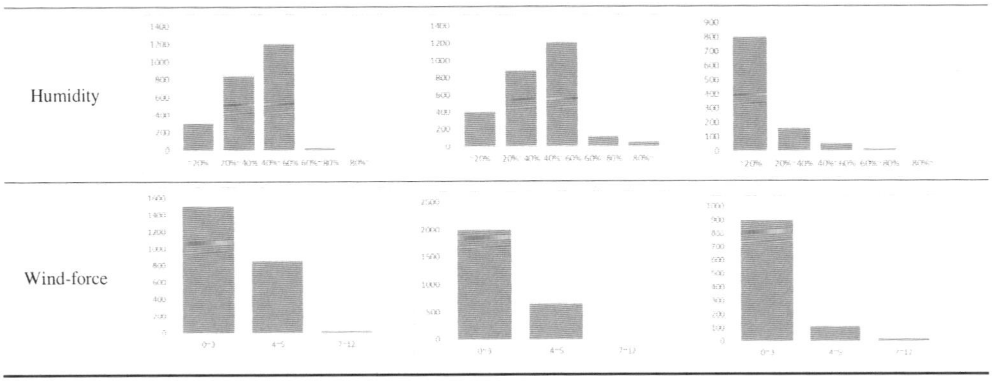

Assuming that the minimum support is 10% and the minimum confidence is 50%, the obtained partial strong association rules are as follows.

Table 4. Cluster matching results

Cluster	Strong association rules	confidence
Climate I	{35kV} →Short circuit	87%
	{Area 10, 220kV, Ordinary load} →Heavy load operation	52%
Climate II	{35kV} →Short circuit	95%
	{Area 10, 35kV, Ordinary load} →Short circuit	82%
	{Area 10, 220kV, Ordinary load} →Heavy load operation	50%
Climate III	{35kV} →Short circuit	86%
	{Area 10, 220kV, Ordinary load} →Heavy load operation	59%

Rule 1: Climate I∧35kV→short circuit, confidence= 87%; climate II∧35kV→short circuit, confidence = 95%; climate III ∧ 35kV → short circuit, confidence = 86%.

This rule shows that the voltage sag that is prone to short-circuit properties at 35kV, especially in climate II, the voltage sag accident at the 35kV monitoring point is almost always caused by a short circuit.

Inference 1 can be obtained by using rule 1 and proportion of sample size of clustering.

Inference 1: 35kV → short circuit, confidence = 90%.

Rule 2: Climate II∧area 10∧35kV∧ordinary load→short circuit, confidence = 82%.

This rule indicates that the voltage sag accidents occurring at the 35kV common load monitoring point in the climate region 2 and in the area 10 are mostly caused by short circuits.

Rule 3: Climate I∧area 10∧220kV∧ordinary load→heavy load operation, confidence = 52%. ClimateII∧area 10∧220kV∧ordinary load→heavy load operation, confidence = 50%. Climate III∧area 10∧220kV∧ordinary load→heavy load operation, confidence = 59%.

The rule shows that the proportion of voltage sag accidents caused by heavy load operation of 220kV ordinary load in area 10 accounts for half, and when the climate is climate III, the proportion of heavy load operation is significantly higher than the other two clusters.

Rule 4: Climate I∧area 10∧220kV∧new energy load→new energy operation, confidence = 90%.

This rule shows that in the voltage sag accidents in the 220kV new energy load in the area 10 under the climate I, the accidents caused by the new energy operation account for 90%.

By analyzing all the strong rules, it can guide the voltage sag investigation to a certain extent, and provide data support for the prevention and treatment of voltage sag.

5. Conclusions

This paper divides the historical voltage sag data by cluster analysis, and obtains several categories of different climatic characteristics. Then, it combines the correlation analysis to analyze each cluster, and puts forward a voltage sag analysis method based on cluster analysis and correlation analysis. The results show that the climate has a certain impact on the voltage sag. In some climates, the voltage sag caused by the short circuit is significantly more than other climates. Moreover, the voltage level, region, load type, week, time and other dimensions in the voltage sag dimension also have a certain impact on the voltage sag event. Most of the voltage sag events at the 35kV monitoring point are caused by the short circuit. The research results can effectively guide the prevention and treatment of local voltage sag and reduce the occurrence of voltage sag.

Acknowledgments

Foundation item: Supported by the Technical Projects of Guangdong Power Grid Limited Liability Company (No. 030600KK52160006)

References

[1] Sun J.J. Voltage sag in modern distribution network and its operational toughness improvement.Industrial Engineering2019,38(3):4-7.
[2] Yang X.D.Study on stochastic estimation of voltage sag and its economic management method,North China Electric Power University2009.
[3] Zhang Y., Yin L.S., Ma R.Q., Liu D.M., Yang C.L. Voltage sag detection method based on complex wavelet transform and effective value algorithm.Electrical Measurement and Instrument2017,54(10):74-79.
[4] Tao S. Study on the influence of voltage sag on reliability of distribution system and its evaluation index. North China Electric Power University,2005.
[5] Luo Y., Qi L.H. Application of association rules in voltage sag analysis. China Science and Technology Information,2013,23:169-171
[6] Xu Z., Mo W.X., Zhang Z., Zhong Q. Cluster analysis of urban power system voltage sag event. Power supply and consumption,2018,35(02):31-35.
[7] Shen X., Yang H.G., Duan C. Data mining analysis method based on gray target theory and cloud model for voltage sag event. Power grid technology,2019,43(02): 722-731.

IOP Conf. Series: Materials Science and Engineering **853** (2020) 012019 doi:10.1088/1757-899X/853/1/012019

Model Predictive Current Control of Dual Inverter Fed Open-End Winding PMSM Based on Fast Vector Selection

Di Wu[1], Rui Zhu[2], Jifeng Chen[2], Jiaquan Hu[2], Guowu Hua[2], Ying Wang[1]

[1]Electronic Engineering Institute in Huaihai Institute of Technology Lianyungang 222005, China

[2]College of Electrical and Power Engineering in China University of Mining and Technology, Xuzhou 221116, China

749805964@qq.com

Abstract. Aiming at the problem of model predictive current for the dual two-level inverter needs to optimize the objective function of all 19 voltage vectors, which makes the calculation of optimization too large for practical application. A MPCC scheme for open-winding pmsm based on voltage vector preselection was proposed. First, the pre-selection of the voltage vector is performed by calculating the reference voltage vector using the deadbeat solution. And establish a voltage vector based objective function to achieve predictive current control and simplify the control algorithm. Next, the switching frequency constraint is introduced into the objective function to select the optimal switching state and reduce the average switching frequency. Finally, the simulation is verified and the effectiveness of the method is proved.

1. Introduction

The open-end winding permanent magnet synchronous motor (OEW-PMSM) can use two different power supplies, with multi-level features, flexible control modes and powerful fault tolerance.[1]. It has broad application prospects in the field of electric vehicles.

For getting good system performance, the high-performance control strategy of OEW-PMSM has always been a research hotspot. At present, DTC and FOC have been widely used. In DTC systems, good dynamic response is obtained, but torque ripple is usually large; in FOC systems, steady state performance is better than DTC, but the inverter has a higher switching frequency and a slower transient response.

In motor drive, MPC has received extensive attention owing to its simplicity and flexibility for multiple constraints [2]. Among them, FCS-MPC [3] uses the characteristics of the finite-switch state of the converter, and combines the system mathematical model to predict the future performance of the system. The objective function is constructed for online optimization, and the required switching state is directly selected. With no additional modulation technique, the parameters are easy to set, and the objective equation can be used to add other control objectives and deal with system constraints.

However, in a dual inverter driven permanent magnet synchronous motor drive system, more switching states and voltage vectors are involved, which adds the complexity of the prediction vector. There are more voltage vectors and multiple control constraints in [4] and [5]. In order to select the optimal voltage vector, more evaluation work on the objective function is required. Therefore, reducing calculation amount becomes an important issue in the new generation of permanent magnet synchronous motor systems. In addition, since some voltage vectors have several different switching

IOP Conf. Series: Materials Science and Engineering **853** (2020) 012019 doi:10.1088/1757-899X/853/1/012019

states, the optimum voltage vector with the least switching loss must be considered.

In this paper, an improved MPC method is proposed for the problem of excessive selection of optimal switching vector selection and high switching frequency in the traditional dual-inverter feed system model predictive current control.

Firstly, mathematical modeling was carried out, and then an improved algorithm was proposed based on the traditional model of predictive current control, which including the calculation of no difference beat voltage, divided into sectors for voltage vector preselection. The optimal voltage vector can be obtained by a single comparison, which obviously reduces the computation of the program. In addition, in order to reduce the average switching frequency, the switching frequency constraint is introduced into the value function to select the optimal switching state. Finally, the correctness of the control strategy studied in this paper is verified by MATLAB, and the simulation results are given.

2. OEW- PMSM

2.1. Model of drive system

The structure of the open-winding PMSM system is shown in Fig.1, where Vdc1 = Vdc2.

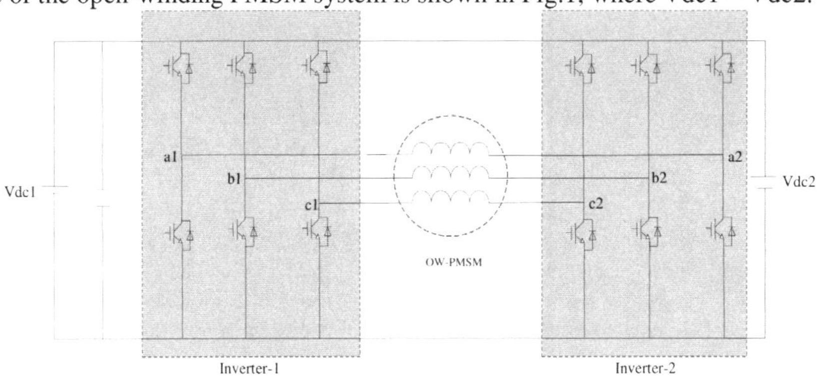

Figure 1. OEW- PMSM system

In this paper, the rotor magnetic field orientation method is adopted. The voltage equation in the dq coordinate system may be expressed as:

$$\begin{cases} u_d = R_s i_d + L_d \dfrac{di_d}{dt} - w_e L_q i_q \\ u_q = R_s i_q + L_q \dfrac{di_q}{dt} + w_e L_d i_d + w\psi_f \end{cases} \tag{1}$$

Where $u_d, u_q, i_d, i_q, L_d, L_q$ are the d- q axis components of the stator voltage, current, and inductance, respectively; ω_e is the rotor electrical speed; ψ_f is permanent magnet flux linkage; R_s is the stator resistance.

The torque can be expressed as:

$$T_e = \frac{3}{2} P_n \left[\psi_f i_q + (L_d - L_q) i_d i_q \right] \tag{2}$$

Where P_n is the number of pole pairs.

2.2. Voltage Vectors of Dual Inverters

The stator voltage can be expressed as:

$$\begin{cases} u_a = V_{a1} - V_{a2} \\ u_b = V_{b1} - V_{b2} \\ u_c = V_{c1} - V_{c2} \end{cases} \tag{3}$$

The voltage vector can be defined as:

$$V_s = \frac{2}{3}\left(u_a + u_b e^{j\frac{2\pi}{3}} + u_c e^{j\frac{4\pi}{3}} \right) \tag{4}$$

In the OEW-PMSM system, each inverter can generate eight switching states. Therefore, there are a total of 64 voltage switching states and 19 voltage vectors. The distribution of all switching states and voltage vectors is shown in fig.2.

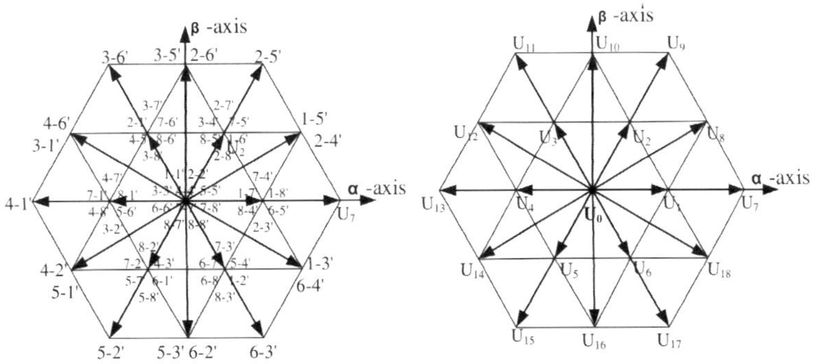

(a) (b)

Figure 2. Space voltage vector plane for dual two-level inverters: (a) all switching states; (b) all different voltage vectors

3. Proposed method

3.1. The basic control algorithm

Based on (1), the stator current can be expressed as

$$\begin{cases} \dfrac{di_d}{dt} = -\dfrac{R_s}{L_d} i_d + \dfrac{L_q}{L_d}\omega_e i_q + \dfrac{1}{L_d} u_d \\ \dfrac{di_q}{dt} = -\dfrac{R_s}{L_q} i_q - \dfrac{L_d}{L_q}\omega_e i_d + \dfrac{1}{L_d} u_d - \omega_e \dfrac{\psi_f}{L_q} \end{cases} \tag{5}$$

Although the traditional first-order forward Euler discrete method is relatively simple, the accuracy is poor[6]. To improve prediction accuracy without significantly increasing computational complexity, In this paper, the second order Euler discrete method is used to discretize equation (5).

$$\begin{cases} i_{dp}^{k+1} = i_d^k + \dfrac{T_s\left(u_d^k - R_s i_d^k - E_d^k\right)}{L_d} \\ i_{qp}^{k+1} = i_q^k + \dfrac{T_s\left(u_q^k - R_s i_q^k - E_q^k\right)}{L_q} \\ i_d^{k+1} = i_{dp}^{k+1} + \dfrac{T_s(-R_s)(i_{dp}^{k+1} - i_d^k)}{2L_d} \\ i_q^{k+1} = i_{qp}^{k+1} + \dfrac{T_s(-R_s)(i_{qp}^{k+1} - i_q^k)}{2L_q} \end{cases} \tag{6}$$

Where i_{dp}^{k+1} and i_{qp}^{k+1} are the predictive correction current; E_d^k and E_q^k are the d-q axis components of the counter electromotive force can be expressed as:

$$\begin{cases} E_d^k = -\omega_e L_q i_q^k \\ E_q^k = \omega_e\left(L_d i_d^k + \psi_f\right) \end{cases} \tag{7}$$

The voltage components in d- q axis can be expressed as:

$$\begin{bmatrix} u_d^{\ k} \\ u_q^{\ k} \end{bmatrix} = \begin{bmatrix} \cos q & \sin q \\ -\sin q & \cos q \end{bmatrix} \bullet \begin{bmatrix} real(V_s) \\ \mathrm{imag}(V_s) \end{bmatrix} \tag{8}$$

To minimize the current vector tracking error as the control target, the objective function is constructed as:

$$g = \left| i_d^{\ k+1} - i_d^{\ ref} \right| + \left| i_q^{\ k+1} - i_q^{\ ref} \right| \tag{9}$$

3.2. Reduce calculation time

For the dual inverter feed system, each time the optimal voltage vector is selected, 19 current predictions are required, and the target function values need to be sorted, resulting in a large amount of computation. In order to achieve a good control effect in the actual project, the selected control period is relatively small, so the requirements for the digital controller are high, which makes the application of the method in engineering practice constrained. To this end, this paper proposes a fast algorithm based on partition judgment. The candidate voltage vector is determined based on the sector in which the deadbeat voltage vector is located. Only a comparison is needed to select the final optimal voltage vector, which reduces the amount of calculation.

Further simplifying equation (6)

$$\begin{cases} i_d^{\ k+1} = i_{d0}^{\ k} + \dfrac{T_s L_d - \dfrac{T_s^2 R_s}{2}}{L_d^2} u_d \\ i_q^{\ k+1} = i_{q0}^{\ k} + \dfrac{T_s L_q - \dfrac{T_s^2 R_s}{2}}{L_q^2} u_q \end{cases} \tag{10}$$

Where

$$\begin{cases} i_{d0}^{\ k} = \dfrac{L_d^2 - L_d R_s T_s + 0.5 T_s^2 R_s^2 i_d^{\ k} + (0.5 T_s^2 R_s - T_s L_d) \mathrm{E_d}}{L_d^2} \\ i_{q0}^{\ k} = \dfrac{L_q^2 - L_q R_s T_s + 0.5 T_s^2 R_s^2 i_q^{\ k} + (0.5 T_s^2 R_s - T_s L_q) \mathrm{E}_q}{L_q^2} \end{cases} \tag{11}$$

If $i_d^{\ k+1} = 0$, $i_q^{\ k+1} = i_q^{\ ref}$ in (10), the deadbeat reference voltage can be obtained as:

$$\begin{cases} u_d^{\ ref} = \dfrac{-i_{d0}^{\ k} L_d^2}{T_s L_d - 0.5 T_s^2 R_s} \\ u_q^{\ ref} = \dfrac{(i_q^{\ ref} - i_{d0}^{\ k}) L_d^2}{T_s L_q - 0.5 T_s^2 R_s} \end{cases} \tag{12}$$

At this point, the objective function can be established as:

$$g = l_d \left| u_d^{\ ref} - u_{di}^{\ k+1} \right| + l_q \left| u_q^{\ ref} - u_{qi}^{\ k+1} \right| \tag{13}$$

Where $u_{di}^{\ k+1}$ and $u_{qi}^{\ k+1}$ are the preselected voltage vector.

$$\begin{cases} l_d = \left| \dfrac{T_s (L_d - 0.5 R_s T_s)}{L_d^2} \right| \\ l_q = \left| \dfrac{T_s (L_q - 0.5 R_s T_s)}{L_q^2} \right| \end{cases} \tag{14}$$

It is divided into 6 parts equally as shown in Fig.3.

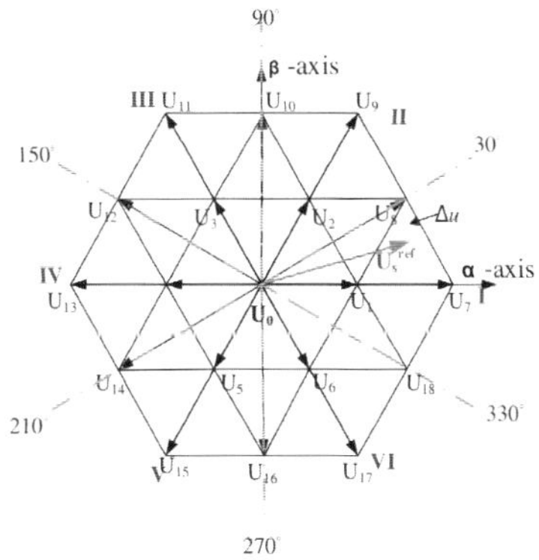

Figure 3. voltage vector sector distribution

For a given deadbeat reference voltage:

$$V_{ref} = [V_a \ V_b \ V_c]^{\mathrm{T}} \tag{15}$$

In order to avoid determining the sector to which the V_{ref} is located, the method shown in Fig .4 is used to determine:

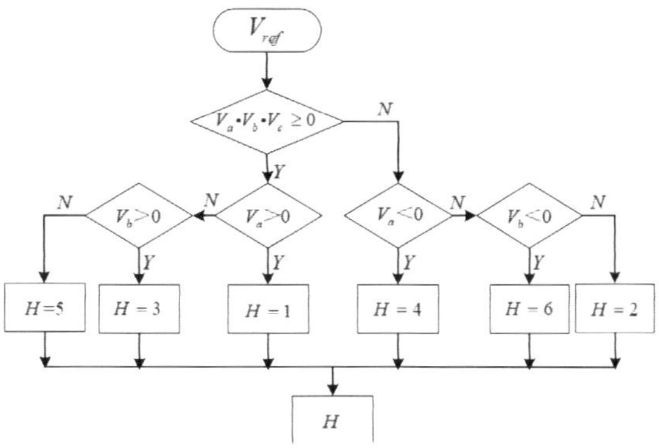

Figure 4. Sector judgment flow chart

It can be seen from (13) that the objective function has been minimized by minimizing the current error to a voltage error.

In Fig. 4, the deadbeat reference voltage vector falls in the first sector, and the voltage difference with the non-zero voltage vector u_8 is obviously the smallest, then

$$\Delta u = \left| u_s^{ref} - u_8 \right| \tag{16}$$

If the zero vector is taken as the selected voltage vector, the magnitude of the error is:

$$\Delta u_0 = \left| u_s^{ref} \right| \tag{17}$$

It is only necessary to compare the two to select the optimal voltage vector.

If Δu is greater than $\Delta \mathrm{u}_0$, a non-zero voltage vector u8 is output, otherwise a zero vector is output.

This method is essentially the same as the traditional FCS-MPC, but the method avoids the calculation of 19 predicted currents, and only needs to make an optimal voltage vector based on the voltage tracking error.

3.3. Minimum switching loss

As shown in Fig. 2, there are redundant switching states which means that the determined optimal voltage vector can have several corresponding switching states. Since different switching states have an effect on the switching frequency, it is necessary to select an appropriate switching state to reduce the switching frequency. According to the principle of minimum switching loss, the number of switching of it in one control cycle is calculated as follows:

$$S=\left|S_{a1}^{k+1}-S_{a1}^{k}\right|+\left|S_{b1}^{k+1}-S_{b1}^{k}\right|+\left|S_{c1}^{k+1}-S_{c1}^{k}\right| \\ +\left|S_{a2}^{k+1}-S_{a2}^{k}\right|\left|S_{b2}^{k+1}-S_{b2}^{k}\right|+\left|S_{c2}^{k+1}-S_{c2}^{k}\right| \tag{18}$$

Where,Sx1(x=a, b, c)=0,1 is the INV1 switching state of U_{opt} ; Sx2(x=a, b, c)=0,1 is the INV2 switching state of U_{opt} ; "k+1" and "k" indicate the dual inverter switch state and current inverter switch state to be applied at the next moment.

3.4. Delay compensation

Considering the actual project, there is a beat delay between the actual output voltage of the digital controller and the command voltage. It is necessary to replace the predicted current $i_{dq}^{P}(k+1)$ of the next moment with i_{dq}^{k} to calculate the ideal voltage vector, and then use the improved algorithm to select the optimal voltage vector.

4. Simulink simulation

In order to verify the effectiveness of the algorithm, the improved algorithm is simulated in Matlab/Simulink environment. The parameters of the OW-PMSM used in the Table 1. The sampling frequency of the control system is 10khz.

Table 1. The parameters of the ow-pmsm

parameter	*value*
Stator resistance R_s	*2.375Ω*
Direct inductance L_d	*5.8mH*
Quadrature inductance L_q	*6.6mH*
DC bus voltage U_{dc}	*400V*
Permanent magnet flux linkage ψ_f	*0.1286wb*
number of pole-pairs P	*4*

4.1. Simulation waveform

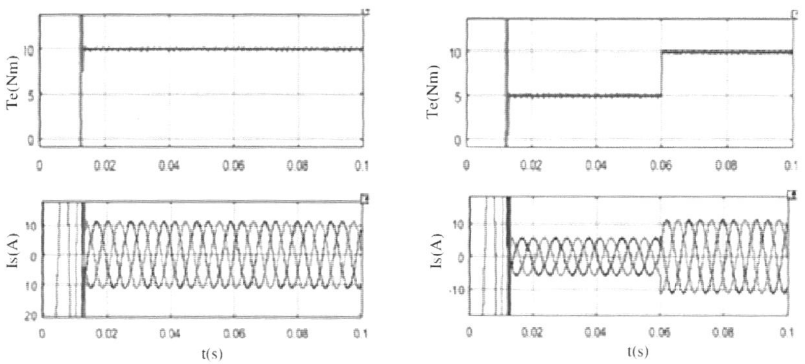

(a) (b)

Figure 5. Simulation waveform:
(a) Steady state operation (b)Dynamic responses

4.2. Simulation results

First, it can be seen that the proposed MPCC produces good current and torque performance of the

steady state performance in Fig. 5(a).
Second, it can be seen that the system has good anti-interference ability with sudden load from 5 Nm to 10 Nm at 0.06 s .The torque response time is short and the tracking time is short about the dynamic behavior in Fig. 5(b).
The fast vector selection MPCC method can indeed select the optimal voltage vector in a more convenient way, so it has higher practicability.

Acknowledgments
The work done in this paper has been funded by the National Natural Science Foundation of China (Grant No.51707078) and Natural Science Foundation of Colleges and Universities in Jiangsu Province (Grant No.15KJD470003).

References
[1] K. Sivakumar, A. Das, R. Ramchand, C. Patel and K. Gopakumar, "A Hybrid Multilevel Inverter Topology for an Open-End Winding Induction-Motor Drive Using Two-Level Inverters in Series With a Capacitor-Fed H-Bridge Cell," IEEE Trans. Ind. Electron., vol. 57, no. 11, pp. 3707-3714, Nov. 2010
[2] Kouro S, Cortes P, Vargas R,et al. Model Predictive Control — A Simple and Powerful Method to Control Power Converters[J]. IEEE Transactions on Industrial Electronics, 2009, 56(6):1826-1838
[3] Vazquez S, Rodriguez J, Rivera M,et al. Model Predictive Control for Power Converters and Drives: Advances and Trends[J]. IEEE Transactions on Industrial Electronics, 2017, 64(2):935-947
[4] B. Zhu, K. Rajashekara and H. Kubo, "Predictive torque control with zero-sequence current suppression for openend winding induction machine," 2015 IEEE Industry Applications Society Annual Meeting, Addison, TX, 2015, pp. 1-7.
[5] S. Chowdhury, P. W. Wheeler, C. Gerada and C. Patel, "Model Predictive Control for a Dual-Active Bridge Inverter With a Floating Bridge," in IEEE Transactions on Industrial Electronics, vol. 63, no. 9, pp. 5558-5568, Sept. 2016.
[6] Su J, Sun D. Simplified MPCC for four-switch three-phase inverter-fed PMSM[J]. Electronics Letters, 2017, 53(16):1108-1109.

IOP Conf. Series: Materials Science and Engineering **853** (2020) 012020 doi:10.1088/1757-899X/853/1/012020

Study on the reactive power coordinated control in hybrid parallel HVDC system

Yan Han, Song Rui, Wang Xuan, Zhao Shichang, Yang Wenli, Luo Min

Power Research Institute of Qinghai Power Grid Company

yh19940103@163.com

Abstract. Hybrid parallel HVDC system adopts line commutated converter high voltage direct current(LCC-HVDC) and voltage source converter high voltage direct current(MMC-HVDC) parallel structure, which combines the advantages of LCC-HVDC and MMC-HVDC. Based on the existing reactive power control of HVDC system, the reactive power coordination control strategy of hybrid parallel HVDC system is innovatively proposed in this paper. Firstly, the reactive power control of LCC-HVDC and MMC-HVDC are studied respectively. And then the voltage stability of LCC-HVDC fed weak AC system is analyzed. Furthermore, a reactive power coordination control strategy between reactive power compensation device of the converter station and MMC-HVDC is proposed. Finally, based on PSCAD/EMTDC simulation platform, a hybrid parallel HVDC simulation system is built. The reactive power coordination control strategy under steady and transient conditions is simulated and analyzed respectively. The analysis results show that the reactive power coordinated control strategy can suppress the voltage fluctuation of converter bus in steady state. At the same time, the reactive power coordinated control strategy can restrain the commutation failure of LCC-HVDC in the fault of receiving power grid and improve its fault recovery ability.

1. Introduction

The energy storage and load in China are inversely distributed. In order to promote the development of new energy sources and optimize the allocation of resources, high voltage direct current (HVDC) transmission technology has entered a booming stage. With the gradual formation of grid structure with AC and DC parallel and multi-infeed direct current system, the safe and stable operation of power grid is facing many problems and challenges[1]. Hybrid parallel DC(HPDC) system integrates the advantages of HVDC system based on line commutated converter (hereinafter referred to as "LCC-HVDC unit") and HVDC system based on voltage source converter (hereinafter referred to as "MMC-HVDC unit"). HPDC system can optimize grid structure and reduce the risk of large-scale power outage.

LCC-HVDC unit is often used to realize long-distance and large-capacity transmission. However, the voltage stability of LCC DC-fed weak AC systems seriously affects the safe and reliable operation of DC projects. MMC-HVDC unit has excellent control performance. It is without reactive power compensation and there is no commutation failure problem[2]. Reactive power control of MMC-HVDC unit can effectively improve the voltage stability level of LCC DC-fed weak AC system. This paper innovatively proposes a reactive power coordination control strategy between reactive power compensation equipment and MMC-HVDC unit by analyzing the slow reactive power control of LCC-HVDC unit and the fast reactive power control of MMC-HVDC unit. Finally, PSCAD/EMTDC simulation analysis is carried out with the HPDC system simulation model , which proves that the reactive power coordination control strategy can suppress the voltage fluctuation of converter buses in steady state. And the fault recovery characteristics of conventional DC system can be effectively improved.

2. Reactive powercontrol of hybrid parallel HVDC system

2.1. Structure of hybrid parallel HVDC system

Hybrid parallel DC system integrates the advantages of LCC-HVDC unit and MMC-HVDC unit transmission system. LCC-HVDC unit is connected by two 12 pulse converters through a smoothing reactor (L_d). Each 12-pulse converter is composed of two 6-pulse converters. The DC grounding electrode is located between the 6 pulse converters on the inverter side. MMC-HVDC unit consists of 2 back-to-back modular multilevel converters(MMC). The schematic diagram of hybrid parallel DC system is shown in Figure 1.

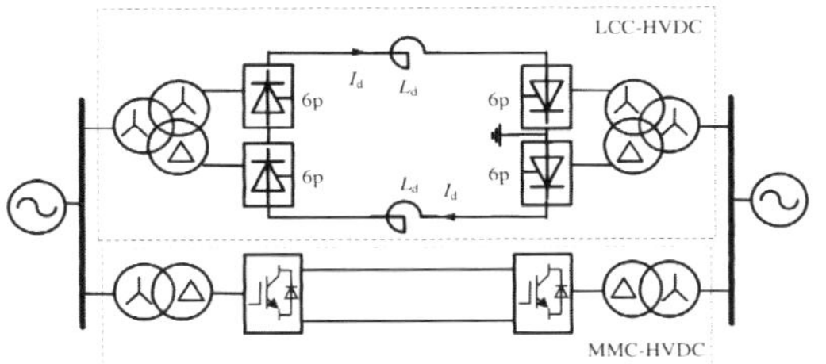

Figure 1. Schematic diagram of hybrid parallel HVDC system.

2.2. Slow reactive power control of LCC-HVDC unit

LCC-HVDC unit needs to consume a lot of reactive power when it is running. According to the requirements of AC/DC system, the reactive power compensation device of converter station is designed to compensate the reactive power needed by converter. Slow reactive power control of LCC-HVDC unit is the switching control of reactive power compensation device[3].

The substitution and filtering control are composed of each sub-function which realizes its specific control function in turn according to the preset priority and criterion. According to the order of the sub-functions of reactive power control from the highest priority 1 to the lowest priority 5, the slow reactive power control logic of converter station is formulated as shown in Figure 2.

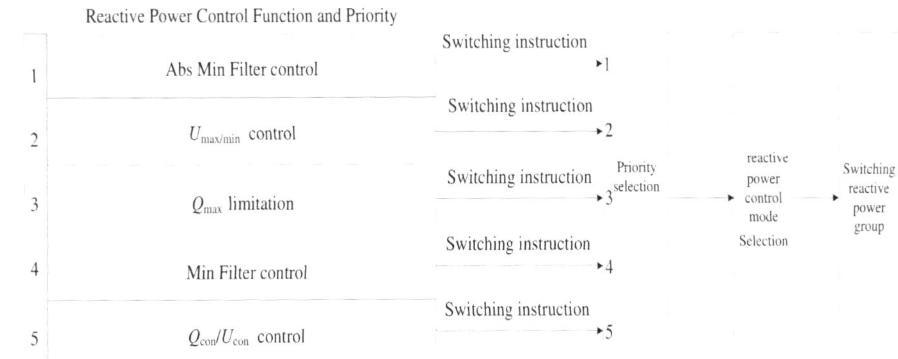

Figure 2. Logic diagram of reactive power control of LCC-HVDC in converter station.

2.3. Fast reactive power control of MMC-HVDC unit

MMC-HVDC unit does not need reactive power compensation. Furthermore MMC-HVDC unit can play the role of Static Synchronous Compensator (STATCOM) when capacity permits to improve the voltage stability of AC system. MMC-HVDC unit control system includes inner-loop control and outer-loop control[4]. The inner-loop control provides two independent controlled variables I_d and i_q for active and reactive control of MMC respectively. The control block diagram of MMC-HVDC unit is shown in Figure 3.

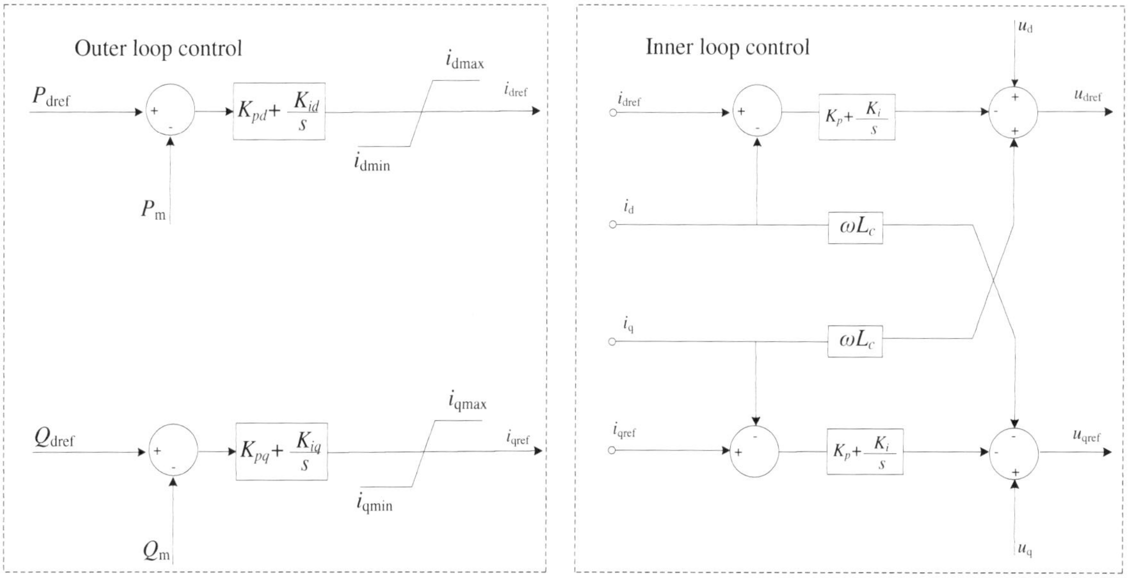

Figure 3. Control block diagram of MMC-HVDC unit.

According to the voltage deviation of converter bus or the switching reactive power deviation of converter station, the outer-loop control of MMC-HVDC unit provides reference current I_{qref} to the inner-loop control through proportional integral(PI) link. And the inner loop control real-time control I_q to achieve no-difference tracking of the voltage of converter bus or the switching reactive power of converter station. The core of reactive power control of MMC-HVDC unit is to realize the dynamic reactive power compensation function of MMC by controlling I_q in real-time.

3. Voltage stability of LCC-HVDC fed weak AC systems

3.1. Reactive power packet configuration and voltage stability of converter bus

When the slow reactive power control of LCC-HVDC unit realizes reactive power grouping switching, the safe and stable operation of the conventional DC unit will be affected. According to GB/T31460-2015 , the transient voltage change rate of converter bus caused by switching reactive power grouping should not exceed 2%, and the steady voltage change rate should not exceed 1%[5]. When the receiving power grid is weak, the voltage fluctuation caused by reactive power grouping switching becomes larger, which will limit the selection of reactive power grouping capacity. However, reducing reactive power grouping capacity will inevitably increase the number of reactive power grouping, increase the area of AC filter field and affect the flexibility and economy of converter station[6] .

3.2. Commutation failure and fault recovery characteristics of LCC-HVDC

When the receiving power grid fails, the voltage amplitude and phase of the converter bus will change, which will affect the commutation process of LCC-HVDC unit and lead to the commutation failure of the inverters[7]. The weaker the receiving power grid is, the more serious the voltage sag of the converter bus will be when the fault occurs and the more likely the commutation failure will be caused[8]. At the same time, the fault recovery of LCC-HVDC units is closely related to the dynamic reactive power compensation provided by the receiving power grid. Therefore, the receiving power grid is required to have sufficient voltage support capability[9].

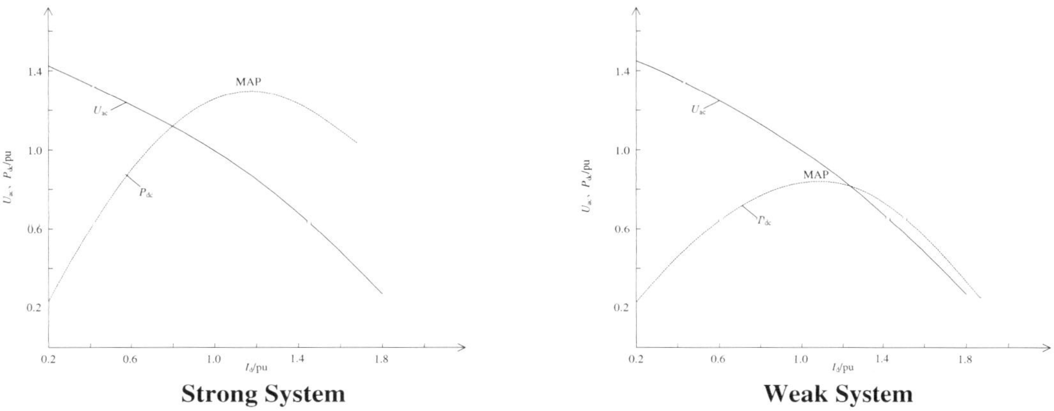

Figure 4. Changes of electrical quantities in AC/DC system with DC current.

4. Reactive power coordinated control strategy in hybrid parallel HVDC system

Aiming at the problem of voltage stability in LCC DC-fed weak AC systems, this paper proposes a novel reactive power coordination control strategy between converter station reactive power compensation device and MMC-HVDC unit to improve the voltage stability of hybrid parallel DC system under steady and transient conditions respectively.

4.1. Steady state reactive power coordinated control strategy

When the converter bus voltage U_{pcc} belongs to (0.9, 1.1), the steady-state fluctuation of the converter bus voltage is determined. According to the coordinated control of the reactive power compensation equipment and the MMC-HVDC unit, the MMC-HVDC unit has priority to regulate the voltage. And the switching times of the reactive power compensation equipment are reduced as much as possible to restrain the voltage fluctuation of the converter bus. The specific control strategy is as follows:

(1)When $U_{pcc}<U_{pcc\_aim}$, the steady-state voltage change rate caused by switching the reactive power compensation equipment in the converter station is not allowed to exceed 1%. Firstly the ability of the MMC-HVDC unit to raise the converter bus voltage by 1% is judged.

$$Q_{MMC\_max\_s} - Q_{MMC} > Q_m \tag{1}$$

In the formula(1), $Q_{MMC\_max\_s}$ is the reactive power output limit of the MMC-HVDC unit in steady state. Q_{MMC} is the actual reactive power output of the MMC-HVDC unit and Q_m is the reactive power limit of 1% increasement of the converter bus voltage in steady state.

In order to calculate Q_m , reactive power required by unit voltage increment of converter bus is represented by introducing MMC-HVDC unit reactive power-voltage control coefficient k. In steady state, the reactive power provided by MMC-HVDC unit is changed in a small range. Reactive power increment ΔQ_{MMCi} and voltage increment ΔU_{ppci} are collected. Assuming that M sets of data are collected, the k value can be expressed as follows:

$$k = \frac{1}{m}\left(\frac{\Delta Q_{MMC1}}{\Delta U_{pcc1}} + \frac{\Delta Q_{MMC2}}{\Delta U_{pcc2}} + \ldots + \frac{\Delta Q_{MMCm}}{\Delta U_{pccm}}\right) \tag{2}$$

$$Q_m = 0.01k \tag{3}$$

If the judgment of formula(1) is established, the control instruction of the reactive power compensation equipment and the MMC-HVDC unit is set to formula(4). The reactive power compensation equipment of the converter station does not operate and the MMC-HVDC unit increases the voltage of the converter bus by 1% or to the target value. After a predetermined time step (3) is entered.

$$\begin{cases} U_{\mathrm{MMC\_ref}} = \min(U_{\mathrm{pcc}} + 0.01, U_{\mathrm{pcc\_aim}}) \\ f_{\mathrm{c\_swi}} = 0 \end{cases} \tag{4}$$

In the formula(4) $U_{\mathrm{MMC\_ref}}$ is the reference voltage value of the converter bus and $f_{\mathrm{c\_swi}}$ is the action instruction of the reactive power compensation equipment in the converter station.
If formula (1) is not valid, it is judged whether the voltage fluctuation of converter bus caused by putting in a maximum capacity reactive power group is satisfied.

$$U_{\mathrm{pcc}} < U_{\mathrm{pcc\_aim}} - \Delta U \tag{5}$$

$$\Delta U = \frac{Q_{\mathrm{cmax}}}{k} \tag{6}$$

If formula (5) is established, step (2) is entered. If it is not satisfied, step (3) is entered.
(2)If the converter station still has a set of reactive power compensation equipment that has not been put into operation , it will be put into. it follows the principle that AC filter should be put in firstly. At the same time, the MMC-HVDC unit ensures that the voltage change rate of the converter bus does not exceed 1% after the reactive power compensation equipment of the converter station is put into operation. And the reactive power control instruction is set to formula(7). After a predetermined time step (4) is entered.

$$\begin{cases} U_{\mathrm{MMC\_ref}} = U_{\mathrm{pcc}} + 0.01 \\ f_{\mathrm{c\_swi}} = 1 \end{cases} \tag{7}$$

If there is no reactive power compensation equipment available in the converter station, the reactive power compensation equipment will not operate. The MMC-HVDC unit adjusts the voltage of the converter bus to a limited extent. The steady-state reactive power limit of MMC-HVDC unit is set to $Q_{\mathrm{MMC\_max\_s}}$. And the reactive power control instruction is set to formula(8). After a predetermined time step (5) is entered.

$$\begin{cases} Q_{\mathrm{MMC\_ref}} = Q_{\mathrm{MMC\_max\_s}} \\ f_{\mathrm{c\_swi}} = 0 \end{cases} \tag{8}$$

(3)Judging whether the flexible DC unit can raise the converter bus voltage to the target value $U_{\mathrm{pcc\_aim}}$.

$$Q_{\mathrm{MMC\_max\_s}} - Q_{\mathrm{MMC}} > (U_{\mathrm{pcc\_aim}} - U_{\mathrm{pcc}}) / k \tag{9}$$

If formula(9) is not valid, step(3) is turned into step(2). On the contrary, the voltage is regulated by a MMC-HVDC unit and the corresponding reactive power control instruction is set to formula(4). After a predetermined time step(4) is entered.
(4)Determine whether U_{pcc} is equal to $U_{\mathrm{pcc\_aim}}$ If $U_{\mathrm{pcc}} < U_{\mathrm{pcc\_aim}}$, step(4) is turned into step(1). If $U_{\mathrm{pcc}} = U_{\mathrm{pcc\_aim}}$, step(4) goes to step(5).
(5)End.

When $U_{\mathrm{pcc}} < U_{\mathrm{pcc\_aim}}$, the flow chart of reactive power coordination control strategy between converter station reactive power compensation equipment and MMC-HVDC unit is shown in Figure 5.

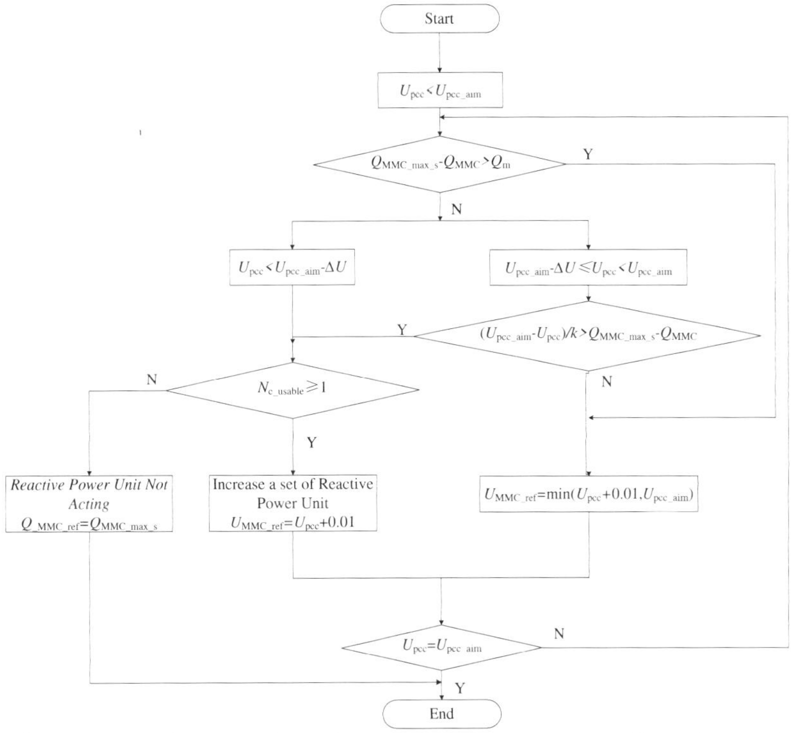

Figure 5. Flow-process diagram of steady state reactive power coordinated control strategy

4.2. Transient state reactive power coordinated control strategy

Transient reactive power coordination control strategy is adopted when the voltage of converter bus drops sharply due to the faults of AC system near the converter station. Voltage changes rapidly in the transient process. Voltage fluctuation of converter buses will be further caused by switching reactive power compensation equipment. Therefore, slow reactive power control of converter stations will be blocked in the transient. MMC-HVDC unit takes the rating voltage of converter bus as reference value. MMC-HVDC unit automatically adjusts the reactive power output according to the reactive power required for voltage regulation, restrains the commutation failure of conventional DC unit and provides dynamic reactive power compensation for LCC-HVDC fault recovery.

4.3. Simulation and verification

Based on the parameters of Luxi back-to-back DC project , the simulation system of HPDC system and weak receiving power grid is established in PSACD/EMTDC simulation software. The reactive power coordination control strategies in steady and transient conditions are simulated and analyzed respectively.In order to verify the effectiveness of the reactive power coordination control strategy, the operation characteristics of HPDC system in steady and transient state are compared and analyzed in two cases. Scheme 1: MMC-HVDC unit does not participate in reactive power control and scheme 2: reactive power coordination control strategy is adopted.

The simulation time is set to 3 seconds .The conveying power of the LCC-HVDC unit increases from 1.0 pu to 1.1 pu after 1 second. It lasts two seconds and recovers later. The waveform of the converter bus voltage is shown in Figure 6. Under both schemes the voltage of converter bus drops. The voltage of converter buses drops to 0.969pu and 0.980pu respectively under scheme 1 and scheme 2 when the conveying power of LCC-HVDC unit increases to 1.1pu. Moreover the voltage fluctuation of converter bus under scheme 2 is smaller than that under scheme 1 during the sudden power increasement of LCC-HVDC transmission. When the power is restored to 1.0 pu, the voltage of the converter bus under the two schemes will be restored to the rated value. The results show that the reactive power coordination control strategy between MMC-HVDC unit and reactive power compensation equipment of converter station can provide reactive power support for weak AC system in steady state, reduce the voltage fluctuation of converter bus. What's more, it will not cause commutation failure on the inverter side.

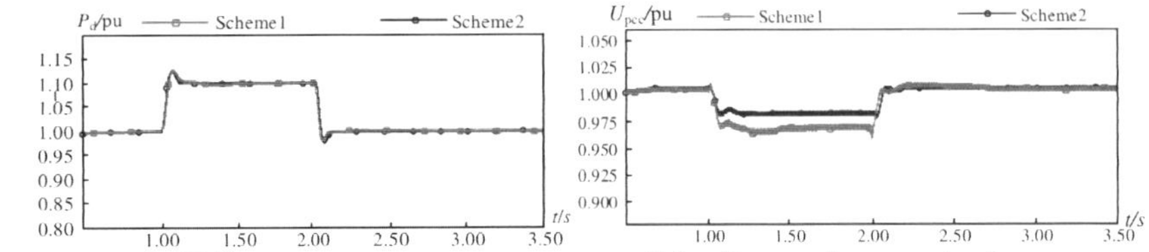

(a)DC power (b)voltage of converter bus

Figure 6. Simulation waveform of steady reactive power coordinated control

Due to the limitation of space, this paper only gives the simulation results of single-phase-to-ground fault in the receiving power grid. The single-phase grounding fault of converter bus lasts 0.1second after 2 seconds. The simulation results are shown in Figure 7. The waveforms from (a) to (d) in Figure 7 are converter bus voltage, LCC-HVDC unit transmission power, DC voltage and turn-off angle.
Because of the short fault time and the long response time of reactive power compensation equipment, no reactive power compensation equipment is put on/off during the fault period. The bus voltages of scheme 1 and scheme 2 at the instant of single-phase grounding of receiving end power grid decrease to 0.76pu and 0.8pu respectively at t=2s. Because the receiving power grid is the weak AC system, the fault easily causes the voltage of the commutation bus droping when there is no dynamic reactive power compensation device. It results in the commutation failure of the LCC-HVDC unit. In severe cases, the LCC-HVDC system will be blocked by oscillation because of the repeated commutation failures. In scheme 2, the MMC-HVDC unit rapidly generates full reactive power during the fault period. The LCC-HVDC unit is compensated by dynamic reactive power to restrain the voltage fluctuation of the converter bus, which improves the oscillation and sag of DC power and DC voltage to a certain extent. After eliminating the fault about 0.2s HPDC system will be restored to the rated value. During the commutation failure of LCC-HVDC unit the DC current Id and turn-off angle γ increase sharply, which leads to reactive power consumption of LCC-HVDC unit increasing. It is necessary to absorb dynamic reactive power from AC system. Therefore, dynamic reactive power compensation has an important impact on the fault recovery characteristics of LCC-HVDC unit.

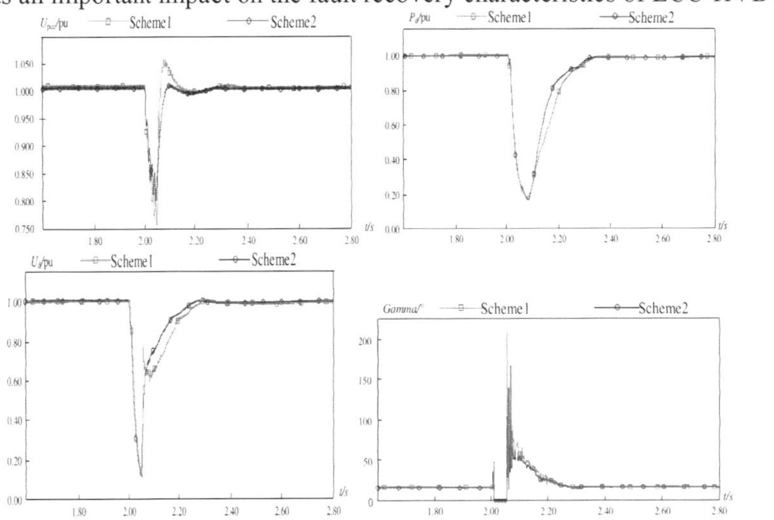

(a)voltage of converter bus (b)DC power (c)DC volitage (d)turn-off angle

Figure 7. Simulation waveform of transient reactive power coordinated control

5. Conclusion

Aiming at the voltage stability of LCC-HVDC fed weak AC system, this paper proposes a reactive power coordination control strategy between reactive power compensation equipment in converter station and MMC-HVDC unit. The effectiveness of this strategy is verified by simulation analysis of Luxi back-to-back DC system. The main conclusions are as follows:

1) Hybrid parallel HVDC system adopts parallel structure of LCC-HVDC unit and MMC-HVDC unit. And reactive power control of MMC-HVDC unit can effectively solve the voltage stability problem of LCC-HVDC fed weak AC system.

2) When the voltage fluctuates under the steady state range, the MMC-HVDC unit is adopted to regulate the voltage through the coordinated control of the reactive power compensation equipment and the MMC-HVDC unit in the converter station. It is useful to reduce the number of reactive power grouping switching and improve the voltage stability of the converter bus.

3) The voltage fluctuation caused by AC system faults will be suppressed through the coordinated control strategy between reactive power compensation equipment of converter station and MMC-HVDC unit. It can reduce the probability of commutation failure ofLCC-HVDC units and improve the fault recovery capability of AC/DC systems.

In summary, the proposed reactive power coordination control strategy for hybrid parallel HVDC system can not only restore the voltage of the converter bus to the target value when the voltage fluctuates steadily but also reduce the risk of commutation failure of LCC-HVDC unit when the AC system fails. Therefore it is beneficial to improve the stability of hybrid parallel HVDC system.

References

[1] ZHOU Baorong, HONG Chao, JIN Xiaoming,et al.Study on the transition from synchronized operation grid to asynchronous operation grid in China southern power grid. Proceedings of the CSEE, 2016, 36(8): 2084-2092.

[2] GUO Chunyi, ZHAO Chengyong, Allan,et al.Risk assessment framework design on flexibility power grid planning and its application prospect[J]. Electric Power Construction, 2012(10): 98-104.

[3] CHEN Huan, WANG Zhen, YANG Zhizhong,et al.Coordinated reactive power control approach for LCC-HVDC and VSC-HVDC in hybrid parallel HVDC system[J]. Power System Technology, 2017, 41(6): 1719-1725.

[4] ZHAO Chengyong, HU Dongliang, LI Guangkai.Analysis of Reactive Power Regulation Characteristics of a New Double-fed HVDC Transmission System[J]. Automation of Electric Power Systems, 2008, 32(21):51-55.

[5] ZHANG Wang, HAO Junfang, CAO Seng,et al.Design of reactive power control for HVDC converter station[J]. Power System Protection and Control, 2009, 37(14): 72-76.

[6] CHEN Peidong, GUAN Lin. Analysis and Modeling of Dynamic Reactive Power Demand and Power Recovery Characteristics of HVDC System[J]. Modern Electric Power, 2017, 34(4): 50-58.

[7] TU Jingzhe, ZHANG Jian, ZENG Bing,et al. HVDC transient reactive power characteristics and impact of control system parameters during commutation failure and recovery[J]. High Voltage Engineering, 2017, 43(7):2131-2139.

[8] MA Leipeng, WEI Gang, LI Qinyu,et al. Reactive power reserve optimization of power system considering HVDC transmission[J]. Power System Protection and Control, 2017(24): 146-151.

[9] YANG Huanhuan, CAI Zexiang, ZHU Lin,et al. Reactive Power Dynamic Characteristics of DC System and Its Impact on Transient Voltage Stability of Received Power Grid[J]. Electric Power Automation Equipment, 2017(10): 86-92.

IOP Conf. Series: Materials Science and Engineering **853** (2020) 012021 doi:10.1088/1757-899X/853/1/012021

A Fiber-Optic Current Sensor Based on Fuzzy PI Control

Zhaowei Diao, Zhongxiao Ji*, Mei Liu, Yifei Wu, Yi Wu, Fei Yang

State Key Laboratory of Electrical Insulation and Power Equipment, Xi'an Jiaotong University, Xi'an 710049, China

jzx1230@xjtu.edu.cn

Abstract. The development of power grid has put forward higher requirements on the performance of current sensors. Traditional current sensors already can not meet these needs of grid metering and protection. This paper proposes a fiber-optic current sensor (FOCS) based on fuzzy PI control strategy which combines fuzzy control with PI control. These parameters of fuzzy PI controller can be adjusted online at different stages of the system response. Simulation results show that the FOCS based on fuzzy PI control has a fast dynamic response and no overshoot, compared with FOCS based on traditional PI control. These advantages of accuracy in metering and rapidity in protection of power grid, can be satisfied simultaneously.

1. Introduction

With the development of the smart grid, especially the construction of high-voltage DC power grid, higher requirements of the current sensor have been put forward for accurate measurement of electric energy and quick protection of circuit breakers. Traditional electromagnetic sensors have been unable to meet these requirements of power grid [1][2]. Compared with the traditional DC current sensor, the fiber-optical current sensor (FOCS) has many technical advantages, which include high measurement accuracy, wide measurement range, excellent insulation performance, anti-electromagnetic interference, and so on [3][4]. It has been widely used in the measurement and protection of various high-voltage DC equipment.

The FOCS with the traditional PI control strategy, is unable to meet these requirements simultaneously between the measurement accuracy in the case of small current and the response speed in the case of large short-circuit current. In order to solve this problem, the FOCS based on the fuzzy PI control strategy is proposed in the paper, by establishing and analyzing the dynamic model of FOCS. Lastly the simulation result shows the effectiveness of this control strategy.

2. Principle and modeling

2.1. Principle of FOCS

The measurement principle of FOCS is Faraday magneto-optical effect [5], and the structure of FOCS is shown in Figure 1[6] [7].

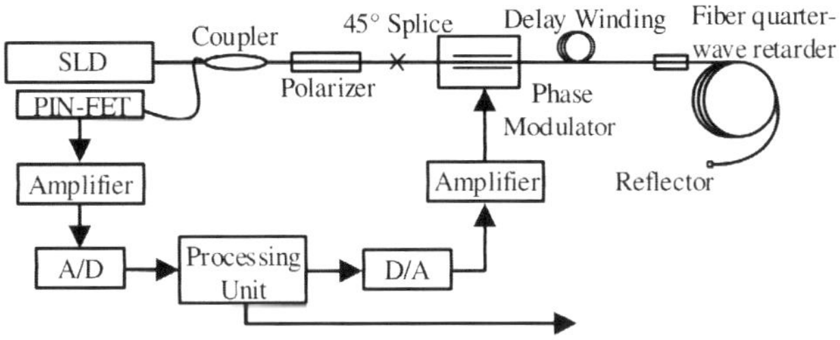

Figure 1. The structure diagram of FOCS

In Figure 1, the SLD source emits light waves, then these light waves pass through the polarizer, the point of 45-degree fusion, the phase modulator, the delay winding of polarization-maintaining fiber, the current sensing fiber in which inducts the Faraday effect by the current. After they are reflected by fiber-optical reflector, these returned light waves with the current information are interfered at the point of 45-degree fusion, and these interference light waves can be shown as:

$$P = \frac{P_0}{2}(1 + \cos\varphi_s) \tag{1}$$

Where P is the light intensity after the interference, P_0 is the light intensity emitted by SLD. φ_s is the phase shift of the Faraday effect, which is expressed as:

$$\varphi_s = 4VNI \tag{2}$$

Where N is turns of the sensing fiber, V is the Verdet constant of the sensing fiber, and I is the current in the conductor.

2.2. Dynamic model of FOCS

By analysing the principle and structure of FOCS, the nonlinear dynamic model which be shown in Figure 2 can be established[8][9].

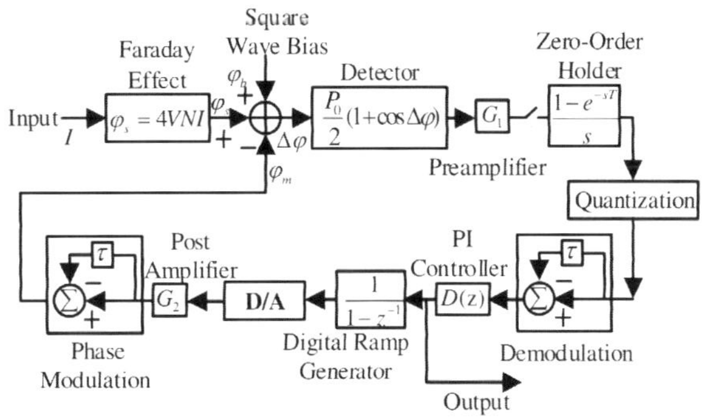

Figure 2. The nonlinear dynamic model of FOCS

The Figure 2 shows a nonlinear closed-loop control model with PI controller. The error signal demodulation, the digital ramp generator and the phase modulation are done in the processing unit of FOCS. By linearizing these parts in the structure of FOCS, this model can be simplified into a discrete linear model which is shown in Figure 3[10].

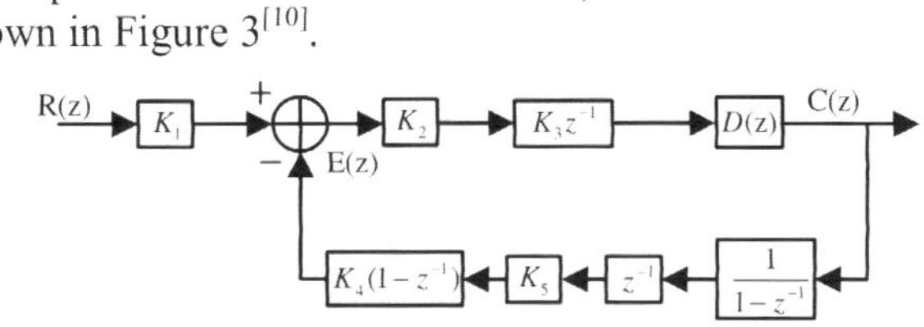

Figure 3. The discrete linear model of FOCS

The Faraday phase shift can be expressed as proportional element K_1. The detector and preamplifier are expressed as proportional element K_2. The sampling, quantization and demodulation process can be replaced by K_3z^{-1}. D/A conversion and the feedback gain corresponding to the post-amplifier are represented by proportional element K_5. The modulation process of a phase modulator is $K_4(1\text{-}z^{-1})$ and $D(z)$ is PI controller. So the closed-loop transfer function of FOCS can be expressed as:

$$\begin{aligned} H(z) = \frac{C(z)}{R(z)} &= \frac{K_1K_2K_3z^{-1}D(z)}{1+K_2K_3z^{-1}D(z)K_4(1-z^{-1})z^{-1}K_5/(1-z^{-1})} \\ &= \frac{K_1K_2K_3z^{-1}D(z)}{1+K_2K_3K_4K_5z^{-2}D(z)} \end{aligned} \tag{3}$$

$D(z)$ can be expressed as:

$$D(\mathrm{z}) = K_p + \frac{K_i}{1-z^{-1}} \tag{4}$$

Where K_p is proportional coefficient and K_i is integral coefficient. The error transfer function is:

$$H_e(z) = \frac{E(z)}{R(z)} = \frac{K_1}{1+K_2K_3K_4K_5z^{-2}D(z)} \tag{5}$$

Given a step current input, such as:

$$R(z) = A/(1-z^{-1}) \tag{6}$$

Where A is the amplitude of step current signal. The steady-state error of the FOCS system with PI control is expressed as:

$$e_{ss} = \lim_{z\to 1}(1-z^{-1})E(z) = \lim_{z\to 1}(1-z^{-1})H_e(z)R(z) = 0 \tag{7}$$

Consequently, FOCS can measure accurately the current in the conductor with these right control parameters of PI controller.

However, these parameters of traditional PI controller are fixed in FOCS. In actual application, FOCS must follow the large change of the current quickly and the steady-state error of FOCS should be zero. The response time and accuracy of FOCS cannot be satisfied simultaneously. For example, when the amplitude of the measured current is small, in order to meet the measurement accuracy and avoid overshoot, we select small parameters which cause a longer response time in FOCS. And when the current changes rapidly such as the short-circuit current, these larger PI parameters are selected to ensure the quick response, which cause the overshoot or oscillation of the system. Therefore, the FOCS with the traditional PI controller cannot meet these requirements of metering and protection at the same time. To solve this problem, we propose the fuzzy PI control strategy of FOCS.

3. Fuzzy PI control

The fuzzy PI controller combines fuzzy control with PI control, which has advantages of simple control, easy implement and parameters to be adjusted online. It takes error E and error change rate EC as input signals of the control system. By fuzzy PI controller, these PI parameters are adjusted at these different stages of the system response according to E and EC. The schematic diagram of fuzzy PI controller is shown in Figure 4.

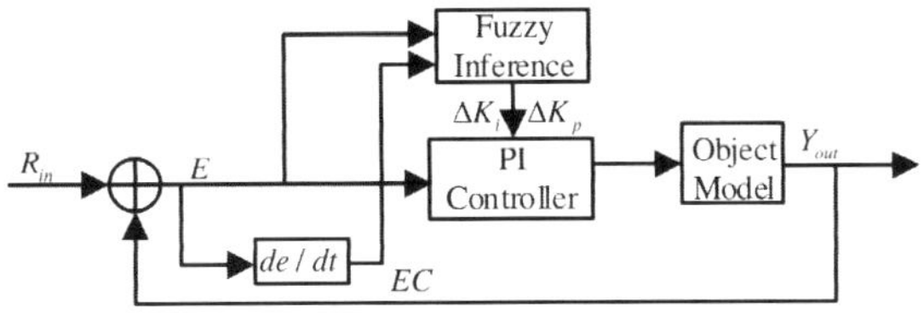

Figure 4. The schematic diagram of fuzzy PI controller

The main part of fuzzy PI controller includes fuzzy input, fuzzy rule base, fuzzy logic reasoning, ambiguity resolution, etc. The basic framework is shown in Figure 5.

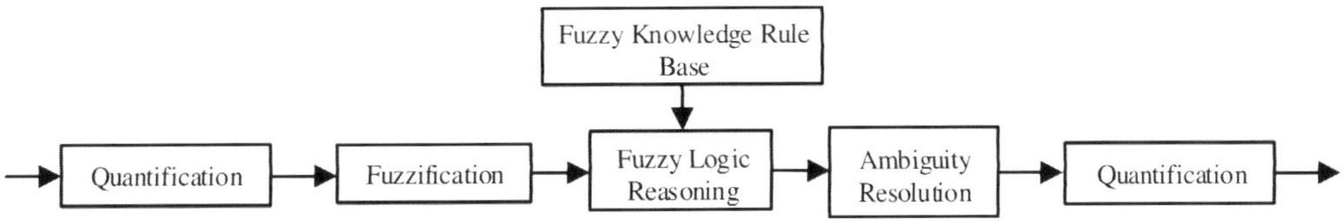

Figure 5. The basic block diagram of fuzzy reasoning

Main steps of fuzzy PI controller are as follow [11][12][13]:

1) Fuzzy subset selection

The fuzzy PI controller has two inputs and two outputs. These input variables are the error E and error change rate EC, these output variables are PI coefficient variable ΔK_p and ΔK_i. Fuzzy field of EC, E and PI coefficients are expressed as:

$$E, EC, \Delta K_p, \Delta K_i = \{-3, -2, -1, 0, 1, 2, 3\} \tag{8}$$

Elements in the fuzzy subset represent negative large, negative medium, negative small, zero, positive small, positive median and positive large, respectively. So the fuzzy subset can be expressed as:

$$E, EC, \Delta K_p, \Delta K_i = \{NB, NM, NS, ZO, PS, PM, PB\} \tag{9}$$

The subordination functions of E, EC and ΔK_p, ΔK_i are shown in Figure 6.

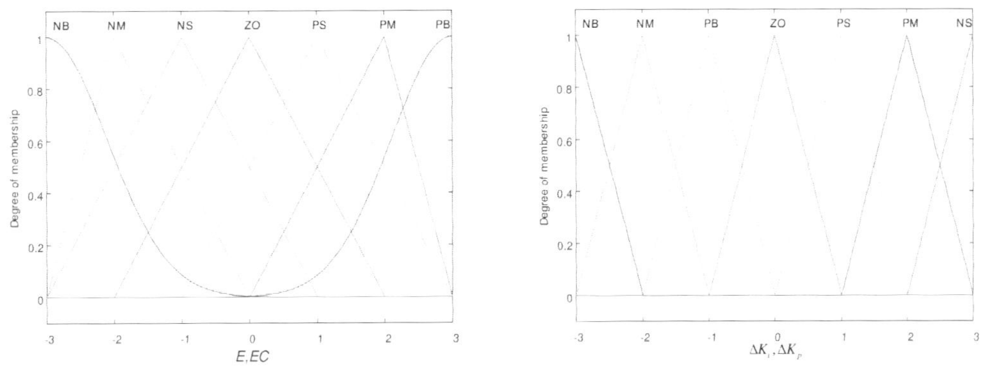

Figure 6. Subordination functions of E, EC, ΔK_p, ΔK_i

2) Establishment of fuzzy rules

The proportionality coefficient affects the response speed of the system. The integral coefficient affects the steady-state accuracy of the system. According to different factors, these parameter setting rules are concluded as shown in Table 1(each cell contains two fuzzy rules, the left is the fuzzy rule of ΔK_p, the right is the rule of ΔK_i).

Table1. Fuzzy control rule library

E	*EC*						
	NB	NM	NS	ZO	PS	PM	PB
NB	PB/NB	PB/NB	PM/NM	PM/NM	PS/NS	ZO/ZO	ZO/ZO
NM	PB/NB	PB/NB	PM/NM	PS/NS	PS/NS	ZO/ZO	NS/ZO
NS	PM/NB	PM/NB	PM/NS	PS/NS	ZO/ZO	NS/PS	NS/PS
ZO	PM/NM	PM/NM	PS/NS	ZO/ZO	NS/PS	NM/PS	NM/PM
PS	PS/NM	PS/NS	ZO/ZO	NS/PS	NS/PS	NM/PS	NM/PB
PM	PS/ZO	ZO/ZO	NS/PS	NM/PS	NM/PM	NM/PB	NB/PB
PB	ZO/ZO	ZO/ZO	NM/PS	NM/PM	NM/PM	NB/PB	NB/PB

3) Making fuzzy judgement

According to input signals *E* and *EC*, the fuzzy judgment is made by centroid method through subordination function, and $\Delta K_p, \Delta K_i$ are obtained:

$$K_p = K_p' + \Delta K_p, K_i = K_i' + \Delta K_i \quad (10)$$

Where K_p and K_i are the final PI coefficient, K_p' and K_i' are the set value of the PI controller, ΔK_i and ΔK_p are the output value of the fuzzy controller.

4. Simulation result and analysis

We used Simulink and Fuzzy Logic Toolbox in Matlab for the model simulation. The simulation model of FOCS was carried out by using two control methods: conventional PI control and Fuzzy PI control. These parameters of FOCS are set as following, the number of sensing fiber turns is 10, the Verdet constant is 1.09×10^{-6}rad/A. The resolution of A/D converter is 14bit, and the reference voltage is 1.25V. The resolution of D/A converter is 16bit, and the reference voltage is 2.5V. The half-wave voltage of the phase modulator is 3.7V. Thus, the system simulation parameters are as follow:

N=10, $V = 1.09\times10^{-6}$ rad/A, $K_1 = 4.36\times10^{-5}$, $K_2 = 1$, $K_3 = 2^{14}/1.25$, $K_4 = \pi/3.7$, $K_5 = 2.5/2^{16}$

In fuzzy PI control, the set value of PI controller parameters are K_p'=0.1 and K_i'=0.6. The traditional PI control has two sets of parameters, one is K_p=0.1 and K_i =0.6, the other is K_p=0.2 and K_i=1.5.

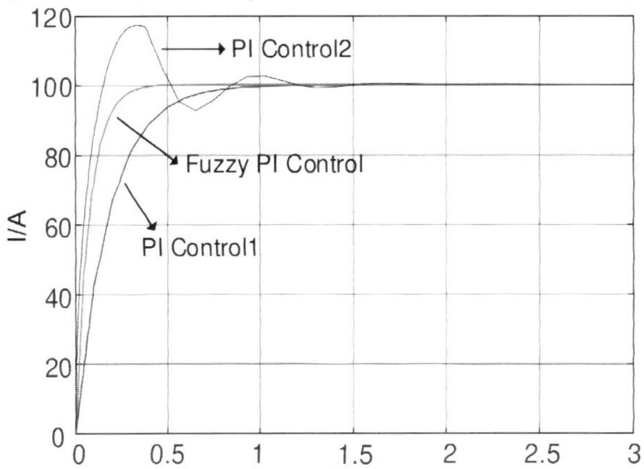

Figure 7. Step response curves of two control strategies

The input signal is a step signal of 100A, and simulation results are shown in Figure 7. PI Control 1 is the curve of small parameters of PI control, and PI Control 2 is the curve of big parameters. Another is the fuzzy PI control. From the Figure 7, we can know that the PI Control 1 has a long rising time about

1ms and settling time 1.5ms, but there is no overshoot in the curve. Meanwhile, PI Control 2 has quick response speed, but it has 20% overshoot, with rising time 0.1ms and the settling time 1.5ms. In comparison, the fuzzy PI control has quick response speed, the rising time is about 0.2ms, the settling time 0.5ms, and no overshoot. The performance of the fuzzy PI control is better than that of the traditional PI control in these simulation curves.

5. Conclusion

A FOCS based on fuzzy PI control strategy is proposed in this paper, which has quick response speed and no overshoot by simulating analysis. It solves the contradiction between the measurement accuracy and the response speed of traditional PI control effectively, which can meet the requirement of grid metering and protection at the same time.

Acknowledgments

This work was supported by National Key R&D Program of China(2018YFB0904700).

References

[1] Klaus Bohnert, Philippe Gabus, Jürgen Nehring. Fiber-optic current sensor for electro winning of metals[J]. Journal of Lightwave Technology, 2007, 25(11): 3602-3609.
[2] PANG Hong-mei, LI Huai-hai, ZHANG Zhi-xin, et al. Research situation of 110 k V smart substation technology[J]. Power System Protection and Control, 2010, 38(6): 146-150.
[3] Y. N. Ning, Z. P. Wang, A. W. Palmer, K. T. V. Grattan, and D. A. Jackson, "Recent progress in optical current sensing techniques," Rev. Sci. Instrum., vol. 66, no. 5, pp. 3097–3111, 1995.
[4] LI Jiu-hu,ZHANG Yu-ping,GU Shi-dong,et al. Application of electronic sensor in digital substation[J]. Automation of Electric Power System,2007(07):94-98.
[5] M. Gaugitsch and H. Hauser, "Optimization of a magneto-optical light modulator-Part I: modeling of birefringence and Faraday effect," in Journal of Lightwave Technology, vol. 17, no. 12, pp. 2633-2644, Dec. 1999.doi: 10.1109/50.809685
[6] WANG Xia-xiao, ZHANG Chun-xi, ZHANG Chao-yang, et al. A new all digital closed-loop fibre optic current sensor[J]. Automation of Electic Power Systems, 2006, 30(16): 77-80.
[7] Zhang Chaoyang , Lei Linxu , Wang Chenghao . Low current measurement error analysis of digital closed-loop fiber optic current sensor[J]. Instrument Technique and Sensor, 2012(10) : 7-10.
[8] WANF-Na,WANG-Quan,SHAO-Xia,XU-Xianyong.Research on modeling and simulation technology of all-fiber current sensor [J]. Journal of hunan university (natural science edition),2011,38(10):44-49.
[9] WANG-Xiaoxia,WANG-Ye,WANG-Xichen,etal.Experimental study on dynamic characteristics of all-fiber current sensor [J]. Power System Protection and Control,2014,42(03):9-14.
[10] HAN-Junliang,GE-Shengmin,SHEN-Yi. Research on modeling and simulation of digital closed-loop fog [J]. Journal of system simulation,2008(04):833-836+881.
[11] WANG-Shuyan,SHI-Yu,FENG-Zhongxu. Research on control method based on fuzzy PID controller [J]. Mechanical Science and Technology for Aerospace Engineering,2011,30(01):166-172.
[12] Li Wei, Yang Lirong. Research on the stability of servo control of hydraulic machine tool slide based on fuzzy PID control[J]. Hydraulic Pneumatics and Sealing, 2018,38 (8):8-12
[13] ZHENG-Meiru. Simulation of fuzzy PID controller [J]. Equipment Manufacturing Technology,2011(04):74-76+88.

Respiratory detection using non-contact sensors

Chen Xinqiang[1], Huang Zhicheng[1], Su Lumei[1,2,a]

[1]School of Electrical Engineering and Automation, Xiamen University of Technology, Xiamen430070, China

[2]Xiaolong Institute of Engineering and Technology, Second floor, Building A, Longzhou, Industrial Park, Longyan364000, China

[a]Corresponding author: sulumei@163.com

Abstract: Current respiratory detection study are mainly based on contact respiratory detection technology which has the limitation that human body should wear corresponding detection instruments. This paper studies three types of non-contact respiratory detection technologies. (1) Kinect-based respiratory detection method, which uses the depth sensor to obtain the depth value of the chest and abdomen region; (2) Ordinary web camera-based respiratory detection method, which applies optical flow to detect the motion information of the chest and abdomen region; (3) Thermal imager based respiratory detection method, which obtains the temperature change information of the nostril region from thermal video.The advantages and disadvantages of these three non-contact respiratory detection technologies are explored according to the experimental results. Kinect-based respiratory detection method is less affected by environmental interference, but the detection accuracy is lower and the detection range is limited. The detection method based on ordinary web camera is more accurate. However, the environmental illumination requirements are high. The detection method based on the thermal imager is not affected by the ambient light source, but the detection has poor real-time performance.

1. Introduction

Respiratory frequency is an important physiological index, which is closely related to the health of the human body. There are many contact respiratory detection technologies that can detect respiratory frequency. The technologies are mainly divided into two categories. One category is a thermistor-based respiratory frequency detecting device. The other category is a chest-belt type breathing detection device based on pressure sensor [1].

Since contact respiratory detection technologies requires the human body to wear the corresponding detecting instrument, it has certain inconvenience. But non-contact respiration detection technology is simpler. People can check the respiratory rate in their daily lives through smartphones, tablets and so on. Therefore, this paper studies three non-contact respiratory detection technologies. First, based on the detection method of the Kinect camera, the built-in depth sensor is used to obtain the depth value of the chest and abdomen region undulation [2]. Second, ordinary web camera-based respiratory detection method, which utilizes optical flow to detect the motion information of the chest and abdomen region [3]. Third, Thermal imager based respiratory detection method, which obtains the temperature change information of the nostril region from thermal video. Such technologies are lower in cost, more feasible, and can be widely promqoted.

2. Non-contact respiratory detection technology

2.1. Non-contact respiratory detection system

The system is mainly divided into three parts. In the first part, three different sensors, Kinect, web camera and thermal imager are used for feature acquisition, respectively, and a data set strongly related to the respiratory signal is obtained. In the second part, since the data sets strongly related to the respiratory signal obtained by the above three detection technologies are time domain information. The time domain data sets need to be converted into frequency domain information by fast Fourier transform (FFT). So that the respiratory signal frequency distribution map can be obtained.

2.2. Three non-contact respiratory frequency detection method

2.2.1. Kinect-based respiratory depth feature extraction

When the human body is breathing, the fluctuation of the chest and abdomen area will cause the depth value of the depth image to change. The depth value is the gray value at each pixel. Let the pixel coordinates of each frame depth image be (x, y). Select a pixel area $(x + w, y + h)$ with width w and height h as the region of interest (ROI). Position the chest and abdomen area to the ROI. And matrix summation of the chest and abdomen depth values obtained in each frame. When the sum is divided by the total number of pixels $w \times h$, the average depth value I_D can be obtained as

$$I_D = \frac{\sum_{k=1}^{w \times h} I_k}{w \times h} \tag{1}$$

where I_k is the depth value of the kth pixel point of each frame of the depth image.
Finally, after obtaining the average depth value of each frame of the depth image, the average depth value of successive frames is taken as a data set, and this data set is used as a strong correlation data set of the respiratory signal.

2.2.2. Image-based respiratory motion feature extraction

The undulation of the chest and abdomen area during breathing corresponds to the change in brightness of the pixels on the image. Using the optical flow method to transform the velocity of the chest and abdomen region into the instantaneous velocity of pixel motion. By grading the image sequence, the motion information of the same gray pixel point between adjacent frames is found.
Set $\mathrm{I}(\mathrm{x}, \mathrm{y}, \mathrm{t})$ and $\mathrm{I}(\mathrm{x} + \delta\mathrm{x}, \mathrm{y} + \delta y, z + \delta z, t + \delta t)$ as two image sequences before and after. Then define an energy function.

$$E(u, v) = \iint \left[\left(I_x u + I_y v + I_t\right)^2 + \alpha^2(P\nabla\mu P^2 + P\nabla v P^2)\right] dx\, dy \tag{2}$$

where $\left(I_x u + I_y v + I_t\right)^2$ is gray scale change factor, and $\alpha^2(P\nabla\mu P^2 + P\nabla v P^2)$ is smoothing factor. The ideal optical flow field should make the two factors as small as possible.
The optical flow value can be obtained by operation.

$$|\lambda_k - \lambda_{k-1}| < thresh \tag{3}$$

where

$$\lambda = I_x\mu + I_y v + I_t \tag{4}$$

2.2.3. Respiratory temperature feature extraction based on thermal imager

When the human body breathes, the hot air exhaled by the nostrils causes the temperature of the nostril area to be greater than the temperature of the surrounding air, thereby causing a change in the thermal characteristics of the nostril area. First, the nostril area is used as the ROI. The temperature of $\delta \times \delta$ pixel points in the vicinity of the nostril region was taken as a strong correlation data set, and the sum of the data sets of each frame was calculated. Secondly, when the sum of the temperature of each frame is divided by the number of pixels, the average temperature I_T can be obtained as

$$I_T = \frac{\sum_{k=1}^{\delta\times\delta} I_k}{\delta \times \delta} \tag{5}$$

where I_k is the temperature value of the kth pixel point of each frame.
Finally, a continuous change in the temperature of the nostril can be obtained, that is, the change law of the respiratory signal can be analyzed.

2.3. FFT-based respiratory rate detection
All strong correlation data sets of respiratory signals are obtained by the above three non-contact respiratory signal detection technologies. They can only reflect the law of change in breathing. However, with the fast Fourier transform FFT, the time domain data set can be converted into frequency domain information, so that the human body's respiratory rate can be observed.

2.4. Prediction of predicts respiratory signals using LSTM
This paper studies a method for predicting respiratory signals based on LSTM. LSTM unit is mainly composed of three "gates", which are input gate, forget gate and output gate [4]. The input gate is used to update the unit status, the forget gate determines what information is discarded or retained, and the output gate determines the output at the current moment. The formula for each "gate" is defined as follows:

$$z = tanh(W_z[h_{t-1}, x_t]) \tag{6}$$

$$i = \sigma(W_i[h_{t-1}, x_t]) \tag{7}$$

$$f = \sigma(W_f[h_{t-1}, x_t]) \tag{8}$$

$$o = \sigma(W_o[h_{t-1}, x_t]) \tag{9}$$

$$c_t = f \cdot c_{t-1} + i \cdot z \tag{10}$$

$$h_t = o \cdot tanhc_t \tag{11}$$

where z, i, f, o, c_t, h_t are input value, input gate, forget gate, output gate, new state and output. W_z, W_i, W_f, W_o are corresponding parameter matrices. σ is a sigmoid function. h_{t-1}, x_t, and c_{t-1} are the output of the previous time, the input of the current time, and the state of the previous time.

3. Experimental results and analysis

3.1. Experimental result

3.1.1. Kinect-based respiration detection
After obtaining the average depth value of each frame depth image, the average depth value of successive frames is taken as a data set, and this data set is used as a respiratory signal strong correlation data set. The experiment we designed was to fix the Kinect camera on the desk, and the tester sat about 1m away from the Kinect camera. During the test, the tester should try to stay still to avoid too much interference during the test. When ready, the tester begins normal breathing.
As can be seen in the Figure 1(a), the depth image data obtained by Kinect is not high precision. Therefore, the obtained waveform diagram does not respond well to the respiratory rate.

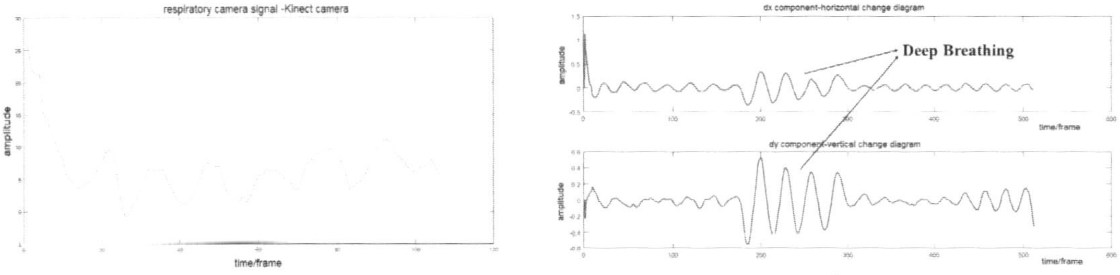

(a) (b)

Figure 1. Respiratory waveform obtained by Kinect camera and ordinary web camera

3.1.2. Respiratory detection based on ordinary web camera

The designed experiment is that the tester and the camera maintain a distance of about 0.5m to test the human body's breathing in the normal breathing mode. Gradient information on the chest and abdomen was collected in the experiment at 512 frames. First, normal breathing is performed, deep breathing is performed after the 180th frame, and normal breathing is resumed after the 300th frame. As shown in Figure 1(b), the amplitude of deep breathing is significantly larger than that of normal breathing, and the waveform is relatively smooth, which reflects the breathing frequency well, and the detection accuracy is high.

3.1.3. Respiratory detection based on thermal imager

The flir thermal imager is used to collect temperature change data in the nostril area, which can be approximated as the change law of human breathing. Align the marked point of the flir thermal imager to the nostril area, and then start to measure the temperature of the nostril area. As shown in Figure 2(a), when the human breathes, the temperature of the gas exchanged in the nostril area changes within the range of 36-40 ° C. And the temperature waveform is zigzag, the flir thermal imaging camera has a low sensitivity to reflecting temperature changes in continuous time.

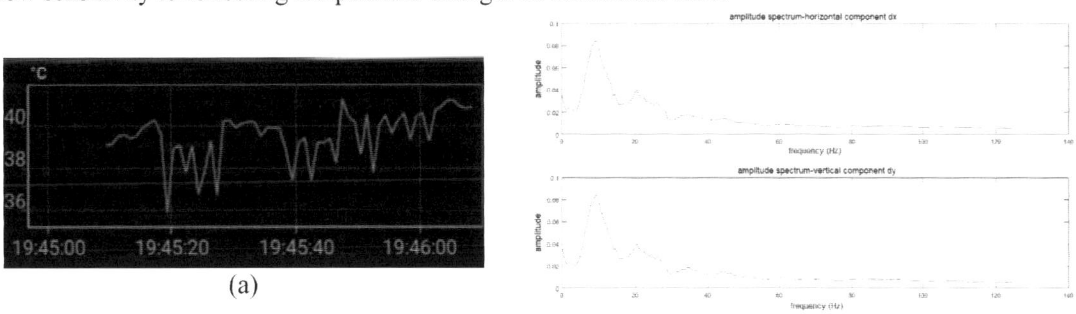

(a) (b)

Figure 2. Temperature information waveform and respiratory frequency spectrum after FFT

3.2. FFT to obtain the respiratory frequency spectrum

By analyzing the respiratory waveforms obtained by the above three non-contact detection methods, Kinect-based and thermal imager based respiratory detection method have low precision, and the detection precision of ordinary web camera-based respiratory detection method is high. Therefore, the FFT transform is performed by selecting a strong correlation data set of the respiratory signals collected by the ordinary web camera. The respiratory data of 512 frames of normal breathing collected by the ordinary camera is used as input data, and the fast Fourier transform FFT is performed to obtain a respiratory signal frequency distribution map. As shown in Figure 2(b), the human body's respiratory frequency is considered to be 0.05 Hz at this time.

3.3. LSTM-based respiratory signal prediction

First collect a strong correlation data set of respiratory signals through an ordinary web camera and input this data set into the LSTM model. Set the number of iterations to 100 and train the LSTM model. The results of respiratory signal prediction are shown in Figure 3. The blue line represents the observed respiratory waveform, and the red is the predicted respiratory waveform. It can be seen from the figure that the repetition rate of the blue curve and the red curve is high, and the root mean square error (RMSE) is 0.13829, which can meet the basic prediction requirements. Therefore, this method is suitable for predicting respiratory signals.

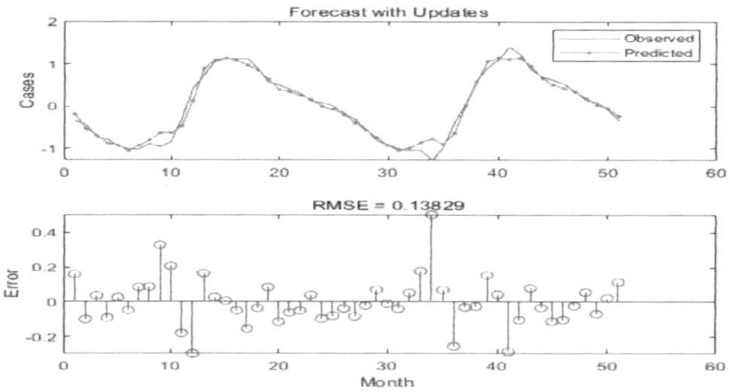

Figure 3. LSTM predicts respiratory signals

4. Conclusion

This paper studied three non-contact respiratory detection methods. The experimental results indicated that they each have their advantages and disadvantages. (1) Kinect-based non-contact breathing detection methods are not easily interfered by the light source environment. However, the detection rate of this method is low, and the requirements on the equipment are high, making it less practical. (2) The detection accuracy of ordinary web camera-based respiratory detection method is higher, and the hardware cost is lower, which is most suitable for practical application promotion. However, its application has certain limitations, and the brightness of the environment is also required. (3) Thermal imager based respiratory detection method is not affected by human motion and environmental light, and is most suitable for detecting respiratory signals of human body in real time. However, the equipment costs are relatively high, and they are also affected by ambient temperature.

Acknowledgments

This work is financially supported by Natural Science Foundation of department of science and technology of Fujian Province(No.2017J01510),Foundation for Science and technology of Longyan(No.2017LY91), Science Research Foundation for Overseas Scholars of Xiamen (No.XRS[2016]314-05).

References

[1] Xiao X Y, Huang S L, Chen S J, et al. Design of a wearable respiration monitoring system based on pressure sensor. Transducer and Microsystem Technologies, 2016.35(02):126-129.

[2] Li B, Xie D, Duan W J, et al. Kinect-based non-contract type breathing detecting system. Transducer and Microsystem Technologies,2017,36(03):107-109.

[3] Kumar M, Veeraraghavan A, and Sabharwal A. DistancePPG: Robust non-contact vital signs monitoring using a camera [J]. Biomedical optics express, 2015, 6(05): 1565–1588.

[4] Zeng M R, Zheng Z S, Luo S. Human behavior recognition combining two-stream CNN with LSTM. Modern Electronics Technique, 2019, (19):37-40.

Prediction of Short-term PV output power Based on PCA-Stacking under different weather conditions

Zhu Xiaofei

Nanjing University of Posts and Telecommunications,Nanjing 210023,China

1870559155@qq.com

Abstract. With the continuous expansion of photovoltaic scale, the accurate prediction of photovoltaic power generation is increasingly important for grid dispatching and grid optimization operations. In this paper, the photovoltaic power generation mainly uses meteorological factors and historical data as the input and output of the neural network. The input quantity is large, the data is redundant, and the network is difficult to converge, which always has a great adverse effect on the accuracy of photovoltaic output prediction. Firstly, different weather types are classified according to the trend graphs of different weather types. Principal components analysis (PCA) is used to analyze less comprehensive features from multiple meteorological factors and reduce the input of predictive models. At the same time, aiming at the problem that the prediction accuracy of a single prediction model such as the existing neural network and wavelet analysis method is limited, the idea and method of integrated learning are introduced, and a short-term prediction method based on Stacking method combined with SVM and Xgboost is proposed. Compared with the single model of SVM and Xgboost, the results show that the proposed method has a significant improvement compared with the accuracy of a single prediction model.

1. Introduction

In recent years, as China attaches great importance to energy conservation and environmental protection, the demand for renewable energy is increasing. Due to its sufficient cleanliness and safety, solar energy makes solar energy one of the fastest growing renewable energy sources in China. Photovoltaic power generation has become a research hotspot in the energy field and has developed rapidly.Meanwhile,With the persistent progress of solar power generation technology and the rapid growth of global installed capacity of photovoltaic power generation, grid-connected solar power generation technology has become an important development direction of solar photo-voltaic power generation [1]. However,photovoltaic power generation exhibits time-varying,volatility and randomness due to objective factors such as solar irradiance and weather, and its uncertainty brings many difficulties to the power grid on safety and economic operation of the grid[2,3].Accurate prediction of photovoltaic power generation is of great significance for grid dispatching and optimization operations.

In term of different classification criteria, the prediction methods of the photovoltaic output power are different [4].According to the classification of prediction process, the PV power prediction method can be divided into direct prediction and indirect prediction; according to the modeling method, it can be divided into physical method prediction and statistical method prediction Currently, neural networks[5-7], such as Elman neural network [8], support vector machines (SVM)[9], autoregressive moving average models (ARAM)[10], Markov chains and others are the most widely used statistical learning methods. According to different prediction time scales[11], classify photovoltaic power prediction methods. they can be divided into ultra-short-term (0 ~ 6 h), short term (6h-1d)[12],long term(one month-one year). Ultra-short-term PV power prediction can provide power transient information; short-term prediction can be used in scheduling, load tracking and forecasting, power market, etc. Medium and long-term PV power forecasting can be used in optical resource assessment, new photovoltaic power plant planning and other fields.

At present, most of the researches on photovoltaic power generation mainly focus on solar energy monitoring system, short-term prediction of the photovoltaic power generation and performance prediction of grid-connected photovoltaic systems. Voyant et al. used exogenous meteorological data as the optimized multi-layer perceptron (MLP) inputs to forecast daily global horizontal irradiance. Through comparing with different forecasting models containing a persistence model, ARIMA and ANN, they claimed the advantages of the proposed methodology[13]. Chen et al. presented an application of SVM to forecast daily solarradiation using sunshine duration in Liaoning province of China. The results illustrated that the SVM method could outperform the traditional forecasting methods [14]. Azimi et al. presented a hybrid solar irradiance forecasting framework combined with a novel cluster selection algorithm, a time-series analysis and MLP neural network for different time horizons [15]. Baser and Demirhan developed an approach using fuzzy regression functions with support vector machine (FRFSVM) to estimate yearly mean daily global horizontal solar irradiance in Turkey [16].

In summary, the existing models have the defects of data processing difficulties, complex model structure, and poor dynamic system identification. The prediction error of machine learning method is about 8% [17], the cloud error is about 26.20%, and the rainy day error is about 43.05% [18]. In the above study, the weather forecast method is limited by the accuracy of the weather forecast system, and the forecasted weather When the parameter error is large, the prediction accuracy will be seriously affected. The machine learning method uses BP neural network to face the problems of local optimization, slow iterative convergence, BP and SVM and its combination model. It is essentially based on single model prediction, and its prediction accuracy is limited, and there is a big improvement space.

Aiming at the existing problems, this paper proposes a short-term output combination forecasting method based on PCA-stacking photovoltaic power station. Considering the non-stationarity of the hourly output power distribution of photovoltaic power plants in the case of sudden weather changes, the data set is first divided by the trend graph analysis of power generation and different weather, and the data is processed by principal component analysis. Feature extracts feature components, compresses data dimensions, reduces data redundancy, obtains principal component components, and introduces the idea of integrated learning. Using Stacking method to combine SVM and Xgboost, use multiple primary learners SVM to predict the predicted samples. Then "generate" a new data set for training the secondary learner Xgbsoost. In this new data set, the output of the primary learner is used as a sample input feature, and finally the secondary training obtained by the secondary Xgbsoost for the SVM. Set training to get the final prediction result.

2. Prediction model algorithm

2.1. Principal component analysis model

The principal component analysis method is one of the most commonly used dimensionality reduction methods. The linear transformation transforms the original data into a set of linearly independent

representations of each dimension, which can be used to extract the main feature quantities of the data, which is often used for dimensionality reduction of high-dimensional data. Keep the feature of the worst contribution of the data set to the greatest extent. The method is mainly to decompose the covariance matrix to obtain the principal components (feature vectors) of the data and their weights.
The principal component expression is.

$$\begin{cases} y_1 = e_{11}x_1 + e_{12}x_2 + e_{13}x_3 + \cdots + e_{1k}x_k \\ y_2 = e_{21}x_1 + e_{22}x_2 + e_{23}x_3 + \cdots + e_{2k}x_k \\ \cdots \\ y_m = e_{m1}x_1 + e_{m2}x_2 + e_{m3}x_3 + \cdots + e_{mk}x_k \end{cases} \quad (1)$$

The paper is the k-dimensional eigenvector corresponding to the i-th eigenvalue of the correlation matrix of the original variable, X is the initial input variable of the k-dimensional and corresponding feature vector $e_1, e_2 \dots e_k$.

2.2. Stacking algorithm

When there is a lot of training data, a more powerful combination strategy is to use the "learning method", which is a combination of another learning device. Stacking is a typical representative of learning method. Here we call the individual learner the primary learning device. The learner used for the combination is called a secondary learner or a meta-learner.
Stacking first trains the primary learner from the initial dataset and then "generates" a new dataset for training the secondary learner. In this new dataset, the output of the primary learner is used as a sample input feature, and the initial The mark of the sample is still treated as a sample mark. Its specific process is shown in Figure 1.

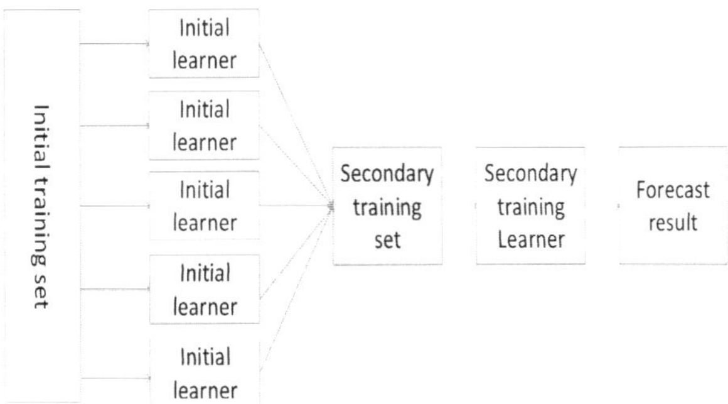

Figure 1. Stacking forecasting process

2.3. Support vector machine

Support vector machine (SVM): The derivation process of the support vector machine is: given training sample$D = \{(x_1, y_1), (x_2, y_2), \dots, (x_m, y_m)\}$,$y_i \in R$,i=1,2,...,n,Through a linear regression function:

$$f(x) = w^T x + b \quad (2)$$

To fit the sample, use the ε-insensitive loss function:

$$|y - f(x)|_\varepsilon = \begin{cases} 0, & |y - f(x)| \le \varepsilon \\ |y - f(x) - \varepsilon|, & |y - f(x)| > \varepsilon \end{cases} \quad (3)$$

Introducing the regularization constant and the relaxation variable and , the optimization problem is:

$$\min_{w,b,\xi_i,\hat{\xi}_i} \frac{1}{2}\|w\|^2 + C\sum_{i=1}^{m}\left(\xi_i + \hat{\xi}_i\right) \tag{4}$$

Introduce the Lagrange multiplier, convert it into a dual form to get the optimal solution. The regression function is:

$$f(x) = \sum_{i=1}^{m}\left(\hat{\partial}_i - \partial_i\right)\kappa(x, x_i) + b \tag{5}$$

Where $\kappa(x_i, x_j) = \emptyset(x_i)^T \emptyset(x_j)$ is a kernel function.

2.4.Xgboost

Xgboost has received a lot of attention due to its excellent learning effect and effective training speed. The derivation process of Xgboost is: for a given training set $T = \{(x_1, y_1),(x_2, y_2),\ldots(x_n, y_n)\}$, define an objective function:

$$Obj(t) = \sum_{i=1}^{n} L\left(y_i, \hat{y}^{t-1} + f_t(x_i)\right) + \Omega(f_t) + constant \tag{6}$$

Constant is a constant, and the regular term $\Omega(f_t)$ is as follows:

$$\Omega(f_t) = \gamma T + \frac{1}{2}\lambda \sum_{i=1}^{T} w_j^2 \tag{7}$$

Where T represents the number of leaf nodes and w_j represents the weight of the jth leaf node.Use the Taylor expansion $f(x + \Delta x) \approx f(x) + f'^{(x)}\Delta x + \frac{1}{2}f''(x)\Delta x^2$ to expand equation (6):

$$Obj(t) \approx \sum_{i=1}^{n}\left[L\left(y_{i,}\hat{y}^{t-1}\right) + g_i f_t(x_i) + \frac{1}{2}h_i f_t^{\,2}(x_i)\right] + \Omega(f_t) + constant \tag{8}$$

Where g_i denotes the first derivative of $L(y_{i,}\hat{y}^{t-1})$ for $\hat{y}^{t-1}$, h_i denotes $L(y_{i,}\hat{y}^{t-1})$ The second derivative of $\hat{y}^{t-1}.L(y_{i,}\hat{y}^{t-1})$ is known as the real value and the residual calculated by the previous function. So the final result is:

$$\sum_{i=1}^{T}\left[G_j w_j + \frac{1}{2}(H_j + \lambda)w_j^{\,2}\right] + \gamma T + C \tag{9}$$

By deriving equal to 0. Can get:

$$w_j = -\frac{G_j}{H_j + \lambda} \tag{10}$$

The simplified formula for bringing w_j into the objective function is as follows:

$$Obj(t) = -\frac{1}{2}\sum_{j=1}^{T}\frac{G_j^{\,2}}{H_J + \lambda} + \gamma T + C \tag{11}$$

After the objective function is simplified, you can see that the objective function of Xgboost is customizable, and only the first and second derivatives of it are used in the calculation. After obtaining the simplified formula, the next step is to select the appropriate splitting feature for the gain of the selected feature.

$$gain(\Phi) = gain(before) - gain(after) = \frac{1}{2}\left[\frac{G_L^{\,2}}{H_L + \lambda} + \frac{G_R^{\,2}}{H_R + \lambda} - \frac{(G_L + G_R)^2}{H_L + H_R + \lambda}\right] - \gamma \tag{12}$$

3. Predictive model construction analysis

First, the training data is divided into four data sets, such as sunny, cloudy, cloudy and rainy days, according to the characteristics of photovoltaic power generation, and then input into the principal component analysis model (PCA) for dimensionality reduction, and the dimensionality-reduced data is used as the prediction sample. Enter 5 primary learning SVMs and use the secondary learner Xgboost to combine the predicted outputs of the primary SVM to arrive at the final prediction. The specific process is shown in Figure 2.

The prediction process of the Stacking learner is shown in Figure 1. The prediction results are input into n primary SVM, and the primary SVM is combined using the secondary learner Xgboost to obtain the final prediction result.

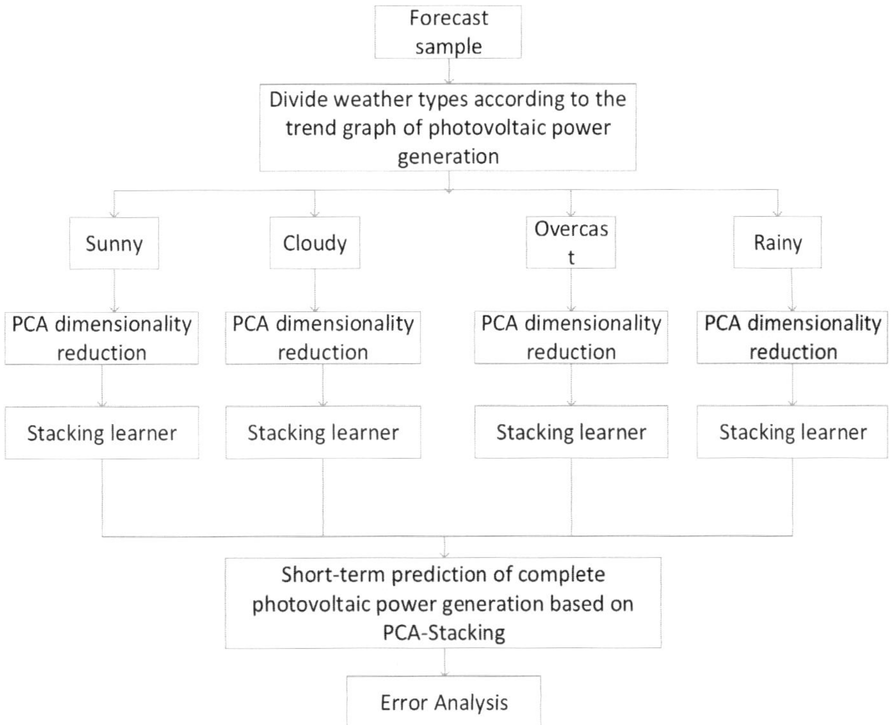

Figure 2. Overall workflow

3.1. Trends in photovoltaic power generation analysis and training sample composition

Photovoltaic power generation is affected by weather, environment and other factors with randomness and volatility, especially under cloudy and rainy weather conditions. The power generation power curve fluctuates drastically, and multiple peaks are generated, which has a great influence on the prediction accuracy of photovoltaic power generation. The sunny day is relatively stable, the curve is gentle, and it is easy to obtain better prediction accuracy, and the power generation on sunny days is much larger than the extreme weather. In cloudy weather, the difference from the sunny curve fluctuated significantly in different time periods, but we finally found that in cloudy weather, the power generated is very close to that of sunny days.

3.2. PCA analysis of meteorological factors

Taking the sunny photovoltaic power generation prediction sub-model as an example, the principal component analysis is carried out on eight original meteorological factors (ambient temperature, environmental humidity, irradiance, wind speed, wind direction, plate temperature, temperature difference, and air pressure) that affect photovoltaic power generation. The contribution rate and cumulative contribution rate of the principal components are shown in Table 1. As can be seen from Table 1, when the principal component is 3, a significant inflection point appears, but the cumulative contribution rate at this time is less than 85%, so four of them are selected. The main component, and a new input variable is calculated using equation (1). Input variables are entered into the Stacking learner for prediction.

Table1. Characteristic value and cumulative contribution rate

Serial number	Initial eigenvalue		
	Total	Contribution rate%	Grand total%
1	5.34	35.20	35.20
2	3.25	26.13	61.33
3	1.68	15.37	76.70
4	1.27	13.12	89.82
5	0.46	5.15	94.97
6	0.27	3.43	98.40
7	0.05	0.84	99.24
8	0.03	0.76	100%

3.3. Data normalization and model evaluation indicators

In this paper, the meteorological data and irradiance data of a PV power plant in Nanjing are used as input samples. The data collection interval is 15 minutes, The normalization method used in this paper will be used to transform the data into intervals. The normalization function is shown in equation (13):

$$\bar{x}_i = \frac{x_{i-}x_{min}}{x_{max}-x_i} \tag{13}$$

In this paper, MAE (mean absolute error) is used to estimate the accuracy of the model. The expression is shown in equation (14):

$$e_{MAE} = \frac{1}{N}\sum_{i=1}^{N}\left|y_{model,i} - y_{actual,i}\right| \tag{14}$$

Where:$y_{model,i}$ is the predicted value of the model, which $y_{actual,i}$ is the measured value.

4. Case analysis

On a sunny day, the amount of clouds is small, and the meteorological conditions such as irradiance and temperature are relatively stable, so the photovoltaic power generation fluctuation is small. Figure 3 shows the comparison of prediction results based on PCA-Stacking model and PCA-SVM, PCA-Xgboost, Xgbsoot and other models on sunny days. The prediction results are shown in Fig. 3. The power generation of the model in this model is very close to the real value, and it is not much different from the PCA-SVM, Xgboost, PCA-Xgboost and other models.

Under cloudy conditions, illumination has become an important factor affecting photovoltaic power generation. Due to the influence of wind power, the thickness and position of clouds, fluctuations in photovoltaic power generation have caused fluctuations. Figure 4 shows the comparison between the model and PCA-SVM, Xgboost, and PCA-Xgboost in cloudy forecast. It can be seen that there are multiple peaks in the real value, and each model has a large deviation at the peak. while other times performed better. overcast and rainy days have less exposure time, so the amount of electricity generated is less than sunny and cloudy. Figures 5 and 6 show the predictions of different models of the two weathers. It can be seen that the predicted values of the model in this model are more realistic. In rainy and bad weather conditions, the solar radiation is very small, so the amplitude of the photovoltaic output power is very small, too.

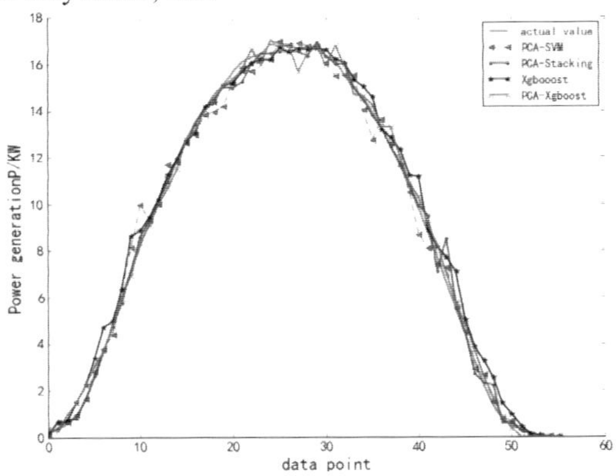

Figure 3. Sunny performance of each model

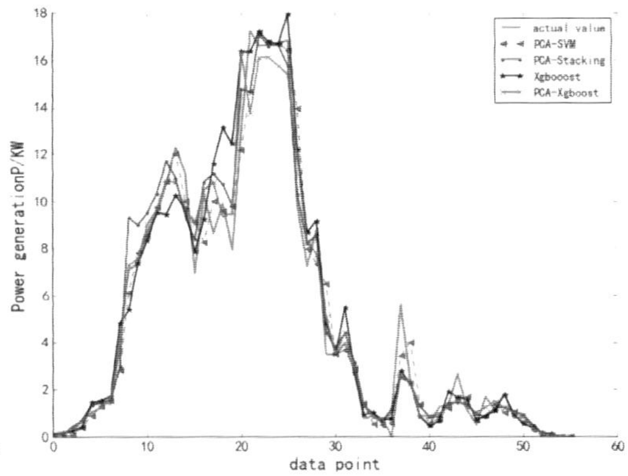

Figure 4. Cloudy performance of each model

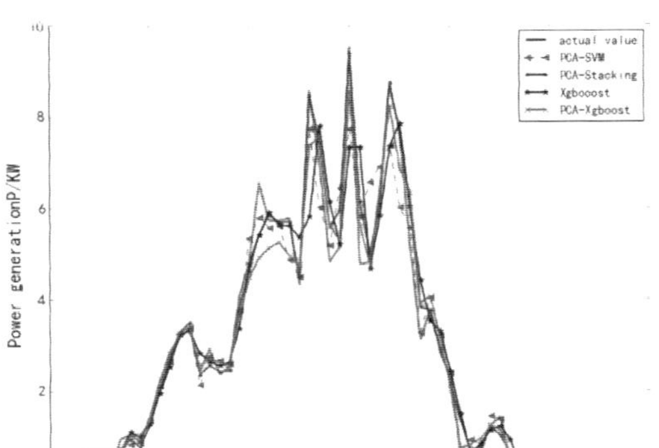

Figure 5. Overcast performance of each model

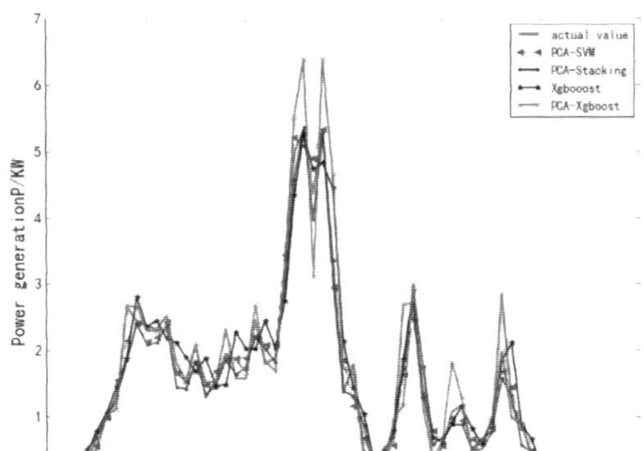

Figure 6. Raindy performance of each model

Table 2 shows the MSE (mean absolute error) values of the four models of PCA-Stacking, PCA-Xgboost, PCA-SVM, and xgboost in four different weather conditions.

Table2. Four model MSE values

	Sunny	Cloudly	Overcast	Rainy
Pca-stacking	0.220	0.679	0.231	0.205
Pca-xgboost	0.282	0.683	0.323	0.216
Pca-Svm	0.339	0.709	0.357	0.269
Xgboost	0.429	0.759	0.442	0.337

From the comparison of Table 2, it can be seen that the PCA-Stacking proposed in this paper has higher precision on sunny days and cloudy days. In the rainy days, it is similar to the prediction performance of other models. At the same time, from the performance of the Xgboost model, the prediction error is worse than other models due to the lack of dimensionality reduction.

Based on the above, the model is superior to other models in the prediction of four weather conditions, and has good prediction accuracy for the extreme weather such as overcast and rainy.

References

[1] Tao Ma, Hongxing Yang, Lin Lu, Jinqing Peng,Pumped storage-based standalone photovoltaic power generation system: Modeling and techno-economic optimization,Applied Energy,Volume 137,2015,Pages 649-659.

[2] D.O. Akinyele, R.K. Rayudu, N.K.C. Nair,Development of photovoltaic power plant for remote residential applications: The socio-technical and economic perspectives,Applied Energy,Volume 155,2015,Pages 131-149.

[3] Arnau González, Jordi-Roger Riba, Antoni Rius, Rita Puig,Optimal sizing of a hybrid grid-connected photovoltaic and wind power system,Applied Energy,Volume 154,2015,Pages 752-762,S.I.Sulaiman,

[4] Wang Fei, Mi Zengqiang, Yang Qi-xun, et al. Power forecasting approach of photovoltaic lant based on ANN and relevant data. J Sol Ener 201233(7):1172

[5] T.K.A. Rahman, I. Musirin, and S.Shaari, “Performance Analysis of Evolutionary ANN for Output Prediction of a GridConnected Photovoltaic System,”World Academy of Science, Engineering and Technology, pp. 1023-1029.

[6] Lee, D., & Kim, K. (2019). Recurrent Neural Network-Based Hourly Prediction of Photovoltaic Power Output Using Meteorological Information. Energies, 12(2), 215.

[7] Wang Fei, Mi Zengqiang, Yang Qi-xun, et al. Power forecasting approach of photovoltaic plant based on ANN and relevant data. J Sol Energy 2012;33(7):1172–7.

[8] Zhong Chunxia. Output power prediction of a photovoltaic system based on similarday algorithm and Elman neural network. J Nanjing Inst Technol (Nat Sci Ed)2016;14(1):42–7

[9] Fei Wang, Zhao Zhen, Zengqiang Mi, Hongbin Sun, Shi Su, Guang Yang,Solar irradiance feature extraction and support vector machines based weather status pattern recognition model for short-term photovoltaic power forecasting,Energy and Buildings,Volume 86,2015,Pages 427-438.

[10] Y. Sun et al., "Short-term prediction model of module temperature for photovoltaic power forcasting based on support vector machine," International Conference on Renewable Power Generation (RPG 2015), Beijing, 2015, pp. 1-6.

[11] Shuang Han. Study on short-term forecasting method of wind power in wind farm.North China Electric Power University; 2008.

[12] Stanley KH, Chow, Lee Eric WM, Li Danny HW. Short-term prediction of photovoltaic energy generation by intelligent approach. Energy Build 2012;55:660–7.

[13] Voyant C, Muselli M, Paoli C, Nivet ML. Optimization of an artificial neural networkdedicated to the multivariate forecasting of daily global radiation. Energy2011;36(1):348–59

[14] Chen JL, Li GS, Wu SJ. Assessing the potential of support vector machine for estimating daily solar radiation using sunshine duration. Energy Convers Manage2013;75:311–8.

[15] Azimi R, Ghayekhloo M, Ghofrani M. A hybrid method based on a new clustering technique and multilayer perceptron neural networks for hourly solar radiation forecasting. Energy Convers Manage 2016;118:331–44

[16] Lima FJL, Martins FR, Pereira EB, Lorenz E, Heinemann D. Forecast for surface solarirradiance at the Brazilian Northeastern region using NWP model and artificialneural networks. Renew Energy 2016;87:807–18.

[17] LEVA S，DOLARA A，GRIMACCIA F， et al. Analysis and validation of 24 hours ahead neural network forecasting of photovoltaic output power[J]. Mathematics and Computers in Simulation，2017，131：88-100.

[18] T. Yamazaki, S. Wakao, Y. Fujimoto and Y. Hayashi, "Improvement of prediction interval estimation algorithm with just-in-time modeling for PV system operation," 2015 IEEE 42nd Photovoltaic Specialist Conference (PVSC), New Orleans, LA, 2015, pp. 1-6

Combination Forecasting Model of Daily Electricity Consumption in Summer Based on Daily Characteristic Meteorological Factors

Ruimiao Wang[1], Xinyu Yao[2], Chunyan Li[3], Bo Hu[3], Kaigui Xie[3], Tao Niu[3], Mushui Li[3], Jian Fu[3], Qingsong Sun[3]

[1]State Grid Chongqing Electric Power Research Institute, Chongqing, China
[2]State Grid Chongqing Power Company, Chongqing, China
[3]State Key Laboratory of Power Transmission Equipment & System Security, College of Electrical Engineering, Chongqing University, Chongqing, China

934910580@qq.com

Abstract. Daily electricity consumption in summer is affected by numerous meteorological factors. In order to improve the accuracy of daily electricity consumption forecasting, a summer daily electricity consumption combination forecasting model based on daily characteristic meteorological factors is proposed in this paper. Firstly, the relationship between various meteorological indexes and daily electricity consumption is analysed considering the coupling effect and accumulative effect of meteorological factors on daily electricity consumption, and a single forecast model of daily electricity consumption is established. Then the meteorological mapping function is designed to evaluate the impact of each meteorological index on daily electricity consumption, and the weight of each single forecasting model is calculated. Finally, a combination forecasting model of daily electricity consumption in summer is established. The parameters of meteorological mapping function are optimized by adaptive training and virtual forecasting, and solved by genetic algorithm. Taking the daily electricity consumption in Chongqing in the summer of 2018 as an example, the results have shown that the proposed combination forecasting model can effectively improve the accuracy of daily electricity consumption forecasting.

1. Introduction

In recent years, with the development of social economy and the improvement of people's living standards, the ownership rate and utilization rate of air conditioners have increased significantly. In many large cities, the use of summer cooling equipment and winter heating equipment has led to a year-on-year increase in the proportion of electricity used by the tertiary industry and residents in urban electricity consumption [1]. Meteorological factors have become the main cause of high load and high electricity consumption. Therefore, in order to improve the accuracy and practicality of urban daily electricity consumption forecast, it is necessary to consider the influence of meteorological factors.

At present, there are relatively few studies on the electricity consumption forecasting considering meteorological factors in the field of load forecasting. In [2], Luo considered a lot of meteorological factors such as temperature, wind speed and precipitation, and established a city electricity consumption prediction model, but ignored the influence of coupling effect generated by different meteorological factors. In [3], Zhong introduced the new meteorological indicator of effective temperature and analysed its relationship with electric load, but did not study the effect of multi-day accumulative effect of meteorological factors. In [4]-[5], authors studied the law of temperature accumulative effect on load and electricity consumption, and established a prediction model based on this law. However, the accumulative effect of meteorological factors also needs to include other meteorological indicators other than temperature.

This paper proposes a summer daily electricity consumption combination forecasting model based on daily characteristic meteorological factors. For the coupling effect and accumulative effect of meteorological factors, various meteorological indicators are used to describe them and the corresponding single-consumption model of electricity consumption is established. According to the meteorological mapping function of each meteorological indicator, the degree of influence on the electricity consumption is evaluated, the weight coefficient of each single prediction model is calculated, and the electricity consumption combination prediction model is established. For the optimal value of the meteorological mapping function parameters, adaptive training and virtual prediction methods are used to optimize them. This paper analyses the daily electricity consumption forecast of Chongqing in the summer of 2018 to verify the accuracy of the proposed model.

2. A single forecasting model considering daily characteristic meteorological factors

2.1. Decomposition of daily electricity consumption

A large number of studies have shown that time and meteorological are the main factors affecting daily electricity consumption in summer [6]. Based on this, it can be considered that the daily electricity consumption is composed of trend electricity and meteorological electricity.

$$E = E_d + E_w + \varepsilon \tag{1}$$

where E_d and E_w are trend electricity and meteorological electricity, respectively, and ε is the error. The trend electricity is determined by the time variable t. The meteorological electricity is determined by the meteorological index vector $\boldsymbol{w}=[w_1,w_2,\ldots,w_n]^{\mathrm{T}}$ considering that the meteorological electricity may be affected by multiple meteorological indexes.

For a group of continuous daily electricity consumption and meteorological index historical data, the daily electricity consumption is decomposed into the trend electricity and the meteorological electricity according to the correlation between the electricity consumption and the meteorological indexes. The relationship between time and the trend electricity and that of meteorological electricity and meteorological indexes is established respectively. The detailed steps are as follows:

1) Dates are numbered according to the time sequence as $1,2,\ldots,N_d$. N_d is the total number of days.
2) Calculate the correlation coefficient ρ between the daily electricity consumption and meteorological indexes [7].

$$\rho = \frac{\sum_{t=1}^{N_d}(E_t - \bar{E})(w_{i,t} - \bar{w}_i)}{\sqrt{\sum_{t=1}^{N_d}(E_t - \bar{E})^2 \sum_{t=1}^{N_d}(w_{i,t} - \bar{w}_i)^2}} \tag{2}$$

   where E_t and $w_{i,t}$ are the daily electricity consumption and the meteorological index i on day t, respectively. $\bar{E}$ and $\bar{w}_i$ are the mean value of daily electricity consumption and that of the meteorological index i, respectively.
3) Remove some days with high correlation between the daily electricity consumption and meteorological indexes until the correlation coefficient ρ is lower than the threshold z_ρ. If $\rho<0$, it will be removed from the day with the lowest meteorological index; otherwise, it will be removed from the day with the highest meteorological index. The days that are reserved constitute the set Ω_1, and the days that are removed constitute the set Ω_2.

4) Since the daily electricity consumption in the set Ω_1 has a low correlation with meteorological indexes, the daily electricity consumption in Ω_1 does not contain the meteorological electricity, that is, the daily electricity consumption is equal to the trend electricity. Regression analysis of trend electricity in Ω_1 is carried out by using one dimensional linear linear model:

$$E_{d,t}=at+b \tag{3}$$

where $E_{d,t}$ is the trend electricity on day t. Constant coefficient a and b are calculated by parameter estimation.

5) Calculate the trend electricity in the set Ω_2 according to the trend electricity regression model of (3), and calculate the meteorological electricity in Ω_2.

$$E_{w,t}=E_t - E_{d,t} \tag{4}$$

where $E_{w,t}$ is the meteorological electricity on day t.

6) A variety of regression analysis models are used to model the variation of meteorological electricity. Each analysis model is as follows:

$$E_{w,t}=\sum_{i=1}^{n}\beta_{1,i}w_{i,t}+\alpha_1 \tag{5}$$

$$E_{w,t}=\sum_{i=1}^{n}\chi_{2,i}w_{i,t}^2+\sum_{i=1}^{n}\beta_{2,i}w_{i,t}+\alpha_2 \tag{6}$$

$$E_{w,t}=\sum_{i=1}^{n}\delta_{3,i}w_{i,t}^3+\sum_{i=1}^{n}\chi_{3,i}w_{i,t}^2+\sum_{i=1}^{n}\beta_{3,i}w_{i,t}+\alpha_3 \tag{7}$$

The different functional relationships between the meteorological electricity and the meteorological index $\boldsymbol{w}$ are described by equations (5)-(7), and select the function with the best fitting effect as the final meteorological regression model of the meteorological electricity.

2.2. Single forecasting model of daily electricity consumption

There are many meteorological indexes describing meteorological conditions, such as: maximum temperature, minimum temperature, mean temperature, humidity, rainfall, wind speed, etc. In order to forecast daily electricity consumption more accurately, it is necessary to extract characteristic meteorological factors that have great influence on daily electricity consumption.

The temperature has the most significant impact on daily electricity consumption in summer. The correlation between humidity and daily electricity consumption is low. However, the coupling effect of humidity and temperature will strongly affect daily electricity consumption. The humidity-temperature index well reflects the coupling effect of humidity and temperature on daily electricity consumption, which is defined as [8]:

$$THI = Temp_C + \frac{1450.8(Temp_C + 235)}{4030-(Temp_C+235)\ln Hmd} - 43.4 \tag{8}$$

where $Temp_C$ is the temperature in centigrade, which is the daily mean temperature; Hmd is the percentage humidity.

In addition to the coupling effect, it is necessary to consider the multi-day accumulative effect of meteorological factors on daily electricity consumption. In order to better quantify the accumulative effect of meteorological factors, a weighted humidity-temperature index is adopted by the US PJM market. There are two commonly used formulas, which are as follows [8]:

$$WTHI_{1,t}=(10THI_t+4THI_{t-1}+THI_{t-2})/15 \tag{9}$$

$$WTHI_{2,t}=(10THI_t+5THI_{t-1}+2THI_{t-2})/17 \tag{10}$$

where THI_t, THI_{t-1} and THI_{t-2} are the humidity-temperature index of day t, yesterday t-1, and day before yesterday t-2, respectively.

According to the above analysis, five groups of characteristic meteorological factors can be extracted. For each group of characteristic meteorological factors, combined with the trend electricity regression model and the meteorological electricity regression model, five single daily electricity consumption forecasting models can be established. The meteorological indexes used in each single forecasting model are as follows:

Single model 1: The lowest temperature $Temp_{min,t}$, the highest temperature $Temp_{max,t}$ and the mean temperature $Temp_{mean,t}$ of the day.

Single model 2: The lowest temperature $Temp_{min,t}$, the highest temperature $Temp_{max,t}$ and the mean temperature $Temp_{mean,t}$ of the day; The lowest temperature $Temp_{min,t-1}$, the highest temperature $Temp_{max,t-1}$ and the mean temperature $Temp_{mean,t-1}$ on yesterday; The lowest temperature $Temp_{min,t-2}$, the highest temperature $Temp_{max,t-2}$ and the mean temperature $Temp_{mean,t-2}$ on the day before yesterday.

Single model 3: The humidity-temperature index THI_t of the day.

Single model 4: The weighted humidity-temperature index 1 $WTHI_{1,t}$ of the day.

Single model 5: The weighted humidity-temperature index 2 $WTHI_{2,t}$ of the day.

3. Combination forecasting model and meteorological mapping function

This section establishes a combination forecasting model of daily electricity consumption, combining the advantages of various single forecasting models [9]. The combination forecasting model can be expressed as follows:

$$\hat{E}_{t+c} = \sum_{j=1}^{5} f_{j,t+c} \hat{E}_{j,t+c} \tag{11}$$

$$\sum_{j=1}^{5} f_{j,t+c} = 1 \tag{12}$$

where $\hat{E}_{t+c}$ is the forecast value of electricity consumption of the combination forecasting model in the future day $t+c$; $\hat{E}_{j,t+c}$ is the forecast value of electricity consumption of the single forecasting model j in the future day $t+c$; $f_{j,t+c}$ is the weight of the single forecasting model j in the future day $t+c$.

When a single forecasting model is determined, the weight of a single forecasting model directly determines the accuracy of the combination forecasting model. It can be seen from the above modeling process that different single forecasting models consider different characteristic meteorological factors. It is not difficult to find that the weight of a single forecasting model actually reflects the degree of influence of the characteristic meteorological factor on daily electricity consumption. Therefore, it is necessary to evaluate the impact of various meteorological indexes on daily electricity consumption.

Since the dimension of each meteorological index is different, it is necessary to map each meteorological index accordingly. Considering that the meteorological indexes used in this paper are continuous variables, we map them by linear functions [10]. The meteorological mapping function of the meteorological index w_i can be expressed as follows:

$$f(w_{i,t}) = a_i w_{i,t} + b_i \tag{13}$$

where a_i and b_t are parameters of meteorological mapping function. In order to limit the range of meteorological index mapping values, $0< a_i <0.1$ and $0< b_i <0.1$ are specified.

According to the mapping values of various meteorological indexes, the degree of influence of meteorological indexes on daily electricity consumption can be obtained:

$$r_{i,t} = \frac{w_{i,t}}{\sum_{i=1}^{N_w} f(w_{i,t})} \tag{14}$$

where N_w is the number of meteorological indexes used for five single forecasting models, with a value of 12.

The calculation formulas of weight of each single forecasting model are as follows:

$$f_{1,t+c} = r_{Temp_{\min},t+c} + r_{Temp_{\max},t+c} + r_{Temp_{mean},t+c} \tag{15}$$

$$f_{2,t+c} = \sum_{k=c-2}^{c} r_{Temp_{\min},t+k} + r_{Temp_{\max},t+k} + r_{Temp_{mean},t+k} \tag{16}$$

$$f_{3,t+c}=r_{THI_{t+c}} \tag{17}$$

$$f_{4,t+c}=r_{WTHI_{1,t+c}} \tag{18}$$

$$f_{5,t+c}=r_{WTHI_{2,t+c}} \tag{19}$$

4. Parameter adaptive optimization of meteorological mapping function

It can be seen from the above analysis that the parameters needed to be determined for the forecasting by using the combination model are divided into two parts: One is the parameters of the single forecasting model itself, which can be calculated by parameter estimation based on historical data. The other is the weight coefficient of each single forecasting model, which essentially needs to determine the parameters of the meteorological mapping function. In this section, an adaptive training method and virtual forecasting idea are used to optimize the parameters of the meteorological mapping function [11].

For convenience of presentation, it is assumed that the forecast date is day t, day t+1~day $t+c$ is the day to be forecasted, and day $t\text{-}N_d$~day t-1 is the historical day. According to the virtual forecasting idea, the historical day is divided into two groups: the day $t\text{-}F_v$ ~day t-1 and day $t\text{-}N_d$ ~day $t\text{-}F_v$-1. Taking the historical data of day $t\text{-}N_d$ ~day $t\text{-}F_v$-1 as the known quantity, the above combination forecasting model is used to forecast the electricity consumption of day $t\text{-}F_v$ ~day t-1. The objective function is to minimize the sum of the absolute value of the relative error between the forecasting value and the actual value of the daily electricity consumption of day $t\text{-}F_v$ ~day t-1. The parameters of the meteorological mapping function are used as the decision variables, and the genetic algorithm is used to solve the optimal value of the parameters. The detailed process of the algorithm is as follows:

1) The electricity consumption of day $t\text{-}N_d$~day $t\text{-}F_v$-1 is decomposed using the decomposition method of daily electricity consumption mentioned in Section 2.1. The parameters of the trend electricity forecasting model and the meteorological electricity forecasting model are calculated by the parameter estimation method.
2) Based on the input data of time and meteorological indexes of day $t\text{-}F_v$ ~day t-1, the electricity consumption of day $t\text{-}F_v$ ~day t-1 is forecasted by each single forecasting model. The forecasting value of single forecasting model j is denoted as $\hat{E}_{j,t-k}$ $(1 \le k \le F_v)$.
3) Two types of chromosomes are generated, one represents the primary coefficient $\boldsymbol{A}$ of the meteorological mapping function, and the other represents the constant coefficient $\boldsymbol{B}$ of the meteorological mapping function. The initial population is formed by several chromosomes.
4) Decoding the information of all chromosomes in the population and obtaining the parameters of meteorological mapping function and the influence coefficients of meteorological indexes, thereby obtaining the weight coefficient $f_{j,t-k}$ $(1 \le k \le F_v)$ of the single forecasting model j, and calculating the fitness value of each chromosome. The fitness function is set to (20).

$$fitness=-\sum_{k=1}^{F_v}\left|\frac{\sum_{j=1}^{5} f_{j,t-k}\hat{E}_{j,t-k}-E_{t-k}}{E_{t-k}}\right| \tag{20}$$

5) Generate new paternal population by selecting, crossing and mutating on chromosomes of the existing population.
6) Determine whether the maximum number of iterations is reached. If it is reached, the calculation is stopped and the result is output to obtain the optimal value of the parameters of the meteorological mapping function; otherwise, turn to step 4).

5. Case studies

In order to verify the effect of the combined forecasting model, the electricity consumption data of Chongqing in the summer of 2018 was selected for analysis. This paper selects August 22th 2018 to August 28th 2018 as the forecast date, and predicts the daily electricity consumption during the week.

In order to effectively decompose electricity consumption, estimate parameters and optimize parameters, you should select as much historical data as possible. This paper uses electricity consumption and meteorological data from May 1th 2018 to August 21th 2018 as historical data. The correlation coefficient threshold z_ρ is set to 0.2.

Table 1 shows the predicted values of various predictive models and the relative errors between predicted value and actual value. As it can be seen from Table 1, the combined prediction model is superior to each single prediction model, and can more accurately predict the daily electricity consumption. Especially for the electricity consumption on August 22th, the relative error of the combined forecast is only 0.30%, and the prediction accuracy is significantly higher than the single prediction. Among the five single prediction models, the overall prediction of single model 2 is the best, and the prediction of single model 3 is good on some days. These reflect that temperature and humidity-temperature index are important meteorological factors affecting electricity consumption.

Table 1. Comparison of predicted value and actual value of electricity consumption in both August 22th 2018 and August 28th 2018 in Chongqing

Date	Actual Electricity Consumption (GWh)	Predicted Electricity Consumption (GWh) /Relative Error (%)					
		Single Model 1	Single Model 2	Single Model 3	Single Model 4	Single Model 5	Combined Model
08/22	306.16	275.20/-10.11	310.82/1.52	274.32/-10.40	284.98/-6.92	290.88/-4.99	307.09/0.30
08/23	285.13	278.66/-2.27	293.88/3.07	281.16/-1.39	275.90/-3.24	278.65/-2.27	291.09/2.09
08/24	298.79	290.64/-2.73	283.89/-4.99	292.05/-2.26	283.56/-5.10	282.92/-5.31	286.09/-4.25
08/25	303.90	306.48/0.85	294.57/-3.07	300.20/-1.22	294.19/-3.19	293.86/-3.30	294.52/-3.09
08/26	300.82	307.73/2.30	309.18/2.78	296.70/-1.37	294.81/-2.00	295.85/-1.65	306.88/2.01
08/27	332.33	322.97/-2.82	323.58/-2.63	308.61/-7.14	303.57/-8.65	304.11/-8.49	320.31/-3.62
08/28	332.38	339.33/2.09	339.51/2.15	326.43/-1.79	320.26/-3.65	319.71/-3.81	336.28/1.17

In order to further compare the various prediction models, Table 2 gives average values and maximum values of the relative error absolute values of the respective prediction models. The average value and maximum value reflect the accuracy and the stability of the forecast, respectively. It is not difficult to find that the average value and maximum value of the relative error of the combined prediction model are 2.36% and 4.25%, respectively, which are the lowest. This proves the correctness of the combined prediction method proposed in this paper. The weight coefficient obtained by the meteorological mapping function can be used to avoid weaknesses and combine the advantages of each single prediction well. In addition, the prediction advantage of single model 2 is more obvious than the other single predictions, which indicates that the accumulative effect of temperature has a huge impact on electricity consumption.

Table 2. Comparison of the relative error absolute value between predicted value and actual value of each prediction model

Type of Relative Error Absolute Value (%)	Single Model 1	Single Model 2	Single Model 3	Single Model 4	Single Model 5	Combined Model
Average Value	3.31	2.89	3.65	4.68	4.26	2.36
Maximum Value	10.11	4.99	10.40	6.92	8.49	4.25

The correlation threshold affects the decomposition of daily electricity consumption, resulting in different prediction accuracy. The results are shown in Table 3. Choosing the appropriate correlation threshold will facilitate the improvement of prediction accuracy.

Table 3. Comparison of combined prediction accuracy under different correlation thresholds

Correlation Coefficient Threshold z_ρ	Average Value of Relative Error Absolute Value (%)
0.15	5.13
0.20	2.36
0.25	2.61
0.30	3.33

6. Conclusions

This paper proposes a combined forecasting model of daily electricity consumption, and synthesizes the effect of meteorological factors. The combined forecasting model considers the effects of multiple meteorological indicators, coupling effect, and cumulative effect. Moreover, this paper designs the meteorological mapping function. The mapping value of meteorological indicators describes the influence degree of meteorological indicators on electricity consumption, and determines the weight coefficient of single prediction model, thus the influence of various meteorological indicators is effectively combined. The example calculation proves that the combined forecasting model can greatly improve the accuracy of power consumption forecasting, and it can provide guidance for power system planning, grid operation and transaction management.

Acknowledgments

The work is supported by the Science and Technology Project of State Grid Chongqing Power Company under the project of "Key technologies and methods of Chongqing social electricity consumption prediction considering the influence of multiple factors".

References

[1] Duan Hailai, Qian Huaisui, "Responses of the Electric Power Consumption to Climate Change in Guangzhou City," Journal of Applied Meteorological Science, vol. 20, no. 1, pp. 80-87, 2009.
[2] Luo Hui, Xu Junchang and Xiao Bo, "Study on the Impact of Weather on Electricity Consumption in Xi'an and Its Application to Mid-Long Term Prediction," Meteorological Monthly, no. 1, pp. 54-60, 2016.
[3] Zhong Lihua, Li Yong and Ye Dianxiu, "The Impact of Composite Meteorological Factors on the Electrical Load in Guangxi," Meteorological Monthly, vol. 34, no. 5, pp. 31-37, 2008.
[4] Cheng Dan, Liu Jing and Guo Chunwei, "Prediction of Electric Loads over Central China Based on Accumulated Temperature Effect," Meteorological Science and Technology, vol. 46, no. 4, pp. 814-821, 2018.
[5] Tian Fenglei, Chen Mengtao and Wang Longlong, "Short-term Electric Power Load Forecasting Based on Accumulated Temperature Effect and Optimized Support Vector Machine," Power Demand Side Management, vol. 20, no. 5, pp. 33-36, 2018.
[6] Li Hanglong, " Short-term Load Forecasting of Power Systems Considering Real-time Weather factors," North China Electric Power University, China, June. 2015.
[7] J. Rodgers, W. Nicewander, "Thirteen Ways to Look at the Correlation Coefficient," American statistician, vol. 42, no. 1, pp. 59-66, 1988.
[8] Kang Chongqing, Xia Qing and Liu Mei, " Power System Load Forecasting," Publisher: China Electric Power Press, pp. 241-244, 2007.
[9] Yang Dasheng, Li Tao and Wu Dajun, "Optimal Combination Forecasting Method for Electricity Consumption in Society," Electrotechnical Application, vol. 32, no. 24, pp. 81-84, 2013.
[10] Kang Chongqing, Cheng Xu and Xia Qing, "A New Unified Approach to Short-term Load Forecasting Considering Correlated Factors," Automation of Electric Power Systems, vol. 23, no. 18, pp. 32-35, 1999.
[11] Gao Feng, Kang Chongqing and Cheng Xu, "Adaptive Training Approaches Considering Impact Factors in Short-term Load Forecasting," Automation of Electric Power Systems, vol. 26, no. 18, pp. 6-10, 2002.

IOP Conf. Series: Materials Science and Engineering **853** (2020) 012025 doi:10.1088/1757-899X/853/1/012025

Active Magnetic Bearing Design and Backstepping-Adaptive Control for High-Speed Rotors

Umar Khayyam, He Boxia*, Shaheer ul Hassan
School of Mechanical Engineering, Nanjing University of Science and Technology, Nanjing, Jiangsu 210094, China.

heboxia@163.com

Abstract. Friction and vibrations hindering high-speed are the most pertinent problems facing the rotating machines. Thus, friction and vibration attenuation are essential in improving the overall performance of turbomachines. Conventionally, high-speed flywheel energy storage systems use Active Magnetic Bearings (AMB) to nullify friction losses and to deal with unbalances but, in the present era with increasing demand for high-speed machinery, the applications of AMBs are proliferating. They are not like the traditional bearings; they generate forces through magnetic fields with no contact between bearing and rotor thus decreasing the friction and providing the ability to counteract imbalances actively. This paper investigates an optimized design methodology along with backstepping adaptive control design of an 8- pole radial AMB used for flywheel energy storage systems. Design optimization using GA, along with analytical design constraints are presented. The design details along with finite element simulations are given to verify the optimization results and system requirements. Bearing parameters then obtained are used to develop a linearized mathematical model of the system, an adaptive back-stepping controller is designed for it to regulate the deviation of the rotating shaft from its equilibrium position.

1. Introduction

Active Magnetic Bearing (AMB) is being extensively used in many rotating machine applications, mainly because of their non-contact motion control characteristics. As a consequence, these properties allow novel constructions, high speeds with the possibility of active vibration control, operation with no mechanical wear, less maintenance and, therefore, lower costs. Due to above-mentioned properties Active magnetic bearing (AMB) is a promising solution for many kinds of potentially promising machines such as turbine engines, helicopter, Agile satellites, vacuum pump, motors and flywheel energy storage systems (FESS) [1-7]. Many researchers have presented the capability of producing a controllable electromagnetic force by AMB, among them, Nickolajsen et al. [8] in 1979 introduced its use for vibration damping for the first time. Many authors have investigated their use for vibration suppression [9-12]. They have demonstrated the use of AMB as between bearings to introduce damping and stiffness into the system and compensate for rotor unbalance; their focus has been on suppressing the vibration by changing the pole currents based on position feedback. An active magnetic bearing is purely a Mechatronics product; The hardware is composed of electromechanical components: electromagnets sensors amplifiers and an information processing part, usually a microcontroller [8]. AMB is inherently an unstable system; therefore, a control algorithm is essential.

In the past many researchers have done work on hardware designs and many researchers have done the controller designs. However, an active magnetic bearing system is a mechatronics product; the design involves multidisciplinary challenges; which makes it a thought-provoking task. Hence, this paper presents a design methodology of 8-Pole Radial AMB for an 8kg high-speed composite flywheel presented in [7].
The developed AMB will be exhibited in the following sections. The second section will present a brief overview of the AMB operation. The third section will show the hardware optimization along with Finite element analysis. The fourth section will elaborate the analytical modelling of a designed AMB depicting the relation between current and force generation in electromagnets. In addition, a control technique that combines the merits of both backstepping and adaptive control laws is presented. Finally, concluding remarks are added to summarize the present research.

2. Operation of an Active Magnetic Bearing (AMB)

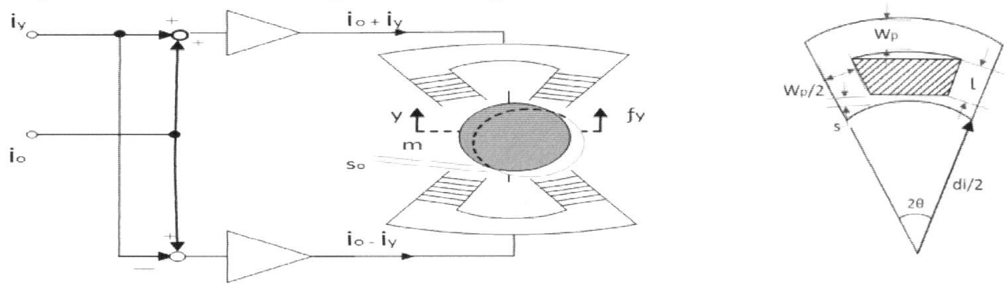

(a) (b)

Figure 1. (a) 1-DOF Radial AMB model in Differential Driving Mode (b) Radial-AMB Design Sketch

Fig. 1 (a) shows a simple 1-DOF AMB magnetic actuator model. Since the principle for vertical and horizontal plane forces is the same, for simplicity, only 1-DOF AMB is considered here. In Fig.1, *i* is the coil current, s_o is a nominal air gap, *N* is the number of coil rounds on the core, *Ap* represents the cross-section area of pole face and the angle of each pole relative to the centreline between the poles is θ_p. The magnetic field generated by the current will create a force in the horizontal axis. According to Ampere's loop law, we have (1), where *H* is the magnetic field, which involves the flux density B.

According to *constitutive Law:* $$B = \mu_o \mu_r H \quad (1)$$

Ampère's circuital law [16]: $$\oint H\, ds = l_{fe} H_{fe} + 2 s_o H_a = NI \quad (2)$$

Combining (1) and (2) will result in a relation (3):

$$B = \mu_o \frac{NI}{\frac{l_{fe}}{\mu_r} + 2s_o} \quad (3)$$

As $\mu_r >> 1$, Hence magnetization of Iron is neglected, resulting in a more simplified form as:

$$B \approx \frac{\mu_o NI}{2s_o} \quad (4)$$

Thus, the energy stored in the airgap is presented as:

$$\mathrm{E} = \frac{1}{2} B_a H_a (A_p 2 s_o) \quad (5)$$

The force acting on the ferromagnetic body ($\mu_r >> 1$) is generated by a change of the field energy in the air gap, as a function of the body displacement. For small displacements *ds,* the magnetic flux $B_a A_p$ remains constant. When the air gap *s* increases by *ds*, the energy *E* in the field increases by an amount *dE.* This energy has to be provided mechanically by electromagnets. Thus, Force *F* is equalled:

$$F = \frac{\delta E}{\delta s} = B_a H_a A_p = \frac{B_a^2 A_a}{\mu_o} = \frac{1}{4} \mu_o N^2 A_p \frac{i^2}{s_o^2} = \frac{K}{4} \frac{i^2}{s_o^2} \quad (6)$$

In the case of a Radial AMB, the forces of both magnetic poles affect the rotor with an angle θ_p . *So,*

$$F = \frac{K}{4} \frac{i^2}{s^2} cos\theta_p \quad (7)$$

IOP Conf. Series: Materials Science and Engineering **853** (2020) 012025 doi:10.1088/1757-899X/853/1/012025

3. Active Magnetic Bearing Design for Flywheel

3.1. Design Optimization

In practice, the design of an AMB encounters several restrictions due to peripheral equipment or actuator itself and these restrictions interfere with one another such that tweaking one parameter will also have an effect on other parameters. Therefore, A compromise between different parameters is to be made in order to reach an ideal solution depending upon the application of AMBs. As part of the design process, this section will deal with the design optimisation of a Radial AMB. AMBs are designed to achieve higher load carrying capacity and compact size. Multi-Objective Genetic Algorithm optimisation strategy is used in Matlab®. The preliminary design conditions used in R-AMB design procedure as follows: Two Radial AMBs are to be used for a Flywheel system of total mass (m_r=10kg) presented in [7]. The nominal air gap is to be kept at (s_o=1mm). Diameter of the shaft is (d_r=40mm). Number of poles (N_p=8) and Silicon iron is to use with Saturation Flux density of (B_{sat}=1.8Tesla). Fringing-leakage factor is assumed to be ($\in = \mathbf{0.8}$) and permeability ($\boldsymbol{\mu_o = 4\pi \times 10^{-7}}$). Along with the initial conditions, optimization method requires a set of constraints which are part of the design methodology adopted in developing the Radial AMB depicted as follows:

The allowable magnetomotive force: $$MMF = 2\frac{s_o B_{sat}}{\mu_o} \quad (8)$$

Required Coil Diameter: $$\emptyset = 2 \times \sqrt{\frac{I_{max}}{\pi J_{cu}}} \quad (9)$$

Minimum Area of Pole: $$Ap \geq \frac{F_{max}\mu_o}{B_{sat}^2 \epsilon (\cos\theta_p)} \quad (10)$$

AMB Pole length and width:

$$w_p \leq 0.8(d_r + 2s_0)\sin\theta_p - 2.5(s_o\cos\theta_p), \quad Lp\ \frac{A_p}{w_p} \quad (11)$$

Required Number of turns: $$\mathrm{N} \leq \frac{\mathrm{MMF}}{I_{max}} \quad (12)$$

Winding space limit: $$A_r < A_w \quad (13)$$

Where A_r and A_w are the required winding space and available winding space respectively. Assume the bulking factor as 80% and most effective winding space as the trapezoid region then the winding space available will be presented by equation (14).

$$A_w = 0.8\left[(d_r + 2s_o + l)\sin\theta_p - w_p\cos\theta_p\right]l\cos\theta_p, \quad A_r = 2N\frac{\pi\emptyset^2}{4} \quad (14)$$

The diameter of the stator is given by: $$D_{st} = d_r + 2(s_o + w_p + 1.15l) \quad (15)$$

Objective of the optimization is to minimize the size of the AMB and maximize the force capability within the constraints from eq. (8-17). Considering an AMB as cylinder, the objective of the optimization is to minimize the total volume and to maximize the force produced and get the optimized value of Number of turns '*N*', current '*I*', Area 'A_p' and Pole width 'w_p'.

$$\text{Cost Function 1: } \min(V) = \frac{\pi D_{st}^2 A_p}{4w_p} \quad (16)$$

$$\text{Cost Function 2: } \max(\mathrm{F}) = \frac{K}{4}\frac{i^2}{s^2}\cos\theta_p \quad (17)$$

Fig.3 shows the optimization results in the form of the Pareto-diagram and Table 1.0 shows design parameters obtained by the preliminary design, GA optimization.

Table 1. Optimization Results

PARAMETER	Current (I_{max})	Number of turns (N)	Area of Pole (A_p)	Width of Pole (W_p)	Length of pole (l)	Stator dia. (D_{st})	Coil dia. ($\emptyset$)	Max. Force (F)
UNITS	A	N	mm^2	mm	mm	mm	mm	N
OPTIMUM VALUE	4	360	300	10	33	138	0.9	140

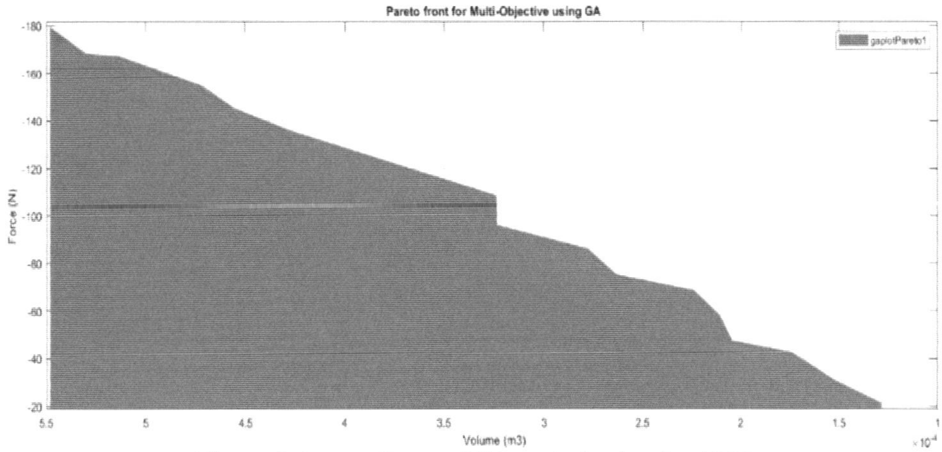

Figure 2. Pareto Front of GA optimization for AMB

3.2. Finite Element Analysis (FEA)

The AMB model based on the parameters presented in Table 1 is designed and magneto-static analysis is done to simulate the magnetic flux density and magnetic force density. In Fig. 3 the top-plots show the flux density distribution and bottom plots show the force density distribution in the designed AMB and rotor lamination.

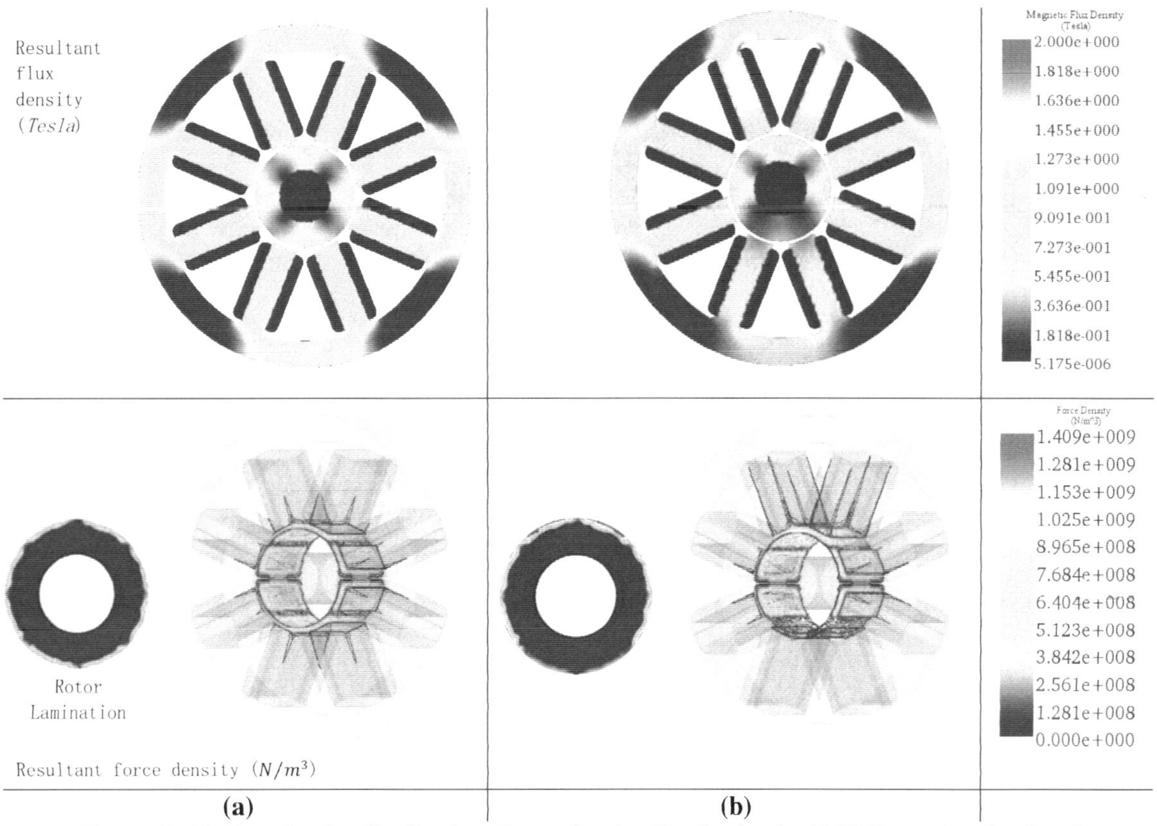

Figure 3. (a) Flux density distribution, Force density distribution in AMB Rotor-Lamination for i_x=0A, i_y=0A (b) Flux density distribution in AMB Force density on Rotor-Lamination for i_x=0A, i_y=2A.

To ensure that the rotor is capable of high rotational speeds, the losses should be minimized. To do so, the magnetic circuit is laminated. Silicon Iron laminations are used both in magnetic bearing and on the rotor. Using laminations on Rotor; in the area where AMB's magnetic field is to be applied results in lower flux loss and helps to avoid excess rotor heating resulting in the long-life span of rotor.
Following the control methodology of Fig 1(a), In the nominal operation when the rotor is at the desired position and reference current i_x and i_y are set to zero, biased current i_o set to 2A then the resulting force-flux distribution is presented in Fig 3(a). It can clearly be seen that the flux is evenly distributed along all axes around the stator and the rotor lamination. Similarly, Fig 3(a) bottom plot presents a detailed view of force density distribution in whole AMB and Rotor lamination. There are attractive magnetic forces between stator poles and rotor lamination, but as amplitudes of these attractive radial forces are the same and the directions are equally distributed, the resultant radial force acting on the rotor is zero. Hence, the rotor will always be stabilized at its equilibrium position unless an external disturbance occurs.
As soon as disturbance occurs and the rotor moves in the negative y-axis. The controller then set i_y to 1A, the new flux and force distribution plots are given in Fig.3(b). The flux density is increased in the positive y-direction side and decreases in the negative y-direction side. The magnetic forces in the two air gaps in the y-direction are no longer equal and the force in the positive y-direction is more significant than that negative y-direction as shown in the force density plot in Fig.3(b). Hence, a radial force F equals to 94-Newton is produced in the positive y-axis. In both of the plots, the maximum magnetic field produced is not more than 1.4 Tesla which is below the saturation point and force density is strong enough to lift the rotor.

4. Controller Design

4.1. Strict Feedback, AMB Model

For designing an Adaptive Backstepping Controller, a mathematical model of AMB in strict feedback form is developed. Fig.1(a) shows the nominal position of the rotor is s_o and m is the rotor's mass, but if a disturbance, F_d occurs and displaces the rotor by an amount s, then according to Newton's law:

$$m\ddot{s} = (F_1 - F_2) + (F_d) \tag{18}$$

Let s_1 and s_2 are now the air gaps between the rotor and stator from top and bottom, respectively. Replacing s in (7) with s_1 and s_2, we can derive the two electromagnetic forces F_1 and F_2:

$$F_1 = \frac{K}{4}\left(\frac{i_1}{s_1}\right)^2,\ F_2 = \frac{K}{4}\left(\frac{i_2}{s_2}\right)^2 \tag{19}$$

By Kirchoff's Voltage Law (KVL), we have:

$$u_1 = Ri_1 + L_s\frac{di_1}{dt} + \frac{K}{2}\frac{d}{dt}\left(\frac{i_1}{s_1}\right),\ u_2 = Ri_2 + L_s\frac{di_2}{dt} + \frac{K}{2}\frac{d}{dt}\left(\frac{i_2}{s_2}\right) \tag{20}$$

Where: u_1 and u_2 are input voltages to control the two currents i_1 and i_2 and $\frac{K}{2}\frac{d}{dt}\left(\frac{i_1}{s_1}\right)$ and $\frac{K}{2}\frac{d}{dt}\left(\frac{i_2}{s_2}\right)$ represents back-electromotive force generated by the air gap flux changes. Nominal states of the systems are represented by (s_o, i_o, μ_o), but as the system will get disturbed, the new states will be presented by (s_1, i_1, μ_1) and (s_2, i_2, μ_2) for top and bottom electromagnets respectively.

$$s_1 = s_o - s,\ s_2 = s_o + s \tag{21}$$

$$i_1 = i_o + i,\ i_2 = i_o - i \tag{22}$$

$$u_1 = u_o + u,\ u_2 = u_o + u \tag{23}$$

Putting (21~23) into (20) and (19) into (18) will result in formulation of a nonlinear AMB model:

$$\begin{cases} \dot{s} = v \\ \dot{v} = \frac{K}{4m}\left(\frac{i_1}{s_o - s}\right)^2 - \frac{K}{4m}\left(\frac{i_2}{s_o + s}\right)^2 + \frac{F_d}{m} \\ \dot{i_1} = \frac{2(s_o - s)}{2L_s(s_o - s) + K}\left[-Ri_1 + \frac{K}{2(s_o - s)^2}vi_1 + u_1\right] \\ \dot{i_2} = \frac{2(s_o + s)}{2L_s(s_o + s) + K}\left[-Ri_2 + \frac{K}{2(s_o + s)^2}vi_2 + u_2\right] \end{cases} \tag{24}$$

Where v is the linear velocity of the rotor. By using Jacobian transformation, (24) can be linearized at its equilibrium states as:

$$\begin{bmatrix} \dot{s} \\ \dot{v} \\ \dot{i} \end{bmatrix} = \begin{bmatrix} 0 & 1 & 0 \\ a & 0 & b \\ 0 & c & d \end{bmatrix} \begin{bmatrix} s \\ v \\ i \end{bmatrix} + \begin{bmatrix} 0 \\ 0 \\ e \end{bmatrix} u + \begin{bmatrix} 0 \\ f \\ 0 \end{bmatrix} F_s \quad (25)$$

Where: $a = \frac{2k_s}{m}$, $b = \frac{2k_i}{m}$, $c = \frac{-k_i}{L_0 + L_s}$, $d = \frac{-R}{L_0 + L_s}$, $e = \frac{1}{L_0 + L_s}$, $f = \frac{1}{m}$, $F_s = F_d$,

$k_s = \frac{K\, i_0^2}{2\, s_0^3}$, $k_i = \frac{K\, i_0}{2\, s_0^2}$ and $L = L_o + L_s$, $L_o = \frac{K}{2s_o}$, $R = \frac{\rho_{cu} l_w}{A_c}$, l_w is the length of winding and A_c the cross-sectional area of conducting wire and the bias voltage caused by the coil resistance R denoted by $u_o = R i_o$.

Parameters used in (25) are evaluated from Table 1 and the values presented in Table 2.

Table 2. AMB Parameter Configuration

Parameter	Force-Displacement constant (K_s)	Force-Current constant (K_i)	Coil Self Inductance (L_s)	Air Gap Inductance (L_o)	Bias-current (i_o)	Disturbances Force (F_d)	Mass per AMB (m)	Resistance (R)
Unit	N/m	N/A	mH	mH	A	N	kg	ohm
Value	97667	49	90	24.4	2	5	4.5	1

For creating a "strict feedback form", (25) is transformed into system shown in (26), (27) and (28).

Where: $x_1 = \frac{s}{b}, x_2 = \frac{v}{b}, x_3 = i$.

$$\dot{x_1} = x_2 \quad (26)$$

$$\dot{x_2} = x_3 + \frac{a}{b} x_1 + \theta = x_3 + \varphi_1(\theta, x_1) \quad (27)$$

$$\dot{x_3} = u' + cbx_2 + dx_3 = u' + \varphi_2 \quad (28)$$

In (19), (20) and (21), the disturbance force and control effort are defined as $\theta = \frac{F_s}{bm} = 1.02$, $u' = eu$, respectively. In (20) and (21), φ_1 and φ_2 are defined as $\varphi_1(\theta, x_1) = \frac{a}{b} x_1 + \theta$ and $\varphi_2 = \varphi_2(x_2, x_3) = cbx_2 + dx_3$

4.2. Back Stepping with Adaptive Disturbance Rejection

From (25) and Table 2, we can calculate the eigenvalues of matrix A, which comes out to be [**112 -766+99*i* -766-99*i***]. Since there is a positive eigenvalue for A, the system is inherently unstable. Therefore, primary control objectives are to stabilize the AMB and to drive the position of the rotor to its equilibrium point in the presence of external disturbance and system uncertainties.

The control goal is to regulate the position of the rotor x_1 in the presence of disturbance to our desired position. The process adopted for designing a backstepping controller for position stabilization is as follows:

- The goal is to be achieved by implanting two virtual controllers α_1 *and* α_2 in (26) and (27) respectively (i.e. $\alpha_1 = x_2$ and $\alpha_2 = x_3$) which will help drive x_1 and x_2 to zero.
- Starting from x_1, CLF (control Lyapunov function) for x_1 will be chosen as $V = \frac{1}{2} x_1^2$ and the virtual control action $\alpha_1 = -k_1 x_1$ will make $\dot{V} = -k_1 x_1^2$ which is negative indefinite, making x_1 asymptotically stable.
- However, practically it is not possible and there is always an error between virtual controller α and real state x.
- We construct an error system whose states will compute the difference between the actual states and virtual controls. Such that:

$$z_1 = x_1,\ z_2 = x_2 - \alpha_1,\ z_3 = x_3 - \alpha_2$$

- Now, the control goal is to asymptotically stabilize all the error system states such that they get eventually driven to zero.
- CLF for states z_1 and z_2 as follow:

$$V_1 = \frac{1}{2}z_1^2 + \frac{1}{2}z_2^2 \tag{29}$$

The derivative of V_1 will be: $\dot{V}_1 = x_1x_2 + z_2\left(x_3 + \varphi_1 - \frac{\partial\alpha_1}{\partial x_1}\dot{x}_2\right)$ (30)

Virtual function x_3 that is to make the derivative of negative definite:

$$\alpha_2 = x_3 = -k_2z_2 - \varphi_1 + \frac{\partial\alpha_1}{\partial x_1}x_2 - z_1 \tag{31}$$

Putting (31) in (30) results in: $\dot{V}_1 = -k_1z_1^2 - k_2z_2^2$

The derivative of V_1 has become negative semi-definite; hence, z_1 and z_2 are eliminated.

- Now for eliminating z_3, new CLF including all the existing errors and displacement is created as:

$$V_2 = \frac{1}{2}z_1^2 + \frac{1}{2}z_2^2 + \frac{1}{2}z_3^2 \tag{32}$$

The derivative of V_2 will be: $\dot{V}_2 = -k_1z_1^2 - k_2z_2^2 + z_3(z_2 + u' + \varphi_2 - \dot{\alpha}_2)$ (33)

Virtual function u' that makes the derivative $\dot{V}_2$ negative definite:

$$u' = -k_3z_3 - z_2 - \varphi_2 + \dot{\alpha}_2 \tag{34}$$

Putting 34 in 33 results in: $\dot{V}_2 = -k_1z_1^2 - k_2z_2^2 - k_3z_3^2$ (35)

Derivative of V_2 has also become negative semi-definite. So, the control goal is achieved.

The above control law u' will work efficiently unless there is no external disturbance, but in the case of high-speed flywheels supported by AMB's, rotor imbalances can cause unknown disturbances. Therefore, we generate an adaptive law to make an estimate disturbance error to zero. The disturbance estimation methodology adopted is as follow:

‘

- Let disturbance be θ, and estimated disturbance be $\widetilde{\theta_1} = \theta - \widehat{\theta_1}$. We have an estimation error. We add the quadratic form of it to (29) and then form a new CLF (36). Positive real numbers γ_i(i=1,2,3) are chosen as adaptive coefficients and the derivative is presented in (37).

$$V_1 = \frac{1}{2}z_1^2 + \frac{1}{2}z_2^2 + \frac{1}{2\gamma_1}\widehat{\theta_2}^2 \tag{36}$$

$$\dot{V}_1 = -k_1z_1^2 - k_2z_2^2 + z_2\tilde{\theta}_1 + z_3z_2 - \frac{1}{\gamma_1}\dot{\widehat{\theta}}_1\widetilde{\theta_1} = -k_1z_1^2 - k_2z_2^2 + z_3z_2 + \widetilde{\theta_1}(z_2 - \frac{1}{\gamma_1}\dot{\widehat{\theta_1}}) \tag{37}$$

Virtual function α_2 that to make the derivative of negative definite:

$$\alpha_2 = -z_1 - k_2z_2 - \varphi_1x_1 - \hat{\theta}_1 + \dot{\alpha} = -(k_1k_2 + \varphi_1 + 1)z_1 - (k_1 + k_2)x_2 - \widehat{\theta_1} \tag{38}$$

Putting (38) in (37), and selecting adaptive law as $\dot{\widehat{\theta_1}} = \gamma_1z_2$ will make derivative $\dot{V}_1$ negative definite assuming the term z_2z_3 will be cancelled in the future.

$$\dot{V}_1 = -k_1z_1^2 - k_2z_2^2 + z_3z_2 \tag{39}$$

- If θ_2 is the estimate of θ_1 and the estimation error is $\widetilde{\theta_1} = \theta_1 - \widehat{\theta_2}$. Then the (32) can be reconstructed as:

$$V_2 = \frac{1}{2}z_1^2 + \frac{1}{2}z_2^2 + \frac{1}{2}z_3^2 + \frac{1}{2\gamma_1}\widetilde{\theta_1}^2 + \frac{1}{2\gamma_2}\widetilde{\theta_2}^2 \tag{40}$$

From (38) computing $\dot{\alpha}_2$ and putting in control law (34) will result in (41):

$$u' = -(k_3 + d)x_3 - \left[k_3\left(k_1k_2 + \frac{a}{b} + 1\right) + k_1\right]x_1 - [k_3(k_1 + k_2) + 1 + cb]x_2 - k_3\widehat{\theta_1} + \dot{\alpha}_2 \tag{41}$$

Differentiating the CLF (V_2) in (40), given the control law (41) and adaptive law as $\dot{\widehat{\theta_2}} = -z_3\gamma_{w2}\frac{\partial\alpha_2}{\partial x_2}$, the derivative of V_2 becomes:

$$\dot{V}_2 = -k_1z_1^2 - k_2z_2^2 - k_3z_3^2 \tag{42}$$

Now, that the derivative of the final CLF is negative semi-definite, which ensures the system is now stabilized at its equilibrium point. The final back-stepping control law is generated by calculating (41):

$$u' = -\left[k_1\left(\frac{a}{b} - 2\right) + k_2\left(\frac{a}{b} - 1\right) + k_3\left(\frac{a}{b} + k_1k_2\right)\right]x_1 - [k_1k_2 + k_2k_3 + k_1k_3 + cb + \varphi_1 + k_1 + k_2 + 3]x_2 - [k_1 + k_2 + k_3 + d]\,x_3 \tag{43}$$

IOP Conf. Series: Materials Science and Engineering **853** (2020) 012025 doi:10.1088/1757-899X/853/1/012025

4.3. Controller simulation results

From Table.2, (24) and using the control law u', we construct the Backstepping controlled AMB system in Simulink. The disturbance is added to the system as a step input at 1.5 seconds and all the initial values of state variables are assumed as zeros. For tuning first ACs γ_1 *and* γ_1 are set to 1. Lyapunov coefficients LCs k_1, k_2 and k_3 are then tuned like a P-controllers. The time response of all three states is then computed as shown below:

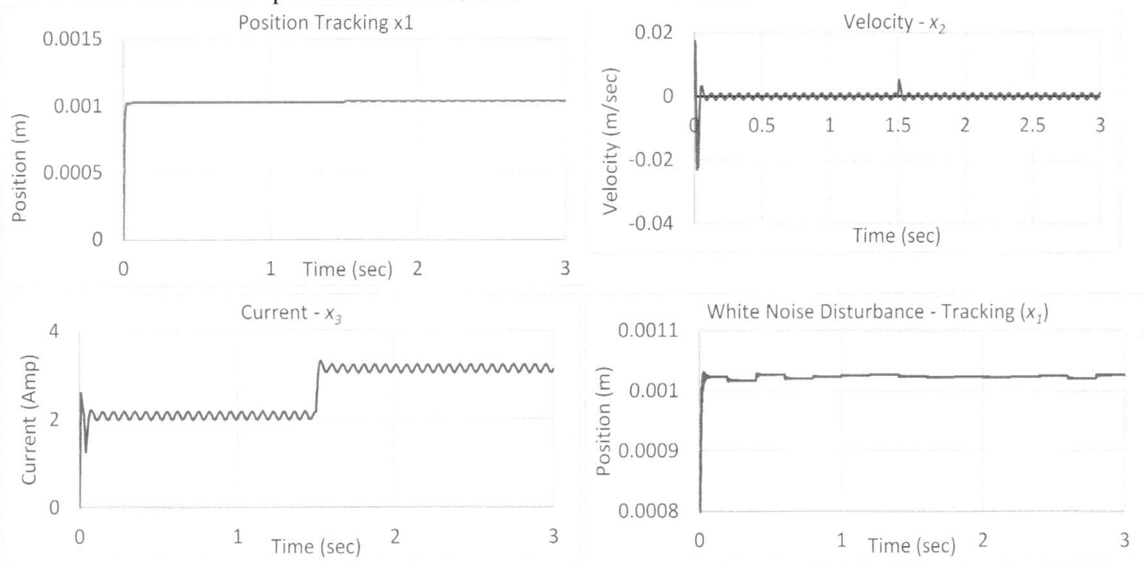

Figure 4. Time response of all three states of AMB

Figure.4 shows the time response of AMB with backstepping adaptive control. It has been seen the maximum tracking error was 0.02mm that can guarantee the rotor will not touch stator. The maximum current drawn as disturbance occur is less than $I_{max,}$ which guarantees that the controller will not force the AMB to cross its saturation point. To check the robustness of the controller white noise with a sample time of 0.20sec and with noise power of 50 is fed as a disturbance Fig.4 bottom left plot shows the tracking response. The maximum deviation from the desired position was no more than 0.07mm which assures that the system is robust and adaptive part is estimating the disturbances accurately.

5. Conclusion

Due to the outstanding characteristics such as complete contact-free support, no wear, no friction, no lubrication and active vibration control; Active magnetic bearing (AMB) Systems are finding more and more industrial applications. Paper presented a complete design methodology for a Radial AMB; Analytical Modelling, design constraints and the resulting GA Optimization was presented. Force and flux density were evaluated to verify the optimized parameters. Based on the parameters obtained, a mathematical model of an AMB was created and back-stepping adaptive control was introduced to regulate the AMB's position, this technique combined both the merits of backstepping and adaptive control law and was able to deal with the parametric uncertainties as well as continuously varying external disturbances. Control objectives were to stabilize the closed-loop system and to regulate the displacement of the rotor in the presence of the disturbance and uncertainties. The Lyapunov method proves the stability of the control system despite the presence of disturbance. Magnetostatic and Control Simulation results, verify the effectiveness and robustness of the AMB system.

Acknowledgments
This work is supported by the National Natural Science Foundation of China, No.51575281 and the Fundamental Research Funds for the Central Universities, No.30916011304. The authors gratefully acknowledge all financial supports.

References
[1] Y. Yuan, Y. Sun and Q. Xiang, "Design and Analysis of a Magnetic Bearings with Three Degrees of Freedom", Chinese Journal of Mechanical Engineering, vol. 32, no. 1, 2019
[2] B. Han, Z. Huang and Y. Le, "Design aspects of a large scale turbomolecular pump with active magnetic bearings", Vacuum, vol. 142, pp. 96-105, 2017. Available: 10.1016/j.vacuum.2016.12.010.
[3] J. Yadav and H. Sahu, "Design, Optimization and Analysis of a Radial Active Magnetic Bearing for Vibration Control", International Journal of Trend in Scientific Research and Development, vol. -3.
[4] H. Bangcheng, Z. Shiqiang, W. Xi and Y. Qian, "Integral Design and Analysis of Passive Magnetic Bearing and Active Radial Magnetic Bearing for Agile Satellite Application", IEEE Transactions on Magnetics, vol. 48, no. 6, pp. 1959-1966, 2012. Available: 10.1109/tmag.2011. 2180731..
[5] Development of a radial active magnetic bearing for high speed turbo-machinery motors 2006.
[6] M. Pichot, J. Kajs and B. Murphy, "Active magnetic bearings for energy storage systems for combat vehicles", IEEE Transactions on Magnetics, vol. 37, no. 1, pp. 318-323, 2001.
[7] Shaheer, H. Boxia and U.Khayyam, "Design and Analysis of flywheel for small scale energy storage systems using different structures and their comparison" International Conference on Power & Renewable Energy", Chengdu, China 2019.
[8] G. Schweitzer, E. Maslen, Magnetic Bearings - Theory, Design and Application to Rotating Machinery, Springer – Verlag, Berlin. 2009.

[9] J. Nikolajsen, R. Holmes, V. Gondhaleker, Investigation of an electromagnetic damper for vibration control of a transmission shaft, Applied Mechanics Group, Proceedings of the Institution of Mechanical Engineers, 193 (1979) 331–336.
[10] X. Wang, C. Han and Y. Yu, "A modified adaptive backstepping method for shaft deflection tracking control of magnetically suspended momentum wheel with nonlinear magnetic torque", Journal of the Franklin Institute, vol. 355, no. 7, pp. 3274-3287, 2018.
[11] L. Dong and S. You, "Adaptive back-stepping control of active magnetic bearings," in 2013 10th IEEE International Conference on Control and Automation, pp. 452–457, Hangzhou, China, 2013

Study on Characteristics of no-load Converter Transformer closing Accessing to High-Cable Rate Network

Tongtong Wang[1], Shengyi Zhu[2], Haiqiang Jin[1], Jun Wen[1]

[1]School of Electrical and Electronic Engineering, North China Electric Power University, Beijing 102206, China.
[2]State Grid Chongqing Electric Power Research Institute, Chongqing 401120, China.

pyyzwtt@163.com

Abstract. When the converter transformer is closed at no load state, the transient voltage is too large, which may cause the transformer core to saturate; meanwhile, the capacitance of the cable line to the ground is much larger than that of the overhead line, which may cause ferromagnetic resonance phenomenon. Taking the parameters of typical transmission line and 500kV converter transformer as an example, the simulation results of Matlab are used to analyze the possible effects of no-load closing of converter transformer under different cable-forming rates and different outlet modes, and put forward targeted measures, which have certain reference value for the future design of the converter station outline.

1. Introduction

Transformer closing refers to the process of connecting the primary side to the grid when the secondary side of the transformer is not loaded [1]. When the DC project is put into operation, the converter transformer's no-load close should first be carried out. 1) Due to the nonlinear excitation characteristics of transformer, a high value of magnetizing inrush current may be generated, which contains a large number of higher harmonic, mainly covers 2nd and 3rd harmonics. The distortion rate can reach 28%, which threatens the safe operation of the power grid [2]. 2) The excitation circuit may be saturated, causing the excitation impedance to drop; thereby causing ferromagnetic resonance; overvoltage and overcurrent are generated, which threaten the safety of the operation personnel.

As the city's electricity consumption increases sharply, overhead lines are no longer meet the demand. Cable lines will be used wider in urban distribution networks in the future due to their high reliability and low operating [3-4]. Compared with the overhead line, cable line is characterized by large capacitance to ground. Therefore, its resonance characteristics are quite different from overhead line [5].

There are no examples of cable outlets in converter stations, and the influence of cable lines is rarely considered in the literature. However, with the increase ratio of the cable line, the research on the cable mode and cable rate is of great significance for the future. This paper first introduces the topology of transmission system, analyzes the mechanism of ferromagnetic resonance, and establishes the harmonic model of converter transformer and transmission line. Based on ±500 kV DC constructions and transmission line parameters. It is studied whether large resonance over-voltage will be generated when commutation and no-load closing under different outlet modes and different cable ratios. Measures for effectively suppressing resonant overvoltage are discussed.

2. Converter station access to the AC system

2.1. No-load converter transformer closing

The converter station in the DC project use 12-pulse converters for each pole, as shown in Figure 1. The 12-pulse converter is composed of two 6-pulse converters (6p) connected in series on the DC side [6]. Each converter is connected to the AC system via the YNy0 and YNd11 type converter transformers [6]. In Figure 1, *ACF* is short for AC filter, *l* is an AC transmission line.

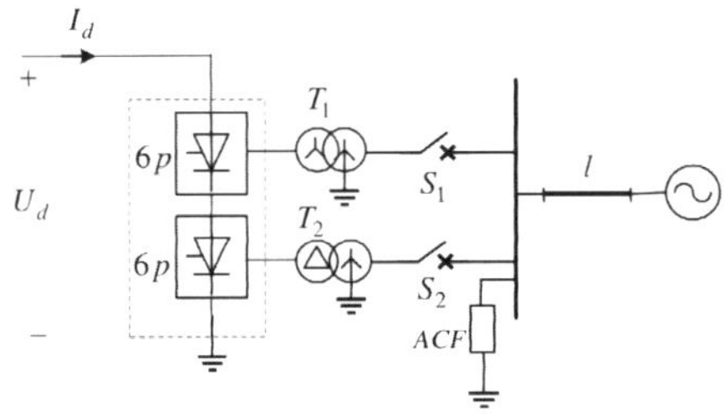

Figure 1. Schematic diagram of the converter station

The *ACF* is out of running when the converter is switched, so its influence doesn't be considered. The no-load closing of the transformer is the first step. When the converters are locked, and the converter transformer charging condition is provided, the three-phase circuit breaker S_1、S_2 is closed.

2.2. Converter station outlet mode

Nowadays, converter stations are connected to the grid via AC overhead lines. In order to investigate the possible impact of the cable line accessing the grid in the future， it sets the two outlet modes:
(1) M1: Connected to the grid by cable lines via overhead lines.
(2) M2: Connected to the grid by overhead lines via cable lines.

2.3. Transmission line harmonic model

This paper defines the cable ratio σ as the ratio of the cable line to the total transmission line. $\sigma = l_c/l$ .The total length of the line is l, the length of the cable line is l_c .

In general, both of the cable line and the overhead line can be represented by a π-type equivalent circuit as shown in Figure 2.

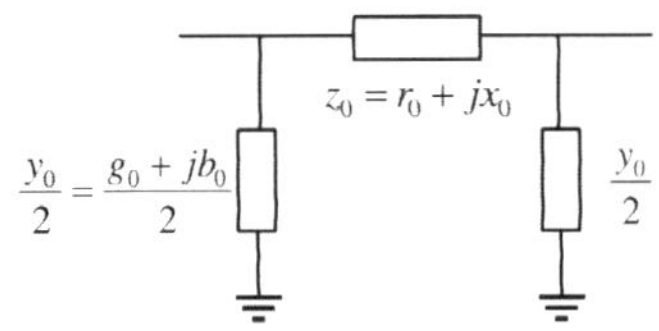

Figure 2. Unit length type π equivalent circuit

In Figure2, r_0, x_0, g_0, and b_0 are the resistance, reactance, conductance, and susceptance of the AC transmission line per unit length. As the leakage current of the line is small, g_0 is usually ignored. The AC transmission line of length l, a three-stage line model is established as shown in Figure 3.

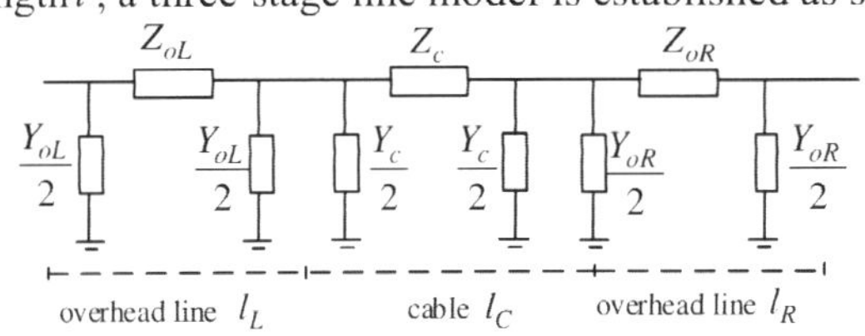

Figure 3. Transmission line topology

Each segment is equivalent by a π-type equivalent circuit. The two outlets can be expressed as

M1: $I_R = 0$, $I = I_L + I_C$

M2: $I_L = 0$, $I = I_R + I_C$

The Thevenin equivalent circuit for the commutation and AC system is shown in Figure 4.

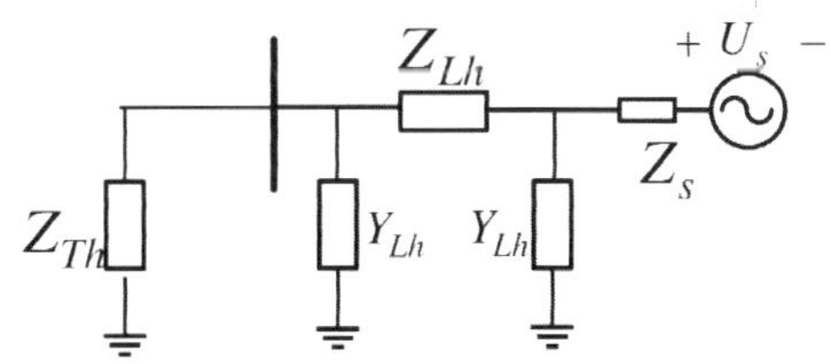

Figure 4. Converter transformer and AC system equivalent circuit

In the figure, Z_{Th} is the harmonic impedance of the transformer; the transmission line topology is equivalent to a π-type equivalent circuit, Z_{Lh} and Y_{Lh} are the harmonic impedance and admittance of the transmission line. The parallel harmonic impedance of the converter transformer and the line is

$$Z_{T-Lh} = \frac{1}{\dfrac{1}{\dfrac{1}{\dfrac{1}{Z_s} + Y_{Lh}} + Z_{Lh}} + Y_{Lh}} \parallel Z_{Th} \tag{1}$$

3. Parallel ferromagnetic resonance caused by no-load close of converter transformer

When the converter transformer is closed, the three-phase high-voltage winding directly connected to the neutral point is instantaneously connected to the transmission network of 500kV and above, causing the saturation of the core and generating ferromagnetic resonance.

The core of converter transformer has non-linear magnetization characteristics, as shown in Figure 5. In the figure, the *ab* segment, the *bc* segment, and after *c* point are respectively a linear segment, a saturated segment, and a deep saturated segment [7]. For the *ab* segment, the field current i_0 is small and ϕ increases linearly with i_0 ; after the saturation segment, i_0 increases sharply with increasing ϕ .

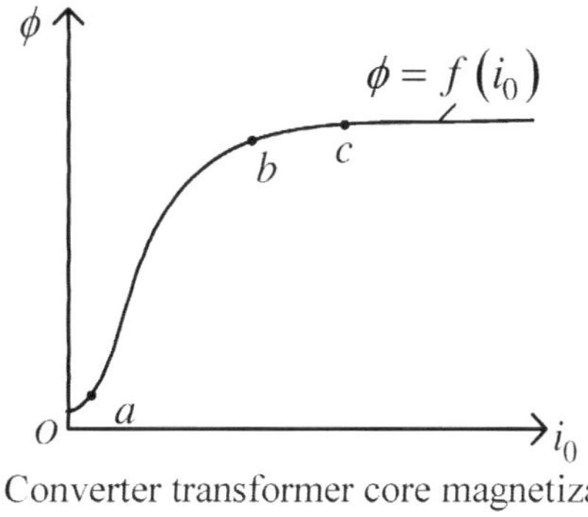

Figure 5. Converter transformer core magnetization curve

Ignoring the impedance of the converter transformer's high-voltage winding, taking the ground-coupling capacitance C of the AC transmission line into account, it is intended to illustrate that the simplified circuit of the converter transformer closing may induce parallel ferromagnetic resonance as shown in Figure.6a. Ignore the equivalent resistance R_s of the AC system, $\Delta I = |I_L - I_C|$, the volt-ampere relationship of the circuit is shown in Figure 6b.

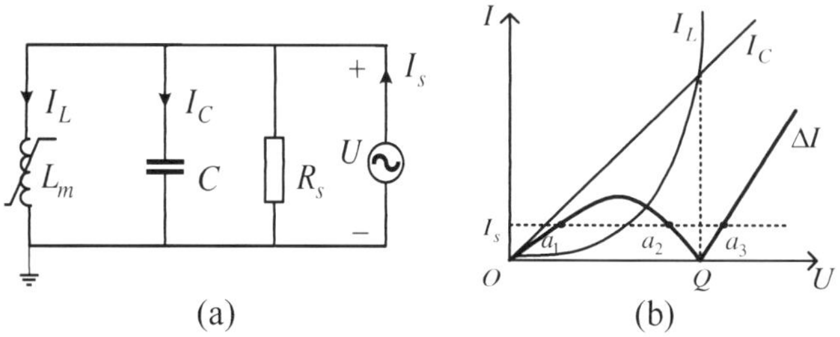

Figure 6. Parallel ferromagnetic resonance description

The curve ΔI intersects the supply current I_s at three points a_1、 a_2 and a_3, a_1 and a_3 are the steady-state point, and a_2 is the imbalance point. Taking a_1 point as an example, if the system causes a rising disturbance of ΔU, the corresponding I_L and I_C increase. But since I_L rises faster, ΔI increases. At this time, I_s is below the ΔI, the voltage cannot be maintained, so the voltage drops, it returns to a_1 point finally; Similarly, when the voltage drops ΔU, as I_L falls faster, the voltage is forced to rise until it stabilizes at a_1. For the a_2 point, the voltage also has a disturbance of ΔU, both I_L and I_C increase, I_C rises faster and ΔI decreases. The voltage continues to increase and finally stabilizes to a_3.

When the converter transformer is closed, the total magnetic flux is very large, so that the core enters the saturation section, and the working point will cross the a_1 point to reach the a_2 point, but the a_2 point cannot run stably, as the L_m is reduced, it forms parallel resonance with the capacitor C. L_m continues to fall, eventually crossing the resonance point Q to reach a new stable point a_3. During this process, the voltage surges, causing the excitation current to increase rapidly, result in an overcurrent phenomenon. Since the equivalent resistance of the system is small, it can be self-sustained after the resonance is excited. This process is called parallel ferromagnetic resonance.

4. Influence of cable rate on no-load closing of converter transformer

4.1. Study System

Take the typical unit parameters of the ±500 kV overhead line and cable line, the unit parameter of the overhead line[8] is $r_o = 0.0163\Omega/km$, $x_o = 0.287\Omega/km$, $b_o = 3.51\times10^{-6}\Omega/km$; The unit parameter of the cable line is $r_c = 0.058\Omega/km$, $x_c = 0.09\Omega/km$, $b_c = 128.3\times10^{-6}\Omega/km$。 The basic parameters of the converter transformer are shown in Table 1 below[9]. Based on Matlab software, the influences of cable rate on converter transformer closing are calculated. Set the cable rate to vary from 0 to 1.

Table 1. Converter transformer excitation parameters

capacity /MVA	Rated voltage /kV	No-load excitation current /%	Short circuit impedance /%
283.7	525	0.102	16

4.2. Influence of different modes on ferromagnetic resonance

In order to study the influence of cable rate on the excitation ferromagnetic resonance under different outlet modes, the transformer is set to close at 0.2s, the simulation lasts 10s, and the supply voltage is 500kV. When the AC transmission line uses overhead lines, the cable rate is 0. The effective value of the excitation current after 1 s is 2.51A, which can be approximated as the normal no-load excitation current value, and no ferromagnetic resonance will occur.

When the cable rate is 1, and the transformer is input, after 1 s, the stable resonance state is reached. The current waveform flowing through the magnetizing inductance and the capacitance to the ground is shown in Figure7. It can be seen that the saturation characteristic of the commutation is obtained when the cable ratio is high. Ferromagnetic resonance occurs; the current waveform is obviously distorted. The maximum excitation current can reach about 6kA, and once the resonance occurs, the amplitude maintains continuous oscillation and does not attenuate.

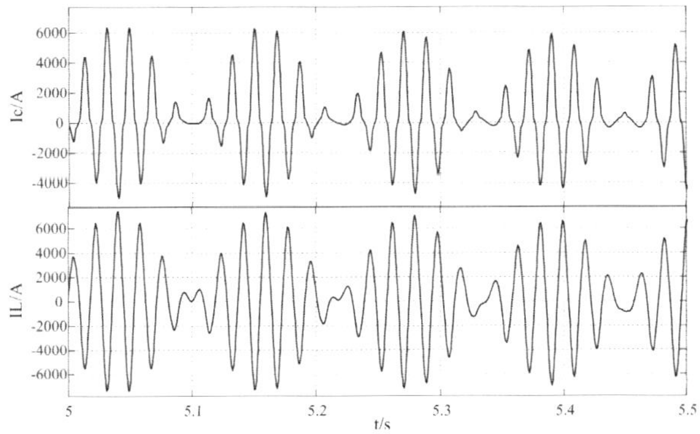

Figure 7. Ferromagnetic resonance current waveform

4.3. Influence of core saturation on ferromagnetic resonance

When the cable rate of the transmission line is different, the closing of transformer may cause different saturation states, which will affect the magnitude of the magnetizing inductance, resulting in different parallel ferromagnetic resonance, and thus the frequency of resonance is affected. The simulation results are shown in Figure 10 below.

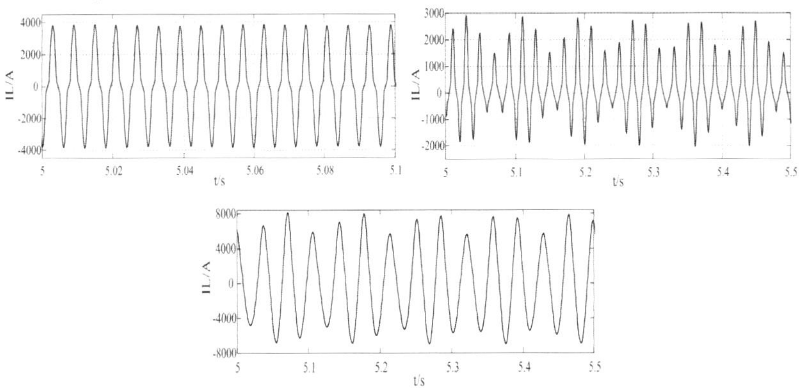

(a) Triple frequency resonance （b） Fundamental resonance (c) 1/2 frequency resonance

Figure 8. Resonance at different cable rates

It can be seen that when the converter transformer core is saturated, the excitation inrush current mainly contains the fundamental wave and the third harmonic. When the line length is short or the cable rate is low (0.5-0.6), as shown in Figure 8(a), the capacitor and inductance parameters may be properly combined to generate triple harmonic resonance; When $\sigma > 0.8$, the line capacitance is large, and the resonant frequency is lower, it is close to 1/2 of the power frequency, as shown in Figure 8 (b). When σ is low or the line is too long, no resonance occurs.

5. Conclusion

（1） The capacitance of the cable line to the ground is larger than that of the overhead line. Therefore, when the converter is connected to cable line directly, the possibility of parallel ferromagnetic resonance is greater. At the same time, the cable rate will have a great impact on the converter transformer closing. $\sigma \leq 0.5$, the parallel ferromagnetic resonance is not easy to occur, $\sigma > 0.5$, the AC transmission line of the cable outlet directly may resonate. Therefore, when the AC grid is erected in the city, it can be considered to minimize the proportion of direct cable outlets.

（2） For the M2 outlet mode, as the cable rate increases, it will cause a change in saturation of the core. In order to prevent the occurrence of ferromagnetic resonance, σ should not exceed 0.5, and the line length should not be too short.

References

[1] Hao Zhiguo, Zhang Baohui, Chu Yunlong, Research on suppression technology of transformer inrush current excitation inrush[J]. High Voltage Apparatus, 2005(02): 81-84.
[2] Zhu Taoxi,Wang Chao,Zhang Xuesong. Analysis of the Influence of No-load Input Converter Transformer on HVDC [J].Automation of Electric Power Systems,2007(23):108-112.
[3] Ding Zhilin, Wen Cao, He Xingqi. Analysis of Ferro- -magnetic Resonance of 500kV Large Transformer Transformer[J]. Electric Power Automation Equipment, 2012,32(02):141-144..
[4] Zhang Dongyi. Study on Inrush Current Characteristics and Suppression Methods of Converter Transformer[D].South China University of Technology,2018.
[5] He Huiwen. Analysis of operation characteristics of cable outlet substation [D]. Huazhong University of Science and Technology
[6] Li Guoqing,Peng Shi,Zhang Shaojie,.A Ferromagnetic Resonance Analysis of Transformer and Shunt Capacitor[J].Power System Protection and Control,2014,42(09):26-32.
[7] Li Zhenqiang, Gu Dingxie, Dai Min. Analysis and suppression method of resonant overvoltage and magnetizing inrush current for UHV no-load transformers[J].High Voltage Engineering, 2012, 38 (02):387-392.
[8] Zhang Siguang, Wang Zimin, Yang Yang. Simulation Analysis of Ferromagnetic Resonance of ±500kV Transformer in Converter Station[J].Guangdong Electric Power,2017,30(03):76-80
[9] Zhang Caijun, Quan Fengyun, Chen Jia. Analysis of overvoltage of cable sheath caused by main transformer no-load closing[J]. High Voltage Apparatus, 2019, 55(02) :164-170.

IOP Conf. Series: Materials Science and Engineering **853** (2020) 012027 doi:10.1088/1757-899X/853/1/012027

Simulation Analysis of Magnetizing Inrush Current of 500kV Transformer

Shengyi Zhu[1], Tongtong Wang[2], Haiqiang Jin[2], Jun Wen[2],

[1]State Grid Chongqing Electric Power Research Institute, Chongqing 401120;
[2]School of Electrical and Electronic Engineering, North China Electric Power University, Beijing 102206

pyyzwtt@163.com

Abstract. The 500kV power grid is the most important high-power AC power system in China. The 500kV transformer is responsible for the regular uploading and sending tasks. It is important electrical primary equipment that connects two or even three different voltage levels. Taking a new 500kV transformer in Chongqing as an example, this paper discusses the influencing factors of 500kV transformer inrush current when connected to the grid through PSCAD/EMTDC software, and puts forward feasible measures to suppress the inrush current.

1. Introduction

The 500kV power grid is the most important power transmission system in China, and it undertakes the task of transporting the power generated by the power generation base to the remote substation. Two main transformers are usually set in a 500kV substation. After the new substation is built or main transformer is overhauled, it will always involve the operation of the main transformer no-load grid connection, that is, the closing of the no-load transformer is a routine operation mode of the substation. The no-load closing of the transformer refers to the process of connecting the primary side to the grid when the secondary side is un-loaded. The process of transformer no-load closing is often accompanied with the magnetizing inrush current, as the transformer core material has nonlinear characteristics, during the closing moment; the excessive grid voltage causes the equivalent inductance of the transformer to decrease sharply, so the excitation current is very high. There are plenty of odd harmonics in this current; the main harmonic order is 3, 5 times, it also contains the even harmonics of 2 times. Overcurrent poses a threat to the safety of electrical equipment and staff. Literature [1-3] analyzed the characteristics of transformer excitation inrush current and its potential hazards such as transformer protection disoperation. Literature [4] showed that serious inrush current would lead to bus voltage distortion, base wave voltage drop, and inverter commutation failure.
At the present, the suppression of inrush current mainly includes the following methods. In literature [5], inrush current is suppressed by effectively measuring transformer remanence and controlling the initial phase Angle at closing time. In literature [6], the suppression effect of closing resistance on inrush current during transformer air charging is verified.
In China, the circuit breaker of 500kV transformer is usually equipped with parallel voltage equalizing capacitor, and there is a capacitive voltage transformer (CVT) connected in parallel on the busbar. The phenomenon of excessive magnetizing inrush current caused by transformer no-load closing has occurred many times [7], causing the voltage oscillation.

In this regard, this paper takes a new 500kV transformer in Chongqing as an example. Through the PSCAD/EMTDC software, the 500kV transformer simulation model with actual parameters is established, and the transformer no-load closing circuit including closing circuit breaker and CVT is set as well. The simulation model discusses the influencing factors of the excitation current of 500kV transformer, and puts forward the feasible pre-control measures of the magnetizing inrush current.

2. Model of 500kV Transformer No-load Closing System

2.1. Introduction of 500kV Substation

The main transformer of TL500kV substation in Chongqing is composed of three single-phase transformers (ODFS-334000/500). The main transformer rated capacity is 334000/334000 /100000kVA, rated voltage is (525/ √3)/(230/ √3±2×2.5%)/36kV, rated frequency is 50Hz[8]. The main transformer's no-load closing circuit is shown in Figure 1. In the figure, the main capacitor of C_{CVT} is 5000pF; the capacitance C_S of the parallel capacitor of the circuit breaker is 2000pF; the other basic parameters of the main transformer are shown in Table 1. The main transformer excitation characteristic curve provided by the manufacturer is shown in Figure 2.

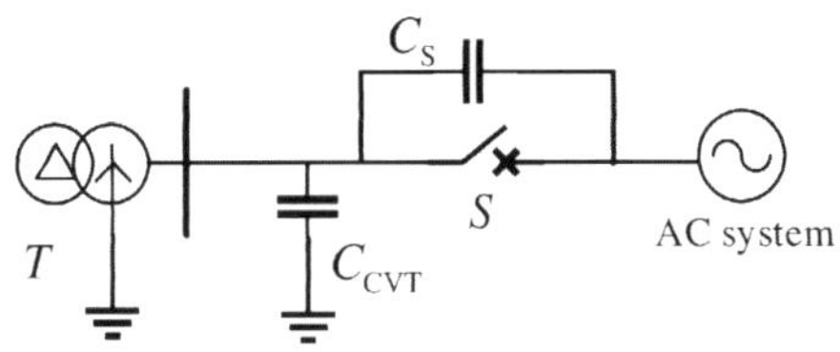

Figure 1. Characteristic curve of transformer excitation

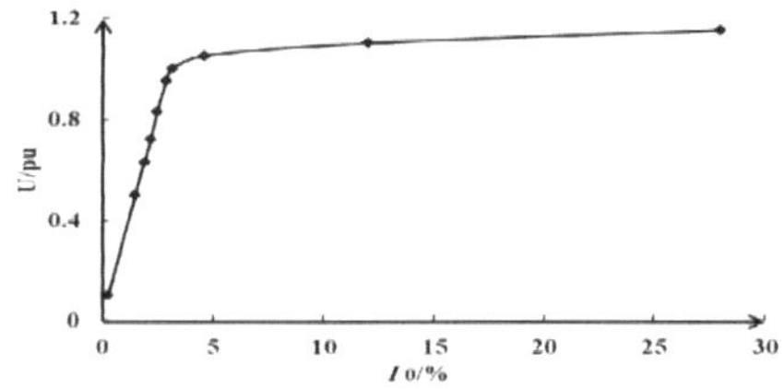

Figure 2. Excitation characteristic curve provided by the manufacturer

Table1. Basic parameters of converter transformer

Rated capacity/MVA	rated voltage/kV	No-load excitation current /A	Short-circuit impedance/%	No-load loss/%	Short-circuit loss/kW
334	525	5.18	15.79	74.55	177.35

2.2. 500kV transformer simulation modelling

2.2.1. UMEC Transformer Model. The UMEC (unified magnetic equivalent circuit) model is a transformer model that takes the magnetic saturation characteristics of the transformer core into account. The model uses a full interpolation method to describe the magnetic saturation characteristics of the transformer, which can model various core structures. The UMEC transformer model provided by PSCAD/EMTDC has the excitation characteristic curve shown in Figure 3.

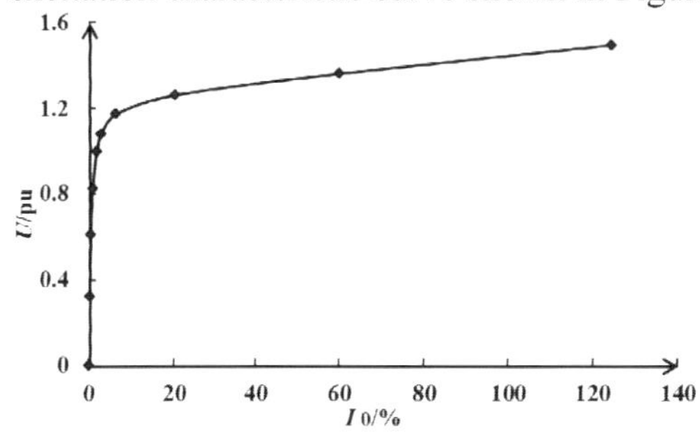

Figure 3. The UMEC excitation characteristic curve in EMTDC

2.2.2. Model verification According to the actual parameters of the TL500kV main transformer provided by the manufacturer, the default parameters of the UMEC model are modified. The modified excitation characteristic curve of the UMEC model is shown in Figure 4. The comparison results are shown in Table 2. It can be seen from Table 2 that the modified UMEC model has a high agreement with the actual model, indicating that the UMEC model can be used for the no-load closing research. .

Table2. Comparison of UMEC model and actual transformer excitation characteristics

Excitation characteristic curve	No-load current					
	0	5%	10%	15%	20%	30%
Default model	0	1.10	1.15	1.18	1.19	1.21
Actual model	0	1.05	1.09	1.11	1.13	1.18

2.3. Substation system model

For the TL500kV transformer no-load closing circuit shown in Figure 1, the simulation takes the CVT detailed model into account. The model parameters are consistent with the actual parameters, that is, the CVT main capacitor is 5000pF, and the 500kV circuit breaker is a double-break circuit breaker, each breaker has a parallel voltage capacitor, which is 2000pF and the system voltage is 525kV.

3. Simulation Analysis of Magnetizing Inrush Current

The simulation system is closed at 0.1s and lasts 0.5s. The following text changes of the CVT, the parallel capacitor of the breaker and the inflection point of the excitation characteristic curve of the transformer to simulate the inrush current. Take the phase A as an example.

3.1. Influence of CVT main capacitor value on magnetizing inrush current

In the transformer closing circuit, the CVT's original capacitor is 5000pF, now it is changed from 1000pF to 15000pF to simulate the magnetizing inrush current. The peak value of this current under different CVT is shown in Figure 4.

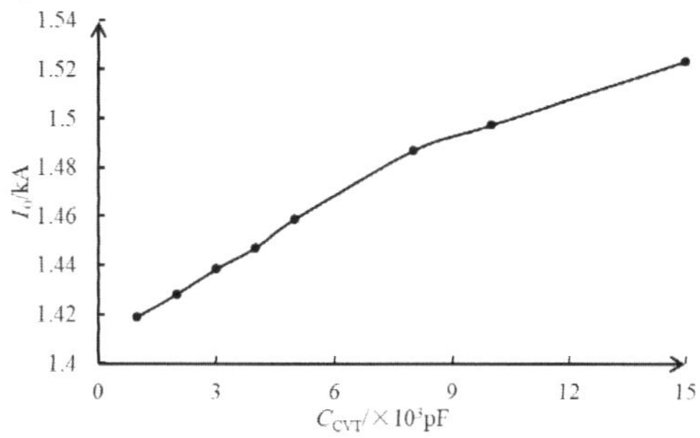

Figure 4. Peak value of magnetizing inrush current under different C_{CVT}

The simulation results show that increasing the capacitance of CVT may lead to an increase in the peak value of the magnetizing inrush current, but the increase is negligible. When C_{CVT}>6000pF, continue to increase the CVT main capacitor, which has no significant effect on the peak value of the magnetizing inrush current.

3.2. Influence of parallel capacitance of breaker on magnetizing inrush current

The parallel capacitor value of the circuit breaker was originally 2000pF. Now it is changed from 500pF to 15000pF to simulate the magnetizing inrush current. The main inrush current waveform is shown in Figure 5 and the peak value of the magnetizing inrush current under different capacitances is shown in Table 3.

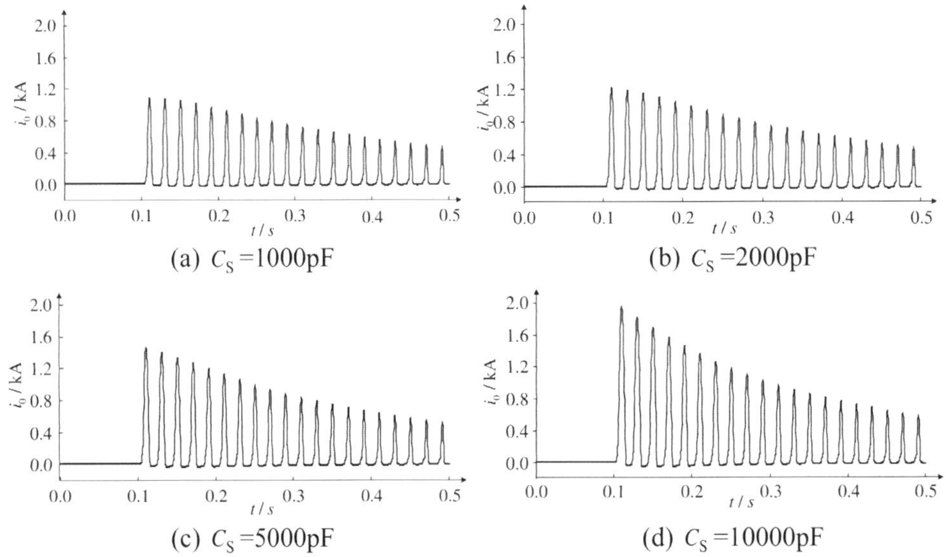

(a) C_S =1000pF (b) C_S =2000pF

(c) C_S =5000pF (d) C_S =10000pF

Figure 5. Simulation waveform of magnetizing inrush current under different C_S

Table3. Magnetizing inrush current peak under different C_S

C_S /pF	Peak value of magnetizing inrush current /kA
500	1.0466
1000	1.0849
1500	1.1236
2000	1.1640
3000	1.2549
5000	1.4587
8000	1.7663
10000	1.9596

The above results show that increasing the parallel equalizing capacitance of the circuit breaker, the peak value of the inrush current will increase rapidly and increase approximately linearly.

3.3. *Influence of the inflection point of the excitation characteristic curve*

The inflection point voltage of the transformer excitation characteristic curve is 1.04 (standard value, the same below). The transformer inflection point voltage is gradually increased from 1.04 to 1.15 at intervals of 0.01, and the inrush current peaks at different inflection point are shown in Figure 6.

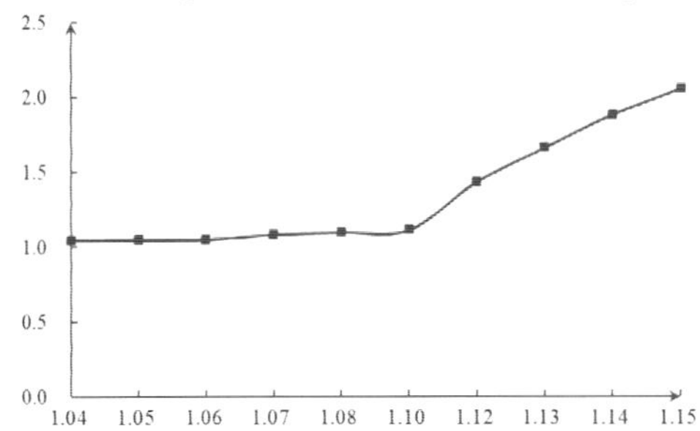

Figure 6. Peak Magnetizing inrush current under different inflection point voltage

The simulation result shows that after the inflection point increasing to 1.06, the peak value of inrush current does not change significantly; when the voltage is continuously increased to 1.07, the peak value of the magnetizing inrush current decreases slightly. When the inflection voltage is raised to 1.10 and higher, the peak value decreases, that is, increase the excitation point voltage means that the saturation condition is hard to satisfy. If the saturation condition is not reached, the excitation current increases linearly with the applied voltage, which does not form magnetizing inrush current. Therefore, as the inflection point voltage increases, the magnetizing inrush current is hard to form.

4. Inhibition measures for magnetizing inrush current

According to the theory and simulation analysis about influence factors of magnetizing inrush current above, the pre-control measures are proposed as followed.

a. Decreasing the value of CVT's capacitor appropriately can reduce the peak value of magnetizing inrush current, but when the value is too large, the inrush current does not change significantly.

b. Reducing the value of parallel capacitor of the circuit breaker can effectively suppress the occurrence of magnetizing inrush current. After reducing the circuit breaker capacitance to 500pF, there is almost no magnetizing inrush current occurring.

c. Increase the saturation voltage of the core. By increasing the inflection point of the excitation characteristic curve, the core is not saturated or just enters the saturation stage when the transformer is closed. The peak value of the magnetizing inrush current is greatly reduced.

5. Conclusion

This paper is aimed at the phenomenon of magnetizing inrush current in Chongqing TL500kV transformer, The influence of the parallel capacitance value of 500kV breaker circuit breaker, the CVT main capacitor value and the inflection point voltage of the transformer excitation characteristic on the peak value of the inrush current are analysed.

Based on the influencing factors above, the measures to suppress the inrush current are proposed, such as reducing the main capacitance of the CVT, reducing the parallel capacitance of the breaker, and changing excitation characteristics, can effectively avoid the occurrence of the inrush current.

References

[1] HAO Zhiguo, ZHANG Baohui, Chu Yunlong, Research on suppression technology of transformer inrush current excitation inrush[J]. High Voltage Apparatus, 2005(02): 81-84.

[2] LI Zhenqiang, GU Dingxie, DAI Min. Analysis and suppression method of resonant overvoltage and magnetizing inrush current for UHV no-load transformers[J].High Voltage Engineering, 2012, 38 (02):387-392.

[3] RAO yufei, ZHANG penghui, LI chenghao. Influence mechanism and evaluation method of excitation inrush current on phase conversion failure of HVDC transmission system [J]. Power System Protection and Control.2019,47(13):54-61

[4] LI Xiaohua,ZHANG Dongyi,WU Lizhu.Analysis of the special characteristics of the inrush current of converter transformer[J].Powe Grid Technology,,2017,41(12):3869-3875.

[5] CHEN chuanjiang, FANG chunen, ZENG junlong. Study on phase selection and closing of no-load transformers with residual magnetism [J]. Power System Protection and Control, 2018,46(16):82-88.

[6] WANG Yan,HUANG Fucheng,ZHAO Shuzhen.Simulation Analysis of Transformer Magnetizing Inrush Current Considering Ferromagnetic Hysteresis[J].Automation of Electric Power Systems,2009,33(15):78-83.

[7] YANG Tongyun， LI Xiaohua， DAI Yangyu. Analysis on Magnetizing Inrush Current Characteristic of Converter Transformer and Its Suppression[J]. Power grids and clean energy,2017,33(01):64-70+78.

[8] CHANG Yong, Analysis on Energizing Inrush Current of No-Load Converter Transformer in 500 kV Gaoling Back-to-Back Converter Station [J].Power System Technology,

IOP Conf. Series: Materials Science and Engineering **853** (2020) 012028 doi:10.1088/1757-899X/853/1/012028

Study on the transformer protection scheme based on double thresholds and waveform similarity

Yu Liang[1] and Fan Yang[2] and Dan Shi[3]

[1]State Grid Hubei Ezhou Electric Power Supply Company, Ezhou Hubei, 436001, China
[2]Maintenance Branch Company of State Grid Fujian Electric Power Co. Ltd., Fuzhou Fujian, 350000, China
[3]State Grid Hubei Xiaogan Electric Power Supply Company, Xiaogan Hubei, 432000, China

935866302@qq.com

Abstract. As the key device of power transmission, the power transformer has higher requirements for the protection sensitivity and reliability. However, the operation time of the differential protection is limit to the data window of the Fourier transform algorithm. In addition, the differential protection cannot recognize the slight transformer turn-to-turn fault. Moreover, it may maloperate when the current transformer (CT) is saturated. To solve above problems, a novel transformer protection scheme based on double thresholds and waveform similarity is proposed. By comparing the current waveform similarity with the double thresholds, the external faults, internal faults and CT saturation scenarios can be effectively identified, while the waveform similarity is obtained by calculating the Hausdorff distance of the current at each side of the transformer. The PSCAD/EMTDC based simulation results show that the proposed protection scheme can operate quickly and the operation time is as low as 5.5ms. In addition, it has high sensitivity and can effectively identify 0.5% or higher turn-to-turn faults, and it will not be affected by CT saturation.

1. Introduction

Transformers play an important role in power transmission and the interconnection of regional power grids. At present, differential protection is the most widely used transformer main protection, which can identify phase-to-phase, phase-to-ground faults and severe turn-to-turn faults [1-2]. The differential protection scheme generally uses a Fourier transform algorithm to extract the fundamental component. To meet the high-precision calculation requirements, the data window length of the Fourier transform algorithm generally takes at least half of the cycle. In addition, CT saturation has been a perennial problem for the differential protection. The fake differential current will emerge when the external fault with CT saturation occurs and the differential protection may maloperate [3-4]. It should be noted that the differential protection is difficult to identify the transformer slight turn-to-turn fault (the number of short circuit turns is less than 3%) [5-6].

Because of its simple principle and high sensitivity to the difference between waveforms, the Hausdorff algorithm has been initially applied in the field of relay protection [7-9]. On the basis of this, this paper further studies the variation rule of Hausdorff distances of the current at each side of the transformer in different scenes such as transformer slight turn-to-turn fault and CT saturation. Then a fast transformer protection scheme based on Hausdorff distance and double thresholds is proposed. Theoretical analysis and simulation tests verify the effectiveness of the proposed protection scheme.

2. The basic protection principle

2.1. Basic principle of waveform similarity recognition based on Hausdorff distance

Assume there are two sets of finite points as $A=\{a_1,a_2,\cdots,a_i,\cdots,a_p\}$ and $B=\{b_1,b_2,\cdots,b_j,\cdots,b_q\}$, the Hausdorff distance between the two sets of points is defined as (1).

$$H(A,B)=\max\left(h(A,B),h(B,A)\right) \tag{1}$$

$$h(A,B)=\max_{a_i\in A}\min_{b_j\in B}\|a-b\| \tag{2}$$

$$h(B,A)=\max_{b_j\in B}\min_{a_i\in A}\|a-b\| \tag{3}$$

Where $\|g\|$ represents the Euclidean distance between two points, and $h(A,B)$ represents the one-way Hausdorff distance from A to B (the same as $h(B,A)$). Equation (4) shows the definition of $h(A,B)$, which is described as follows: for each point in set A, the nearest point can be found in set B.

$$\|a_i-b_j\|<\|a_i-b_k\|(1\le k\le q\text{且}k\ne j) \tag{4}$$

The two way Hausdorff distance called $H(A,B)$ is the larger one between $h(A,B)$ and $h(B,A)$, which is used to measure the maximum dissimilarity between two point sets. The larger Hausdorff distance between the two waveforms is, the greater difference between them is, and vice versa.

2.2. The characteristics of Hausdorff distance of current at each side of transformer under different conditions

2.2.1. Transformer normal operation or external fault without CT saturation. Figure 1 shows the schematic diagram of the currents collected by CT of transformer protection. Considering the different connection modes at each side of the transformer, the measured current on the star side should be transformed in the microcomputer protection device as (5) before comparing the waveform similarity of the currents at both sides:

$$\begin{aligned}\dot{I}_{yA}&=-N(\dot{I}_{yA}-\dot{I}_{yB})\\ \dot{I}_{yB}&=-N(\dot{I}_{yB}-\dot{I}_{yC})\\ \dot{I}_{yC}&=-N(\dot{I}_{yC}-\dot{I}_{yA})\end{aligned} \tag{5}$$

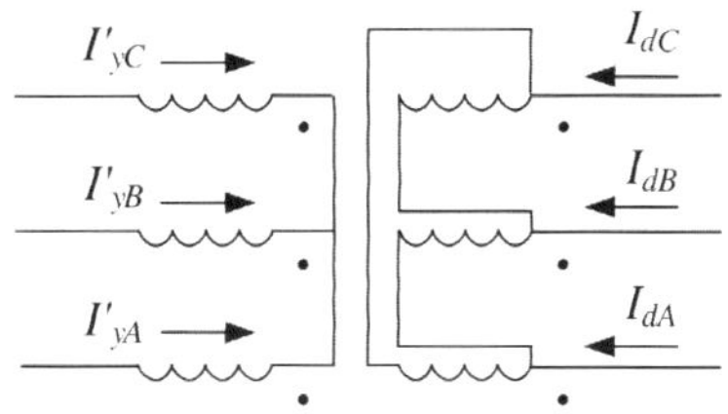

Figure 1. Currents on both sides of the transformer.

According to Kirchhoff's current law, the transformed currents satisfy (6).

$$i_{yA}=i_{dA},i_{yB}=i_{dB},i_{yC}=i_{dC} \tag{6}$$

It can be obtained that transformed currents of the star side $\dot{I}_{yA}$, $\dot{I}_{yB}$ and $\dot{I}_{yC}$ are exactly the same with the triangle side current $\dot{I}_{dA}$, $\dot{I}_{dB}$ and $\dot{I}_{dC}$ respectively when the transformer is under normal operation or external fault. Therefore, the corresponding Hausdorff distance is zero.

2.2.2. Transformer internal fault. When an internal fault occurs at the transformer, the original three-phase symmetrical state of the transformer will be broken. Then, the star side currents obtained from (5) and triangle side currents will no longer meet (6).

2.2.3. Transformer external fault with CT saturation. When a transformer external fault occurs, the involved CT may be saturated due to the influence of DC component and CT parameters. Then the CT works alternately in saturated and unsaturated state, and there is always a period of time for CT to work in the linear transmission area. Considering that the length of the data window of Hausdorff distance algorithm can be adjusted to a small one, the two waveforms in the window are almost the same as long as the data window is within the linear transformation area, and the Hausdorff distance will be near zero. The above characteristics can be used to identify the external faults with CT saturation.

3. The novel transformer protection scheme based on double thresholds and waveform similarity

According to the analysis in the section 2, the Hausdorff distances under above three conditions can be correspondingly divided into the following three categories when a short data window is selected (1/8 cycle (2.5ms) is selected in this paper):1)the waveforms on both sides of the transformer are the same, then the Hausdorff distance is near 0 for a long time; 2)the waveforms on both sides of the transformer are almost opposite, then the Hausdorff distance is at a high level for a long time; 3)there exists a certain time when Hausdorff distance drops to around 0 in one cycle. Figure 2 shows the diagram of three types of Haudorff distances discussed above.

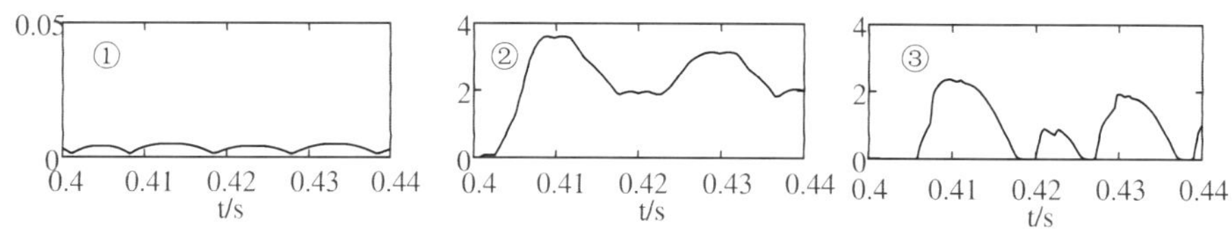

Figure 2. Diagram of three types of Hausdorff distances.

According to the above analysis, the setting method of double thresholds is designed as follows:

1) Low threshold H_{set1}

The low threshold H_{set1} is set as the starting threshold. When the Hausdorff distance is larger than the threshold H_{set1}, it is determined that there is an internal fault or external fault with CT saturation occurs. The setting principle of H_{set1} shall be based on the maximum value (denoted by H_{max1}) of Hausdorff distance in the cases of transformer normal operation and external fault with CT unsaturation. That is to say, $H_{set1}=H_{max1}*K_{rel1}$, where K_{rel1} represents reliability and is set as 1.15~1.3.

2) High threshold H_{set2}

The high threshold H_{set2} is used for further identification between internal fault and external fault with CT saturation. Its setting principle is based on the maximum Hausdorff distance under the condition of transformer external fault with CT saturation, and the reliability coefficient K_{rel2} shall be considered. That is to say, $H_{set2}=H_{max2}*K_{rel2}$. Once the Hausdorff distance is larger than the threshold value, it is determined that a serious internal fault occurs, and the circuit breaker shall be tripped immediately. When the Hausdorff distance is between the above two threshold values, it is judged that slight internal fault or external fault with CT saturation occurs. At this time, move the data window forward in single sample point and continue to calculate the Hausdorff distance for one cycle. When the Hausdorff distance falls below the low threshold value during this period, it is judged an external fault with CT saturation occurs, and protection criterion shall be blocked. If the Hausdorff distance is always between two threshold values within one cycle, it is judged a slight internal fault occurs and the circuit breaker will be tripped.

4. Case study

Figure 3 shows the simulation model built in PSACD/ EMTDC. The transformer parameters are as follows: the ratio is 220/110 kV and the capacity is 100 MVA. The leakage reactance is 0.2 p.u and copper loss is 0.02 p.u. The power voltage is 220 kV and the equivalent system impedance is $Z = 27.5\angle 85^{\circ}\,\Omega$. The sampling frequency is 4 kHz. The high and low thresholds are set as 2.5 and 0.05, respectively.

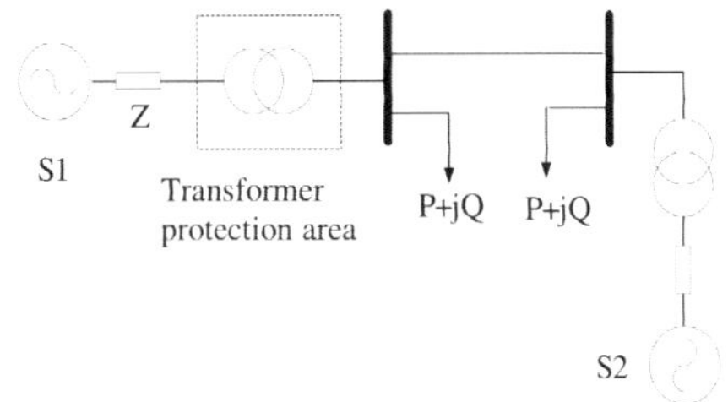

Figure 3. 220kV/110kV transformer model.

4.1. Slight turn-to-turn fault

Assume a slight 0.5% turn-to-turn fault occurs at the phase A winding at 0.4s. Under this condition, phase A current waveform on both sides of the transformer and corresponding Hausdorff distance is shown in Figure 4. The simulation results show that the maximum distance of Hausdorff is 0.16, which is far below the high threshold value of 2.5, so the protection device will continue to judge in one cycle. During this period, the Hausdorff distance is always higher than the low threshold value of 0.05, so the protection scheme determine that transformer internal fault occurs. Then the protection will operate at 0.42s.

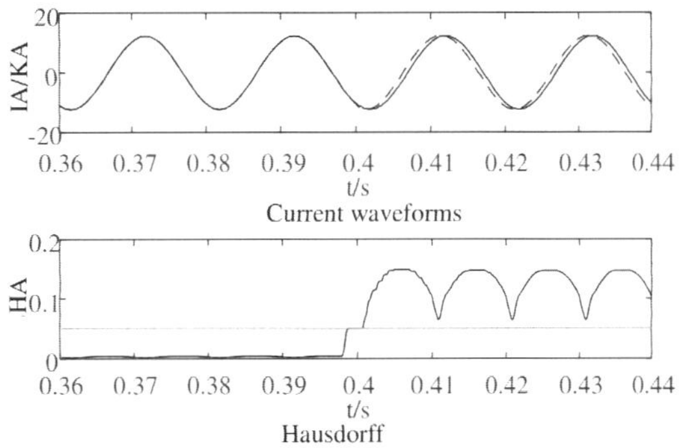

Figure 4. Phase A current waveforms and Hausdorff distance when slight turn-to-turn fault occurs.

If differential protection is adopted, the differential current is too small to exceed the threshold and cannot identify the fault.

4.2. External fault with CT saturation

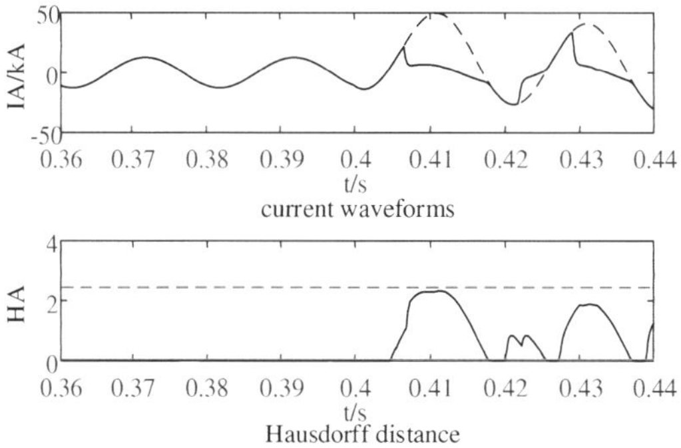

Figure 5. Phase A current waveforms and Hausdorff distance when external fault with CT saturation occurs

It is assumed that a phase A grounded fault occurs at the first end of the transmission line at 0.4s and the CT is saturated. Figure 5 shows the phase A current waveforms and Hausdorff distance. It can be seen that CT starts to saturate at 4ms after the occurrence of the fault and the Hausdorff distance increases to the low threshold value of 0.05. In the following one cycle after the occurrence of the fault, the Hausdorff distance is always lower than the high threshold value of 2.5. The Hausdorff distance drops to the low threshold value at 0.417s. Thus, the protection scheme identifies that a transformer external fault with CT saturation occurs and does not issue a trip command.

5. Conclusion

To solve the problem that the traditional transformer differential protection cannot deal with slight turn-to-turn faults and CT saturation, a novel transformer protection scheme based on double thresholds and Hausdorff distance is proposed. The validity and superiority of the proposed scheme are verified by PSCAD/EMTDC based simulation tests. The basic conclusions are as follows:

1. The proposed scheme has a strong CT saturation tolerance ability, and its operation speed is faster than that of the traditional differential protection in identifying serious transformer internal faults;
2. The proposed scheme has a strong ability to response to turn-to-turn fault. It can identify the slight turn-to-turn faults whose short circuit turn ratio is as low as 0.5% within one week after the occurrence of the fault.

References

[1] D. Zacharias and R. Gokaraju, "Prototype of a Negative-Sequence Turn-to-Turn Fault Detection Scheme for Transformers," in IEEE Transactions on Power Delivery, vol. 31, no. 1, pp. 122-129, Feb. 2016.

[2] Jun Guan, Zhiguo Hao, Wenzhe Chen, Zhiyuan Liu and Xiaojun Yu, "Novel power transformer differential protection scheme based on improved short-window algorithm," 2016 IEEE PES Asia-Pacific Power and Energy Engineering Conference (APPEEC), Xi'an, 2016, pp. 830-834.

[3] T. Zheng, T. Huang, Y. Ma, Z. Zhang and L. Liu, "Histogram-Based Method to Avoid Maloperation of Transformer Differential Protection Due to Current-Transformer Saturation Under External Faults," in IEEE Transactions on Power Delivery, vol. 33, no. 2, pp. 610-619, April 2018.

[4] F. Costa and R. Prado, "A Wavelet-Based Transformer Differential Protection With Differential Current Transformer Saturation and Cross-Country Fault Detection," 2018 IEEE Power & Energy Society General Meeting (PESGM), Portland, OR, 2018, pp. 1-1.

[5] L. M. R. Oliveira and A. J. M. Cardoso, "Comparing Power Transformer Turn-to-Turn Faults Protection Methods: Negative Sequence Component Versus Space-Vector Algorithms," in IEEE Transactions on Industry Applications, vol. 53, no. 3, pp. 2817-2825, May-June

2017.
[6] O. Ozgonenel and S. Karagol, "Maximum overlap discrete wavelet based transformer differential protection," 2017 25th Signal Processing and Communications Applications Conference (SIU), Antalya, 2017, pp. 1-4.
[7] A. Wang, X. Sun, X. Zhou, and W. Hu, "Difference squared hausdorff distance based medical image registration," in Proc. Chin. Control Decis. Conf., Mianyang, China, 2011, pp. 4270–4272.
[8] L. Chen et al., "Similarity Comparison Based High-Speed Pilot Protection for Transmission Line," in IEEE Transactions on Power Delivery, vol. 33, no. 2, pp. 938-948, April 2018.
[9] ZHAO Hang, LIN Xiangning, YU Kun, et al. A High-speed Protection Scheme for HVDC Transmission Line Based on Hausdorff Distance Comparison[J]. Proceedings of the CSEE, 2017.

Research on Dynamic Relationship between Urban Rail Transit and Power Consumption

Liu Zhe[1], Du Fengqing[1] and Liu Xu[2*]

[1]State Grid Shanghai Electric Power Research Institute, Shanghai, Shanghai, China
[2] North China Electric Power University, Baoding, Hebei, China

527532129@qq.com

Abstract. Aiming at the problem of insufficient feature quantity and lack of correlation between feature quantity and load, based on time series analysis method, ARIMNA model is used to fit the load forecasting problem. The study found that the urban orbit data has a good autocorrelation. Compared to the Linear Regression method, when the ARIMA model is used, the algorithm can predict the trend of urban rail transit load data better.

1. Introduction

Forecasting is the premise of decision making and provides guidance for planning and scheduling [1]. In the field of power systems, one of the important tasks of the power sector and the dispatching department is to make accurate predictions of the load. In general, accurate prediction of medium and long-term load [2] is conducive to long-term planning of power systems, and fuel purchase plans can be formulated in advance. Similarly, short-term load forecasting can be a guide for short-term power generation planning.

Load forecasting can be generally classified into long-term forecasts, medium-term forecast, short-term forecast for forecasting daily load, and ultra-short-term forecast. At present, the development of the power industry tends to be stable. Foreign research's hotspots mainly focus on short-term load forecasting [3-4]. In China, industrial adjustment has a great impact on power demand [5]. Therefore, the study of load forecasting is a combination of long-term and short-term.

Since the development of load forecasting, many scholars at home and abroad have proposed many prediction methods based on their own research [6]. In literature [7], the support vector machine method is used to perform regression analysis on the load data and the grid search method is used to find the optimal parameters to predict the load data. Another way to predict the load is to combine multiple prediction methods. In [8], Markov chain was used to select two prediction models that satisfy the accuracy from several models. Using this method to filter the model can improve the prediction accuracy. Since the development of artificial intelligence, neural networks have also begun to be used in load forecasting. In [9], the ridge regression-BP neural network was used to predict the load on the premise of using the ridge map to screen the indicators and achieve high precision. With the development of big data, the research hotspot of load forecasting is transformed into parallel processing of data using related technologies of big data [10]. The study found that when the amount of data increases, the prediction accuracy is also improved [11- 12].

The above research has laid the foundation for the application of data science in power systems, as well as the relationship between load data and factors such as region and climate. However, the above research does not reveal the trend of the load in time. When faced with the problem of insufficient feature quantity and insufficient correlation between feature quantity and load, the prediction is not very accurate. The study found that the seasonal characteristics of the load data are obvious and have a linear relationship with time. Based on this, the time series analysis method is used to analyze the load. The energy consumption data of Shanghai Line 6 was selected as the research object, and the time series model was established. The study found that the energy consumption data has a high degree of autocorrelation, and the fitted curve is high when performing load forecasting.

2. Sequence Prediction Model

The sequence method divides the continuous load into a quantifiable sequence, and then uses the sequence research method to predict the value of the next segment by fitting the previous sequences. The load forecasting abstract expression of the sequence model is shown in equation (1):

$$y = \mathrm{f}(\mathrm{S}, \mathrm{X}, t) \tag{1}$$

In the load model, X represents the m factors that affect the load, X recorded as $\mathrm{X} = [x_1, x_2, \cdots, x_m]^T$; t represents the time serial number; S is a coefficient matrix, sort by the order of t from small to large, when the load model is a first-order equation $y = a + bt, \mathrm{S} = [a, b]^T$.

The source of data in load forecasting has two aspects, one is the influencing factor data obtained from the economic and meteorological departments, and the other is the load data obtained from the power sector. The collected data is divided into a training data set and a test data set. The composition of the training sample is the value of the influencing factor y_t and the magnitude of the load obtained over the known time period $t(n + 1 \le t \le N)$. The composition of the predicted sample is the value of the influencing factors $\mathrm{X}_t = [x_{1t+1}, x_{2t+2}, \cdots, x_{mN}]^T$ obtained in the future time period $t(n + 1 \le t \le N)$.The coefficient matrix S is estimated by the above data and the value of $y_t(n + 1 \le t \le N)$ is determined.

After obtaining the parameter estimation value $\hat{\boldsymbol{S}} = [\hat{s}_1, \hat{s}_2, \cdots, \hat{s}_k]^T$, the load model can be fitted or predicted. Substituting $\hat{\mathrm{S}}$ into equation (1) can be obtained:

$$\hat{y}_t = \mathrm{f}(\hat{\mathrm{S}}, \mathrm{X}_t, t), t = 1, 2, \cdots, n, \cdots, N \tag{2}$$

The fitted residual is:

$$u_t = y_t - \hat{y}_t, t = 1, 2, \cdots, n \tag{3}$$

The goal of model optimization is to minimize the value of $\mathrm{u_t}$ by adjusting the value of $\hat{\mathbf{S}}$. There are many methods for parameter optimization. The most classic methods are least squares method and intelligent algorithms represented by ant colony algorithm and particle swarm algorithm.

3. Time Series

A time series is a type of sequence that refers to a sequence of numbers arranged in chronological order. Due to the existence of fluctuations and error terms, the changes are widely present in the time series. Time series is especially good at dealing with changes in trends and periodic composition. Time series analysis is the process of regression analysis and prediction of the obtained data. Commonly used methods are autoregressive (AR), autoregressive moving average (ARMA), and vector autoregressive model (VAR).

From the idea of time series, the predicted value $\mathrm{y_t}$ can be written as a linear multiplication of the constant coefficient with the known value and added to the offset. Define the p-order AR model as shown in equation (4):

$$y_t = \varphi_1 y_{t-1} + \varphi_2 y_{t-2} + \cdots + \varphi_p y_{t-p} + a_t \tag{4}$$

Where p represents a p-order function, $\varphi_i, i = 1, 2, \cdots, p$ represents a constant coefficient, and the offset a_t is a value of white noise at a predicted time t.

The MA model is to study the relationship between predicted values and noise sequences. The q-order MA model is defined as shown in equation (5):

$$y_t = a_t - \theta_1 a_{t-1} - \cdots - \theta_q a_{t-q} \tag{5}$$

Where q is a q-order function, $\theta_i, i = 1, 2, \cdots, p$ represents a constant coefficient, offset $a_{t-i}, i = 0, 1, \cdots, p$ is the white noise at the prediction time ti value.

Combining the AR model with the MA model, a more general model ARMA model is obtained, as shown in equation (6):

$$y_t - \varphi_1 y_{t-1} - \varphi_2 y_{t-2} - \cdots - \varphi_p y_{t-p} = a_t - \theta_1 a_{t-1} - \cdots - \theta_q a_{t-q} \tag{6}$$

The ARMA model is applied to a smooth time series. In the load forecasting, the load data shows the non-stationary period characteristics. In order to meet the conditions of the ARMA model, a differential operation can be added before the model fitting to remove the periodicity and non-stationarity.

The steps of time series analysis are shown in Figure 1.

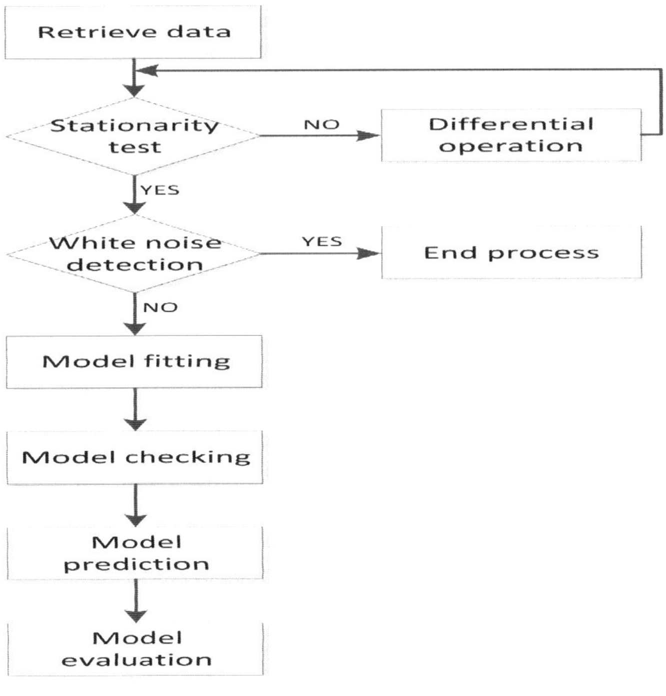

Figure 1. Time Series Analysis Steps

As can be seen from Figure 1. Firstly, the energy consumption data of the subway operation is acquired, and the missing value and the error value are modified. Then, the obtained data is subjected to stationarity detection. If the data is not stable, a difference operation is performed, and the data for performing the difference operation is again subjected to stationarity detection, and this step is repeated until the data is smooth. White noise detection is then performed. Model fitting is performed on non-white noise data, and the commonly used fitting model in time series analysis is ARIMA. Finally, data prediction and evaluation of the model are performed.

4. Examples

As an important livelihood project of urban infrastructure construction, urban rail transit reflects the strategic positioning and competitiveness of urban development. Taking Shanghai Metro as the research object, analyzing the electricity consumption model of urban rail transit and forecasting demand, it can provide methods and models for rail transit power consumption forecast of new lines in Shanghai. The Shanghai Metro was built in 1990. By December 2018, there were 16 subway lines in Shanghai, with a total of 415 stations with a total mileage of 705 kilometers.

This section combines passenger flow and annual operating mileage data to analyze the changes in energy consumption of Line 6. The specific data is shown in Table 1:

Table 1. Shanghai Rail Transit - Energy Consumption of Line 6 for Nearly 7 Years

Years	Passenger Flow (10,000 people)	Annual Operating mileage (10,000 kilometers)	Electricity Consumption (10,000 degrees)
2011	8711.613	1383.411	6853.1
2012	9568.014	1431.229	7145
2013	10622.05	1603.415	8080.8
2014	11598.74	1934.383	11441.2
2015	12970.33	2020.484	12008
2016	14126.9	2045.305	14370
2017	14683.95	2033.448	14040

In order to make the data image smoother, the First-order difference operation is performed on the original data. The calculated data is shown in Figure 2.

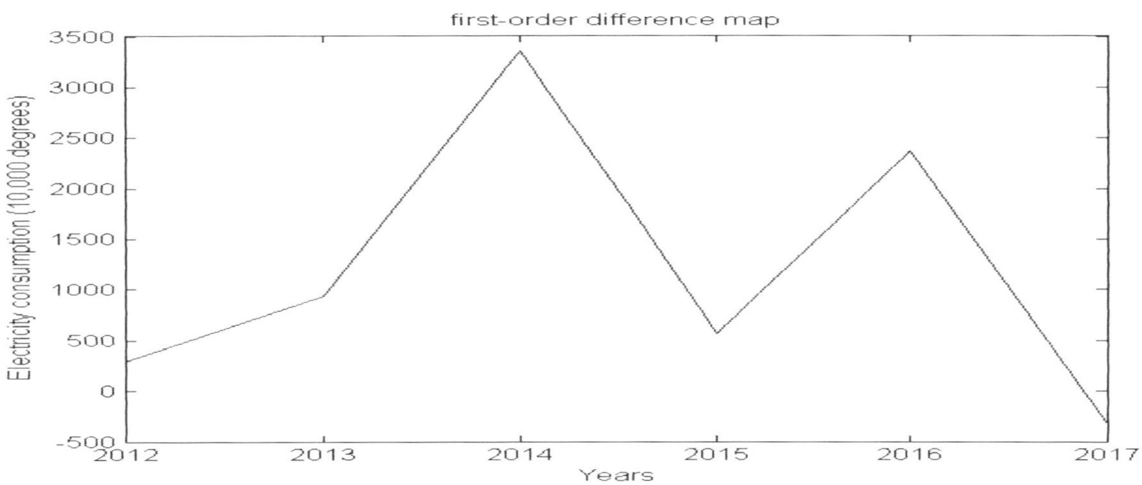

Figure 2. First-order difference map

White noise detection is performed on the differential data. The results are shown in Table 2.

Table 2. White noise monitoring results

Significant Level	*P* Value
1.82327468	0.17692398

Observed in Table 2, the P value is significantly smaller than the significance level, and the data is non-white noise.

The above model is used to predict the data and predict the data from 2015 to 2017. The results are shown in Figure 3.

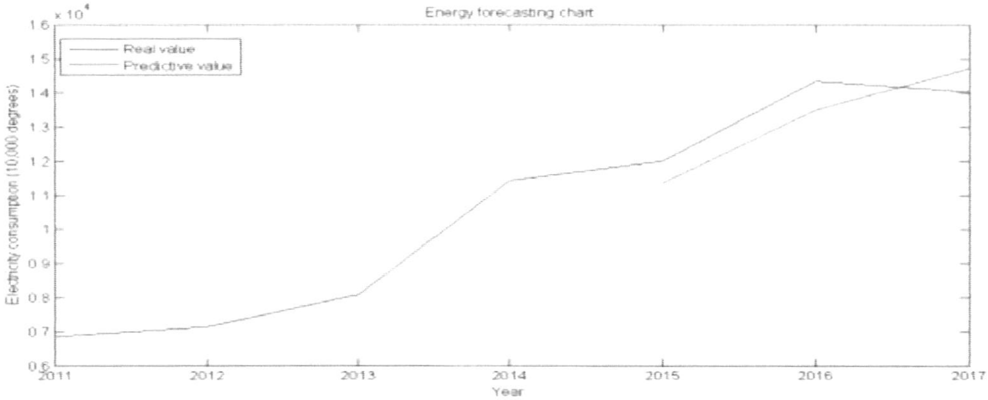

Figure 3. Energy forecasting chart

The blue line represents the raw data and the red line represents the predicted value. Errors are introduced to analyze the prediction results. The values are shown in Table 3.

Table 3. Real and predicted values

Year	Real value	Linear Regression Model Predictive Value	ARIMA Model Predictive Value
2015	12008	11390	11358
2016	14370	12816	13508
2017	14040	14242	14732

It can be seen from the calculation that the error value of the load prediction by the linear regression model is 2837644, and the error value of the load prediction by the ARIMA Model is 1644408. The accuracy of load forecasting with time series is higher.

5. Conclusion

This paper takes Shanghai Metro data as the research object and time series as the analysis model to predict the energy consumption data. The study found that the urban orbit data has a good autocorrelation. Compared to the Linear Regression method, when the ARIMA model is used , the algorithm can predict the trend of urban rail transit load data more better.

Acknowledgments

Project Supported by State Grid Shanghai Electric Power Research Institute.

References

[1] Kang Chongqing, Xia Qing, Zhang Boming.Review of Power System Load Forecasting Research and Discussion on Its Development Direction.Automation of Electric Power Systems, 2004, (17):1-11.
[2] Zhang Zhi, Du Yanling, Cui Huijun, Wang Yang, He Zhe, Lai Xiaowen. Intelligent mid-long electricity load forecast method considering associated factors . Power System Protection and Control, 2019, 47(02): 24-30.
[3] Barakat E H,Al-rashed S A. long Range Peak Dernand Foreeasting Under Conditions of High Growth.IEEE Transon Power Systems.1992,7(4):1483-1486.
[4] Kandil M S, El-Debeiky S M, Hasanien N E.Overview and Comparison of Long-term Foreeasting Teehniques for a Fast Developing Utility: Part1.Eleetric Power Systems Researeh,2001,58(1):11-17.
[5] Yang Fangyuan,Shi Yuchao,Hou Yucheng.Correlation Analysis of Electricity Consumption Index and Load Forecasting Considering Adjustment of Industrial Structures. Electrical

Engineering,2017,(05):19-23+29.
[6] Zhou Chao, Xing Wenyang, Li Yulong. A Survey of Power System Load Forecasting Methods. Journal of Power Supply,2012(06):32-39.
[7] Wang Ning,Xie Min ,Deng Jialiang, Liu Mingbo, Li Jialong, Wang Yi, Liu Sijie.Mid-long term temperature-lowering load forecasting based on combination of support vector machine and multiple regression. Power System Protection and Control,2016,44(03):92-97.
[8] Zhang Dongliang, Yan Jian, Li Xiaobo, Ren Xiaoda, Zhang Jinzhong, Zhang Fulai.Mid-long term load forecasting based on Markov chain screening combination forecasting models. Power System Protection and Control,2016,44(12):63-67.
[9] Wen Ruiying，Wang Hongyong. A Forecasting Method of Controller's Workload Based on Ridge Regression—BP Neural Network. Journal of Transportation Systems Engineering and Information Technology,2015,15(01):123-129.
[10] Wang Dewen, Sun Zhiwei. Big Data Analysis and Parallel Load Forecasting of Electric Power User Side. Proceedings of the CSEE,2015,35(03):527-537.
[11] Lei Zhengxin, Han Bei, Nie Meng, Zhen Ying, Wang Keyou, Li Guojie. Research and application of multi-node load forecasting algorithm under the environment of distribution network's big data. Power System Protection and Control,2016,44(23):68-78.
[12] Xu Yuan, Cheng Qianshan，LI Yang，Zhang Hao，Yu Wei，He Bing. Mid - long Term Load Forecasting of Power System Based on Big Data Clustering. Proceedings of the CSU - EPSA, 2017, 29(08):43-48.

A Non-intrusive home load identification method based on adaptive reinforcement learning algorithm

Haoyang Li

School of Electrical and Electronic Engineering, North China Electric Power University, Beijing , China

1586934450@qq.com

Abstract. At present, the load data collection of residential users mainly starts from the lower acquisition frequency, so the non-intrusive home load identification method based on low frequency sampling has attracted wide attention. However, low-frequency sampling has a low recognition accuracy when the training data set is small. Therefore, a non-intrusive home load identification method based on adaptive KNN reinforcement learning algorithm is proposed. The method firstly analyzes the state of the electrical appliance by KNN to obtain the initial HMM model, and then solves the HMM model by adaptive KNN reinforcement learning algorithm to obtain the optimal state transition strategy. This method reduces the model pair data. The dependence improves the recognition accuracy of the model and the adaptability to new data. Finally, the experimental verification is carried out by the low frequency data set AMPds. The results show that the method improves the state recognition accuracy of the electrical appliance and enhances the adaptability of the algorithm to new data.

1. Introduction

With the development of smart grids, power systems are increasingly demanding data interaction and information mining. Non-intrusive load monitoring can provide real-time and accurate user power load status information, which is essential for power system planning, load forecasting and market regulation, and has broad development prospects in the field of load monitoring and energy analysis [1]. The non-intrusive load monitoring system provides key data support for the two-way interaction between the power grid and the user, deep mining of user information, and analysis of household energy consumption by mining user load data [2-3]. The main idea of non-intrusive load monitoring is to collect and analyze information such as voltage, current or power through sensors installed at the power supply inlet, extract corresponding electrical characteristics, and then analyze the operating state of the user's internal load [4-5].

Non-intrusive load monitoring has low economic cost and strong practicability, so it has attracted widespread attention in the field of smart grid, and many scholars have studied the NILM method. Wu et al. [6] first constructed the de-mixing matrix through the whitening total current signal, and then proposed a non-intrusive home load identification method based on the negative entropy maximization criterion. Finally, the recognition results were analyzed by the constructed evaluation function. Sun et al. [7] proposed a non-intrusive home load identification method based on dynamic adaptive particle swarm optimization algorithm. The load characteristics of power feature and total harmonic distortion coefficient are used as the objective function of dynamic adaptive particle swarm optimization algorithm. Breschi et al. [8] first designed a set of jump models to describe the consumption behavior of each appliance, and then proposed a load identification method based on the dynamic programming filtering algorithm. Qi et al. [9] takes the steady-state current and the steady-state voltage as load characteristics, and then combines the local average decomposition and model matching to identify the internal electrical state of the user. Lin et al. [10] proposed a non-intrusive load identification method based on quadratic programming in order to improve the accuracy of recognition.

At present, most of the non-intrusive load monitoring research is based on high-frequency sampling, but high-frequency sampling has high hardware requirements and high economic costs. Moreover, the load data collection of residential users mainly starts from the lower acquisition frequency, so high-frequency sampling is not conducive to The promotion of non-intrusive load monitoring. And existing models based on low-frequency sampling also have problems such as low recognition accuracy and poor system stability. This paper proposes a non-intrusive home load recognition method based on adaptive KNN reinforcement learning algorithm. The adaptive KNN reinforcement learning algorithm is used to solve the HMM model, which reduces the dependence of the model on a large amount of data and improves the model's Identify accuracy and adaptability to new data.

2. Research Methods

2.1. Non-invasive household load identification system

The key technologies implemented by the non-intrusive home load identification system include: Load monitoring and Identification technology, Information communication technology and Data deep mining technology. Load monitoring and Identification is the real-time status of the user's internal load through the information analysis at the user's entrance. The information communication technology is to realize the two-way interaction between the power grid and the user. And the Data deep mining technology can further mine the user's power load data information. As shown in Figure 1, the non-intrusive home load identification system collects user real-time voltage and current information at the user entrance, obtains the user's internal load information by analyzing the characteristics of the load, and then uploads the user load information to the data center through the communication network. After the data center obtain the user side information, the user load information is further analyzed and mined to obtain the user's power usage rules and abnormal power usage behaviors. And the user side and the power grid side are provided with support for flexible interaction services such as energy efficiency management and demand response.

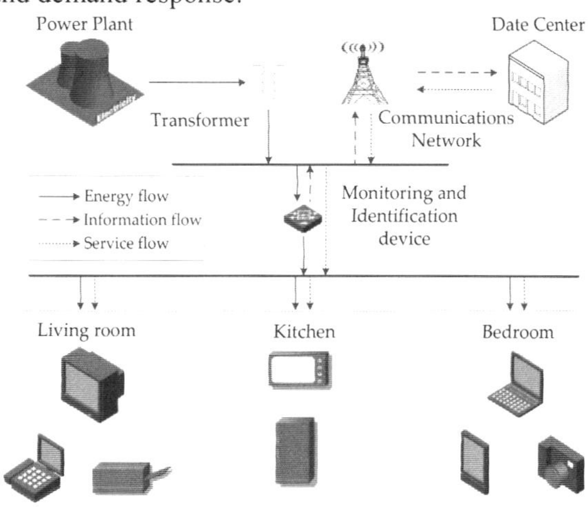

Figure 1. Non-invasive home load identification system

2.2. NILM method based on adaptive reinforcement learning

2.2.1. Reinforcement Learning. Reinforced learning obtains Enhanced signals in the interaction with the environment, and learns by strengthening the signal in a "trial and error" way. Enhanced signals is an evaluation of the current action in the environment, rather than letting the reinforcement learning system to perform the correct action. Because the external environment provides very little information, the reinforcement learning system must learn on its own experience. In this way, the reinforcement learning system gains knowledge in the interaction with the environment, and then dynamically adjusts the parameters to improve the action strategy to adapt to the new environment [11]. The decision-making process of reinforcement learning can be seen as a process of Markov decision, which can be simply expressed as $\varphi = (S, A, P, R)$, S is a finite state set, A is a finite set of actions, P is the transition probability matrix, and R is the return of the completed action. Reinforcement learning mainly evaluates the value of each state or action through the state value function and the action value function, then selects the strategy. The status value function represents the expected return from the current state in accordance with the current strategy, that is $V(s) = \mathrm{E}(R_t | S_t = s)$. The state behavior value function represents the expectation return of the current strategy after performing an action from the current state, that is $Q(s,a) = \mathrm{E}(R_t | S_t = s, A_t = a)$. In the function $R_t = r_{t+1} + \gamma r_{t+2} + \gamma^2 r_{t+3} + \ldots + \gamma^{T-t-1} r_T$, it indicates the cumulative return from time t [12]. In this paper, the optimal state value function $V^*(\mathrm{s})$ and the optimal action value function $Q^*(\mathrm{s,a})$ are calculated by the Bellman optimal equation. The function is as follows:

$$V^*(s) = \max_a \sum_{s'} P_{ss'}^a \left[R_{ss'}^a + \gamma V^*(s') \right] \tag{1}$$

$$Q^*(s,a) = \sum_{s'} P_{ss'}^a \left[R_{ss'}^a + \gamma \max_{a'} Q^*(s', a') \right] \tag{2}$$

In the function, $P_{ss'}^a$ indicates the probability that execution action a will transition from state s to s'. $R_{ss'}^a$ indicates the return of execution action a from state s to s'.

In the process of Agent interaction with the environment, the value function is updated. In this paper, the classic Sarsa algorithm is used to update the value function. The Sarsa primitive value function update formula is:

$$Q(s,a)_{t+1} = Q(s,a)_t + \alpha(R_{ss'}^a + \gamma Q(s',a')_t - Q(s,a)_t) \tag{3}$$

This paper first analyzes the load state through KNN algorithm and establishes a home HMM model with M loads, that is $\lambda = (X, Y, T, O, \pi)$. X is a finite set of states, Y is a finite set of observations, T is a state transition probability matrix, O is the observation matrix, and π is the initial probability vector. Based on the HMM model established in this paper, the optimal state value function and the optimal action value function can be rewritten as:

$$V^*(x_i) = \max_{y_i} \sum_{x_j} T_{ij} \left[O_j + \gamma V^*(x_j) \right] \tag{4}$$

$$Q^*(x_i, y_i) = \sum_{x_j} T_{ij} \left[O_j + \gamma \max_{y_j} Q^*(x_j, y_j) \right] \tag{5}$$

Then the corresponding value function update formula can be rewritten as:

$$Q(x_i, y_i)_{t+1} = Q(x_i, y_i)_t + \alpha\left[O_j + \gamma\sum_{x_j} T_{ij} Q(x_j, y_j)_t - Q(x_i, y_i)_t\right] \tag{6}$$

2.2.2. Adaptive KNN Reinforcement Learning Algorithm(AD-KNN-RL). The adaptive KNN reinforcement learning algorithm proposed in this paper combines the KNN algorithm with reinforcement learning, classifies the state of the load through KNN, establishes the state space, and then updates it through reinforcement learning. The reinforcement learning is learned through the enhanced signal of the environment, and When the new data is input, the parameter improvement action plan can be dynamically adjusted to adapt to the environment, so the prior knowledge is less dependent. The adaptive KNN reinforcement learning algorithm is divided into two parts: state space learning and value function learning. The state space learning determines whether to add the new state to the new state space representative point by judging whether the minimum distance between the newly appearing state and the existing discrete state is greater than the distance threshold. The value function learning calculates the corresponding action value function according to the known degree of the current state, and selects the action according to the value and moves to the next state; at the same time, calculates the corresponding update amount according to the known degree of the next state. The algorithm in this paper estimates the state behavior values of k neighbors in the current state. The algorithm in this paper uses the state behavior values of the k nearest neighbors of the current state x_i to estimate $Q(x_i, y_i)$.

Assume that the set of k-nearest neighbors of the state x_i is expressed as:

$$K = \{x_1, x_2, \ldots, x_k\} \tag{7}$$

Then calculate the distance between k neighbors $x_1, x_2, x_3, \ldots, x_k$ and the current state x_i :

$$d = d(x_i, x_j), x_j \in K \tag{8}$$

Calculate the distance set:

$$D = \{d(x_i, x_1), d(x_i, x_2), \ldots, d(x_i, x_k)\} \tag{9}$$

Finally, calculate the weight set and the neighbor point contribution ratio set:

$$W = \left\{ w_1, w_2, \ldots, w_k \,\middle|\, w_j = \frac{1}{1 + \left(d(x_i, x_j)\right)^2}, 1 \le j \le k \right\} \tag{10}$$

$$G = \left\{ g_1, g_2, \ldots, g_k \,\middle|\, g_j = \frac{w_j}{\sum_{j=1}^{k} w_j}, 1 \le j \le k \right\} \tag{11}$$

Then the optimal state value function and the action value function can be rewritten as:

$$V^*(x_i) = \max_{y_i} \sum_{x_j} g_j \left[O_j + \gamma V^*(x_j)\right] \tag{12}$$

$$Q^*(x_i, y_i) = \sum_{x_j} g_j \left[O_j + \gamma \max_{y_j} Q^*(x_j, y_j)\right] \tag{13}$$

Then the corresponding value function update formula can be rewritten as:

$$Q(x_i, y_i)_{t+1} = Q(x_i, y_i)_t + \alpha\left[O_j + \gamma\sum_{x_j} g_j Q(x_j, y_j)_t - Q(x_i, y_i)_t\right] \tag{14}$$

2.2.3. NILM method based on adaptive KNN reinforcement learning algorithm. The algorithm first inputs the historical data of each power load and analyzes the load state through KNN algorithm, then generates the initial HMM model. It solve the optimal state transition strategy through the adaptive KNN reinforcement learning algorithm. The algorithm flow is shown in Figure 2, the detailed steps are as follows:
Step1: Input each load power data and initial parameters, perform cluster analysis by KNN algorithm, and generate an initial HMM model.
Step 2: Input the total load power data through the preprocessing, and calculate the weight set and the contribution ratio set of the neighbor points by using equations (10) and (11).
Step3: According to the current state, calculate the corresponding state value function and action value function through equations (12) and (13), and then select and execute the action through the greedy strategy to move to the next state.
Step4: Perform state space learning, and update the value function according to formula (14).
Step5: Change the next state to the current state and return to Step 3 until the last observation value ends.
Step6: Get the optimal state transition strategy and the status of each load.

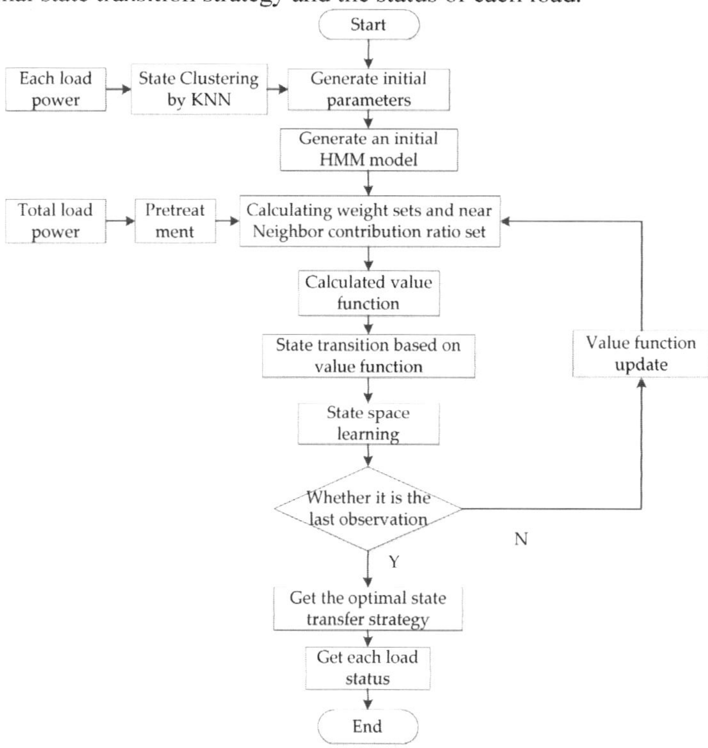

Figure 2. Algorithm flow chart

3. Experimental Results and Analysis

3.1. Data Preparation

The experimental data set used in this paper is the public low-frequency data set AMPds [13]. Eight electrical power data were selected from the AMPds data set to train the model, including: Lamps, Dishwashers, HVAC, Refrigerators, Heat Pumps, Televisions, Washing Machines, Dryers. These eight kinds of electrical appliances have various operating modes and different powers, covering various types such as resistance type, motor type and switching power supply type, and have certain representativeness. The total table data is equal to the sum of the power values of the eight electrical devices at each moment. In this paper, state recognition accuracy and power squared error are selected as non-intrusive load decomposition evaluation indicators [14-15]. At the same time, three algorithms of HMM, KNN and genetic algorithm (GA) are selected as the comparison algorithm [16-18].

(1) State recognition accuracy

$$Acc = \frac{S_{correct}}{S_{correct} + S_{incorrect}} \times 100\% \tag{15}$$

In the formula, $S_{correct}$ is the number of states that identify the correct state, and $S_{incorrect}$ is the number of states that identify the error.

(2) Power squared error

$$RSE = \frac{\sum_{i=1}^{T}\left(P_i - \tilde{P}_i\right)^2}{\sum_{i=1}^{T} P_i^2} \tag{16}$$

In the formula, P_i is the actual power value at the ith time, and $\tilde{P}_i$ is the estimated power value at the ith time.

3.2. Results and Analysis

This article is divided into two scenarios for experimentation. Scene 1 is five kinds of electrical appliances, and Scene 2 is eight kinds of electrical appliances, to analyze the influence of the number of electrical appliances on the algorithm. Scene 1 selects appliances as Lamps, Dishwashers, HVAC, Refrigerators, Heat Pumps. And the training data set is a week of data randomly selected from the AMPds data set.

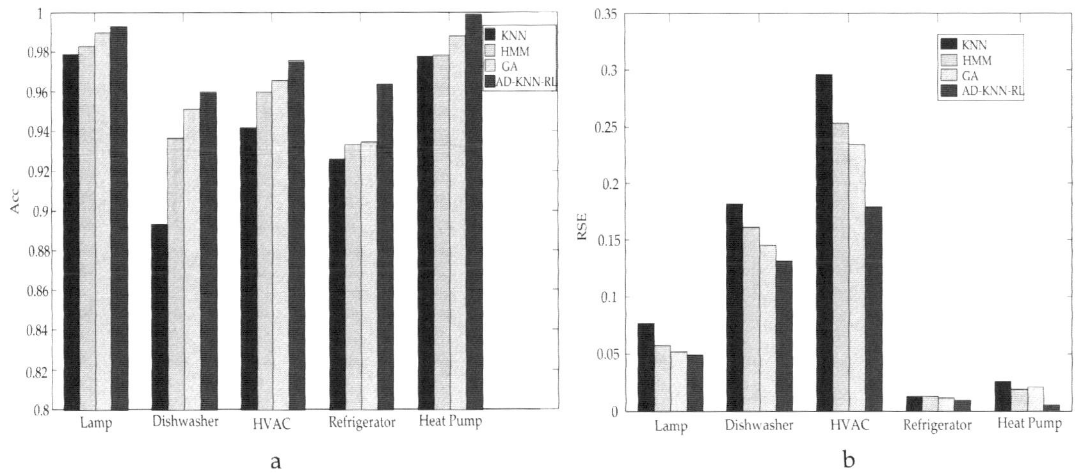

Figure 3. Scene 1 State recognition accuracy comparison and Power square error comparison

Figure 3 a is a comparison of the recognition accuracy in the Scene 1. It can be seen from the figure that the accuracy of the state recognition of the electrical appliances with simple state changes such as lamps and heat pumps is higher than that of other electrical appliances.The accuracy of electrical identification is generally low for multi-state continuous changes.Relatively speaking, the proposed algorithmthe of the recognition accuracy in this paper can reach 95%, and the recognition performance of the algorithm is good. Among them, the HMM's state recognition accuracy is the worst, which is related to the training data set for one week. The HMM model requires a large amount of prior knowledge [22]. In contrast, the KNN and GA state recognition accuracy has improved, but Obviously, AD-KNN-RL better and the dependence on data is lower.

Figure 3 b is a comparison of the power squared error. It can be seen that the power squared error of the algorithm is also improved. Compared with the other three algorithms, the power square error of the five types of electrical appliances is reduced, further illustrating the superiority of AD-KNN-RL .
Scene 2 added Televisions, Washing Machine and Dryer, which will have certain influence on the recognition accuracy of the algorithm. It can be seen from Figure 4 a and b that the recognition accuracy of the original five kinds of electrical appliances has decreased. The recognition accuracy needs to be improved. Relatively speaking, the recognition accuracy of AD-KNN-RL is the best, and the downward trend is relatively flat, and AD-KNN-RL has better adaptability to new data.

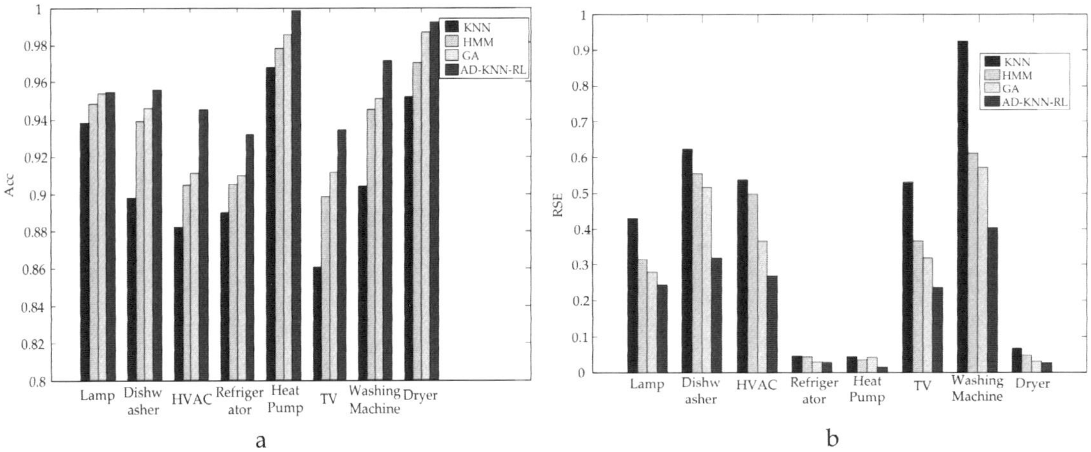

Figure 4. Scene 2 State recognition accuracy comparison and Power square error comparison

4. Conclusion

Based on the low-frequency sampling, this paper proposes a non-intrusive home load identification method based on adaptive KNN reinforcement learning algorithm. The model can identify the running state of the home load from the low-frequency mixed power signal and realize the non-intrusive load identification. The low-frequency data set AMPds is used to verify the decomposition effect of the algorithm. It can be seen from the comparison of two non-intrusive load decomposition evaluation indicators of state recognition accuracy and power square error. This method can obtain higher load recognition accuracy when the training data set is smaller, and reduce dependent of the model's prior data. And when adding new home load data, the system performance can be stable and adaptable to new data. However, the accuracy of the recognition of the multi-state and continuously changing electrical appliances needs to be improved.

Acknowledgments

This work was financially supported by National Natural Science Foundation of China, grant number 51777068.

References

[1] Mueller J A; Kimball J W. Accurate Energy Use Estimation for Nonintrusive Load Monitoring in Systems of Known Devices[J]. IEEE Transactions on Smart Grid, 2016:1-1.
[2] Wang Shouxiang; Sun Zhiqing; Liu Zhe. Co-scheduling strategy of home energy for smart power utilization [J]. Automation of Electric Power Systems, 2015, 39(17):108-113.
[3] Hoyo-Montano J A; Pereyda-Pierre C A; Tarin-Fontes J M , et al. Overview of Non-Intrusive Load Monitoring: A way to energy wise consumption[C]// 2016 13th International Conference on Power Electronics (CIEP). IEEE, 2016.

[4] Vardakas J S ; Zorba N ; Verikoukis C V . A Survey on Demand Response Programs in Smart Grids: Pricing Methods and Optimization Algorithms[J]. IEEE Communications Surveys & Tutorials, 2015, 17(1):152-178.

[5] Cheng Xiang ; Li Linzhi; Wu Hao, et al. A survey of the research on non-intrusive load monitoring and disaggregation [J]. System Technology, 2016, 40(10):3108-3117.

[6] Wu Xin; Wang Zhen. A non-intrusive decomposition algorithm for resident power load based on negative entropy estimation [J]. Power System Technology, 2017, 41(3):931-937.

[7] Sun Yi; Zhang Lu; Zhao Honglei, et al. A Non-Intrusive Household Load Monitoring Method Based on Dynamic Adaptive Particle Swarm Optimization Algorithm [J]. Power System Technology,2018,42(06):1819-1826.

[8] Breschi V. ; Piga D. ; Bemporad. Jump model learning and filtering for energy end-use disaggregation [J]. IFAC-Papers On Line,2018(1).

[9] Qi Bing ; Liu Liya ; Wang Lili . Identification Algorithm for Appliance Load Based on LMD and Model Matching [J]. Automation of Electric Power Systems, 2017(22):74-80.

[10] Shunfu Lin ; Lunjia Zhao; Fangxing Li ; Qingqiang Liu ; Dongdong Li ; Yang Fu .A nonintrusive load identification method for residential applications based on quadratic programming[J]. Electric Power Systems Research, 2016, 133:241-248.

[11] Sutton R S, Barto A G. Reinforcement learning: An introduction[M]. MIT press, 2018.

[12] Hirotaka Hachiya; Takayuki Akiyama ; Masashi Sugiayma; Jan Peters. Adaptive importance sampling for value function approximation in off-policy reinforcement learning[J]. Neural Networks,2009,22(10).

[13] Makonin S; Popowich F; Bartram L , et al. AMPds: A Public Dataset for Load Disaggregation and Eco-Feedback Research[C]// Electrical Power and Energy Conference (EPEC), 2013 IEEE. IEEE, 2013.

[14] Johnson M J , Willsky A S . Bayesian nonparametric hidden semi-Markov models[J]. Journal of Machine Learning Research, 2013, 14(1): 673-701.

[15] Cao H, Liu S, Wu L, et al. Achieving Differential Privacy against Non-Intrusive Load Monitoring in Smart Grid: a Fog Computing approach[J]. 2018.

[16] Makonin S; Popowich F; Bajić I V, et al. Exploiting HMM Sparsity to Perform Online Real-Time Nonintrusive Load Monitoring[J]. IEEE Transactions on Smart Grid, 2016, 7(6):2575-2585.

[17] Song Xufan; Zhou Ming ; Tu Jing ; Li Gengyin . Non-intrusive Load Monitoring Method Based on k-NN and Kernel Fisher Discriminant [J]. Automation of Electric Power Systems,2018,42(06):73-80.

[18] Xu Qingshan ; Lou Oudie; Zheng Aixia ; Liu Yujun . A Non-Intrusive Load Decomposition Method Based on Affinity Propagation and Genetic Algorithm Optimization[J]. Transactions of China Electrotechnical Society,2018,33(16):3868-3878.

IOP Conf. Series: Materials Science and Engineering **853** (2020) 012031 doi:10.1088/1757-899X/853/1/012031

Research on the Civil-Military Integration Logistics Support System of Warships

Xiaobai Liu[1,*], Fei Shao[1], Xingxing Zhang[2]

[1]Shanghai Department, China Ship Development and Design Center, Shanghai, China

[2] High Vocational and Technical College, Shanghai Dianji University, Shanghai, China

mail-fox@163.com

Abstract. This paper had researched the construction of the Chinese Navy warship logistics support system, on the basis of drawing lessons from foreign military work experience. It had pointed out that, the institution building of Chinese Navy warship civil-military integration logistics support should be carried out from three aspects: establishing whole life-cycle logistics support mechanism, perfecting market competition mechanism, establishing the joint logistics support mechanism. At the same time, it had suggest that we should pay more attention to improving the research and development (R&D) capability, peacetime and wartime transfer capability and quality management capability of civilian enterprises. So it would to improve the quality and efficiency of Chinese Navy civil-military integration logistics support.

1. Introduction

In recent years, the contradiction between warship logistics support demand and Chinese Navy internal logistics support capability is becoming increasingly prominent, with the extensive service of modern naval warships and onerous combat readiness cruise tasks of their implementing. In this condition, the model of independent logistics support in Chinese Navy, which has been implemented for half century, has been difficult to provide fast, accurate and high-quality logistics support services for batch, high-strength and whole life using warships, either could not meet the needs of easy support and well support [1-2]. Therefore, it has great significance to improve the support capability, reduce the support cost and enhance the operational effectiveness of warships by learning from the successful experiences of foreign military, that was reforming the logistic support system and utilizing the advantages of civil industry department including scientific and technological strength, information resources, human resources and logistics network, which could provide logistics support services for Chinese Navy warships.

On the basis of foreign military experices, this paper had researched on Chinese Navy warship logistics support system construction, and put forward suggestions on institutional-building and capacity-building of the civil-military integration logistics support, which had the great significant for both improving the logistics support capacity and reducing the logistics support cost of warship.

2. The current situation of foreign army civil-military integration logistics support

The concept of civil-military integration logistics support was first proposed by the United States. In 1994, the U.S. Congressional Bureau of Technology Assessment had issued the 'Potential Assessment of Civil-Military Integration', which calls for the combination of defense industry and civil industry, to form a unified national science and technology innovation system [3]. The so-called 'civil-military integration logistics support' refers to making full use of professional military and civilian support resources under the unified planning and guidance of the state, for enable the weapon to maintain or restore to a good operational readiness to ensure the military's requirements [4].

After more than 20 years development, the U.S. had established a relatively perfect civil-military integration defense science, technology and industry system. According to the research report of the U.S. Congress, the civil-military integration logistics support had saved at least 30 billion US dollars annually for the Ministry of Defense, equivalent to more than 20% of the weapon acquisition funds [5]. At present, the U.S. civil-military integration logistics support has following features:

1) Introducting civil technology to supply support. The U.S. military is good at using mature and advanced civilian technology to provide supply support for weapons. Such as the signal processor of F-35 joint fighter had been provided by Mercury Systems Company, and the defense airborne reconnaissance antenna was provided by Harris Company. It had great improved the scientific and technological level of U.S. weapons by utilizing advanced technological method in the civilian field.

2) Introducting commercial company to armament logistics. In the logistics transportion support field, the U.S. Army had introduced Wal-Mart, Actuate Corporation and other civilian retail giants to participate in it actively. By utilizing the visual logistics transportion support resources of professional logistics companies, the spare parts, maintenance tools and other armament materials can be transported to the global battlefields accurately and rapidly, and the benefits of weapon logistics support can be maximized.

3) Introducing design and construction units to provide weapon maintenance support. The U.S. Army had introduced design and construction units to participate in weapon maintenance support actively. For example, in the Iraq war, the maintenance support tasks of U.S. Army's communication system, armed helicopters and ground weapon system were all accomplished by the original design and construction units, as shown in figure 1. It had't only improved the efficiency of weapon maintenance support, but also reduced the pressure of rush repair by the military internal logistics support force greatly.

Figure 1. The rush repairment of Tank by U.S. construction units

In order to change the deformity of military development in the former Soviet Union era, Russia had promulgated a series of laws and documents, such as the 'Law on the Conversion of National Defense Industrial Army to Civil Affairs', aiming at promoting the level of civil-military integration logistics support, and promoting cooperation between military enterprises and civilian enterprises, military enterprises and foreign enterprises vigorously.

France, Germany, Italy and others were committed to establishing the cooperation framework of EU defense science and technology integration support, for carrying out joint civilian-military technology development, extending from civilian technical cooperation to military technical cooperation.

Japan had adopted the means such as developing civilian military industry and establishing civil-military integration support service company to promote the development of civil-military dual-use technology and industry, so that to advance the process of civil-military integration support strategy for the Japan Self-Defense Force weapons, as shown in figure 2.

Figure 2. The maintenance support for Japanese armored vehicle by Mitsubishi Corporation

3. The institution building of Chinese Navy civil-military integration logistics support

3.1. The whole life-cycle logistics support mechanism

Influenced by traditional thinking, the combat and support capabilities of Chinese Navy warships have not developed synchronously at present, relevant support capability can not been formed after the warship service. Therefore, in order to meet the needs of high-strength and whole life-cycle support, it is necessary to establish a whole life-cycle logistics support mechanism for the warships. For the maximization of operational effectiveness and the minimization of logistics support cost, it is necessary to consider the relevant support issues synchronously or even in advance at all stages from project demonstration to warship retirement, and constantly revise the ship maintenance plan, resource allocation plan, retirement disposal plan and so on.

Hence, in order to maintain the industrial departments to provide whole life-cycle support services, the civil industrial departments should be actively involved in all the stages of warship development as an important part of the civil-military integration logistics support. And the civil industrial departements shoud be used to design the use and maintenance support plans in line with national conditions, and the plans should be constantly revised with the construction, service and retirement of warships.

3.2. The market competition mechanism

In recent years, the PLA has gradually adopted the centralized procurement and competitive procurement mechanism in the weapons procurement [6], and formed a relatively perfect market economy competitive procurement mechanism, which has made a significant improvement in the development level and procurement cost of weapons. Therefore, the competitive bidding mode should also be introduced in the process of civil-military integration logistics support of warships, for perfecting the market competition mechanism. Neither threshold nor wide absorption should be set, but quality and procedure should be guaranteed.

Especially, the network information platform such as Chinese Army Weapons Procurement Information Network, which has been popularized in recent years, should be used widely and actively. It is for releasing the requirement information of weapon's logistics support activity, invitating the civil departments participate in weapons logistics support services extensive. Encouraging the benign market competition among different service providers and mobilizing the enthusiasm of local enterprises and civiliian enterprises to participate in the weapons logistics support, that could not only prevent the market monopoly, but also could improve the support service quality continuously.

3.3. The joint logistics support mechanism

With the Liaoning and the Chinese first domestic aircraft carrier in service continuous and establishment of the Strategic Support Forces, it has become an important future development direction of establishing carrier formation group, air-sea cooperative operation group and forming systematic combat capability. As a result, the multi-service integrated joint logistics support has become a hot topic in the reform of Chinese army's logistics support institutional in recent years, and a joint logistics support force of PLA had been established specially.

Therefore, the civil-military integration logistics support of Chinese Navy warships can not be separated from the macro-structure of the overall support mechanisem reform. Exploring the multi-service joint logistics support mode based on the future war scenarios such as aircraft carrier formation operations, integrated air-sea operations and so on, and building a multi-service civil-military integrated joint logistics support framework, which could realize the civil-military integrated joint logistics support for the sea, air, sky and shore. These has provide a fast and accurate integrated support service for the multi-service joint operations under the condition of informationization, and is an important supporting force for the national strategy of Chinese Navy.

Figure 3. The Joint Logistics Support Command Center of U.S. Army

Figure 4. The Joint Logistics Support Network Center of Japan Self-Defense Force

4. The capability building of Chinese Navy civil-military integration logistics support

4.1. Research and development (R&D) capability of civilian enterprises

Today, the more and more local enterprises and civilian enterprises are willing to participate in the warships logistics support, which has a positive significance for improving the logistics support efficiency and reducing the whole life-cycle support cost. Yet, owing to the characteristics of small scale, lack of accumulation and pursuit of profits, either local enterprises or civilian enterprises have limited investment in the research of new materials, new technologies and new processes, and even more difficult to sustain basic research. They are difficult to meet the logistics support development trend of informationization, intellectualization and integration, which will inevitably restrict the development of warships logistics support. Therefore, more policy support should be given to local enterprises and civilian enterprises, and the incubation platform of new technology should be provided to improve their independent research and development (R&D) capabilities. In this way, it will promote the renewal of logistic support technology for Chinese Navy, and realize the Complementing each other and advancing side by side of military and civil technology development.

4.2. Peacetime and wartime transfer capability of civilian enterprises

Recent decades, logistic support tasks of Chinese Navy had been carried out in a relaxed and peaceful environment, especially in the industrial department and civilian enterprises, which had not experienced the test of real war. It is still unknown whether they can cope with the future high-intensity and information-based maritime war. In order to form the civil-military integrated logistics support capability to meet future maritime war, it is necessary to plan ahead and pay attention to improving the peacetime and wartime transfer capability of civilian enterprises.

For example, the 'Logistics Militia Supplementary Program' military exercises held by the U.S. Army in 1999, had organized thousands of weapon support provider to participate in the live exercises, which had enabled the civilian enterprises to gain valuable battlefield experience for making up many loopholes in the wartime support services. Hence, the Chinese Navy should learn from this mode of introducing civilian enterprises to participate in military exercises, so as to accumulate wartime logsitics support experience and exercise the psychology in front of the battlefield, so that the peacetime and wartime transfer capability of civilian enterprises could be formed rapidly.

4.3. Quality management capability of civilian enterprises

As the purpose of civil-military integration logistics support is to provide efficient and high-quality logistics support services for weapon. Hence, it is necessary to improve the quality management capability of civilian enterprises. On one side, we can draw lessons from the experience of weapon's logistics support quality management of U.S. Army, and develop a quality management system based on ISO-9000 series and GJB9001 series under the leadership of Chinese Navy. The local enterprises and civilian enterprises, which had participated in logistic support, should be included in the unified quality management system, so as to facilitate quality supervision and control.

On the other side, the choice of logistics support service providers should be strickly. The comprehensive consideration should be takingf out from technical strength, business scope, holding background, past performance and other aspects, so as to improve the warship logistics support service quality from the source, and avoiding workshop-style individual enterprises mixed up among them.

Addition, the third-party supervision mechanism could be introduced in logistics support according to the design and development experience of warships. The original warship development units, military internal support departments, China Classification Society (CCS) and other professional supervision organizations could be invited to participate in the whole process quality management of warship logistics support services, for improving the support quality continuously.

5. Conclusion

The civil-military integration logistics support of Chinese Navy should actively draw lessons from foreign successful experience, and build up the institution from aspects of establishing whole life-cycle support mechanism, improving market competition mechanism and establishing the joint logistics support mechanism. And then carry out the capacity building from improving R&D capability, peacetime and wartime transfer capability, quality management capability and so on, so that for improving the quality and efficiency of logistics support for Chinese Navy warships.

References

[1] Crino, S. T. (2007) Lean six sigma for supply chain management as applied to the amry rapid fielding initiative. In: 2007 1st Annual IEEE System Conference. Honolulu. pp.28-57.

[2] Miao, L. H., Shu, Z. P., Liu, X. L., et al. (2014) Building the Security System of Theater Equipped Civil-military Integration. Sichuan Ordnance Journal, 35(6): 71-74.

[3] Zhang, W., Yang, X. L. (2017) Construction of Discourse System of Military-civilian Merging with Chinese Characteristics — Out of Misunderstangding of Comparative Study Between China and USA. Journal of Beijing Institute of Technology (Social Science Edition), 19(3):1-7.

[4] Etzkowitz, H., Dzisah, J. (2008) Unity and Diversity in High-tech Growth and Renewal: Learning from Boston and Silicon Valley. European Planning Studies, 16(8):1009-1024.

[5] Liu, X. B., Liang, H. (2016) Research on Civil-military Integration Support of Chinese Warship Equipment. Journal of Naval University of Engineering, 13(1):59-63.

[6] Cai, L. Y., Wang, K., Zhou, Y. C. (2013) Military-civil Task Allocation Method in Military-civil Integration Maintenance Support. Journal of Academy of Armored Force Engineering, 27(3):13-17.

A method of automatic generation of daily work report for power dispatching based on RoBERTa

Jie LI[1], Tao ZHANG[2], Hao LI[2], Jixiang LU[2,3], Jinjun LU[2,3], Ziqi LI[2]

[1] State Grid Jiangsu Electric Power Co., LTD
[2] Technology Research Center of NARI Group Corporation, Nanjing, China
[3] State Key Laboratory of Intelligent Power Grid Protection and Operation Control, Nanjing, China

Lijie_tzgd@sina.com,
{zhangtao11,lihao5,lujinjun,lujixiang,liziqi}@sgepri.sgcc.com.cn

Abstract. Power dispatching operators need to read a large number of power dispatching logs every day and arrange them into daily work reports. It takes them a lot of time to do information extraction and transfer that information into work reports. In order to reduce the workload of power dispatching operators, we propose an intelligent method for automatic generation of daily work reports in this paper by adopting RoBERTa, which is an improved version of BERT. With the outstanding performance of RoBERTa in semantic analysis and entity recognition, the workload of power dispatching operators is reduced approximately 50% through our proposed method, which means the efficiency is greatly improved.

Keywords. Natural Language Processing, Information Extraction, Named Entity Recognition, Pattern Matching, RoBERTa Introduction

1. Introduction

Reading a large number of complicated logs generated from the power dispatching system and transferring them into daily work reports according to corresponding regulations is an essential part of the daily work for power dispatching operators. According to our preliminary investigation, it takes an experienced operator approximately two hours to read power dispatching logs and write daily work reports. In order to help power dispatching operators save time on log reading and information transferring, we propose an intelligent method for automatic generation of daily work reports based on corresponding regulations in this paper by utilizing some technologies and models of NLP based on deep learning [1, 2]. Nevertheless, similar to other NLP tasks, the biggest challenge in our task is the shortage of massive training data, since there are a lot of duplicate contents in power dispatching logs. It is hard to augment text data of logs as dispatching operation contents are strongly professional. Therefore, we choose transfer learning to solve this problem through benefiting from training on enormous amount of un-annotated text corpora, getting pre-trained general-purposed language model and fine-tuning the pre-trained model with specific sparse text data [3].
To make our proposed method more accurate, we choose RoBERTa [4] for semantic analyzing by using pre-trained general-purposed language encoders and fine-tuning them in power dispatching logs.

The experimental results show that the workload is reduced more than 50% by using our proposed method.

Our contributions can be summarized as follows:

1. To the best of our knowledge, what we present in this paper is the first work to exploit power dispatching logs contextualized embedding with RoBERTa pre-trained model, and improve word matching performance with fine-tuning.
2. We redesign a new model for named entity recognition(NER) by employing RoBERTa. It creates a state-of-the-art accuracy for named entity recognition task with the model RoBERTa-BiLSTM-CRF.

2. Related work

A lot of work has been done to make computers reach the level of human beings in terms of understanding human language. Word2vec is introduced and popularized by Mikolov et al. [5] on word embedding process. However, Word2vec cannot dynamically alter the word representation based on the contextual information. BERT, published by Google in late 2018, is a new method of pre-training language representations which obtains state-of-the-art results on a wide array of NLP tasks. Guo et al. [6] worked on reading answer from Database without SQL form, they propose method to learn from question and search result directly in tables with BERT based model. Zhang et al. [7] used BERT model on text summarization task successfully. Recently, the authors of BERT have released an updated version of BERT, which is called Whole Word Masking. The whole word masking mainly mitigates the drawbacks of original BERT that all the WordPiece tokens are masked altogether if a masked WordPiece token belongs to a whole word. It explicitly forces the model to recover the whole word in Masked Language Model (MLM) pre-training task instead of recovering WordPiece tokens [8]. RoBERTa, an improved recipe for training BERT models, is capable of matching or exceeding the performance of all of the post-BERT methods [9]. RoBERTa is the model that we adopted in our proposed method as it proves upon the published BERT results on both GLUE and SQuAD in terms of controlling for training data.

3. Implementation

3.1. Overview of the method

In this paper, we propose an intelligent method of handling power dispatching logs based on NLP technologies. The proposed method is able to assist operators to arrange daily work reports and improve working efficiency. As shown in Figure 1, the whole process is composed of three steps.

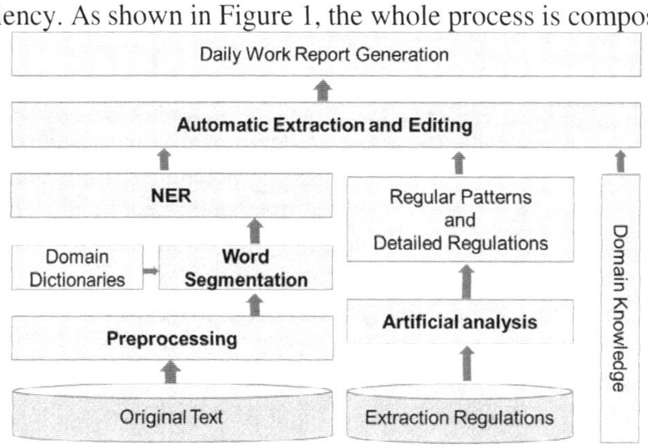

Figure 1. The whole process of the intelligent method.

Step 1: We pre-process the original logs to organize them into records with timestamps. Key contents that we extract from the logs are "Dispatching Operation", "Electrical Overhaul", "Power Grid Accident", "Electrical Defect", "Start-up and Shutdown" and "Operation Mode Adjustment".

Step 2: We analyze the regulations, extract mode elements such as objects, conditions and events from the regulations and form regular patterns composed of mode elements. For example, if we need to parse the sequence "Start-up of devices with the voltage level of 220kV and above in a certain province", we would come up with a pattern "Status + Device + Voltage Level + Province". In order to enhance accuracy, detailed regulation is adopted to define types for those words extracted from daily work reports. The operation function is constructed based on the corresponding type.

Step 3: We utilize NLP technologies to do word segmentation on power dispatching logs based on domain dictionaries. Dictionary based word type annotation is carried out at the same time during word segmentation. Besides, NER is used to get the pattern element annotation of each word. With the annotations of word type and pattern element, sentences are matched into different patterns. In particular, when dealing with those logs lack of conditions, we make supplement by querying domain knowledge. Finally, we extract contents that meet the rules and classify them into different categories through objects. Contents with the same object are merged into one record.

3.2. Key Technologies

3.2.1. Chinese Word Segmentation

We use Jieba to cut specific text sequences into segmented words. The accuracy of word segmentation is highly depended on the dictionary for Chinese. Therefore, we have done a lot of work building a comprehensive and reliable dictionary. During the work of dictionary building, we have summarized that major device types of power dispatching include substation, power station, line, etc. The names of specific devices are provided by dispatching departments of electric power companies.Also, We have done some research on abbreviations and aliases of those standard device names.

3.2.2. Word Based Name Entity Recognition

Name Entity Recognition (NER) is one of the most essential NLP tasks, aiming at identifying entities and laying the groundwork for entity relation extraction. After researching on BERT, we come up with the conclusion that the model pre-trained with Whole Word Masking enjoys better performance than that pre-trained with WordPiece Masking for Chinese. As a result, we adopt word based NER models. The classical NER model is Word2vec-BiLSTM-CRF. As what is shown in Figure 2(a), the model architecture of Word2vec-BiLSTM-CRF contains three layers.

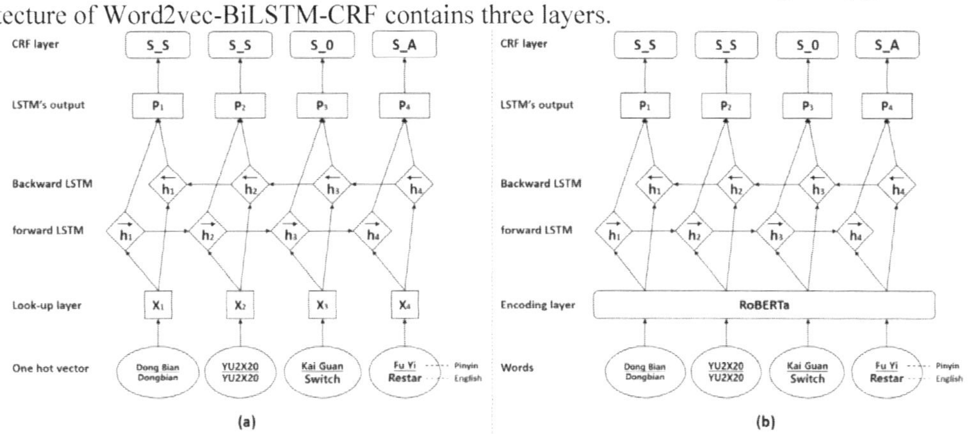

Figure 2. The model architecture of Word2vec-BiLSTM-CRF and RoBERTa-BiLSTM-CRF.

The first layer is a looking-up layer, in which each word x_i of a sentence is transferred from a one-hot vector into a low density word vector through pre-training or randomly initialized embedding matrices. The second layer is a BiLSTM layer in which sentence features are automatically extracted. After the feature extraction, the hidden layer state vectors are mapped from m dimensions to k dimensions through a linear layer after dropout where k is the number of labels of marked data sets. Thus, sentence features are successfully extracted as $P \in R^{n*k}$, where P_{ij} is regarded as the score for the mapping from category of the word x_i to the j-th label. The third layer is a CRF layer used for sentence-level sequence labelling. The parameter of the CRF layer is a (k+2)*(k+2) matrix A. A specific member A_{ij} is the score for the transferring from the i-th label into the j-th label. Therefore, the previous label can be used when marking a position.

In terms of Word2vec, the biggest drawback is its poor effect on ambiguous words. For example, the word "apple" is only represented as one vector in Word2vec despite of the context. Thus, it may be figured out as a kind of fruit or a name of a hi-tech company. Compared with Word2vec, RoBERTa, the abbreviation of Robustly Optimized BERT Pre-training Approach, is an upgraded version of BERT and capable of representing different vectors for different meanings. Thus we propose a new model with the combination of BiLSTM, CRF and RoBERTa for NER tasks. The architecture is shown in Figure 2(b).

4. Experiment

4.1. Data Preparation

Limited by data confidentiality, we use approximately 4000 sentences from power dispatching logs of two weeks provided by the dispatching department of a provincial electric power company. Those sentences are segmented words by words and the type of each word is marked. As shown in Figure 3, in terms of marking for word type, "Substation" and "Line" are jointly marked as "S_S", "Switch" is marked as "S_O", and "Status" is marked as "S_A". In order to prevent over fitting, data augmentation is adopted. We implement data augmentation by replacing words, which means that an entity is randomly replaced by another entity of the same area with the same voltage level and the same type. Hence, the data is expanded 20 times.

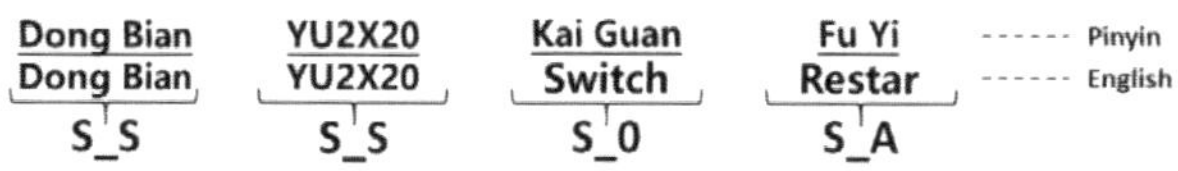

Figure 3. Annotated case for pattern matching.

4.2. Word Segmentation

We make use of Jieba to do word segmentation on the expanded data set, and compare the precision of word segmentation based on dictionaries with that without any dictionary. The precision is calculated as the ratio of the number of correct words segmented by the machine to the total number of manually segmented words. The result is shown in Table 1.

Table 1. Word segmentation precision.

The way of word segmentation	Precision (%)
With dictionaries	91.3
Without dictionaries	70.8

4.3. NER

For the experiment of NER, we divide the expanded data into the training data and the test data in a ratio of 8:2 after data augmentation is done. We make use of three models for NER training in three experiments. The results are shown in Table 2.

It can be seen from the experimental results that RoBERTa-BiLSTM-CRF enjoys better performances than Word2vector-BiLSTM-CRF. Besides, RoBERTa-BiLSTM-CRF with fine-tuning is better than that without fine-tuning. We come up with two viewpoints after analyzing the results. First of all, Word2Vec does not do well in distinguishing named entities of the same word in different sentences, which makes it easy to make mistakes when marking the words. Moreover, RoBERTa, a pre-trained model for Chinese, is trained based on common language. Therefore, a fine-tuned one is better for the corresponding specific domain.

Table 2. NER training and testing results.

Category	Index	Word2vec (%)	Roberta without fine-tuning (%)	Roberta with fine-tuning (%)
training	Precision	86.7	90.8	95.3
	Recall	81.9	85.6	90.5
	F-measure	84.23	88.12	92.84
testing	Precision	85.5	89.3	94.5
	Recall	80.2	83.9	89.8
	F-measure	82.77	86.52	92.09

5. Conclusion and Future Work

The old way of generating daily work reports is to extract information from power dispatching logs through regular expressions. The biggest shortcoming of this way is that it takes a lot of manpower to summarize regular rules before regular matching. In order to solve this problem, we propose an intelligent method to generate daily work reports of power dispatching based on NLP technologies in this paper. RoBERTa is adopted as the semantic analytic model in our proposed method, which reaches a state-of-the-art accuracy of particular domain NLP task. Our proposed method significantly improves efficiency by reducing approximately 50% of the workload and ensures that all the contents complied with rules will be extracted even though there is redundancy.

In terms of future work, we need to focus on some advanced technologies such as data correcting and knowledge reasoning to improve the accuracy of object element recognition and reduce redundant data. When accuracy is improved and data redundancy is solved, our proposed method shall be more universally adaptive.

Acknowledgments

This work is supported by National Key Development Plan Program (No.2016YFB0902600) and State Grids Science and Technology Program (Research on Key Technologies of Knowledge Graph in Power Grid Fault Disposal), and partially supported by Collaborative Innovation Center of Novel Software Technology and Industrialization. This work is supported by Science and Technology Program of NARI Technology Development Limited Company (Research and Application of Key Technologies in the Construction of Power Dispatching Knowledge Graph).

References

[1] Zhou Niancheng, et, al. Developing Status and Prospect Analysis of Deep Learning in Smart Grid. [J]. Automation of Electric Power Systems, 2019,43(04):180-197

[2] Shan Xin, et al. Analysis of Key Technologies for Artificial Intelligence Applied to Power Grid Dispatch and Control [J]. Automation of Electric Power Systems, 2019,43(01):49-57

[3] Cheng Lefeng, et al. Machine Learning for Energy and Electric Power Systems State of the Art and Prospects [J]. Automation of Electric Power Systems, 2019,43(01):15-43

[4] Devlin J, Chang M W, et al. Bert: Pre-training of deep bidirectional transformers for language understanding [J]. arXiv preprint arXiv:1810.04805 (2018).

[5] Mikolov T, Sutskever I, Chen K, et al. Distributed representations of words and phrases and their compositionality[C]//Advances in neural information processing systems. 2013: 3111-3119 (2013).

[6] Guo T, Gao H. Table2answer: Read the database and answer without SQL [J]. arXiv preprint arXiv:1902.04260 (2019).

[7] Zhang H, Gong Y, Yan Y, et al. Pretraining-Based Natural Language Generation for Text Summarization [J]. arXiv preprint arXiv:1902.09243 (2019).

[8] Cui, Yiming , et al. "Pre-Training with Whole Word Masking for Chinese BERT." (2019).

[9] Liu, Yinhan , et al. "RoBERTa: A Robustly Optimized BERT Pretraining Approach." (2019).

Recursive-Biased-Generalized-Inverse Method for Thrust Allocation for Dynamic Positioning Vessels with Non-Rotatable Thrusters and Force Constraints

Bo Zhao, Guoqing Xia

College of Automation, Harbin Engineering University, Harbin, China

bo_zhao@hrbeu.edu.cn, xiaguoqing@hrbeu.edu.cn

Abstract. This paper presents a novel method for thrust allocation for DP vessels with non-rotatable thrusters and force constraints. It uses a modified cost function of the thrust allocation problem for non-rotatable thruster and without force constraints, such that the force constraint can be handled. The proposed method runs in a recursive way to minimize the allocation error. The proposed method is different from the known methods in the literature. The completeness and continuity of the proposed algorithm are studied to verify the effectiveness of the proposed method.

1. Introduction

Thrust allocation (TA), which is control allocation (CA) on thruster systems of marine crafts, is a difficult problem. CA is a frequently encountered problem in motion control system with redundant actuators. If a system is equipped with redundant actuators, since a control command can be achieved in various ways by different actuators. The CA problem is to calculate and select the best control among all feasible solutions. [1] has systematically reviewed the up-to-date technologies in control allocation problem.

DP vessels are the vessels, which can keep their position and heading by only using their thrusters in the presence of wind, wave, and ocean current ([2], [3]). They have been widely used in ocean oil and gas exploitation, offshore engineering, oceanic research, and other marine operations. Due to the possible catastrophic consequence of failures of DP vessels, DP control system is required to be redundant by international regulations. Therefore, it is necessary to have more than three thrusters installed on a DP vessel. However, it is common for a DP vessel to have even more thrusters. As the DP vessel is to be regulated at a certain position and heading, which is in 3 degree-of-freedom, the thruster system is over-actuated. This means the thrust allocation problem have to be solved in DP system design. [2] details the thrust allocation problem.

This paper considers the *constrained linear control allocation* problem in [1], which equals to the problem of thrust allocation for dynamic positioning (DP) vessels with non-rotatable thrusters and force constraints in [2]. This paper proposed recursive-biased-generalized-inverse method for this problem. To the knowledge of the authors, the proposed method is different from any existing method. Comparing with redistributed pseudo-inverse method, daisy chaining method and direct allocation method, the proposed method is complete. Comparing with methods based on mathematical programming, this method is more efficient.

This paper introduces the problem of TA in Section 2 and explains the proposed method in Section 3. The properties of the proposed method are analyzed in Section 4. Section 5 gives the conclusion.

2. Thrust allocation problem

Tunnel thrusters, azimuth thrusters, and main propellers are the common thrusters used for DP vessels. Defining the body-frame of a vessel as Figure 3, the general force and moment produced by a thruster can be written as

$$\begin{aligned}\boldsymbol{\tau} &= \boldsymbol{t}(\alpha)f \\ &= \begin{bmatrix}\cos(\alpha) \\ \sin(\alpha) \\ l_x\sin(\alpha) - l_y\cos(\alpha)\end{bmatrix} f,\end{aligned} \tag{1}$$

where α is the angle of the thrust force, which is measured from x-axis to the force vector, f is the magnitude of the thrust, and l_x, l_y are the location of the thruster. The resulting force and moment of all the thrusters of a DP vessel is a collection of (1), such as

$$\boldsymbol{\tau} = \boldsymbol{T}(\boldsymbol{\alpha})\mathbf{f}, \tag{2}$$

where $\boldsymbol{T} = \begin{bmatrix}\boldsymbol{t}(\alpha_1) & \dots & \boldsymbol{t}(\alpha_n)\end{bmatrix}$ is named thrust configuration matrix following [2], $\boldsymbol{\alpha} = \begin{bmatrix}\alpha_1 & \dots & \alpha_n\end{bmatrix}$, $\boldsymbol{f} = \begin{bmatrix}f_1 & \dots & f_n\end{bmatrix}$, $\alpha_1, \dots, \alpha_n$ and $f_1, \dots, f_n$ are the angles and magnitudes of the forces produced by thrusters, respectively.

Usually, the control command of a DP controller can be expressed as

$$\boldsymbol{\tau}_c = \begin{bmatrix}X & Y & N\end{bmatrix}^\top, \tag{3}$$

where X, Y, N are the force command in surge, sway direction and moment command in yaw, respectively. In TA problem, the thrusters are to be commanded by the TA algorithm to produce a resulting force which follows the generalized force command $\boldsymbol{\tau}_c$, such as $\boldsymbol{\tau} = \boldsymbol{\tau}_c$. A TA algorithm can be written as a mapping

$$\begin{bmatrix}\boldsymbol{\alpha} & \boldsymbol{f}\end{bmatrix}^\top = \boldsymbol{U}(\boldsymbol{\tau}_c), \text{ subject to } \boldsymbol{\tau} - \boldsymbol{T}(\boldsymbol{\alpha})\boldsymbol{f} = \boldsymbol{0}, \tag{4}$$

where $\boldsymbol{U}(\cdot)$ is the TA mapping, and $\boldsymbol{\tau}_c$ is the generalized force to be allocated.

Since the force produced by thrusters is limited, a virtual control $\boldsymbol{\tau}_c$ may not be able to be allocated to a given group of thrusters. This is usually referred as force constraint and can be expressed as $f_{min} < f < f_{max}$.

Azimuth thrusters and rudders are subject to angular constraints. However, we only consider tunnel thruster and azimuth thrusters with a fixed angle in this paper.

3. Recursive-Biased-Generalized-Inverse Method

3.1. Biased-generalized-inverse (BGI) method

The original idea of thrust allocation using generalized inverse is described in [2]. The problem can be defined as

$$\begin{aligned} \boldsymbol{J} &= \min_{\boldsymbol{f}} \left\{ \boldsymbol{f}^\top \boldsymbol{W} \boldsymbol{f} \right\} \\ &\text{subject to} : \boldsymbol{\tau} - \boldsymbol{T}\boldsymbol{f} = \boldsymbol{0} \end{aligned} \tag{5}$$

where $\boldsymbol{W}$ is a weighting matrix. This formula is adopted from [2]. But it is only suitable for solving control allocation problem for *non - rotatable thrusters without force constraint*, which is an ideal case in TA.

Consider the following modification to the cost function.

$$\begin{aligned} \boldsymbol{J} &= \min_{\boldsymbol{f}} \left\{ (\boldsymbol{f} - \boldsymbol{f}_0)^\top \boldsymbol{W}_1 (\boldsymbol{f} - \boldsymbol{f}_0) + \boldsymbol{f}^\top \boldsymbol{W}_2 \boldsymbol{f} \right\} \\ &\text{subject to} : \boldsymbol{\tau} - \boldsymbol{T}\boldsymbol{f} = \boldsymbol{0} \end{aligned} \tag{6}$$

where $\boldsymbol{W}_1$ and $\boldsymbol{W}_2$ are positive definite diagonal weighting matrices, $\boldsymbol{f}_0$ is a parameter which will be discussed in the following. The second term in the cost function is the same as the one in (6), which represent the cost of the usage of thrusters. The first term represents the deviation of the control force to the parameter $\boldsymbol{f}_0$.

The first term in the cost function (6) can serve different practical purposes.

- Consider a discrete time DP control system. If $\boldsymbol{f}_0$ is the control force of the thrusters of a DP vessel at the last time step, and $\boldsymbol{f}$ is the control force at current step. Minimizing $\|\boldsymbol{f}-\boldsymbol{f}_0\|$ is to reduce the change of the thrust force. Because the change rate of the thrust force is limited due to physical constraint, it is meaningful to have this term in the cost function.
- The term can be used to avoid allocated control forces exceeding the limits. For example, assuming $\boldsymbol{\tau}_{c,0}$ is allocated as $\boldsymbol{f}_0$ by the BGI algorithm using cost function (6). But assume that some of the elements of $\boldsymbol{f}_0$ exceed the corresponding bounds. For instance, assume $f_k > f_{k,max}$ and other control forces are within their corresponding bound. Then by choosing $\boldsymbol{f}_0 = [0,\ldots,0,f_{k,max}-f_k,0,\ldots,0]^\top$, where $f_k - f_{k,max}$ is the kth element. In this case, because of the first term in the cost function, the kth element in the control force allocated using cost function (6) will be moved to the direction which reduces the size of the excess.

The optimization problem (6) can be solved by Lagrange Multipliers. Assuming $\boldsymbol{W}_1$ and $\boldsymbol{W}_2$ are symmetric. Consider the Lagrangian

$$L(\boldsymbol{f},\boldsymbol{\lambda}) = (\boldsymbol{f}-\boldsymbol{f}_0)^\top \boldsymbol{W}_1 (\boldsymbol{f}-\boldsymbol{f}_0) + \boldsymbol{f}^\top \boldsymbol{W}_2 \boldsymbol{f} + \boldsymbol{\lambda}^\top (\boldsymbol{\tau} - \boldsymbol{T}\boldsymbol{f}) \tag{7}$$

where $\boldsymbol{\lambda} \in \mathbb{R}^3$ is a vector of Lagrange multipliers. Consequently, differentiating the Lagrangian L with respect to $\boldsymbol{f}$ yields

$$\frac{\partial L}{\partial \boldsymbol{f}} = 2(\boldsymbol{W}_1+\boldsymbol{W}_2)\boldsymbol{f} - 2\boldsymbol{W}_1\boldsymbol{f}_0 - \boldsymbol{T}^\top\boldsymbol{\lambda} \tag{8}$$

Let $\frac{\partial L}{\partial \boldsymbol{f}} = 0$, yields

$$\boldsymbol{f} = \frac{1}{2}(\boldsymbol{W}_1+\boldsymbol{W}_2)^{-1}\left(\boldsymbol{T}^\top\boldsymbol{\lambda} + 2\boldsymbol{W}_1\boldsymbol{f}_0\right) \tag{9}$$

Next

$$\boldsymbol{\tau} = \boldsymbol{T}\boldsymbol{f} = \frac{1}{2}\boldsymbol{T}(\boldsymbol{W}_1+\boldsymbol{W}_2)^{-1}\boldsymbol{T}^\top\boldsymbol{\lambda} + \boldsymbol{T}(\boldsymbol{W}_1+\boldsymbol{W}_2)^{-1}\boldsymbol{W}_1\boldsymbol{f}_0 \tag{10}$$

Then

$$\boldsymbol{\lambda} = 2\left(\boldsymbol{T}(\boldsymbol{W}_1+\boldsymbol{W}_2)^{-1}\boldsymbol{T}^\top\right)^{-1}\left(\boldsymbol{\tau} - \boldsymbol{T}(\boldsymbol{W}_1+\boldsymbol{W}_2)^{-1}\boldsymbol{W}_1\boldsymbol{f}_0\right) \tag{11}$$

Then

$$\begin{aligned}\boldsymbol{f} &= \boldsymbol{W}_{12}^{-1}\boldsymbol{T}^\top\left(\boldsymbol{T}\boldsymbol{W}_{12}^{-1}\boldsymbol{T}^\top\right)^{-1}\left(\boldsymbol{\tau} - \boldsymbol{T}\boldsymbol{W}_{12}^{-1}\boldsymbol{W}_1\boldsymbol{f}_0\right) + 2\boldsymbol{W}_{12}^{-1}\boldsymbol{W}_1\boldsymbol{f}_0 \\ &= \boldsymbol{T}^\dagger\boldsymbol{\tau} + \left(2\boldsymbol{I} - \boldsymbol{T}^\dagger\right)\boldsymbol{W}_{12}^{-1}\boldsymbol{W}_1\boldsymbol{f}_0,\end{aligned} \tag{12}$$

where $\boldsymbol{W}_{12} = \boldsymbol{W}_1 + \boldsymbol{W}_2$ and $\boldsymbol{T}^\dagger = \boldsymbol{W}_{12}^{-1}\boldsymbol{T}^\top\left(\boldsymbol{T}\boldsymbol{W}_{12}^{-1}\boldsymbol{T}^\top\right)^{-1}$.

3.2. Recursive-biased-generalized-inverse (Rec-BGI) method

In this section, the BGI method is modified using cost function (6) and run in a recursive way to handle the constrained control allocation problem for non - rotatable actuators.

The algorithm is as following.

Require: $\boldsymbol{\tau}_c$ - the generalized force to be allocated.
Require: $\boldsymbol{T}$ - the thrust configuration matrix.
Require: $\boldsymbol{f}_0$ - the initial point.
Require: $\boldsymbol{W}_1, \boldsymbol{W}_2$ - weighting matrices.
Require: $\boldsymbol{b}_l = [b_{l,1}, \dots b_{l,p}]^\top, \boldsymbol{b}_u = [b_{u,1}, \dots b_{u,p}]^\top$ - lower and upper boundary of the control force.
Require: i_{max} - the maximum times of iteration.
Ensure: $\boldsymbol{f}$ - allocated control force.

```
function REC-BGI(τ_c, T, f_0, W_1, W_2, b_l, b_u, i_max)
    i = 0
    while i ≤ i_max do
        i++
        f = BGI(τ, f_0)
        if b_l ≤ f AND f ≤ b_u then break the While loop
        else
            for i_f = 1 : p do
                f_{ex,i_f} = 0
                if b_{l,i_f} > f_{i_f} then
                    f_{ex,i_f} = f_{i_f} − b_{l,i_f}
                end if
                if f_{i_f} > b_{u,i_f} then
                    f_{ex,i_f} = f_{i_f} − b_{u,i_f}
                end if
            end for
            f = f − f_ex
            f_0 = f_0 − f_ex
        end if
    end while
end function
```

4. Analysis and Verification of Rec-BGI

4.1. Completeness

In this section, the completeness of Rec - BGI is studied with examples. If a TA algorithm is complete, it means that the algorithm is capable to allocate any attainable generalized control command to the feasible region of the thrusters. Furthermore, the completeness of a TA algorithm means the degree of the TA algorithm being complete, i.e. in which subset of the attainable set, the TA algorithm is complete.

The completeness of the proposed algorithm is studied associated with a give thrust configuration. Using the example ship in Appendix, considering T4 and T5 output force in $-\pi/4$ rad and $\pi/4$ rad, respectively, the study of the completeness gives the results in Figure 1.

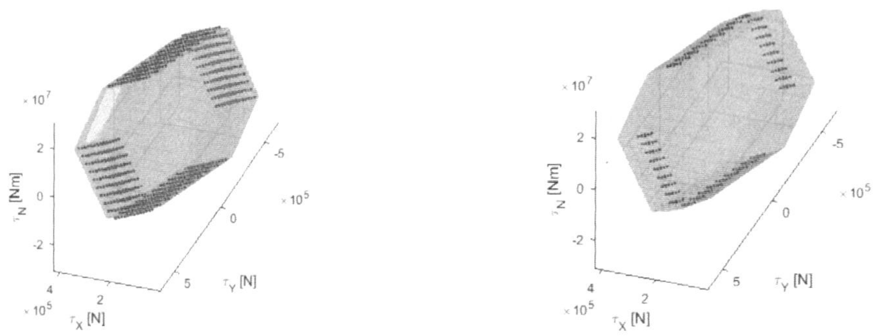

(a) (b)

Figure 1. The thrusters T4 and T5 are pointing at $-\pi/4$ rad and $\pi/4$ rad, respectively.

In Figure 1, the grey surface is the attainable set, which is the set that can be researched by the thrusters. Red scattered spots are the generalized control commands τ_c which cannot be allocated by the proposed method. The maximum number of iterations is 100 in (a) and 1000 in (b).

From Figure 1, it can be concluded that the Rec-BGI can allocate almost any generalized force in the attainable set except for the generalized force close to the boundary. Results also show that the completeness increases with the number of iterations. From the above results, it can be concluded that the completeness of the Rec-BGI algorithm is high.

4.2. Continuity

The continuity of the solution is an important aspect of TA algorithms. The continuity of the solution means that given generalized force $\boldsymbol{\tau}_0$ and corresponding TA solution $\boldsymbol{f}_0$, such that $\boldsymbol{f}_0 = TA(\boldsymbol{\tau}_0)$, the TA shall satisfy that

$$\lim_{\boldsymbol{\tau} \to \boldsymbol{\tau}_0} TA(\boldsymbol{\tau}) = \boldsymbol{f}_0 \tag{13}$$

The BGI algorithm is continuous, if the constraints are not considered. To show this, take the derivative of the TA solution (12), such as

$$\frac{\partial \boldsymbol{f}}{\partial \boldsymbol{\tau}} = \boldsymbol{W}_{12}^{-1}\boldsymbol{T}^\top \left(\boldsymbol{T}(\boldsymbol{W}_{12})^{-1}\boldsymbol{T}^\top\right)^{-1}, \tag{14}$$

which is bounded. This means that if the BGI can find a solution in the feasible region, the solution is continuous.

The continuity of Rec-BGI is to be discussed in different cases.

CASE 1: Both $\boldsymbol{f}_1$ and $\boldsymbol{f}$ are in the attainable set. In this case, Rec-BGI leads to the solution being exactly the same as BGI, and then is continuous.

CASE 2: Consider the situation that $\boldsymbol{f}_1$ and $\boldsymbol{f}_2$, which are the allocation of $\boldsymbol{\tau}_1$ and $\boldsymbol{\tau}_2$ from Rec-BGI. Assuming that $\boldsymbol{\tau}_1$ and $\boldsymbol{\tau}_2$ are close enough such that they can be allocated with the same bias term $\boldsymbol{f}_0$. Analyzing the difference, resulting

$$\frac{\Delta \boldsymbol{f}}{\Delta \boldsymbol{\tau}} = \frac{\boldsymbol{f}_1 - \boldsymbol{f}_2}{\boldsymbol{\tau}_1 - \boldsymbol{\tau}_2} = \boldsymbol{W}_{12}^{-1}\boldsymbol{T}^\top \left(\boldsymbol{T}(\boldsymbol{W}_{12})^{-1}\boldsymbol{T}^\top\right)^{-1} \tag{15}$$

which means that the Rec-BGI method is continuous.

5. Conclusion

This paper presented a novel method for thrust allocation for DP vessels with non-rotatable thrusters and force constraints. The main difference comparing with the literature is that the force constraints is handled with an additional bias term in the cost function. The detailed Rec-BGI algorithm is proposed.

The completeness and continuity study of the proposed algorithm shows that the proposed algorithm can allocate almost any generalized control and keeps the change of the allocated force small if the change of the input is small. This makes the proposed algorithm very practical and robust in practice.

6. Appendix - Example vessel

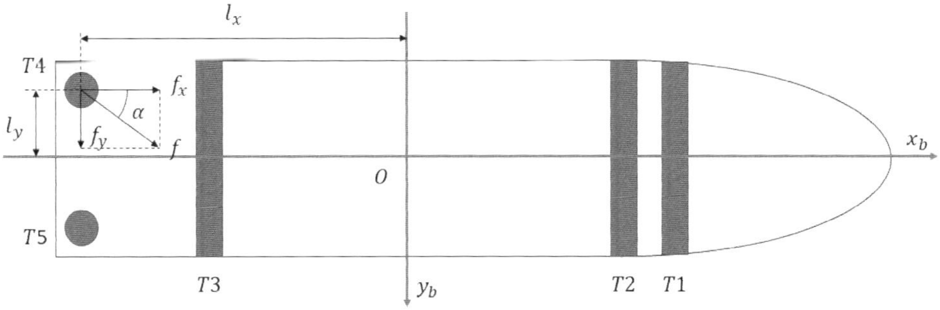

Figure 2. The configuration of the thrusters

Table 1. Thruster setup.

Thruster	Location [m]		Thrust Limit [kN]
	x	y	
T1	50	0	150
T2	40	0	150
T3	-40	0	150
T4	-65	-5	300
T5	-65	5	300

Refer to Figure 2, the vessel in this example is adopted from a real rescue vessel with DP system. The vessel is equipped with three tunnel thrusters to the bow, and two azimuth thrusters to the aft, as shown in Figure 2. The setup of the thrusters is listed in Table1.

Acknowledgments
This work is supported by the 7th Generation Ultra Deep Water Drilling Platform (Ship) Innovation Project, the Fundamental Research Funds for the Central Universities (HEUCFJ180404), and the National Natural Science Foundation of China (Grant No.51879049) research on green and safe mooring-assisted dynamic positioning technology for all-weather deepwater platform.

References

[1] Tor A. Johansen and Thor I. Fossen. Control allocation|A survey. Automatica, 49(5): 1087-1103, may 2013.
[2] Thor I. Fossen. Handbook of Marine Craft Hydrodynamics and Motion Control. John Wiley & Sons, Ltd, 1 edition edition, 2011.
[3] Asgeir J. Sorensen. A survey of dynamic positioning control systems. Annual Reviews in Control, 35(1):123-136, apr 2011.

IOP Conf. Series: Materials Science and Engineering **853** (2020) 012034 doi:10.1088/1757-899X/853/1/012034

Kalman Filter Based Sensor Fusion for Altitude Estimation of Aerial Vehicle

Muhammad Asad Bashir, Fahad Mumtaz Malik, Zeeshan Ali Akbar and Muhammad Uzair

Department of Electrical Engineering, National University of Sciences & Technology (NUST) College Of Electrical & Mechanical Engineering, Rawalpindi ,Pakistan

asadbashir1993@gmail.com, malikfahadmumtaz@yahoo.com, itszeeshanaliakbar@gmail.com, uzair_ahmed2003@hotmail.com

Abstract. Stabilization of Aerial Vehicles requires altitude estimation. In this paper sensor fusion method is developed for aerial vehicles which focuses on estimation of the altitude. Altitude measurement using sensors is too noisy and biased so Kalman filter based sensor fusion is used to estimate altitude. The proposed method is divided into two parts: attitude estimation and altitude estimation. The method is implemented using IMU and barometer sensors. The results show that altitude is estimated accurately and errors are modelled in few seconds after initialization.

1. Introduction

Sensor fusion is a technique which combines data from different sensor to get an optimized/ best estimates/ results. In air navigation/ aerial vehicles different sensors are used, data from sensors are usually noisy and biased. To achieve better results sensor fusion technique is applied. In this paper a method is proposed to get vertical height of aerial vehicles using MEMS (micro electro mechanical system) sensor I M U (inertial measurement units) accelerometer, gyroscope and barometer. However, the data retrieve from the inertial measurement units and barometer are noisy, due to which vertical height estimate or orientation/ attitude estimation is not easy to achieve. If only relay on inertial measurement unit sensors data to get vertical height/altitude, it may suffer from integration drift which cause positioning error. Changes in atmospheric pressure by unpredictable atmospheric condition and rapid changes in altitude of aerial vehicles cause the barometric altimeter data to be noisy and their accuracy is poor. Therefore, to stabilize the position and orientation of IMU sensors it is customary to use fusion method [1]. Applications of the sensor fusion algorithm include vertical trajectory determination in navigation, control of robotic systems, fall detection and sports. [2]

To acquire vertical height position barometric altimeter is used with 3 Axis accelerometer and 3 Axis gyroscope. Hardware part consists of MPU 6050 and BMP 180 barometer, these sensors are employed on a quadcopter to extract real-time data using Arduino DUE. MPU 6050 (6-DOF) uses I2C or (I^2C) interface with Arduino. Such real-time data is used to estimate orientation. BMP 180

Figure 1. Arduino DUE, MPU 6050 and BMP 180 sensors

uses I2C or (I^2C) interface with Arduino to measure the barometric pressure. Barometer based algorithm obtains the height change using the observed barometric pressure change [3] by using Equation (11).

The IMU-barometer fusion method consists of two parts: first one is orientation /attitude estimation and the other is altitude /height estimation. Kalman filter (KF) and Extended Kalman filter (EKF) are used to measure attitude and altitude of aerial vehicles by collecting the requisite data from raw accelerometer and gyroscope. Attitude measured by IMU and barometer is used to estimate vehicles altitude/ height. This paper presents the algorithm for real experiment, the proposed methodology, a collection of available data sources and data processing techniques to provide better navigation solution [4].

2. Mathematical Modelling

2.1. System Dynamics

A causal, linear and time-invariant system can be described in state-space form using the following format;

$$\vec{x}_{t+1} = A.\vec{x}_t + B.\vec{u}_t + \vec{w}_t \tag{1}$$

$$\vec{y}_{t+1} = C.\vec{x}_{t+1} + \vec{v}_{t+1} \tag{2}$$

$$\vec{x}_t = \begin{bmatrix} \hat{\phi}_t \\ b_{\dot{\phi}_t} \\ \hat{\theta}_t \\ b_{\dot{\theta}_t} \end{bmatrix} \tag{3}$$

$$\vec{u}_t = \begin{bmatrix} \dot{\phi}_{Gt} \\ \dot{\theta}_{Gt} \end{bmatrix} \tag{4}$$

$$\vec{z}_t = \begin{bmatrix} \hat{\phi}_{At} \\ \hat{\theta}_{At} \end{bmatrix} \tag{5}$$

Where, $\vec{x}_t$ is a state vector, $\vec{u}_t$ is input vector and $\vec{z}_t$ is measurement vector. A , B and C are system matrices: A relates the current states to the next states, B relates inputs to the next states, and C relates the system states to the measured states. $\vec{w}_t$ is additive process and $\vec{v}_t$ is measurement noise, b is the gyro bias with respect to ϕ and θ at time t .The system matrices may be designed on following considerations:

Varying gyro bias at every instant in time causes drift so this should be eliminated, to get an updated estimate of angular position the bias is subtracted from transformed angular rates of changed measured by the gyros.

IOP Conf. Series: Materials Science and Engineering **853** (2020) 012034 doi:10.1088/1757-899X/853/1/012034

The general state space form is:

$$\vec{x}_{t+1} = \begin{bmatrix} 1 & -\Delta t & 0 & 0 \\ 0 & 1 & 0 & 0 \\ 0 & 0 & 1 & -\Delta t \\ 0 & 0 & 0 & 1 \end{bmatrix} \vec{x}_t + \begin{bmatrix} \Delta t & 0 \\ 0 & 0 \\ 0 & \Delta t \\ 0 & 0 \end{bmatrix} \vec{u}_t + \vec{w}_t \tag{6}$$

$$\vec{y}_{t+1} = \begin{bmatrix} 1 & 0 & 0 & 0 \\ 0 & 0 & 1 & 0 \end{bmatrix} \vec{x}_{t+1} + \vec{v}_t \tag{7}$$

2.2. Accelerometer Model

The accelerometer measures gravitational acceleration by resolving forces through basic trigonometry. Estimate of pitch and roll can be calculated as follows:

$$\hat{\theta}_A = \text{Arctan}\left(\frac{A_x}{\sqrt{A_y^2 + A_z^2}}\right).\frac{180}{\pi} \tag{8}$$

$$\hat{\phi}_A = \text{Arctan}\left(\frac{A_y}{\sqrt{A_x^2 + A_z^2}}\right).\frac{180}{\pi} \tag{9}$$

Where A_x, A_y, A_z are Raw acceleration measurements along each axis $\left(\frac{m}{s^2}\right)$.

2.3. Gyroscope Model

The gyroscope measures angular rates of change around each axis with respect to the body frame. Assuming noise-free measurements, we could integrate these raw measurements to get the angular position relative to the body frame. If we were to integrate this, we would add up all the small errors due to noise, which then would accumulate over time and become unbounded in magnitude. This is commonly known as gyroscopic drift.

We cannot simply integrate the angular rates of change given by the gyroscope to get our roll, pitch, and yaw angles in the inertial frame. The rates of change of Euler angles depend on more than one body frame angular velocity. To get the rates of change in the inertial frame, we need to transform our states using the following matrix.

$$\begin{bmatrix} \dot{\phi} \\ \dot{\theta} \\ \dot{\psi} \end{bmatrix} = \begin{bmatrix} 1 & \sin\phi\tan\theta & \cos\phi\tan\theta \\ 0 & \cos\phi & -\sin\phi \\ 0 & \sin\phi\sec\theta & \cos\phi\sec\theta \end{bmatrix} \begin{bmatrix} G_x \\ G_y \\ G_z \end{bmatrix} \tag{10}$$

To get the inertial attitude, we will use this transformation in a sensor fusion algorithm.

Where G_x, G_y, G_z are Raw gyroscope measurements about each axis $(\frac{\deg}{s})$, usually referred to as p, q, and r respectively. θ, ϕ -Roll angle (rotation about X axis in degrees) and pitch angle (rotation about Y axis in degrees) respectively.

2.4. Barometer Model

Barometer pressure sensor is used to find altitude with change in atmospheric pressure. The barometric Altimeter are used in most of aerial vehicles. The pressure is measured in Pascal then converted to change in height in meters by [5]

IOP Conf. Series: Materials Science and Engineering **853** (2020) 012034 doi:10.1088/1757-899X/853/1/012034

$$h_t = 44330 \cdot \left(1 - \left(\frac{P}{P_0}\right)^{0.19}\right) - h_{init} \tag{11}$$

P_0 is the standard pressure given as 101.325 kPa and P is the measured pressure in Pascal. Errors related to barometric height estimates are corrected through barometric errors [6].

3. Sensor Fusion

Sensor fusion is a technique of combining data from multiple sensors to minimize the amount of uncertainty in order to provide improved and best estimate of the state. Here the uncertainty minimization means the estimate would be more accurate and more reliable. Both the sensors being used are not of ideal accuracy. Accelerometers provide good results when the system is in rest state and gyroscopes are fine for short intervals only, but over longer periods of time, both sensors will provide lesser reliable estimates of pitch and roll. Therefore, sensor fusion method is used to mitigate the estimation uncertainties.

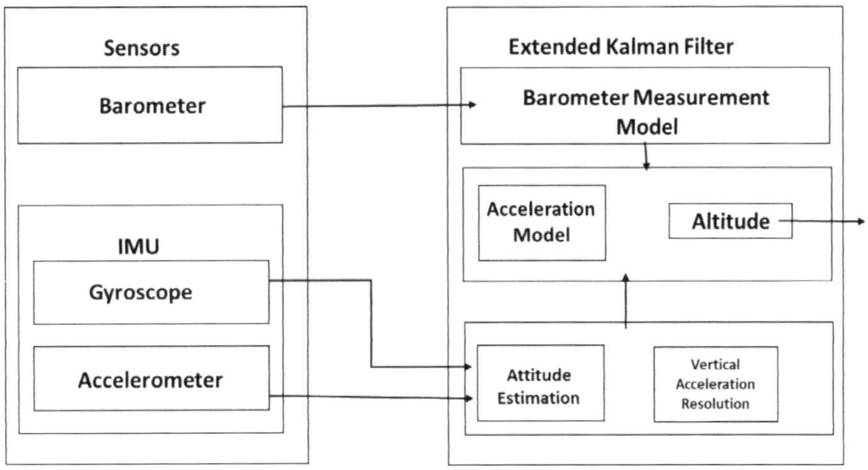

Figure 2. Block diagram for Fusion Algorithm

3.1. Kalman Filter

The Kalman Filter is also a Linear Quadratic Estimator. It is a state estimator which is optimal in the sense that it tries to minimize a quadratic cost function. Kalman Filter equations are divided to two group of equations: Prediction and Update.

Prediction:

$$P = A.P.A^T + Q \tag{12}$$

$$\vec{x}_{t+1} = A.\vec{x}_t + B.\vec{u}_t \tag{13}$$

Update:

$$S = C.P.C^T + R \tag{14}$$

$$K = P.C^T.S^{-1} \tag{15}$$

$$\vec{x}_{t+1} = \vec{x}_{t+1} + K.\tilde{y}_{t+1} \tag{16}$$

$$P = (I - K.C).P \tag{17}$$

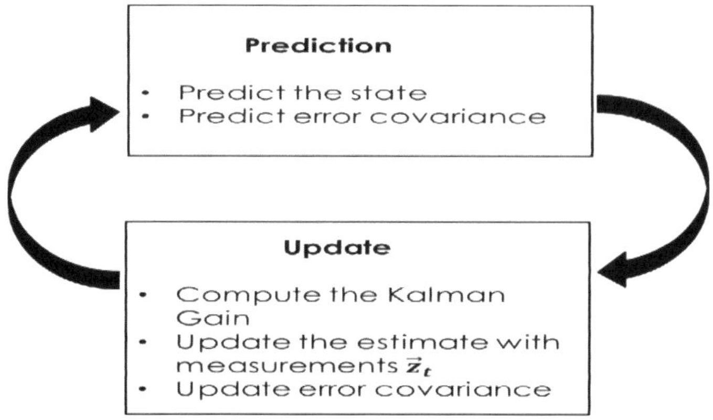

Figure 3. Block diagram for Kalman Filter algorithm

3.2. Attitude Estimation

The attitude can be estimated by fusion of integrated angular velocities and acceleration measurements [7]. It is a 3D orientation of the IMU with respect to the Earth coordinates. In the controlling of aerial vehicles, it is compulsory to have correct feedback states, but there are some practical impediments associated with IMU sensors like accelerometer and gyroscope. Gyroscopes measure angular rate of change but not the angular position. Similarly, accelerometers measure gravitational acceleration but not the linear acceleration. Measurements computed by the sensors are noisy and biased. Body-frame states need to be transformed to the inertial reference frame.

3.3. Altitude Estimation

Attitude estimation resolves acceleration vector into vertical acceleration which is used for estimating the IMU barometer system's Altitude. The state vector for Altitude estimation consists of altitude, accelerometer biases and acceleration scale factor [8].

$$x_t = \begin{bmatrix} h_t & v_t & b_{1\times3} & s_{1\times3} \end{bmatrix}^T \tag{18}$$

Accelerometer bias errors and scales factors cause the errors in Altitude estimation using IMU-Barometer systems.

$$a_t = s.\alpha_t + b + n_t \tag{19}$$

Where a_t, α_t, s, b, n_t represents acceleration measurement , acceleration input ,scale factor, bias , additive noise at time t, respectively. System state is estimated on every update of acceleration. Acceleration measurements corrected by bias and scale estimation.

$$\dot{h}_t = v_t \tag{20}$$

$$\dot{v}_t = r_t(diag(s_t)^{-1}(a_t - b_t)) + g \tag{21}$$

Where $\dot{b}_t = 0$, $\dot{s}_t = 0$ and v_t is velocity of vehicle at time t, gravitational acceleration is given as g and r_t is the third row of rotation matrix.

4. Results

The system is implemented on MATLAB based computer program where real data is collected from sensors using Arduino mounted on Ariel vehicle. We acquired different data sets and performed experiments several number of times.

In the first part, data is used to estimate Attitude of vehicle, which results in pitch, and roll angle of the vehicle in degrees. Kalman filter is used to fuse the data from accelerometer and gyroscope to find out best estimate of the pitch and roll angle.

The EKF is used for combining IMU-Barometer to estimate the vertical height of aerial vehicle. The measurements from sensors which are affected by noise are used in attitude estimation and with EKF

IOP Conf. Series: Materials Science and Engineering **853** (2020) 012034 doi:10.1088/1757-899X/853/1/012034

algorithm process these characterize the noise to accurately estimate the Altitude. Accuracy of estimates may be increased with the implementation of EKF.
MATLAB simulation results are shown here:

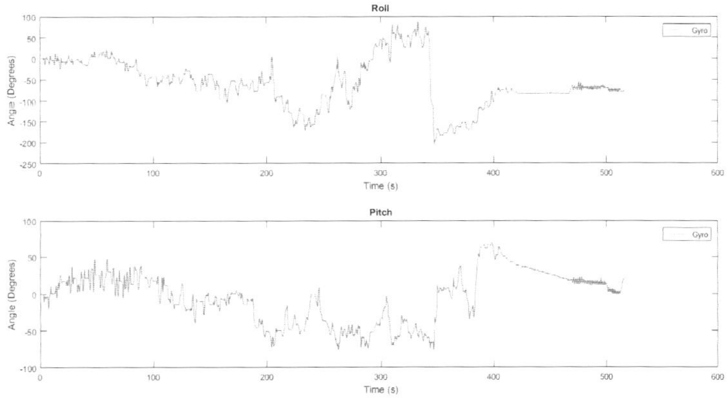

Figure 4. Roll and pitch angle(degrees) for Attitude estimation using raw Gyro values

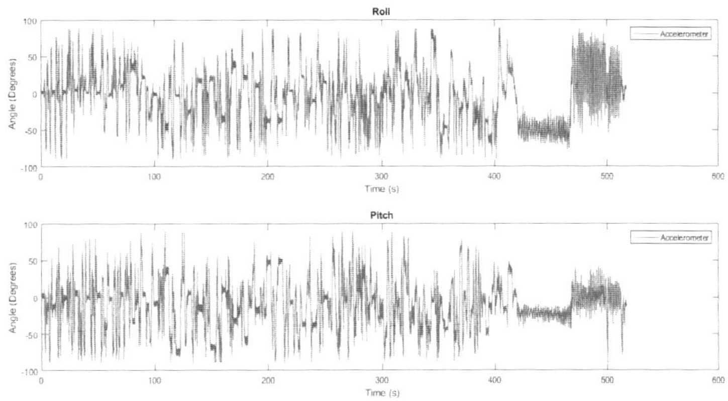

Figure 5. Roll and pitch angle(degrees) for Attitude estimation using raw Accelerometer values

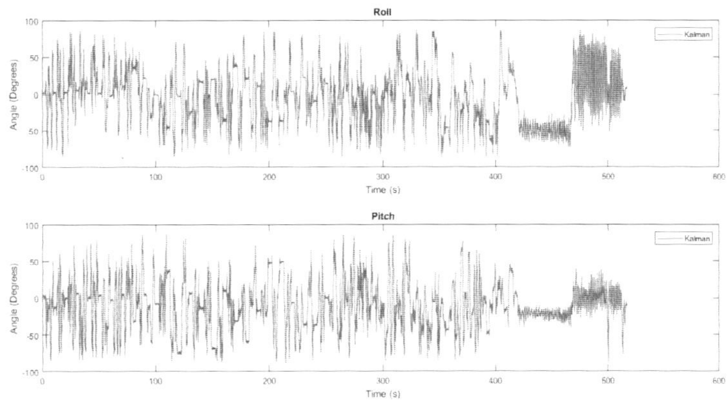

Figure 6. Roll and pitch angle(degrees) for Attitude estimation by Kalman filter

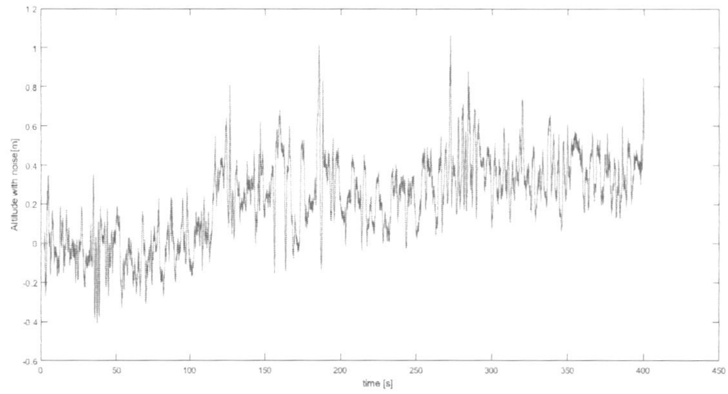

Figure 7. Measured barometric data in meters

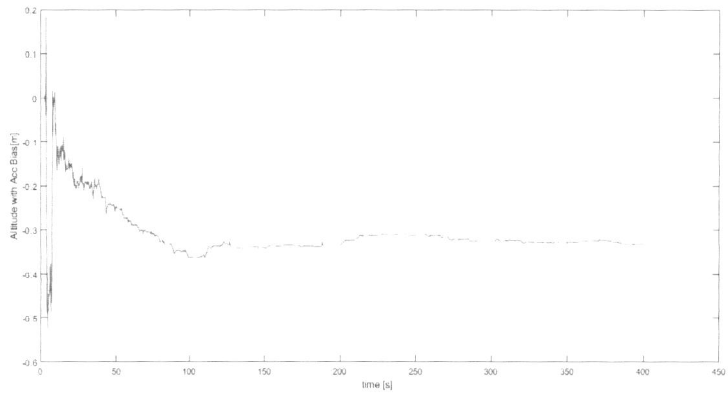

Figure 8. Altitude(meters) with Acceleration bias

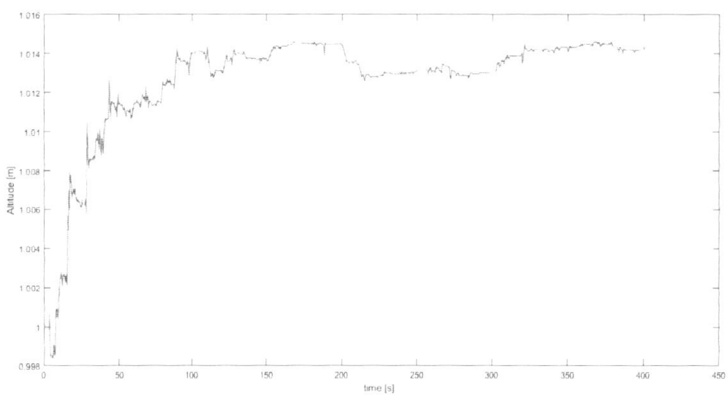

Figure 9. Altitude (meters) estimated

5. Conclusion

In this paper we proposed an algorithm for sensor fusion based on IMU and barometer for Altitude estimation. The Kalman filter is used as states estimator. The Kalman filter performs a method of Altitude estimation with the help of IMU and barometer together. The barometer is used with IMU based attitude estimation to get better results. The proposed method is based on attitude estimation using IMU sensors and EKF which use the linear vertical component of acceleration combined with barometric data. The simulation results show that the errors are modelled within a few seconds. Going forward, this algorithm will be used to find out velocity of vehicle.

6. References

[1] Sabatini, A., & Genovese, V. (2014). A sensor fusion method for tracking vertical velocity and height based on inertial and barometric altimeter measurements. Sensors, 14(8), 13324-13347.

[2] Zihajehzadeh, Shaghayegh, et al. "Integration of MEMS inertial and pressure sensors for vertical trajectory determination." IEEE transactions on Instrumentation and Measurement 64.3 (2014): 804-814.]

[3] Xu, Zhengyi, et al. "A robust floor localization method using inertial and barometer measurements." 2017 International Conference on Indoor Positioning and Indoor Navigation (IPIN). IEEE, 2017.

[4] García, Jesús, Jose Manuel Molina, and Jorge Trincado. "A Methodology for Design and Analysis of Sensor Fusion with Real Data in UAV Platforms." (2018).

[5] Tanigawa, Makoto, et al. "Drift-free dynamic height sensor using MEMS IMU aided by MEMS pressure sensor." 2008 5th Workshop on Positioning, Navigation and Communication. IEEE, 2008.

[6] Bevermeier, Maik, et al. "Barometric height estimation combined with map-matching in a loosely-coupled Kalman-filter." 2010 7th Workshop on Positioning, Navigation and Communication. IEEE, 2010.

[7] Hyyti, Heikki, and Arto Visala. "A DCM based attitude estimation algorithm for low-cost MEMS IMUs." International Journal of Navigation and Observation 2015 (2015).

[8] Son, Youngbin, and Seyoung Oh. "A barometer-IMU fusion method for vertical velocity and height estimation." 2015 IEEE SENSORS. IEEE, 2015.

A Non-Linear Improved Double-Integral Sliding Mode Controller (IDI-SMC) for Modeling and Simulation of 6 (DOF) Quadrotor system

Irfan Ahmad[1], Muwahida Liaquat[2] owais zarin khan[3] and Fahad Mumtaz Malik

[1]Department of Electrical Engineering College of Electrical & Mechanical Engineering National University of Sciences and Technology(NUST) Islamabad, Pakistan

[2]Department of Electrical Engineering College of Electrical & Mechanical Engineering National University of Sciences and Technology(NUST) Islamabad, Pakistan

[3]Department of Electrical Engineering College of Electrical & Mechanical Engineering National University of Sciences and Technology(NUST) Islamabad, Pakistan

irfankhandgk11@gmail.com

Abstract. Due to recent technological achievements, modeling and simulation of Unmanned Aerial Vehicles (UAVs) such as Quadrotor contracted great potential. In this paper, the Improved Double-Integral Sliding Mode Controller (IDI-SMC) scheme is presented to cope with the problem like chattering and tracking. Under-actuated six Degree Of Freedom (DOF) Quadrotor model is formulated through Newton-Euler-Lagrange (NEL) equations. The idea of design (IDI-SMC) is to implement Double Sliding Mode control with proper coefficient scheduling for switching surface with Hurwitz stability criteria. Furthermore, introducing the strong condition of the sliding surface made the results and Simulations became more efficient with Hurwitz stability for both translational and rotational motion of the 6 DOF Quadrotor system is achieved to show the effectiveness of the proposed design controller.

1. Introduction

In recent decades, the importance of unmanned air vehicles (UAVs) in intelligence robot system engineering, especially in control system engineering, has been increased manifold. Compared to other traditional UAVs, Quadrotor optimization is an interesting area and plays a very vital role in flight control due to several advantages [1]. The concept of Quadrotor was first given by Breguet-Riched in 1907. The quadrotor has been extensively considered in many domains. Quadrotor has numerous military and civil applications for having advantages of its smaller size, quadrotor used in military operations like surveillance, trajectory planning, navigation, searching and rescuing. Quadrotor can also be used in civil tasks like aerial photography, production in agriculture plants, surface, and transmissions line inspection, accurate poisoning and altitude control.

The contributions of researchers with several control techniques have been proposed for controlling of Quadrotor. For Implementation and simulations various linear, nonlinear and robust controllers have been studies. PID is used by Bouabdallah et al [2] for controlling Quadrotor which only deals with little disturbance. Li a Young also proposed a PID controller in the pole place technique considering with low speed of propellers [3]. For inner loop (poison) and outer loop (altitude) control, Hongping Liu used PID and PD named PID-PD controller respectively but It needs to online gain regulation [4]. Bilinear modeling and attitude control of quadrotor, Chee Hwee [5] Seah used a robust bilinear controller. To avoid noise, Model Reference Adaptive Control (MRAC) and PID control have been used [6]. Recently, Aws Abdulsalam comes with a Nonlinear PID controller for Quadrotor to ensure Hurwitz stability [7]. Linear parameter varying LPV-LQR control techniques are proposed by Carlos Trapiello which only preferred for a small quadrotor [8]. For the control and design of translational position and heading, T.L Wang proposed H^{∞} and a state feedback controller respectively [9] with the existence of little noise and small algorithm error. Zhi and Li [10] proposed a Lyapunov based SMC algorithm for a quadrotor.
For trajectory tracking, a robust SMC controller integrated with Human Machine Interface (HMI) has been implemented by M.Reinoso [11]. High and varying payload causes quadrotor to unstable and creates wind disturbance. An Adaptive fractional Sliding Mode Control technique based upon Backstepping has been simulated by Mohsen [12]. This scheme has the ability to successfully attenuate wind disturbance and cope with undesirable fluctuations in control angles and poisons due to its inertia. Through results and study, the ideal path trajectory has been achieved with minor transient changes. Recently, the discrete-time Sliding Mode Control technique (DSMC) has been proposed by J.Xiong for attitude and position control after continuous to discrete transformation [13]. Integral Backstepping Sliding Mode Controller (IBS-SMC) with an external disturbance procedure has been applied [14]. The result shows this algorithm robust to disturbance as well as a good tracking problem. Recently, Feedback Linearization [FL] and Backstepping techniques have been proposed by M.A-Vvallego [15]. First, by reducing the system model by FL and after implementing Leader-Follower (LF) based and integrated with Backstepping controller which resulted asymptotic stability for a quadrotor.To achieve asymptotically stability for desired states trajectories nonlinear feedback controller is applied with neglecting high order Lie derivatives [16]. A strong robust based sliding mode and fuzzy controllers for inner loop and Feedback linearization for outer loop has been introduced [17].
In this paper, Improved Double Integral Sliding Mode control (IDI-SMC) is presented which strongly robust to uncertainties with proper gain scheduling and to tackling chattering. In several controller coefficients and gains are designed by simulations. Here introducing proper gain called switching surface coefficient by using of Eigenvalue which based upon Hurwitz criteria. Furthermore, for a sliding mode controller, the proper and strong sliding surface is introduced to ensure the surface remains to zero so that trajectories stay on it.
The rest of this paper has been arranged as follows. In section 2, modeling of simulation derived through Newton's Euler equations. Mathematical modeling of three different control techniques has been presented in sections 3. In section 4, results verified through simulations have been obtained. In last, the conclusion and final discussion have been discussed briefly.

2. Mathematical Modeling

In this section, the Quadrotor model was developed by using Newton-Euler formulations. A Cross like the structure of Quadrotor having four propellers with a control system placed at the center shown in Fig [1]. The power of the propeller supplied by DC motor. Two of them (1, 3) propellers move the same direction (anti-clockwise) and the other two move in clockwise directions to neutralize or cancel the effect of thrust force. A 6 DOF [ϕ, θ, ψ, x, y, z] Quadrotor model is basically Under-actuated in nature with six states and only four inputs which make it difficult to flight control [2]. The system is sub-divided into two parts which are rotationally and transnationally parts having states roll (ϕ), pitch

IOP Conf. Series: Materials Science and Engineering **853** (2020) 012035 doi:10.1088/1757-899X/853/1/012035

(θ) and yaw (ψ) with position (x y z) respectively. The control diagram altitude (x y z) and heading and attitude with heading are shown in Fig [1]. Tracking, stabilizing, and orientation of Quadrotor can be achieved by thrust force and angular torque.
The quadrotor model is divided into two parts. Rotational dynamics roll, yaw, and pitch (ϕ, θ, ψ) and transnational dynamics (x,y,z) have been formulated around X, Y, and Z axes respectively. The rotation matrix R is given by

$$R = \begin{pmatrix} c\theta c\psi & s\phi s\theta c\psi & c\phi s\theta c\psi + s\phi s\psi \\ c\theta s\psi & s\phi s\theta s\psi + c\theta c\psi & c\phi s\theta s\psi - s\theta c\psi \\ -s\theta & s\phi c\theta & c\phi c\theta \end{pmatrix} \quad where \quad c\theta = \cos(\theta) \text{ and } s\theta = \sin(\theta) \tag{1}$$

2.1. Rotational Equations Dynamics

Rotational equations roll, pitch and yaw (ϕ, θ, and ψ) have been derived from body inertia frame (x, y, and z) with
the help of N-Euler equation the total torque is given as:

$$I\dot{w} + w \times Iw + M_G = \tau \tag{2}$$

where
I = Diagonal inertia matrix, w is Angular velocity , M_G is Gyroscopic moments and τ *is* Total torque

The inertia matrix contains diagonal entries with their product is equal to zero having moments of inertia I_{xx}, I_{yy} and I_{zz}

$$I = \begin{pmatrix} I_{xx} & 0 & 0 \\ 0 & I_{yy} & 0 \\ 0 & 0 & I_{zz} \end{pmatrix} \tag{3}$$

From these moments rotation is produced from rotors and creates a force called aerodynamics forces or simply called lift forces. The forces and moments are shown in equation (4)

$$M^j = \frac{1}{2}\rho A C_T r^2 \Omega_{j^2} \Rightarrow K_m \Omega_{j^2} \ ; \ F^j = \frac{1}{2}\rho A C_T r^2 \Omega_{j^2} \Rightarrow K_f \Omega_{j^2} \tag{4}$$

Where K_f and K_m are the constants, ρ is Air density, C_T is Aerodynamic coefficient and Ω_{j^2} is j^{th} Angular velocity. The total moments about x,y, and z-axis has been calculated as

$$\tau = \begin{bmatrix} M_x = lK_f(\Omega_4^2 - \Omega_2^2) \\ M_y = lK_f(\Omega_1^2 - \Omega_2^2) \\ M_z = K_m(\Omega_1^2 - \Omega_2^2 + \Omega_3^2 - \Omega_4^2) \end{bmatrix} \tag{5}$$

2.2. Translational Quadrotor Equations Dynamics

Translational equations of motions can be derived from using Newton's second law according to earth inertia frame

$$\sum \vec{F} = m\vec{a} \Rightarrow m\ddot{r} = \begin{bmatrix} 0 \\ 0 \\ mg \end{bmatrix} + RF_B \tag{6}$$

Where $r = [x\, y\, z]^T$ is Quadratic distance, m is Quadrotor mass while F_B is a force without gravitational. Let we have a state vector X and input vector U can be defined as:

$$X = \begin{bmatrix} x_1 & x_2 & x_3 & x_4 & x_5 & x_6 & x_7 & x_8 & x_9 & x_{10} & x_{11} & x_{12} \\ \Downarrow & \Downarrow & \Downarrow & \Downarrow & \Downarrow & \Downarrow & \Downarrow & \Downarrow & \Downarrow & \Downarrow & \Downarrow & \Downarrow \\ \phi & \dot{\phi} & \theta & \dot{\theta} & \psi & \dot{\psi} & x & \dot{x} & y & \dot{y} & z & \dot{z} \end{bmatrix}^T \quad \left\{ \begin{array}{l} U_1 = K_f(\Omega_1^2 + \Omega_2^2 + \Omega_3^2 + \Omega_4^2) \\ U_2 = K_f(-\Omega_2^2 + \Omega_4^2) \\ U_3 = K_f(\Omega_1^2 - \Omega_3^2) \\ U_4 = K_f(\Omega_1^2 - \Omega_2^2 + \Omega_3^2 - \Omega_4^2) \end{array} \right\} \tag{7}$$

Substituting equations (7) to (5) through mapping state vector X we get total torque from rotational and translational equations of motion and simplifying we get State Space model shown in Eq. (9)

$$\begin{array}{ll} \ddot{\phi} = b_1U_2 - a_2x_4\Omega_r + a_1x_4x_6 & \ddot{x} = \dfrac{-U_1}{m}(\sin x_1 \sin x_5 + \cos x_1 \cos x_5 \sin x_3) \\ \ddot{\theta} = b_2U_3 - a_4x_2\Omega_r + a_3x_2x_6 & \ddot{y} = \dfrac{-U_1}{m}(\cos x_1 \sin x_5 \sin x_3 - \cos x_5 \sin x_3) \\ \ddot{\psi} = a_5x_2x_4 + b_3U_4 & \ddot{z} = g - \dfrac{U_1}{m}(\cos x_1 \cos x_3) \end{array} \tag{8}$$

3. Controller Design

The proposed scheme has been implemented in three different phases for modeling and simulation of Quadrotor. First, one Sliding mode control which deals with uncertainties but produces chattering reaching to steady-state. To deal this further leads to building a double integral sliding mode controller in which Chattering successfully has been reduced. The proposed scheme has been implemented in three different phases for modeling and simulation of Quadrotor. Furthermore, for proper gain scheduling parameterization of sliding surface has been introduced by comparing sliding surface constants gains to the system given Eigenvalue to achieve Hurwitz stability this technique named Improved Double Integral Sliding Mode Controller (IDI-SMC).

$$\left\{ \begin{array}{l} \dot{x}_1 = \phi = x_2 \\ \dot{x}_2 = \dot{\phi} = b_1U_2 - a_2x_4\Omega_r + a_1x_4x_6 \\ \dot{x}_3 = \theta = x_4 \\ \dot{x}_4 = \dot{\theta} = b_2U_3 - a_4x_2\Omega_r + a_3x_2x_6 \\ \dot{x}_5 = \psi = x_6 \\ \dot{x}_6 = \dot{\psi} = a_5x_2x_4 + b_3U_4 \\ \dot{x}_7 = z = x_8 \\ \dot{x}_8 = \dot{z} = g - \dfrac{U_1}{m}(\cos x_1 \cos x_3) \\ \dot{x}_9 = x = x_{10} \\ \dot{x}_{10} = \dot{x} = \dfrac{-U_1}{m}(\sin x_1 \sin x_5 + \cos x_1 \cos x_5 \sin x_3) \\ \dot{x}_{11} = y = x_{12} \\ \dot{x}_{12} = \dot{y} = \dfrac{-U_1}{m}(\cos x_1 \sin x_5 \sin x_3 - \cos x_5 \sin x_3) \end{array} \right\} \tag{9}$$

Let general state space equation is:

$$X(t)^{(n)} = f(x,t) + g(x,t)u \tag{10}$$

Where $X(t)$ the state is vector and u is control input. For designing of sliding mode controller, two steps are necessary. Step 1: The choice of the sliding surface and step 2: The verification of control law.
Generally, we have a sliding surface like in equation (21).

$$S(x) = ce + \dot{e} \tag{11}$$

Where c is a positive constant value and e is an error which is $e = x - x_d$. For control and verification, Lyapunov function and from theory $V(s)$ should be pos2itive and $\dot{V}(s)$ should be negative definite for stability criteria described below

$$V(s,x,t) = \frac{1}{2}S^2(x,t) \text{ and } \dot{V}(s,x,t) = \dot{V}(s) = S\dot{S} - < \eta|s| \tag{12}$$

Where η, K_1 and K_2 are a positive value (K_2, η, K_1>0). After this it obtained two control law called switching and equivalent terms:

$$u = u_{eq} + u_s \; and \; u_s = -K_1 sign(s) - K_2(s) \; where\, sign(s) = \begin{cases} -1 & if \;\; s<0 \\ 1 & if \;\; s>0 \end{cases} \tag{13}$$

3.1. Sliding Mode Controller for Quadrotor

Now for controlling of row angle let we have to design errors and Sliding surface

$$V(s,x,t) = \frac{1}{2}S^2(x,t) \text{ and } \dot{V}(s,x,t) = \dot{V}(s) = S\dot{S} - < \eta|s| \tag{14}$$

From equation (14) we get control law for row angle which is given in equation (25).

$$U_2 = \frac{1}{b_1}\left[K_1 sign(s) + K_2(s) + c_1(\dot{\phi}_d - \dot{\phi}) + \ddot{\phi}_d + a_2\dot{\theta}\Omega_r - a_1\dot{\theta}\dot{\psi}\right] \tag{15}$$

Similarly, for control inputs for Pitch, Yaw and Altitude $\left[U_3, U_4, U_1\right]$ have been given in the equation. (16).The
the phases of the sliding surface have been shown in Figure-2.

$$\begin{aligned} &= \frac{1}{b_2}\left[K_1 sign(s) + K_2(s) + c_1(\dot{\theta}_d - \dot{\theta}) + \ddot{\theta}_d + a_4\dot{\phi}\Omega_r - a_3\dot{\phi}\dot{\psi}\right] \\ &= \frac{1}{b_3}\left[K_1 sign(s) + K_2(s) + c_1(\dot{\psi}_d - \dot{\psi}) + \ddot{\psi}_d + a_5\dot{\phi}\dot{\theta}\right] \\ &= \frac{m}{\cos\phi\cos\theta}\left[K_1 sign(s) + K_2(s) + c_1(\dot{z}_d - \dot{z}) - \ddot{z} + g\right] \end{aligned} \tag{16}$$

A double integral sliding mode controller has been used in this technique for the removal of the steady-state error. To tackle chattering across steady-state integral control also has been preferred. To drive control law mathematical modeling has been given below.

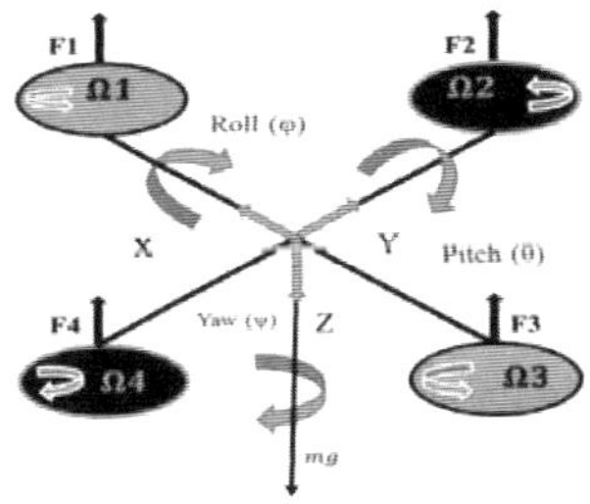

Figure 1. Configuration and schematic of the quadrotor.

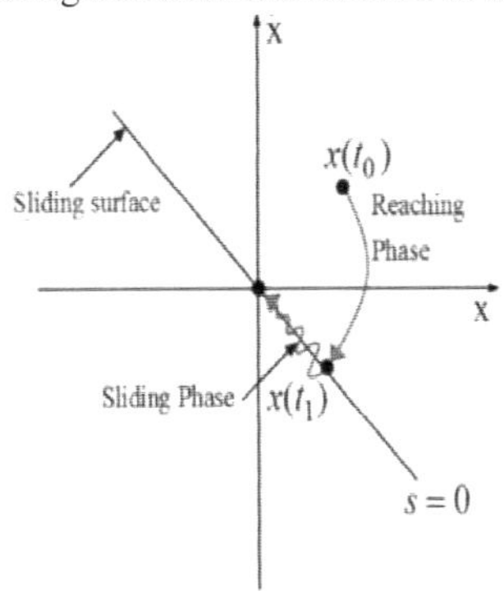

Figure 2. Sliding Phases

Let for Roll angle we can drive control law for Roll angle with their error.

$$\dot{x}_1 = x_2 \Rightarrow \dot{x}_2 = x_4 x_6 a_1 - x_4 \Omega_r a_2 + b_1 U_2 \qquad e_1 = x_1 - x_{1ref},\ \ e_2 = \int e_1 dt\ \ e_3 = \int\left\{\int e_1 dt\right\} dt \quad \therefore \ddot{e}_3 = \dot{e}_2 = e_1 \tag{17}$$

Sliding Surface is given below

$$\dot{S}_1 = (c_1+1)(x_2 - \dot{x}_{1ref}) + (c_2+1)e_1 + c_3 e_2 + x_4 x_6 a_1 - x_4 \Omega_r a_2 + b_1 U_2 - \ddot{x}_{1ref} \tag{18}$$

Also $\dot{S}_1 = -K_1 sign(s) - K_2(s)$ (19)

Now from Eq. (18), (19) a control law has been developed for Double Sliding Mode Controller shown in Eq. (20).

$$U_2 = \frac{1}{b_1}\left[-(c_1+1)(x_2 - \dot{x}_{1ref}) - (c_2+1)e_1 - c_3 e_2 - x_4 x_6 a_1 + x_4 \Omega_r a_2 + \ddot{x}_{1ref} - K_1 sign(s) - K_2(s)\right] \tag{20}$$

Similarly,for control inputs for Pitch, Yaw and Altitude $[U_3, U_4, U_1]$ have been given in equations (21)

$$U_3 = \frac{1}{b_2}\left[-(c_1+1)(x_4 - \dot{x}_{3ref}) - (c_2+1)e_1 - c_3 e_2 - x_2 x_6 a_3 - x_2 \Omega_r a_4 + \ddot{x}_{3ref} - K_1 sign(s) - K_2(s)\right]$$

$$U_4 = \frac{1}{b_3}\left[-(c_1+1)(x_6 - \dot{x}_{5ref}) - (c_2+1)e_1 - c_3 e_2 - x_2 x_4 a_5 + \ddot{x}_{5ref} - K_1 sign(s) - K_2(s)\right] \tag{21}$$

$$U_1 = \frac{m}{\cos x_1 \cos x_3}\left[-(c_1+1)(x_8 - \dot{x}_{7ref}) - c_3 e_2 - (c_2+1)e_1 - g + \ddot{x}_{7ref} - K_1 sign(s) - K_2(s)\right]$$

3.2. Determination of Switching Surface coefficient

To determine the switching surface coefficient S and $\dot{S}$ should be taken as zero. The sliding surface and errors have given below

$$S = a_1 e_1 + a_2 e_2 + a_3 e_3 \qquad e_1 = x_1 - x_{1ref},\ \ e_2 = \int e_1 dt\ \ e_3 = \int \left\{ \int e_1 dt \right\} dt \tag{22}$$

When S taken as zero we get error

$$e_1 = -\frac{a_2}{a_1} e_2 - \frac{a_3}{a_1} e_3 \ \Rightarrow -\frac{a_2}{a_1} \int x_1 - x_{1ref} dt - \frac{a_3}{a_1} \int \left\{ \int x_1 - x_{1ref} dt \right\} dt \tag{23}$$

Considering $y_1 = e_3, y_2 = e_2$ and $y_3 = e_1$ we get

$$\dot{y}_1 = y_2 \Rightarrow \dot{y}_2 - \frac{a_2}{a_1} y_2 - \frac{a_3}{a_1} y_1 \tag{24}$$

After linearizing with Tayler series expansion around equilibrium point we get set of the equation as

$$\begin{bmatrix} \dot{y}_1 \\ \dot{y}_2 \end{bmatrix} = \begin{pmatrix} 0 & 1 \\ -\frac{a_3}{a_1} & -\frac{a_2}{a_1} \end{pmatrix} \Rightarrow \lambda^2 + \frac{a_2}{a_1} \lambda + \frac{a_3}{a_1} \tag{25}$$

According to Hurwitz's stability criteria, a matrix is stable if real parts of Eigen-values lie on the left half-plane. The coefficient of the surface will be obtained by using $\lambda I - A$ comparing with desirable stable points such as $(\lambda + 1)(\lambda + 2)$ and assuming $a_3 = 1$ we get a sliding surface coefficient.

$$\frac{a_2}{a_1} = 3,\ \frac{a_3}{a_1} = 2,\ a_3 = 1 \tag{26}$$

4. Results and Simulations

In this section, Simulations of proposed scheme controller and other controller has been done in the MATLAB /SIMULINK environment. In Fig. (3), the graphs of the roll (ϕ), pitch (θ) and yaw (ψ) with Altitude have been shown to differentiate these controllers respectively Green, blue and red line shown for SMC, Double Integral SMC and Improved Double Integral Sliding Mode Control for Row, angels respectively. As from Figure Simple Sliding Mode controller shows overshoot with a steady-state error and it produces chattering. In the blue, the chattering is successfully removed by the improved double integral sliding mode controller. The graph which is for IDI-SMC has been shown in the red line. The effectiveness of the proposed graph. The reference trajectory for a roll (ϕ), pitch (θ) and yaw (ψ) are taken as 5° where Altitude reference is taken as 2°. In all cases, it has been seen that the SMC controller is good but due to the sliding surface it produces chattering around steady-state. As the integral is preferred for reducing steady-state error to achieve asymptotically stability so applying double integral SMC the chattering has been successfully reduced In the presence of three different control techniques, IDI-SMC based trajectories show better performance.

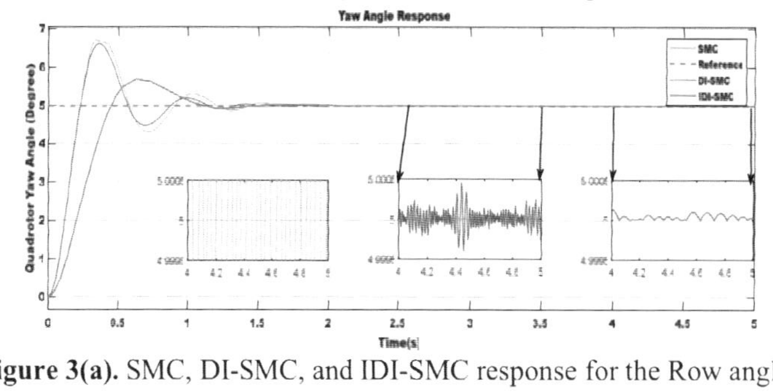

Figure 3(a). SMC, DI-SMC, and IDI-SMC response for the Row angle

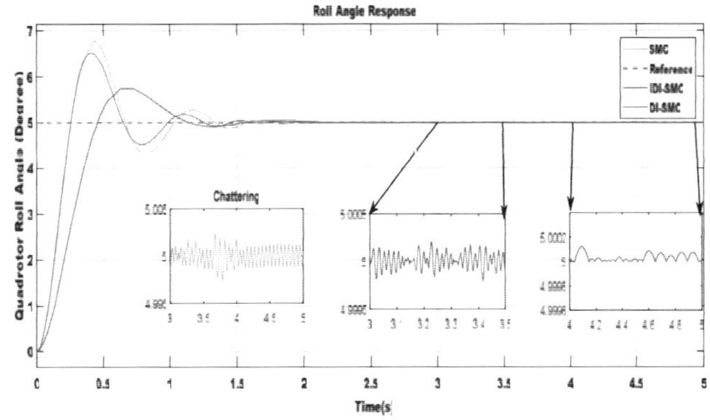

Figure 3(b). Responses of Yaw angle

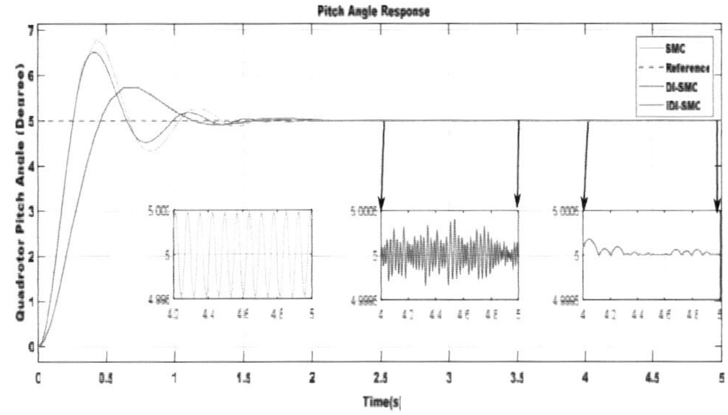

Figure 3(c). Responses of Pitch angle

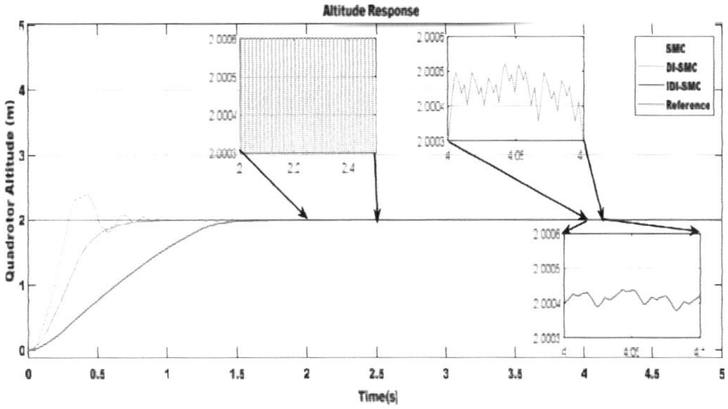

Figure 3(d). Responses of Altitude angle angle

The performance of altitude control has been shown in Fig.3(d) which shows that if Euler angles are not controlled directly, altitude still remains stable. These Schemes have been verified by these simulations. As the integral is preferred for reducing steady-state error to active asymptotically stability so applying double integral SMC the chattering has been successfully reduced. Furthermore, after the parameterizations of sliding surface gain, comprehensive results have been achieved. It can be clearly seen that all states of the quadrotor are converged to their desired states while all control errors converge to zeros shown in Fig (4).

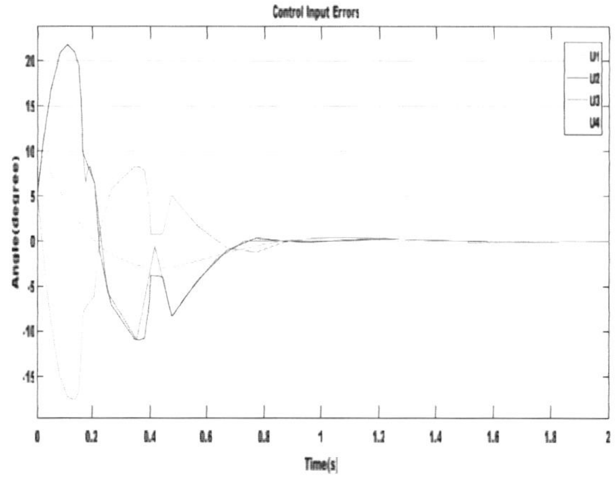

Figure 4. Control inputs and errors

Table 1. Quadrotor Parameters

Parameter	Value	Unit
I_{XX}	7.5e-3	kg.m^2
I_{YY}	7.5e-3	kg.m^2
I_{zz}	1.4e-2	kg.m^2
l	0.24	M
J_r	6e-5	kg.m^2
m	0.65	kg
K_f	3.13e-5	Ns^2
K_m	7.5e-7	Nm s^2
K_t	diag(0.1,0.1,0.15)	Nm s

5. Conclusion

In this paper, we presented an efficient controller technique for a quadrotor. Firstly, start with the development of a dynamic model we design Sliding Mode Controller for a quadrotor. Steady-state errors of the control system have been eliminated by augmented improved double integral action taking the strong condition of the sliding surface. Furthermore rather than taking assumption we introduce parameterization technique through Eigen-values made this scheme valuable. In this way, strong controller techniques have been developed. The simulations results proved satisfactory results addressed by IDI-SMC and proved the effectiveness of the proposed scheme. By further improvements and designing of parameters of sliding surface, the control effort could be less and control will be more valuable.

References

[1] Lee, Sangheon, Dipak Kumar Giri, and Hungsun Son. "Modeling and control of quadrotor UAV subject to variations in center of gravity and mass." In 2017 14th International Conference on Ubiquitous Robots and Ambient Intelligence (URAI), pp. 85-90. IEEE, 2017.

[2] Bouabdallah, Samir, Andre Noth, and Roland Siegwart. "PID vs LQ control techniques applied to an indoor micro quadrotor." In 2004 IEEE/RSJ International Conference on Intelligent

Robots and Systems (IROS)(IEEE Cat. No. 04CH37566), vol. 3, pp. 2451-2456. IEEE, 2004.

[3] Li, Jun, and Yuntang Li. "Dynamic analysis and PID control for a quadrotor." In 2011 IEEE International Conference on Mechatronics and Automation, pp. 573-578. IEEE, 2011.

[4] Liu, Hongping, and Guanbin Gao. "Dynamic modeling and analyzing for a novel X-Quadrotor." In 2017 3rd International Conference on Computational Intelligence & Communication Technology (CICT), pp. 1-4. IEEE, 2017.

[5] Seah, Chee Hwee, Isonguyo J. Inyang, and James F. Whidborne. "Bilinear modelling and attitude control of a quadrotor." IFAC-PapersOnLine 50, no. 2 (2017): 193-198.

[6] Zeng, Yong, Qiang Jiang, Qiang Liu, and Hua Jing. "PID vs. MRAC control techniques applied to a quadrotor's attitude." In 2012 Second International Conference on Instrumentation, Measurement, Computer, Communication and Control, pp. 1086-1089. IEEE, 2012.

[7] Najm, Aws Abdulsalam, and Ibraheem Kasim Ibraheem. "Nonlinear PID controller design for a 6-DOF UAV quadrotor system." Engineering Science and Technology, an International Journal (2019).

[8] Liu, Shida, Zhongsheng Hou, and Jian Zheng. "Attitude adjustment of quadrotor aircraft platform via a data-driven model free adaptive control cascaded with intelligent PID." In 2016 Chinese Control and Decision Conference (CCDC), pp. 4971-4976. IEEE, 2016.Trapiello, Carlos, Vicenç Puig, and Bernardo Morcego. "Position-heading quadrotor control using LPV techniques." IET Control Theory & Applications 13, no. 6 (2019): 783-794.

[9] Wang, Haibo, Zheng Li, Hongyun Xiong, and Xiaohong Nian. "Robust H∞ attitude tracking control of a quadrotor UAV on SO (3) via variation-based linearization and interval matrix approach." ISA transactions 87 (2019): 10-16

[10] Li, Zhi, Xin Ma, and Yibin Li. "Robust tracking control strategy for a quadrotor using RPD-SMC and RISE." Neurocomputing331 (2019): 312-322.

[11] Reinoso, Manuel J., Luis I. Minchala, Paul Ortiz, Darwin F. Astudillo, and Diego Verdugo. "Trajectory tracking of a quadrotor using sliding mode control." IEEE Latin America Transactions 14, no. 5 (2016): 2157-2166.

[12] Vahdanipour, Mohsen, and Mahdi Khodabandeh. "Adaptive fractional order sliding mode control for a quadrotor with a varying load." Aerospace Science and Technology 86 (2019): 737-747.

[13] Xiong, Jing-Jing, and Guobao Zhang. "Discrete-time sliding mode control for a quadrotor UAV." Optik-International Journal for Light and Electron Optics 127, no. 8 (2016): 3718-3722.

[14] Jia, Zhenyue, Jianqiao Yu, Yuesong Mei, Yongbo Chen, Yuanchuan Shen, and Xiaolin Ai. "Integral backstepping sliding mode control for quadrotor helicopter under external uncertain disturbances." Aerospace Science and Technology68 (2017): 299-307

[15] Vallejo-Alarcón, M. A., R. Castro-Linares, and M. Velasco-Villa. "Unicycle-type robot & quadrotor leader-follower formation backstepping control." IFAC-PapersOnLine 48, no. 19 (2015): 51-56.

[16] He, Ze Fang, and Long Zhao.Internal model control/backstepping sliding model control for quadrotor trajectory tracking.In 2017 IEEE 2nd Information Technology, Networking, Electronic and Automation Control Conference (ITNEC), pp. 1254-1258. IEEE.

[17] Lee, Hyeonbeom, and H. Jin Kim. "Robust control of a quadrotor using takagi-sugeno fuzzy model and an lmi approach." In 2014 14th International Conference on Control, Automation and Systems (ICCAS 2014), pp. 370-374. IEEE, 2014.

IOP Conf. Series: Materials Science and Engineering **853** (2020) 012036 doi:10.1088/1757-899X/853/1/012036

Research on Adaptive Reclosing Discriminant Method Based on LMD Sample Entropy

Liqun Shang[1], Yuan Zhang[1] and Chenguang Yang[1]

[1]College of electrical and control engineering of Xi'an University of Science And Technology, Xi'an, China

1224765156@qq.com

Abstract. Based on the adaptive time-frequency analysis characteristics of local mean decomposition and the nonlinear quantization ability of sample entropy, an adaptive reclosing discriminant method based on LMD sample entropy is proposed. Firstly, the local mean decomposition method is applied to decompose the original signal of the transmission line fault into several product function components (PF component). Then calculate the sample entropy of the first three PF components to achieve feature quantization of the PF component. Finally, the sum of the sample entropy of first three PF components is used as the eigenvalue to distinguish the fault type of the transmission line. Obtained from the simulation results, the method can identify the type of fault.

1. Introduction

The fault rate of the transmission line is the highest in the whole power system component [1]. The reclosing device can quickly restore normal power supply and effectively improve the stability of power system operation. So it is necessary to judge whether the fault nature of the line is permanent or transient before reclosing action. Electric power researchers at home and abroad have done a lot of research on adaptive reclosing [2-4]. In terms of the type of fault nature, the method of using the steady-state power-frequency voltage of the fault phase and the non-fault phase accounts for the majority [5].

The secondary arc voltage waveform distortion of the transient short-circuit ground fault is very serious, contains a large amount of high-frequency components, and the content of each frequency component is unstable and changes with time. In this paper, permanent fault and transient fault are distinguished according to the complexity of fault phase voltage signal after tripping. The local mean decomposition (LMD) is used to decompose the fault signal into PF components with physical meaning, and the sample entropy is used to calculate the complexity of PF components as eigenvalues to distinguish the fault properties.

2. Analysis of Arc Characteristics

According to the period of fault arc generation, the transient short circuit grounding fault arc can be classified as: the primary arc from the occurrence of the fault to the protection action on both sides of the line and the secondary arc from the protection action to the complete extinguishing of the arc.

A large amplitude short-circuit current flows through the primary arc. Due to the continuous energy support of the system power supply, the arc length of the arc is relatively stable. Moreover, the arc drop is mainly concentrated on the arc column, and the pressure drop at the arc angle is negligible.
There is a repeated process of reburning and extinguishing of the secondary arc. Only when the actual arc voltage is no longer greater than the arc reburning voltage can the secondary arc be completely extinguished [6].
The voltage waveforms of the primary arc and the secondary arc are shown in Fig.1 and Fig.2.

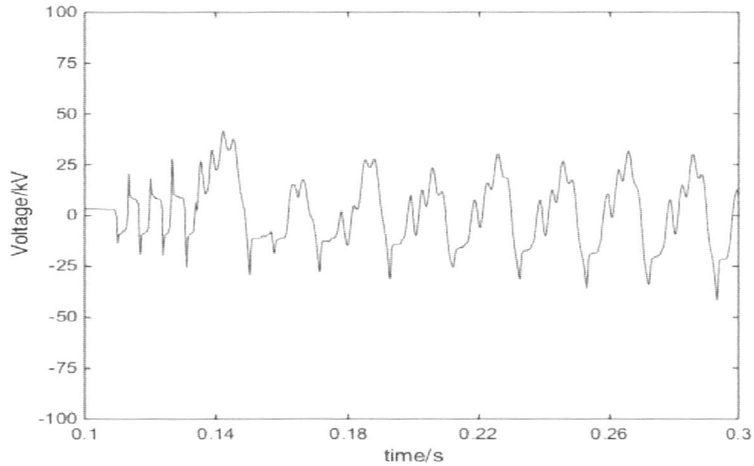

Figure 1. Voltage waveform of secondary arc

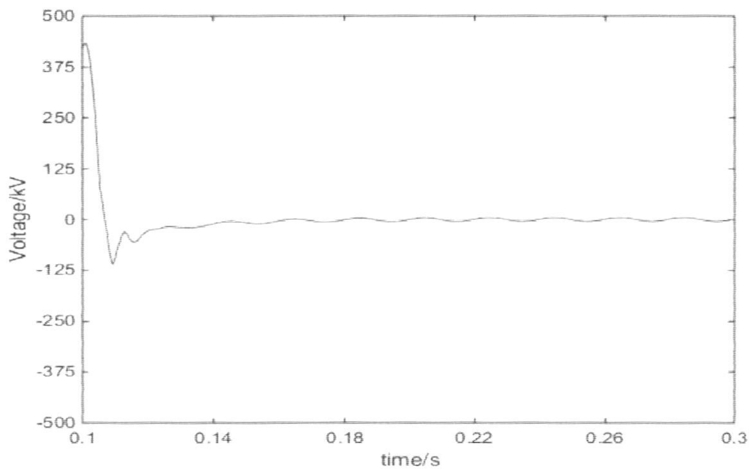

Figure 2. Voltage waveform of primary arc

3. Simulation of single-phase ground fault

Using the EHV transmission line with a length of 478.9km and a voltage of 750kV as a practical example of simulation.

(1) The single-phase grounding transient fault of transmission line is simulated by ATP/EMTP simulation software, which is shown in Fig.3. The waveform of the fault phase terminal voltage is shown in Fig.4.

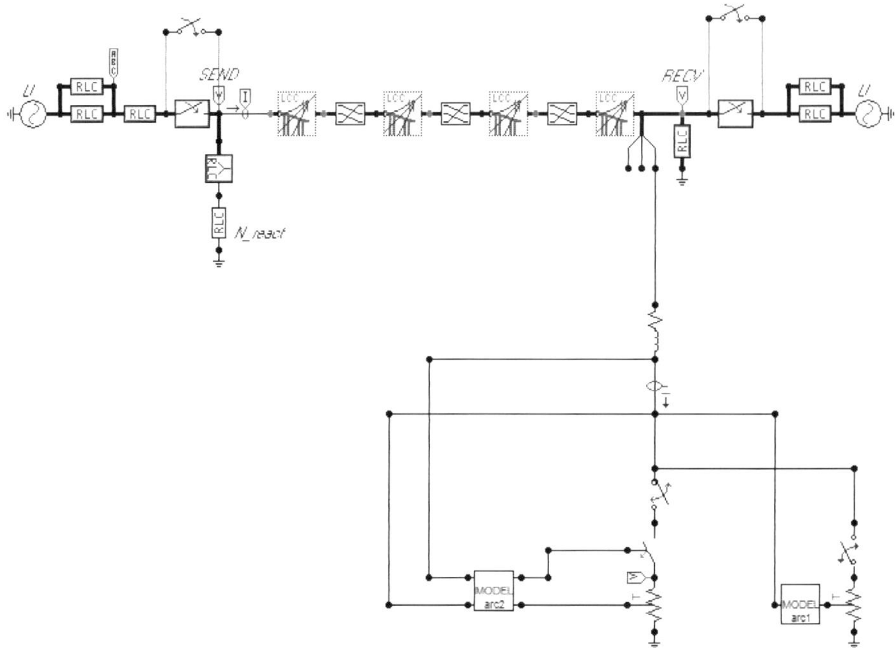

Figure 3. Simulation Model of transient single-phase grounding short-circuit fault

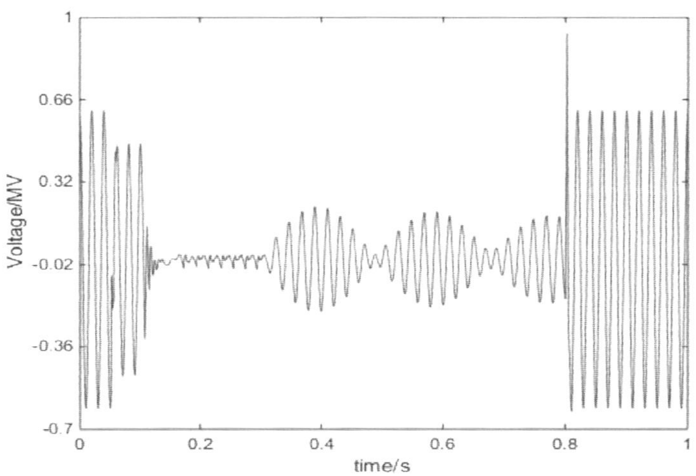

Figure 4. The waveform of the fault phase terminal voltage

The fault occurs in 50ms, and the circuit breaker at both ends jumps off in 100ms. The arc extinguishing time is 300 ms, and the reclosing time is 800 ms, sampling frequency is 10kHz.A phase is taken as the fault phase, and the phase voltage waveform of the transmission line in different places and in the case of different grounding resistance is simulated.

(2) The single-phase grounding permanent fault of transmission line is simulated by ATP/EMTP simulation software, as shown in Fig.5. The setting fault occurs in 50ms, and the circuit breaker at both ends jumps off at 100ms, and the sampling frequency is 10kHz. The waveform of the fault phase terminal voltage is shown in Fig.6.

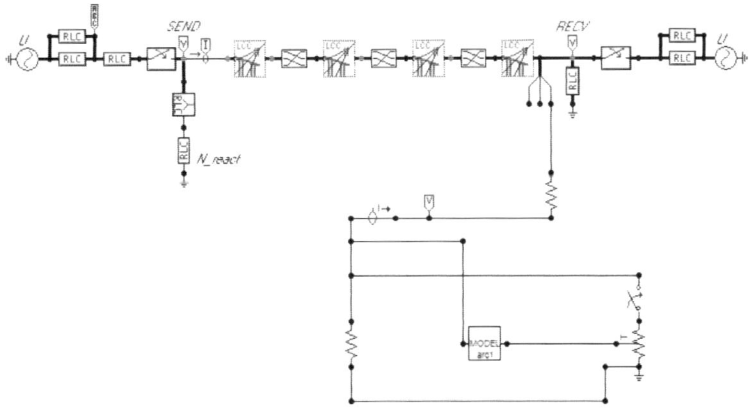

Figure 5. Simulation Model of permanent single-phase grounding short Circuit Fault

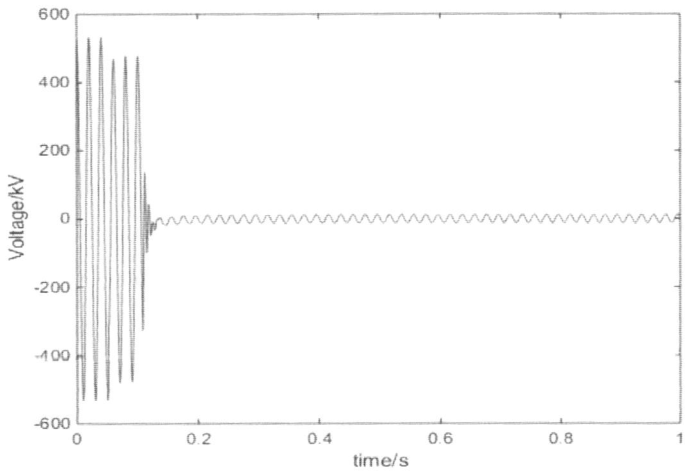

Figure 6. The waveform of the fault phase terminal voltage

4. Theoretical background

4.1. Local mean decomposition

The basic idea of LMD method is to adaptively decompose complex nonlinear nonstationary signals into finite product function (PF) components.

For signal x (t), the LMD decomposition steps [7] are as follows.

(1) By calculating all the local extreme points of x(t), the local mean m_i and envelope estimation values a_i of two adjacent extreme points n_i and n_{i+1} can be obtained as follows.

For all local mean values, the local mean function $m_{11}(t)$ is obtained by moving average method, and the envelope estimation function $a_{11}(t)$.

(2) Remove $m_{11}(t)$ available $h_{11}(t)$ from signal x (t), then decoupling $h_{11}(t)$.

$$s_{11}(t) = \frac{h_{11}(t)}{a_{11}(t)} \tag{1}$$

Repeat steps (1)-(2) until $s_{1n}(t)$ is a pure frequency modulation signal, that is, the envelope estimation function $a_{1(n+1)}(t)$ of $s_{1n}(t)$ is equal to 1.

(3) At the end of the iteration, the envelope signal $a_1(t)$ can be obtained.

$$a_1(t) = a_{11}(t)a_{12}(t)\cdots a_{1n}(t) = \prod_{k=1}^{n} a_{1k}(t) \tag{2}$$

In the formula, k is the number of iterations.

(4) The first PF component of the signal x (t) is

$$P_{F1}(t) = a_1(t)s_{1n}(t) \tag{3}$$

The instantaneous amplitude is $a_1(t)$ and the instantaneous frequency is

$$f_1(t) = \frac{1}{2\pi}\frac{d(\arccos(s_{1n}(t)))}{dt} \tag{4}$$

(5) Repeat steps (1)-(4) until $u_k(t)$ monotone

At this point, the original signal $x(t)$ can be represented as

$$x(t) = \sum_{i=1}^{k} P_{F1}(t) + u_k(t) \tag{5}$$

4.2. sample entropy

Sample entropy (SE) is an improvement of approximate entropy algorithm, and the accuracy of the result is better than that of approximate entropy. The more complex the analysed sequence, the larger the corresponding SE value. On the contrary, the higher the self-similarity of the sequence, the smaller the SE value. The sample entropy needs only a small amount of data to quantitatively analyse the self-similarity and complexity of time series data, so it has been widely used in the field of engineering [8]. At present, most scholars use local mean decomposition method and information entropy in rotating machinery fault diagnosis, but less in transmission line fault diagnosis.

5. Algorithm simulation analysis

It can be seen from a large number of simulated voltage waveforms that the single-phase grounding fault occurs in the transmission line, the grounding resistance and the fault occurrence point are different, and the voltage waveform collected at the transmission end and the receiving end will be different from that of the circuit breaker at the first arc stage and at the moment when the circuit breaker is disconnected, but the secondary arc stage will not be affected by the difference between the fault point and the transition resistance after the circuit breaker has been disconnected for two power frequency cycles. Therefore, in order to avoid the influence of the transient over-voltage when the circuit breaker is open, the fault signal of the transient fault and the permanent fault is collected at 10000 Hz after two power frequency cycles. A total of 18 sets of data were collected.

Using LMD to decompose the fault signal, the decomposition results of one set of data of transient and permanent faults are shown in Fig.7 and Fig.8.

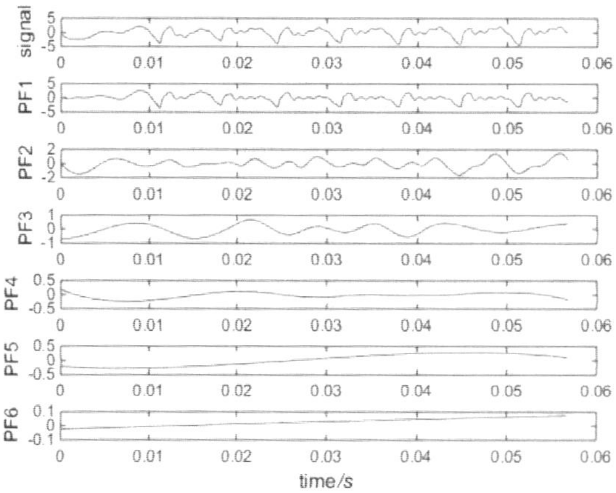

Figure 7. The LMD decomposition results of permanent fault signals

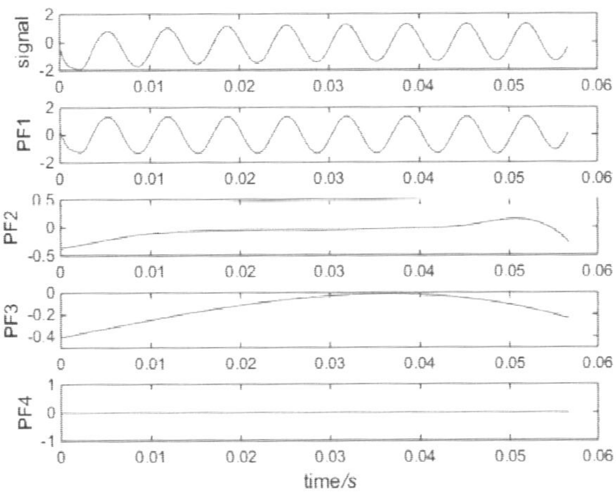

Figure 8. The LMD decomposition results of transient fault signals

From the results of LMD decomposition, it can be seen that the information of fault signal is mainly concentrated in the first three PF components, so the first three PF components are selected as the analysis object. Each group of fault signals is decomposed by LMD, the sample entropy of each PF component is obtained, and the sum of the sample entropy of the first three PF components is used as the characteristic value. Table 1 lists the sample entropy and entropy sum of the first three PF components of each group of fault signals.

Table1. Sample entropy values of different fault signals.

Transition resistance /Ω	Occurrence point	Fault type	The sample entropy of the first three PF components	The sum of sample entropy
0	Line head	Transient Fault	0.7987 0.4947 0.0918	1.3852
		Permanent Fault	0.0042 0.0038 0	0.008
	Line midpoint	Transient Fault	0.9821 0.4867 0.1282	1.597
		Permanent Fault	0.3285 0.0356 0.0087	0.3728
	Line end	Transient Fault	0.6578 0.5643 0.1084	1.3305
		Permanent Fault	0.3453 0.0682 0.0060	0.4195
50	Line head	Transient Fault	0.6970 0.4814 0.0543	1.2327
		Permanent Fault	0.0893 0.0057 0.0025	0.0975
	Line midpoint	Transient Fault	0.7189 0.4179 0.0518	1.1886
		Permanent Fault	0.3152 0.0311 0.0008	0.3471
	Line end	Transient Fault	0.6193 0.4875 0.1418	1.2486
		Permanent Fault	0.2354 0.0124 0.0035	0.2513
100	Line head	Transient Fault	0.5917 0.5226 0.0152	1.1295
		Permanent Fault	0.2319 0.0094 0.0058	0.2471
	Line midpoint	Transient Fault	0.5731 0.3894 0.1064	1.0689
		Permanent Fault	0.3243 0.0416 0.0111	0.3770
	Line end	Transient Fault	0.5811 0.4975 0.0762	1.1548
		Permanent Fault	0.2319 0.0094 0.0058	0.2471

It can be seen that the LMD sample entropy of the line fault signal under transient fault is higher than that in the permanent fault state, because the fault of the secondary arc voltage waveform is very serious when the transient fault occurs. It contains a large number of high-frequency components, and the content of each frequency component is unstable. As time changes, the more information the fault signal contains, the larger the entropy value of the signal sample.

It can be seen from Table 1 that when a single-phase ground fault occurs in a transmission line with a shunt reactor, the sum of sample entropy of different fault types is different and there is a certain gap between them. The threshold is set to 1. When the sum of sample entropy is greater than the threshold, it is determined to be a transient fault. And when it is less than this threshold, it is determined to be a permanent fault. Therefore, the sum of sample entropy can be used to distinguish between transient faults and permanent faults. Fig. 9 is the curve of sample entropy sum for transient and permanent faults in transmission lines at different locations and different transition resistances.

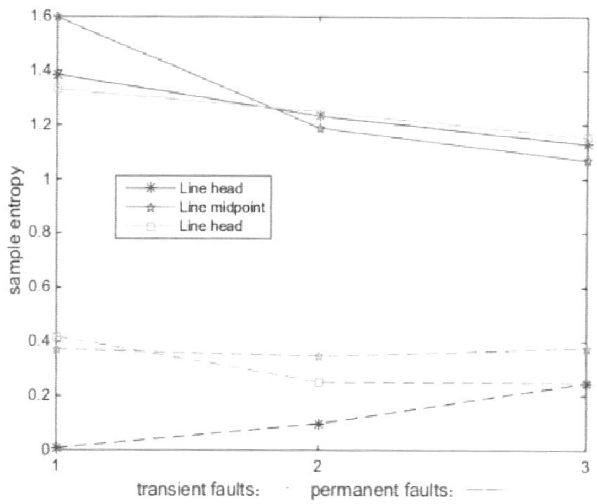

Figure 9. The curve of sample entropy sum under different conditions for different faults

6. Conclusion

Based on LMD algorithm and sample entropy theory, a fault feature extraction method for transmission line based on LMD sample entropy is proposed. The fault feature extraction of single-phase grounding short-circuit signal of transmission line is carried out by using this method. The feasibility of the proposed method is verified by the simulation results, and the following conclusions are drawn.

(1) The signal characteristics of single-phase grounding short circuit of transmission line extracted by LMD sample entropy can be used to judge the fault type of transmission line and accurately reflect the fault type.

(2) LMD sample entropy has strong signal representation ability and can be applied to feature extraction of complex signals.

References

[1] He Jiali, Song Congju. Principles of Power System Relay Protection. Updated Edition. Beijing: China Electric Power Press,2004,258.

[2] Fan Yue, Shi Wei. Steady-state phase angle method for distinguishing single-phase grounding fault types of ultra-high voltage lines [J]. Relay,1999,27(3):23-26.

[3] Li Bin, Li Yongli, Huang Qiang, etc. Study on phase criterion of single-phase adaptive reclosing [J]. Power system Automation,2003,27(22):41-44.

[4] Zhao Qingming, Li Bin. A new phase criterion for single-phase adaptive reclosing based on voltage compensation principle [J]. Power system Protection and Control,2010,38(13):50-54.

[5] Zhang Yin Yi, Yan Pingli, Chen Xue. Research on single-phase adaptive reclosing based on artificial neural network [J]. Electrical switch,2008,46 (3):33-36.

[6] Lin Xin， He BaiNa， Xu JianYuan. Secondary arc spectrum characteristics on EHV transmission lines [J]. High Voltage Engineering,2009,35(8)： 1891-1895.

[7] Yang Dechang, Tang Wei, qu Ruiqian, etc. Extraction of low frequency oscillatory parameters based on improved local mean decomposition [J]. Proceedings of the Chinese Journal of Electrical Engineering,2013,33(4):135-140.
[8] Achmad Widodo,Min-Chan Shim,Wahyu Caesarendra,Bo-Suk Yang. Intelligent prognostics for battery health monitoring based on sample entropy[J]. Expert Systems With Applications,2011,38(9).

3-D Localization of UAV and Detection based on Harmonics Index and Spectral Entropy Criteria

Muhammad Amjad Iqbal[1], Zhao Zhao*, Xu ZhiYong, Saad Ur Rehman

School of Electronic and Optical Engineering, Nanjing University of Science and Technology, Nanjing 210094, P. R China

engramjad@njust.edu.cn, zhaozhao@njust.edu.cn

Abstract. The popularity of drones or unmanned aerial vehicles (UAVs) will be increased enormously in future smart cities due to their usage in several domains. Significantly it's an emerging research to localize and timely detect such kinds of objects to avoid threats. This paper considered to deploy an array of 4-microphones for acoustics localization enhancement and UAV detection optimization. For high accuracy of acoustics localization in 3-D, the cumulative algorithm of GCC-PHAT and least square (LS) based on TDOA is implemented to estimate acoustic direction of arrivals (DOAs). However, detection optimization is achieved by a spectral framework, where harmonics index estimation effectively determined sound and noise range in each frame, while spectral entropy featured noise and sound correlation, the least possible value of entropy verified the presence of UAV and optimize its detection. Moreover, experimental results confirmed the efficiency of proposed techniques, our system has robust performance for localization as it featured less than 5% errors, while detection accuracy is more than 75%.

1. Introduction

UAV is widely using in several commercial applications, on another hand it might be used for illegal purposes which violate social norms, so objective is to detect unauthorized drones. Target tracking with a camera using an image processing technique requires a high-resolution cameras which is not cost-effective [1]. Radars and millimeter waves have an advantage for surveillance of flying objects, even don't get any affect by environmental factors [2]. Although, due to the small cross-section of UAV, radar can hardly pick up these small objects. Radio-frequency approach to detect UAV which has excellent results for long-range detection [3], but it hinders signal of other devices (Wi-Fi). Recently, machine learning approaches, linear predictive coding (LPC) and MFCC are used for speech signal and drone detection [4]. Regardless, due to misclassification and week feature identification of multipath signals, this solution is not suitable in real-time detection.

An emerging topic is a search of targets by its acoustic signature which is typically done by using a set of microphone array [5], which is cost-effective as it uses a large range of detection, correspondingly it hinder noises to maximize accuracy. For the passive method of localization, mostly two modus operands are used in the literature, the Angle of arrival (AOA), and time difference of arrival (TDOA) [6, 7]. TDOA scheme is recommended for high accuracy localization, as it's more efficient in both cases of 2-D and 3-D [8]. For powerful estimation of TDOA, GCC should be filtered with a known technique called phase-amplitude transform (PHAT), and hence resulting in the algorithm as GCC-

PHAT [9, 10]. This algorithm has several advantages; a well-performing attitude in reflective geography and more interesting is its sharp spectrum due to effective use of weighting function. Cleansed method is helpful to eliminate reflected waves, and to obtain those waves which are considered only direct waves and paving most of the spatial information [11]. For the sake of optimization in localization, least square (LS) method defines signal properties to find the location of target in 3-D effectively [12]. LS algorithm can do weight updates for multiple frames simultaneously, and each update does not need traditional resampling.

To detect target sound, several matrices can be selected including harmonics, and spectral entropy. Since drone sound contains harmonics features which are an integer multiple of its fundamental frequency [13]. Harmonics index estimation gives the range of target sound and noise information in each frame. The spectral entropy method [14] is used for the first time as this feature is widely helpful for detection of desired target signal. It can also de-noise the Doppler frequency signal successfully and ensure for the most part to gain coherence of information and maximum SNR.

In this paper, the design of acoustics localization and UAV detection is simple and fast in searching by using an array of 4-microphones. Array is designed to rectify the sound of 32-kHz with as low as possible mixing of interference. Section-2 presents localization and enhancement of acoustic signal. TDOA with GCC-PHAT is adapted to find time delays and the position of the sound. For peaks conversion into angles, Moore-Penrose (LS) technique is used. Localization approximation results are derived in the form of distance, elevation, and azimuth angles. The cleansing is introduced to identify direct waves, moreover, delay and sum beam-forming is applied to enhance localized signal. Section-4 elaborates the detection of UAV based on beam-formed vector, where harmonics indexes are evaluated to distinguish noise and sound range. To optimize detection of UAV, spectral entropy is implemented with different frame lengths. Conclusions with possible future advancements are briefed in section-5.

2. Localization and Enhancement of Acoustic signal in space

Microphone array elements are placed in spherical coordinates in 3-dimensional spacing whose coordinates are $M_i(x_i, y_i, z_i)\ where,\ i = 1,..,N$ as shown in Figure (1). GCC-PHAT and TDOA are employed to obtain geographical information, while 3-D view of acoustic emissions determined by LS method. To enhance localized signal, cleansing method with appropriate Hamming window is used, however for spatial filtration Delay and sum beamforming is used.

2.1. Generalized Cross-Correlation Phase Transformation (GCC-PHAT)

GCC-PHAT [9, 10] computes time delay τ obtained by TDOA and here function assumes that signals at microphone under observation and reference microphone are coming from the same source. GCC-PHAT finds the location of the peaks of cross-correlation between the observatory signal and reference signal. Relationship between the source signal and impulse response is represented by the following expression, $S_m(\omega) = Y(\omega) * H_m(\omega)$ while GCC of two microphones is derived as:

$$Z(\tau) = \frac{1}{2\pi}\int_{-\infty}^{\infty} S_{m1}(\omega) * S_{m2}(\omega)\ e^{j\omega\tau}\, d\omega \quad (1)$$

$$Z(\tau) = \frac{1}{2\pi}\int_{-\infty}^{\infty} Y_1(\omega)Y_2^*(\omega)H_{m1}(\omega)H_{m2}^*(\omega)\ e^{j\omega\tau}\, d\omega \quad (2)$$

The weighting function is $H_{m1}(\omega)H_{m2}^*(\omega) = \psi_{m1m2}(\omega)$ designed to optimize performance criteria. Substituting weighting function into (2) GCC-PHAT will be of the form:

$$Z(\tau) = \frac{1}{2\pi}\int_{-\infty}^{\infty} \psi_{m1m2}(\omega)\, Y_1(\omega)Y_2{}^*(\omega)\ e^{j\omega\tau}\, d\omega \quad (3)$$

Weighting is used to emphasize the power of a directly incoming signal by suppressing noise power.

2.2. Sound Source DOA formulation based on the Least Squares Method.

UAV in space is positioned at $(D\cos\theta\sin\theta, D\cos\theta\sin\varphi,\ D\sin\varphi\ )$ where 'θ, φ, D' are elevation, azimuth and distance from the origin of the reference coordinate. Apply an expression for a 4-element array, by using time difference information between $i_{th}\ and\ j_{th}$ array elements, while $(r_i - r_j)$ is path distance of microphone.

$$\begin{bmatrix} \boldsymbol{r}_2 - \boldsymbol{r}_1 \\ \vdots \\ \boldsymbol{r}_N - \boldsymbol{r}_1 \\ \boldsymbol{r}_3 - \boldsymbol{r}_2 \\ \vdots \\ \boldsymbol{r}_N - \boldsymbol{r}_{N-1} \end{bmatrix} \begin{bmatrix} \frac{\cos\theta\cos\varphi}{v} \\ \frac{\sin\theta\cos\varphi}{v} \\ \frac{\sin\varphi}{v} \end{bmatrix} = \begin{bmatrix} \tau_1 - \tau_2 \\ \vdots \\ \tau_1 - \tau_N \\ \tau_2 - \tau_3 \\ \vdots \\ \tau_{N-1} - \tau_N \end{bmatrix} \tag{4}$$

Where '*N*' represents the number of elements in an array and 'v' is the speed of sound. The optimal parameters θ, φ can be estimated by Moore-Penrose method [15]. To solve a problem in the least square error sense, in general it is an exact solution to solve an over-determined problem.

$$\boldsymbol{Y}.\boldsymbol{x} = \boldsymbol{\tau} \tag{5}$$

'$\boldsymbol{Y}$' is the structure of an array, '$\boldsymbol{x}$' is direction vector of sound and, '$\boldsymbol{\tau}$' is a time delay; LS method is

$$\widehat{\boldsymbol{x}} = \boldsymbol{Y}^{\dagger}\boldsymbol{\tau} = \left(\boldsymbol{Y}^T\boldsymbol{Y}\right)^{-1}\boldsymbol{Y}^T\boldsymbol{\tau}. \tag{6}$$

$\boldsymbol{Y}^{\dagger}$ is Moore-Penrose inverse matrix of $\boldsymbol{Y}$ is given as:

$$\boldsymbol{Y}^{\dagger} = \begin{bmatrix} \boldsymbol{u}_1 \\ \boldsymbol{u}_2 \\ \boldsymbol{u}_3 \end{bmatrix} \tag{7}$$

Thus least square estimates of azimuth and pitch angles of source in the far field are obtained as:

$$\widehat{\boldsymbol{x}} = \begin{bmatrix} \frac{\cos\theta.\cos\varphi}{v} \\ \frac{\sin\theta.\cos\varphi}{v} \\ \frac{\sin\varphi}{v} \end{bmatrix} = \boldsymbol{Y}^{\dagger}\boldsymbol{\tau} = \begin{bmatrix} \boldsymbol{u}_1^T.\boldsymbol{\tau} \\ \boldsymbol{u}_2^T.\boldsymbol{\tau} \\ \boldsymbol{u}_3^T.\boldsymbol{\tau} \end{bmatrix} \tag{8}$$

$$\hat{\vartheta}(\tau) = \tan^{-1}\left(\frac{u_1^T.\tau}{u_2^T.\tau}\right) \tag{9}$$

$$\hat{\varphi}(\tau) = \tan^{-1}\left(\frac{u_3^T}{\sqrt{\left(u_1^T.\tau\right)^2 + \left(u_2^T.\tau\right)^2}}\right) \tag{10}$$

The LS algorithm has less computational cost, and reduces the error for DOA estimation in the form of elevation and azimuth, which are obtained under real-time scenario.

2.3. DOA Improvement by Cleansing of Multi-path Effect

Peaks of GCC-PHAT function sometimes do not certainly match with accurate source location due to the multipath effect. For this concern, cleansing process is helpful to remove non-directional bins, to ensure that the microphones haven't encountered with unnecessary information. To obtain better results of cleansing, Hamming window is employed as illustrated in (11):

$$C\hat{Z}(\tau) = \int_{-\infty}^{+\infty} \frac{W(f)\otimes\hat{Y}_{12}(f)}{|W(f)\otimes\hat{Y}_{12}(f)|} * e^{j2\pi ft} \tag{11}$$

Here $\otimes$ refers to convolution operator and assume that additive noises are independent and also uncorrelated with source signal, while $w(f)$ is an appropriate hamming window.

2.4. Delay and Sum Beam-forming (DSB)

Beamforming is a technique to enhance a sound source in the desired direction [16]. To emphasise that to what extent the desired signal is affected by reflections and other sources. Simple delay-sum beamforming is carried out, where delays are inserted after each microphone to compensate path-delay. Later on, all samples of a relative microphone are delayed or advanced, to get a booster version of sound to form an acoustic beam in particular direction by following equation:

$$Y(k) = \sum_{k=1}^{j} \alpha_k X_k(n - t_j) \tag{12}$$

X_k represents the number of microphones, t_j corresponds to relative time delay and $Y(k)$ is a delay and sum beamforming output.

IOP Conf. Series: Materials Science and Engineering **853** (2020) 012037 doi:10.1088/1757-899X/853/1/012037

3. Spectral matrices for detection of UAV

To detect UAV, it's necessary to collect useful spectral information. Several matrices can be selected, however in our case, harmonic index estimation, and spectral entropy are included.

3.1. Harmonic Frequency Index Estimation

UAV propellers generate complex sound and its acoustics feature fosters many harmonics [13], while harmonics are round about integer multiple of a fundamental frequency, which is helpful to distinguish drone sound. The time domain signal after beamforming of UAV is divided into frames. Apply FFT that data set is sparse in frequency domain especially, as energy is concentrated in harmonic structure. Find 1st fundamental in the spectrum between F_L & F_R and search for '*N*' number of harmonic indexes as follows $\left[(\frac{n}{n-1}F_{n-1} * F_{interval}) - F_T\,,\ (\frac{n}{n-1}F_{n-1} * F_{interval}) + F_T\right]$ where *n=2*, and F_T is threshold frequency, and Output, $F_1, \ldots, F_N$.

3.2. Spectral Entropy-based detection Optimization

Entropy is to encode signal according to its consistency as it featured noise and sound correlation [14]. The output results in the aforementioned step is under the condition of intense noise which may have an apparent error. Therefore to optimize detection, entropy-based method has been reported. The least possible value of entropy indicates the presence of UAV, smaller variations validate the stability of target sound. Calculate the section of each spectral component in the form of energy which would be lean as a signal energy probability at a certain spectrum point.

$$p_x = \frac{S(f_x)}{\sum_{k=1}^{N} S(f_k)}, x = 1, \ldots, M \qquad (13)$$

Here $S(f_x)$ is the energy of f_x, and p_x is related probability density, while '*M*' are points of FFT frequency components. Proportionally, spectral entropy for each UAV sound frame is stated as:

$$E_x = -\sum_{K=1}^{M} p_x * \log p_x \qquad (14)$$

If a frame consists of only noise then the value of entropy will approach to **1**, if it is containing noise with the target signal information, then entropy value will lead to **0**.

4. Experimental Validations

Figure (1) shows experimental setup. Performance of GCC-PHAT based on TDOA was empirically estimated between reference microphone and microphone under observation. To evaluate the performance of GCC-PHAT, the Moore-Penrose LS method is utilized to estimate DOAs. According to the GCC-PHAT result, as shown in Figure (2.a), the estimation performance of DOAs exhibits some non-directional bins and it has several ups and downs. After cleansing, non-directional bins were removed, result shown in Figure (2.b) is near to ground truth which is definitive and only focuses the direct waves coming from UAV, as DOAs are not drastic and mostly it contains target sound.

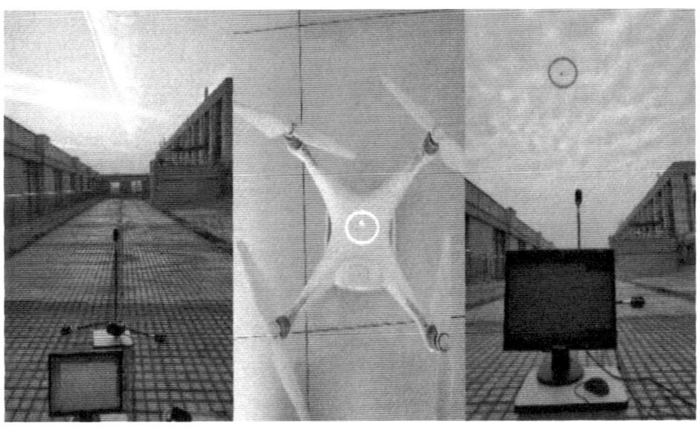

Figure 1. Field experiments using microphone-array.

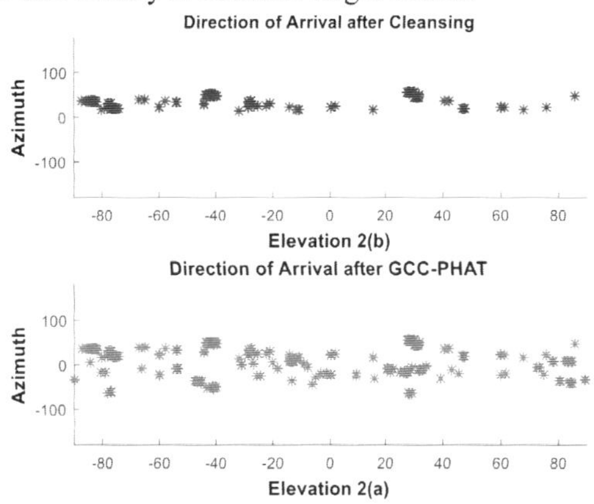

Figure 2. GCC-PHAT result & cleansing of reflections.

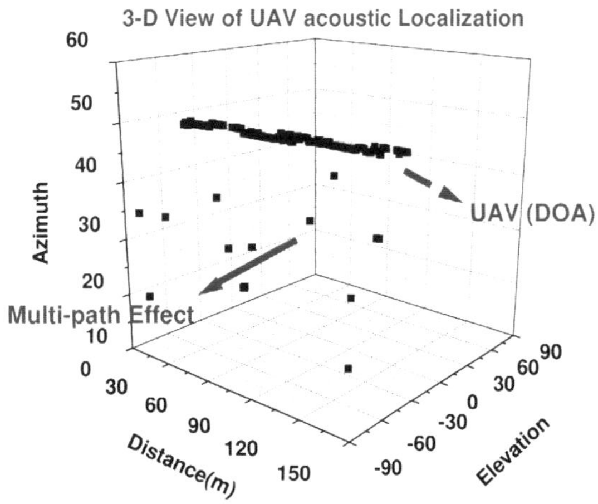

Figure 3. Acoustic localization in 3-D space.

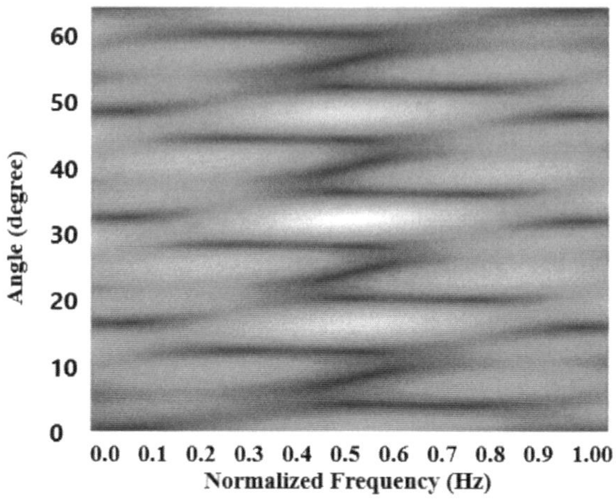

Figure 4. Delay and sum beamforming at 30°.

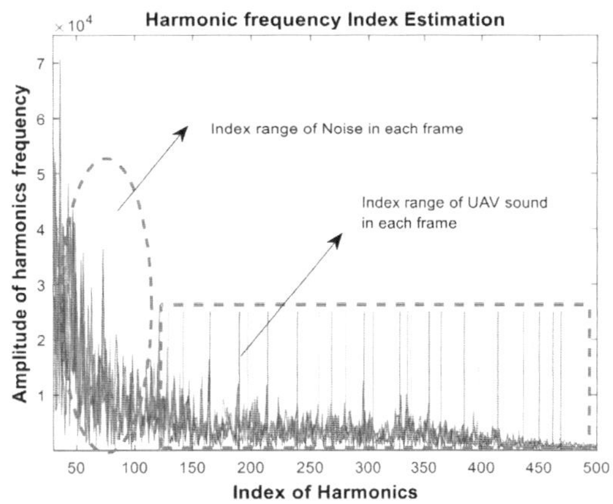

Figure 5. Harmonics index estimation.

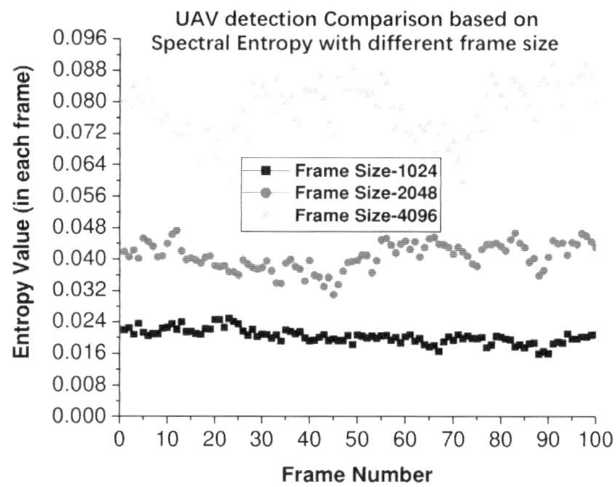

Figure 6. Spectral entropy comparison of UAV signal with different frame lengths.

DOAs of sound in space of 3-D is scrutinized in Figure (3) where drone flying movement is expressed and easily examines the accuracy of DOAs. Some multi-path bins appeared which might be caused by the collision of sound to building walls. In this experiment, the localization task achieved 95% accuracy with ground reality, which in turn will have a maximum success detection rate.

Figure (4) shows a beam pattern, which has an effect of boosting the sound signal at the desired angle, while unwanted signals are combined in a quite unpredictable manner. Beamforming is applied to highlight that, to what extent target sound is deteriorated by noises. UAV is searched at an angle of 30°. The aim is to achieve the main-lobe with minimum reflections at the desired angle. It is noticed that interferences are lower, depression is quite a bit, and noise is not interfering main-lobe signal and its behaviour is stable, side-lobe regions are [10 - 25; 35 - 50] degree.

In Figure (5) index-wise estimation of harmonic frequency is presented. Background noises are mainly existing in the indexes range of 0-120, which arrives before harmonic components of UAV sound. Indexes of the sound signal obtained where peak frequencies were detected and it will arise in a harmonic correlation of fundamental frequency which is distinguishable. This feature gives a range of noise and UAV sounds in each frame to detect a target easily.

For the evaluation of spectral entropy, a specific set of isolated UAV sound is applied with three different frame lengths of 1024, 2048 and 4096. Length selection of frames is an important parameter. As it is shown in Figure (6) when we choose window size smaller it gives better results in the form of low entropy value. The number of samples in a frame, consequently the correlation between sound and noise is less. Slight variation of entropy in adjacent frames is a significant advantage. If the frame size is not appropriate, many samples will be lost and it will seed an error in the detection of UAV. Detection analysis for blocks of different frames is examined based on inverse RMSE relative to SNR. It is concluded that frames with less detection rate are sound of UAV having low potential, while most of the other frames exhibiting the high success detection rate, which indicates that these frames have a high potential of the target sound.

5. Conclusions

This work demonstrates UAV localization and detection, using an array of 4-microphone elements. Several problems on SSL have been addressed based on its pros and cons, we deliberated to improve localization, and summarized a novel strategy to detect UAV. GCC-PHAT with TDOA extracted the phase information to get the position of UAV and LS technique obtain DOA successfully. It has been empirically manifested that, localization has outperformance for DOA estimation, as accuracy is 95%. To detect UAV, representative algorithms of harmonics, and spectral entropy are applied. As long as simulation is quick enough to execute processing and evaluation was done passively, results of harmonics frequency index estimation and spectral entropy are realistic. Noise and sound ranges are extracted based on harmonic indexes, the gain of the UAV signal is directional for most of the frames, and spectral entropy of UAV sound is estimated with minimum entropy value, which depicts that data under observation contains target sound, as there is a very slight variation in entropy behaviour. This research provides many new openings, specifically the improvement for real-time detection pursuit. Thereafter, it will be an opportunity to distinguish if the certain UAV is viable to threat or either it is a functional flight [17]. Acoustic signature of drone which has (malicious) payload is different from a UAV that is just having a simple flight. Tracking will be plus point to extend the validity of the above work.

Acknowledgments

This work was supported by the National Natural Science Foundation of China [grant number 61401203]; the Fundamental Research Funds for the Central Universities [grant number 30918012203]; Equipment Pre-Research Program of China [grant number 41413010401].

References

[1] A. Rozantsev, V. Lepetit, and P. Fua.: 'Detecting Flying Objects using a Single Moving Camera', IEEE Transactions on Pattern Analysis and Machine Intelligence, 2016, vol. 39, pp 879 – 892.

[2] F. Fioranelli, M. Ritchie, H. Griffiths, et al.: 'Classification of Loaded/Unloaded MicroDrones Using Multistatic Radar', Oct. 2015, IET Electronics Letters, vol. 51, no. 22, pp 1813–15.

[3] M. Haluza and J. Cechak.: 'Analysis and Decoding of Radio Signals for Remote Control of Drones', Proc. IEEE New Trends in Signal Processing, , Slovakia, Oct. 2016, pp 1–5.

[4] Muhammad Zohaib Anwar, Zeeshan Kaleem, Abbas Jamalipour.: 'Machine Learning Inspired Sound-Based Amateur Drone Detection for Public Safety Applications', IEEE Transactions on Vehicular Technology, 2019, Vol. 68, pp 2526 – 2534.

[5] Caleb Rascon, Ivan Meza.: 'Localization of sound sources in robotics: A review', Robotics and Autonomous Systems, ELSEVIER, Oct 2017, vol.96, pp.184–210 .

[6] Yang, M., Jackson, D.R., Chen, J., et al.: 'A tdoa localization method for nonlineof-sight scenarios', IEEE Trans. Antennas Propag., 2019, 67, (4), pp. 2666-2676.

[7]. Zheng, Y., Sheng, M., Liu, J., et al.: 'Exploiting aoa estimation accuracy for indoor localization: a weighted aoa-based approach', IEEE Wirel. Commun. Lett., 2019, (1), pp. 65-68.

[8]. Xianyu Chang, Chaoqun Yang, Junfeng Wu et al.: 'A Surveillance System for Drone Localization and Tracking Using Acoustic Arrays', 2018 IEEE 10th (SAM) Workshop.

[9]. J. Xie, H. Tao, X. Rao, and J. Su.: 'Passive localization of mixed far field and near-field sources

without estimating the number of sources', Research gate, Sensors, 2015, vol. 15, pp. 3834–3853.

[10]. Ali Pourmohammad, Seyed Mohammad Ahadi.: 'Real Time High Accuracy 3-D PHAT-Based Sound Source Localization Using a Simple 4-Microphone Arrangement', 2012, IEEE System Journal, vol.6, pp. 455 – 468.

[11]. Sangmoon Lee, Youngjin Park2 and Youn-sik Park.: 'Cleansed PHAT GCC based Sound Source Localization', International Conference on Control, Automation and Systems IEEE, 2010 Oct. 27-30.

[12]. Szymon Wozniak and Konrad Kowalczyk: 'Passive Joint Localization and Synchronization of Distributed Microphone Arrays', IEEE Signal Processing Letters, Vol. 26, no. 2, February 2019.

[13]. Hannes Pessentheiner, Martin Hagmüller, Gernot Kubin.: 'Localization and Characterization of Multiple Harmonic Sources', IEEE/ACM Transactions on Audio, Speech, and Language Processing, 2016,vol. 24(8), pp. 1348-1363.

[14]. Shufen Qi, Jinghong Zhao, Xiaojie Liu.: 'Mill Sound Detection Based On Signal Singular Entropy Noise Reduction', Chinese Control and Decision Conference (CCDC) IEEE, Yinchuan China, may 2016, pp.3766 – 3769.

[15]. Xu Zhi Young, Zhao Zhao.: 'Perturbation sensitivity analysis of azimuth estimation of the acoustic source for the planar microphone array', Journal of Xidian University, 2017, doi:10.3969/j.issn.10012400.

[16]. David Kurc, Vaclav Mach, Kristian Orlovsky et al.: 'Sound Source Localization with DAS Beamforming Method Using Small Number of Microphones', International Conference on Telecommunications and Signal Processing (TSP), 2013 IEEE, pp. 526 – 532.

[17]. Francesco Fioranelli, Matthew Ritchie, Hugh Griffiths.: 'Multistatic human micro-Doppler classification of armed/unarmed personnel', IET Radar, Sonar & Navigation, 2015, Volume: 9, pp. 857 – 865.

IOP Conf. Series: Materials Science and Engineering **853** (2020) 012038 doi:10.1088/1757-899X/853/1/012038

Design of power well cover wireless monitoring system based on freeRTOS and NB-IoT technology

Wei Wang[1], Tairong Xie[2,*] Xianhua Hu[2]

[1]Guangzhou Power Supply Bureau Co., Ltd; 2. Guangdong Yuanpeng Network Technology Co., Ltd

*Corresponding author.E-mail addresses: 756148845@qq.com(Tairong Xie)

Abstract. Power well cover monitoring system（PWC-MS） based on the embedded microprocessor ARM Cortex-M3 with FreeRTOS is proposed in this paper. The software functions of the terminal are divided into four tasks, namely, watchdog refresh and flag monitoring task, system status light task, non-debug sensor mode task, and debug sensor mode task. The priority levels of tasks depend on real-time requirement. The temperature and humidity sensor RS-WS-N01-2-2 and the dual-axis tilt sensor ADXL345 are adopted in the terminal and the sampling data is sent to the IoT (Internet of Things) platform via the NB-IoT communication module BC35-G. The software of application layer in the IoT platform is developed based on the Spring Cloud and database technology with Java language. User can get the real-time data through the Web page. Experiments show that the PWC-MS works well and the real-time alarm can be realized immediately.

Keyword: FreeRTOS; ARM; power well cover; wireless monitoring system; NB-IoT

1. Introduction

Currently, the construction of smart grid is in full swing. The urban pipeline is an important part of urban planning and construction. Gradually, it is a trend that the overhead lines are transformed into underground pipeline [1]. Nowadays, a large number of high-voltage cable well covers in the UPG are still under the management of manual inspection. The manual inspection is not only heavy workload, but also is hard to respond power well abnormal information quickly [2]. These cause a great security risk to vehicular and pedestrian, which brings a great challenge to the safety of UPG and citizens[3].

How to carry out real-time monitoring on well covers and deal with the abnormal situation quickly to guarantee the safety of the citizens and power assets is an urgent problem to be solved for the electric power department.

Meanwhile, FreeRTOS is completely free even in commercial scenarios [4]. It analyzes the principles of FreeRTOS and μc/OS-II kernels. It turns out that FreeRTOS has unlimited tasks and more flexible scheduling strategies [5]. There are also many researches on IoT monitoring applications based on FreeRTOS, including intelligent power monitoring, downhole multi-parameter recorders and smart home solutions [6].

Therefore, in order to solve monitoring management, safety problems in daily operation of power well covers in the urban power tunnel and make it easy for maintainer to obtain power wells condition such covers states and detailed surroundings information in the cable wells. The power well cover monitoring system (PWC-MS) with embedded microprocessor STM32 and real-time operating system FreeRTOS by employing NB-IoT wireless communication network for remote data transmission is proposed in this paper, which can monitor the power well states, realize early warning and ensure power and road safety.

2. Terminal hardware design of PWC-MS

The data processing of the power well cover monitoring system is based on the data uploaded by the monitoring system terminal to the IoT platform. Therefore, the most important part of the system is to realize real-time, accurate collection of sensor data and upload it to the cloud server. In addition, considering the large distribution of power wells, a large amount of data is distributed in various hidden and complex surroundings such as basements, equipment rooms, streets, underground pipelines, tunnels, markets, and building rooms. This means that the power well monitoring system terminal must work for a long time while accurately collecting data in real time.
Based on the above requirements, the hardware design of the power well cover monitoring system terminal is based on the STM32L151C8 microprocessor. The hardware structure block diagram of the power well cover monitoring system terminal is shown in Figure 1.

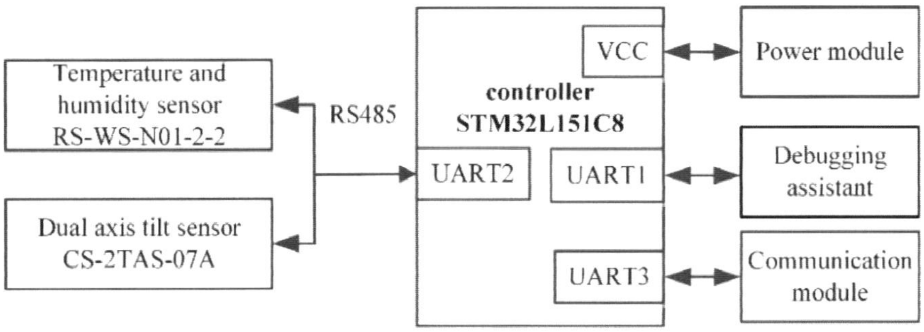

Figure 1. PWC-MS terminal hardware structure block diagram

2.1 Sensor selection

Temperature and Humidity Sensor(THS): The type SHT20 is adopted as THS. Humidity accuracy 3% RH. Humidity range: 0 ~ 100% RH. Temperature accuracy: ±0.3 °C. Temperature range: -40 ~ 125 °C. Working power consumption: 4.8uA.
Dual Axis Tilt Sensor(DATS): The type ADXL345 is adopted as DATS. The ADXL345 is with high resolution, up to 13-bit resolution and variable measurement range sensitivity +/-2g, +/-4g, +/-8g, +/-16g up to 3.9mg/LSB. In addition, it can measure the tilt angle change of less than 1.0°. Most importantly, the ADXL345 possess ultra low power consumption of 40~145uA and is only 0.1uA during standby mode.

2.2 Communication module

The BC35-G module is adopted as the NB-IoT wireless communication module, which is characterized by ultra-low power consumption and ultra-high sensitivity. The power consumption is 3.6μA during power-saving mode, the working power consumption is 6mA, and the sensitivity is -129dBm±1dB. The module adopts LGC package which is easier to be welded and suitable for mass production. Using Quectel enhanced AT command and supporting Chinese Datagram Protocol (UDP) connection, BC35-G module enable accomplish data transmission and realize the interaction between cable well states information and IoT cloud platform.

2.3 Power circuit design

The power is a basic module that provides a stable energy supply to the system. It is mainly divided into two modes. One is battery powered and the other is powered by a 5V adapter.

2.4 Power circuit design

The power is a basic module that provides a stable energy supply to the system. It is mainly divided into two modes. One is battery powered and the other is powered by a 5V adapter.

The first type power supply is powered by a lithium battery. The lithium battery provides 3.7V voltage through the capacitor filter, the SS34 diode outputs 3.5V voltage to the BC35 chip, and the other part through buck circuit outputs VCC_3V3 to the other modules of the circuit.

The second power supply mode is powered by a 5V adapter. The adapter provides a 5V voltage-stabilizing circuit to modulate output 3.7V, and then forms the entire power supply system through the same circuit as the first power supply. In addition, when using external 5V power supply, the circuit automatically switches to the battery charging mode. By connecting the external 5V power supply to the battery charging control chip, the battery charging control chip monitors the battery voltage and charges the battery.

3. Terminal software design of PWC-MS

The software design of the PWC-MS consists of two parts. One is the terminal software design of PWC-MS and the other is the software design of application layer

In the PWC-MS, the terminal needs to complete a plurality of tasks such as system initialization, data acquisition, data communication, and system operation status indication. In order to ensure the real-time performance of the system, make more reasonable and effective use of CPU and transplant the FreeRTOS. So the task needs to be designed into several separate parts to be completed by PWC-MS. FreeRTOS is transplanted via STM32CubeMX visualization tool in this paper. After configuring the clock, serial port, I/O port and other parameters with STM32CubeMX, corresponding Keil project can be easily generated. Compared to traditional one, generating code with the STM32 CubeMX configuration is simpler, more convenient, and more errorless.

3.1 Main process of terminal software design

The terminal of the PWC-MS is mainly to collect the power well covers dip angle, temperature and humidity datum, then package them into data packets and upload them to the IoT platform through the NB-IoT wireless network. The main flow chart of the software is illustrated in Figure 2.

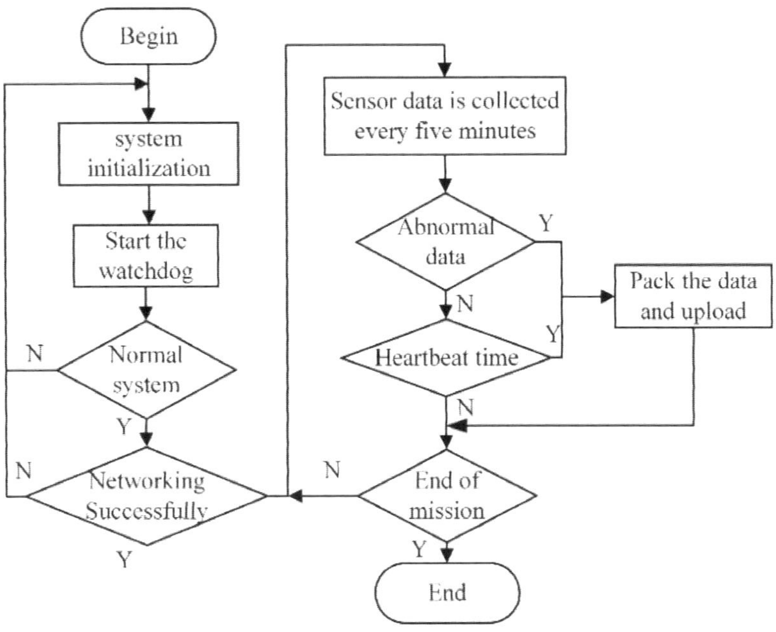

Figure 2. Software basic flow chart

The above flow chart can be described as: initializing the STM32L151C8T6 chip and FreeRTOS after the system is power-on, and then starting the watchdog timer to provide protection for the system; Then the BC35-G module dials the network (detect the SIM card status, module status, query IMSI and IMEI number, attached network, etc.). If the BC35-G module network is unsuccessful less than 10 times, BC35-G will continue attach the network. Otherwise resetting the BC35-G module and redialling again. After the network is successfully attached, the sensors start collecting datum.
Initially, the sensors datum are collected as the initial value sample based on the Grubbs criterion. Then, the sensor datum are collected every 5 minutes and judge theirs integrity. If datum are integrity, then move on next step, otherwise the system collect datum anew. After confirming sensor datum (dip angle, temperature and humidity) are in range of threshold or not. If datum exceed threshold value, the datum will be reported to IoT cloud platform through the BC35-G wireless communication module immediately, otherwise they will be uploaded every 30 minutes.

3.2 Task division and function design
The terminal program design of PWC-MS mainly includes four tasks: watchdog timing refresh, flag position monitoring, system status light, non-debug sensor mode and debug sensor mode. The specific functions are based on FreeRTOS embedded real-time operating system for task scheduling.

3.3. Application layer software design
In this system, the terminal data is uploaded to the IoT platform through the NB-IoT network. The platform is in the cloud and is carried in the server. Externally, the cloud platform only implements pure data processing services, does not involve interface implementation, and provides a unified API interface, so that each control platform can develop interfaces according to its own platform characteristics, and does not affect the implementation of functions. In order to realize the interaction between the user and the platform, the database needs to be designed on the server side to archive and store the data of the IoT platform. In order to realize the visualization of the data, the front-end technology is used to design the web interface, and the data that has been archived and stored is displayed in the form of a graph on the webpage side.
The server-side application project designed in the application layer is developed in Java language. Combining Spring Cloud technology and database technology, a Web-based monitoring platform has been developed.

4. Experiments
The physical map of the PWC-MS terminal is shown in Figure 3. The PWC-MS is packaged and installed on the simulated power well cover. After the system is power-on, the initial data is automatically collected ten times as the original parameters to obtain the initial values of the dip angle, temperature and humidity. After 3 hours of testing, open the client to observe the data flow. As shown in Figure 4-5, the platform can display various environmental data transmitted from the terminal of the PWC-MS in real time.

Figure 3. Physical map of the PWC-MS terminal

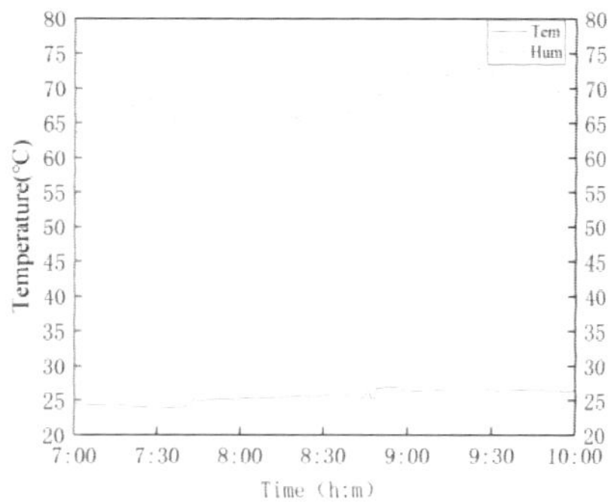

Figure 4. Terminal temperature and humidity data flow

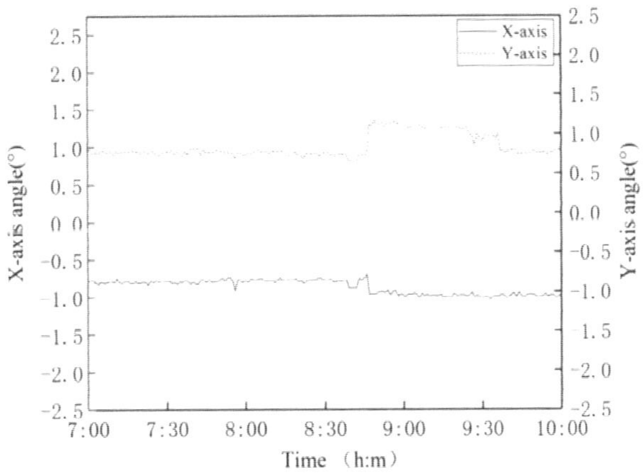

Figure 5. Terminal dip angle data flow

The tests show that the terminal device can stably upload the sensors datum to the IoT platform, and users can view the data flow of sensors through the web page. Therefore, The PWC-MS based on NB-IoT and FreeRTOS can monitor environment information of the power well in real time. As a result, PWC-MS reaches expected effect and lays the foundation for realizing remote information monitoring of large-scale power well covers in urban areas.

5. Conclusion

PWC-MS based on NB-IoT and FreeRTOS embedded real-time operating system is proposed in this paper. The terminal system function is divided into four tasks including watchdog refresh, flag monitoring tasks, system status light tasks, non-debug sensor mode tasks and debug sensor mode tasks. After testing, the system shows stable operation and good real-time performance. What's more, PWC-MS can accurately collect data in real time and upload it to the IoT cloud platform correctly. As a data acquisition terminal, its software design structure is clear and easy to rewrite, so PWC-MS can be applied to other industrial backgrounds as a node in the sensor network.

Acknowledgments

This work was supported in part by Guangzhou Science and Technology Planning Project under Grant 201902020003, in part by China Southern Power Grid Science and Technology Project under Grant GZJKJXM20170032

References

[1] Xinyang Yuan. Application Research of Distribution Network Planning in Urban Power Grid Construction [J]. Low Carbon World,2017(12):35-36.
[2] Man Chen. Urban power integrated pipe gallery monitoring system linkage process [J]. Sichuan Architecture, 2018,38,180(02):242-243.
[3] Wei Zhao. Current Status and Challenges of Smart Grid Big Data Processing Technology [J]. Electronic Technology & Software Engineering, 2017(17):185-185.
[4] Yue Yang, Haoran Shao, Suofei Zhang. FreeRTOS and μC/OS-III Kernel Analysis and Selection Research [J]. Science & Technology Vision, 2018, 246(24):16-18.
[5] Yingshu Liu, Ke Yang, Weihang Ding. Intelligent power monitoring system based on FreeRTOS [J]. Electronic Design Engineering, 2017(21).
[6] Yingjian Hu. Downhole multi-parameter recorder based on TMS570 and FreeRTOS [J]. Microcontrollers & Embedded Systems, 2017(12):45-48.

The optimized layout strategy of ring oscillator network for Trojan detection

Lian Yang, Wenxiu Yang, Huan Li

University of Electronic Science and Technology of China; University of Electronic Science and Technology of China; University of Electronic Science and Technology of China

yanglian@uestc.edu.cn; 201821010626@std.uestc.edu.cn; photon_lee@uestc.edu.cn

Abstract. The method of Trojan detection based on the RON (Ring Oscillator Network) has been widely proposed and has a good effect. However, the method needs to integrate a RON on the chip to sense the voltage drop generated by the Trojan. Currently, the number of ROs (Ring Oscillator) required for a RON has not been discussed, which has a large impact on the additional hardware overhead. Therefore, based on the RON for Trojan detection, we propose an optimized layout strategy that can effectively control the number of ROs integrated into the circuit. By measuring the sensing range that can be perceived by a single RO, the sensing radius of the RO is obtained; then, ROs are placed in the circuit in a way of regular hexagonal lattice to clarify the number of ROs required in the network. Determining the appropriate number of ROs not only reduces the hardware overhead required for on-chip integration, but also ensures the reliability of Trojan detection, ensuring seamless coverage of the circuit with as few ROs as possible.

1. Introduction

Due to the separation of IC (Integrated Circuit) design and manufacturing, IC products are facing new security threats [1]. An attacker can insert a hardware Trojan at the design or manufacturing stage of an IC [2] to achieve the purpose of leaking circuit information, denial of service and so on [3]. There are endless methods of Trojan detection, most of which use the side channel information to distinguish the circuits with or without Trojans [4]. Since the ROs are distributed inside the circuit, they are able to generate signatures more accurately than other side-channel methods.

After the method of Trojan detection based on the RON was proposed, some research based on the method also appeared. But there has been no discussion of the number of ROs required in a RON. Such a discussion is obviously necessary, because how to reduce the resource overhead is an important issue. In this paper, we propose an optimized layout strategy for a RON. By optimizing the layout method and determining the suitable number of ROs, the entire circuit or system can be seamlessly covered with as few ROs as possible.

2. Related work

In the past research work, the researchers put forward lots of methods about RON to detect hardware Trojans.

2.1. The methods of the Trojan detection based on RON
Zhang *et al*. first proposed the technique of detecting hardware Trojans by integrating RON on the on-chip in [5]. They pointed out that RO effectively reduces the influence of noise on the side channel information. The method can implement Trojan detection in the test stage by analyzing the data of ROs. In addition, the RON can also be applied to the real-time defense system of Trojan detection, which alerts when the Trojan is activated, thus providing the last strong barrier for the hardware circuit [6-7].

2.2. The layout of RON
In the past research, there are two common layouts: "Mesh" layout [8-9] and "X" layout [10]. However, neither of the two layouts considers the sensing range of the RO, and the seamless coverage of the sensing range cannot be realized, so that the correct rate of the Trojan detection cannot be reliably guaranteed. Wu *et al*. pointed out that the circles of the same radius completely can cover a plane and the optimal layout with the farthest distance between the centers is a hexagonal grid [11]. This ensures that as few circles as possible are used to seamlessly cover a plane, providing a strong basis for our approach.

3. Our method
We will explain our approach in three ways: determining the coverage radius of the RO, the layout strategy, and determining the number of ROs.

3.1. Determining the coverage radius of the RO
Assuming that the frequency of an RO is F. Due to the actual measurement and the influence of noise, for a circuit without Trojans (H_0), F should satisfy the Gaussian distribution with mean u_0 and variance σ_0. Similarly, for circuits infected with Trojans (H_1), the frequency F should satisfy the Gaussian distribution with mean u_1 and variance σ_1.
The Trojan analog circuit is placed at a position away from the RO by $i(i=1,2,\ldots)$ slice. At this time, the frequency values of the RO are recorded as a Gaussian distribution with mean u_1^i and variance $u_1^i(=\sigma_0)$. The coverage radius of the RO is Equation 1.

$$r = \max\left(\left\{i \mid \left|\mu_0 - \mu_1^i\right| \geq \sigma_0\right\}\right) \tag{1}$$

3.2. The layout strategy
Due to the size of circuit is limited, we will discuss the boundary condition of the circuit. The size of the circuit C is $x \times y$, the measured coverage radius of the RO in the previous section is r.

3.2.1. ROs are laid out according to regular hexagon grid division.
The circuit is subjected to regular hexagonal meshing with the lower left corner of the circuit as the origin. The regular hexagon here is an inscribed regular hexagon of a circle of radius r. Here, a regular hexagon is placed laterally so that the lower left corner is aligned with the origin, and the bottom edge is aligned with the x-axis. Based on the regular hexagon, the adjacent regular hexagon can be found, and then the adjacent regular hexagon is used as a reference. Find the regular hexagons adjacent to them, and so on, until the circuit is completely covered. If the ROs are placed in the center of these grids, seamless coverage of the circuit can be achieved.
The problem is that after the regular hexagonal mesh is divided, some of the grid's center position will be beyond the circuit range. For the position of the edge ROs, adjustments need to be made to make it inside the circuit.

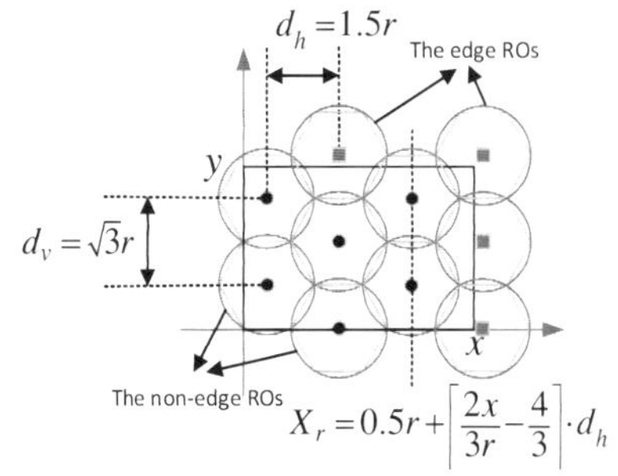

Figure 1. The schematic of Non-edge and edge ROs.

To facilitate description, some symbols are defined and are labeled in Figure 1. Here, the x-coordinate of non-edge ROs in the rightmost column is denoted as $Xr=0.5r+\left\lceil \frac{2x}{3r}-\frac{4}{3} \right\rceil \cdot d_h$. The distance between the ROs is recorded as $d_v=\sqrt{3}r$, and the distance between the two columns is recorded as $d_h=1.5r$. Since the number of non-edge ROs of odd and even columns may be different, the number of odd and even non-edge ROs is $N_o=\left\lfloor \frac{y}{d_v}-\frac{1}{2} \right\rfloor+1$ and $N_e=\left\lfloor \frac{y}{d_v} \right\rfloor+1$, and the number of columns of odd and even columns is $M_o=\left\lceil \frac{x}{2d_h}-\frac{7}{6} \right\rceil+1$ and $M_e=\left\lceil \frac{x}{2d_h}-\frac{5}{3} \right\rceil+1$.

Here, the position of the non-edge ROs are inside the circuit, which are fixed in the following discussion, and the position of the odd column non-edge ROs are given as Equation 2.

$$\left(0.5r+m_o\cdot 2d_h, 0.5d_v+n_o\cdot d_v\right) \tag{2}$$

Where the range of m_o is from 0 to M_o-1 and the range of n_o is from 0 to N_o-1.
Similarly, the position of the even column non-edge ROs are Equation 3.

$$\left(2r+m_e\cdot 2d_h, n_e\cdot d_v\right) \tag{3}$$

Where the range of m_e is from 0 to M_e-1 and the range of n_e is from 0 to N_e-1.

3.2.2. Adjusting the position of the edge ROs.
The edge ROs have exceeded the range in which the circuit can be laid out, so the position of the edge ROs need to be re-adjusted.
Situation 1: The edge ROs exist outside the upper boundary of the circuit.
Given this case, the position coordinates of the upper edge RO after adjustment are as shown in Equation 4.

$$\begin{cases}\left(0.5r+m_o\cdot 2d_h, y\right), & 0<x-X_r\le 0.5r, 0<y\ \%\ d_v< 0.5d_v\\ \left(2r+m_e\cdot 2d_h, y\right), & 0<x-X_r\le 0.5r, 0.5d_v<y\ \%\ d_v<d_v\end{cases} \tag{4}$$

Situation 2: The edge ROs exist outside the right edge of the circuit.
Give adjusted position coordinates of the right edge ROs as Equation 5.

$$\begin{cases}\left(X_r+0.5r, 0.5d_v+n_o\cdot d_v\right), & 0.5r<x-X_r\le r, y\ \%\ d_v=0 \text{ or } 0.5d_v\\ \left(X_r+r, 0.5d_v+n_o\cdot d_v\right), & r<x-X_r\le 1.5r, y\ \%\ d_v=0 \text{ or } 0.5d_v\end{cases} \tag{5}$$

3.3. Determining the number of ROs
We can give the number of ROs required to completely cover the circuit C of the size $x\times y$ using the ROs of radius r. The situation can be reduced to Equation 6 when calculating the number of ROs.

$$Num_{RO} = \begin{cases} M \cdot N, & 0 < y \% d_v \le 0.5 d_v \\ M \cdot \left(\left\lfloor \frac{N}{2} \right\rfloor + 1 \right) + (M+1) \cdot \left(\left\lceil \frac{N}{2} \right\rceil - 1 \right), & others \end{cases} \tag{6}$$

Where M is equal to $\left\lceil \frac{y}{d_v} \right\rceil$ and N is equal to $\left(\left\lceil \frac{x-r}{d_h} \right\rceil + 1 \right)$.

4. Experimental results

4.1. Perceptual radius and number of ROs

For comparison, Reference [8] uses a 4×2 ROs array (the low-density layout), a similar 4×3 ROs array (the high-density layout) is used in [9], and the literature [10] uses 32 ROs which were distributed on the diagonal of the circuit (the X-shaped layout). Table 1 summarizes the size of the underlying circuit, the measured coverage radius of the RO, and the number of ROs required by our method.

Table1. The coverage radius and quantity statistics of ROs.

The name of the base circuit	The size of circuit (Slice)	Perceptual radius of RO(Slice)	The number of ROs
AES	30×30	7	12
BasicRSA	18×18	4	13
Memctrl	26×26	7	9
Wb_conmax	40×48	11	10
MC8051	32×32	9	7

The number of ROs required for our method is less than the low-density layout in the MC8051 basic circuit, and more ROs are required in the other test circuits than the low-density type. In addition, compared with the high-density layout, the method in this paper uses more ROs in two of the test circuits, four test circuits use less RO, and the remaining four test circuits have the same number of ROs as the high-density type.

4.2. The accuracy of Trojan detection

The accuracy of Trojan detection is shown in Figure 2.

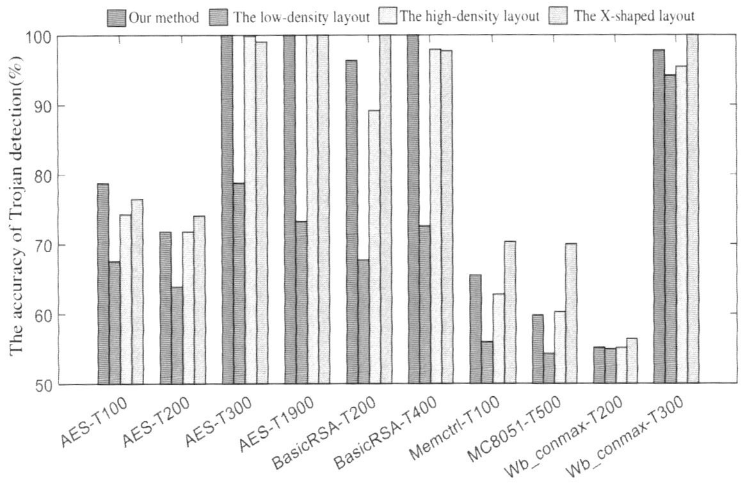

Figure 2. The accuracy result of Trojan detection.

As shown in Figure 2. It can be found that the results of the low-density layout are lower than the other three methods in all the test circuits. In addition, the high-density layout is also inferior to the method proposed in this paper. Beyond that, the methods proposed in this circuit with more than 1% of Trojans have the accuracy of over 95%. In addition to this, because our method can realize the full range sensing of the circuit, the detection rate of the Trojan is relatively stable, while other methods have coverage gaps, and the accuracy of the Trojan changes with the position of the Trojan. In actual situations, the location of the Trojan is unknown, which reflects the advantages of our approach.

5. Conclusion

In this paper, an optimized method for the number of ROs in the network is proposed. The method first gives the actual measurement method of the sensing range. On this basis, we propose a RON layout method based on regular hexagonal meshing. This method can achieve seamless coverage of the circuit. The method proposed in this paper was verified by experiments on 10 test circuits, and the number of ROs and Trojan detection effects in the layout used in other literatures were compared. Experiments show that the proposed method can guarantee the correct rate of Trojan detection with fewer ROs.

References

[1] M. Tehranipoor, C. Wang. Introduction to hardware security and trust[M]. New York: Springer, 2011, 339-364.
[2] B. Shakya, T. He, H. Salmani, *et al.* Benchmarking of hardware Trojans and maliciously affected circuits[J]. Journal of Hardware and Systems Security, 2017, 1(1): 85-102.
[3] M. Tehranipoor, C. Wang. Introduction to hardware security and trust[M]. New York: Springer, 2011, 325-338.
[4] D. Agrawal, S. Baktir, D. Karakoyunlu, *et al.* Trojan detection using IC fingerprinting[C]. 2007 IEEE Symposium on Security and Privacy (SP'07), Oakland, 2007: 296-310.
[5] X. Zhang, M. Tehranipoor. RON: An on-chip ring oscillator network for hardware Trojan detection[C]. 2011 Design, Automation & Test in Europe, Grenoble, 2011: 1-6.
[6] C. Bao, D. Forte, A. Srivastava. On application of one-class SVM to reverse engineering-based hardware Trojan detection[C]. International Symposium on Quality Electronic Design, Santa Clara, 2014, 47-54.
[7] D. Forte, C. Bao, A. Srivastava. Temperature tracking: An innovative run-time approach for hardware Trojan detection[C]. International Conference on Computer-Aided Design, San Jose, 2013, 532-539.
[8] S. Kelly, X. Zhang, M. Tehranipoor, *et al.* Detecting hardware trojans using on-chip sensors in an asic design[J]. Journal of electronic testing, 2015, 31(1): 11-26.
[9] X. Zhang, M. Tehranipoor. RON: An on-chip ring oscillator network for hardware Trojan detection[C]. 2011 Design, Automation & Test in Europe, Grenoble, 2011: 1-6.
[10] C. Lamech, R. M. Rad, M. Tehranipoor, *et al.* An experimental analysis of power and delay signal-to-noise requirements for detecting Trojans and methods for achieving the required detection sensitivities[J]. IEEE Transactions on Information Forensics and Security, 2011, 6(3): 1170-1179.
[11] C. W. Wu, D. Verma. A sensor placement algorithm for redundant covering based on riesz energy minimization[C]. 2008 IEEE International Symposium on Circuits and Systems, Seattle, 2008: 2074-2077.

Humanoid Robot Torso Motion Planning Based on Manipulator Pose Dexterity Index

Baoshi Cao, Kui Sun*, Yikun Gu, Minghe Jin and Hong Liu

State Key Laboratory of Robotics and System, Harbin Institute of Technology, Harbin, China

sun.kui@163.com, cbob.robot@gmail.com

Abstract. Humanoid robot operation task is mainly executed with arm-torso system. The torso system can effectively increase humanoid robot arm operation range and vision range of the humanoid robot, but it also increases the difficulty of the humanoid robot motion planning. Humanoid robot torso increases humanoid robot arm operation range by changing the pose of the robot arm base and it also reconstruct humanoid robot arm workspace pose dexterity. Humanoid robot torso position is optimized with manipulator pose dexterity in different humanoid robot arm sub-workspace. For different operation tasks, the indicators to be considered for the position of the torso joint include: whether the robot arm can grasp the target position, whether the specified motion trajectory can be reached, and whether it has higher dexterity in specified motion trajectory. In this paper, we consider higher manipulator pose dexterity sub-workspace is better operation position, and we introduce a scheme to optimize torso joint position.

1. Introduction

Humans learn to grasp object when they grow up, they grasp object in dexterous orientation with motion of torso and manipulator subconsciously. And also we can know which place is easy to reach and which are difficult to approach, it seems that we have a map/model of our arm's capabilities and dexterity. To a human robot, a similar map of manipulator capability and dexterity is meaningful [1]. The information can be used in robot design stage to extend its workspace or maximize robot dexterous region of workspace, and it could apply in humanoid torso motion planning.

There are many different manipulator dexterity description index, such as, Klelin et al, [2] use ratio of maximum Jacobian matrix singular value and minimum Jacobian matrix singular value as condition number to optimize manipulator joint angles. When the condition number has an optimal value of 1, the configuration has been termed isotropic. With the goal to obtain a global isotropy design parameter, Stocco et al. [3] optimized the ratio of the maximum and the condition number in whole workspace to obtain a global version of the condition number. Yoshikawa et al. [4] quantify the ease of arbitrarily changing the position and orientation of the end-effector with manipulability ellipsoid. Zacharias et al. emphasized the importance of being able to easily find robot target configurations that solve the subtask and facilitate the subsequent process of finding paths in a task planning process in paper [5]. Then they proposed a capability map to analysis manipulator workspace [1]. They focus on manipulator directional structures, by positioning of the torso, manipulator could reach target in highest reachability index. Also, A pose dexterity map is created with pose dexterity index in our previous article [6].

The basic function of the humanoid robot motion planning is to allocate the system motion resources, coordinate the resource allocation with time and space, form the optimal or better motion output, and realize the input task [7]. The basic operational task form of the robot can be divided into two types: continuous trajectory tracking and realization of discrete directional target constraint tasks. The continuous trajectory tracking problem needs to solve the inverse solution planning problem, while the other one needs to plan the smooth continuous joint motion trajectory that satisfies the discrete position constraint in the feasible variable space. The two types of tasks are different, but in the same way, the smooth joint continuous trajectory of the time base is required as the motion output or task realization method. The trajectory planning for redundant manipulators is mainly dominated by the former, while the other one is mostly used for large-scale mobile trajectory planning of mobile platforms.

In the multi-degree-of-freedom robot system motion planning, redundant degree of freedom or the solution is not unique, the unique solution must be determined by the performance index function optimization. The weighted sum mode combination performance index function to achieve multi-objective decision-making to obtain a unique solution, which is the most common multi-objective decision-making method. However, the application of weighting and performance functions is inherently flawed [8], because the objective functions have the effect of each other[9]. This makes the adjustment strategy of the weight parameter design one of the determinants of the performance of the weighted objective function algorithm. When the combined objective function needs to be two, the adjustment weight strategy is simple, but when the combined objective function needs to be three or more The incompatibility caused by different unit dimensions, different rate of change, and different ranges of values of the objective function [10] makes it difficult to design a weight adjustment strategy with unclear physical meaning. In this paper, a torso motion planning method with manipulator dexterity index is proposed to optimize humanoid torso trajectory.

The remainder of this paper is organized as follows. Section II details our HIT humanoid robot. In section III, the dexterity map with different pose dexterity index range is proposed. In section IV, high pose dexterity index zone transform with torso movement, section V humanoid robot torso position optimized with manipulator pose dexterity index data, section VI introduces experiment to verify the proposed algorithm and section VII concludes this paper.

2. Robot systems

HIT humanoid robot, as shown in Figure 1, is designed with "from up to down" sequence and consists of an anthropomorphic upper body which is mounted on a wheeled mobile robot. The upper body carries a 3-DOF binocular head [11], two extremities, each of them consists a 7-DOF humanoid arm [12] and a 15-DOF five-fingered dexterous hand [13], and a torso with two joints [14]. The wheeled mobile robot is employed as a first solution to provide mobility to robot when main focus of design was on the stable operation of upper body, future developments will concentrate on a new lower limbs of mobility. The initial specifications for design of the robot is to be a versatile platform for research on replicating human in special environments. Therefore, it should have an anthropomorphic kinematic configuration for research on bi-manual grasping in addition of a human-like appearance. More detail is presented in [15].

3. The dexterity map with different pose dexterity index range

Humans grasp an object most in a dexterous orientation when they could reach the object in many orientations, it facilitates the following operations and in case of some unexpected matters. The same operation criterion is in point for HIT humanoid robot, whose design is oriented at human model. In general, each manipulator has different orientation dexterity in smaller subspace due to its configuration and D-H parameters. We show that manipulator capabilities result in end-effector pose dexterity specific to sub-workspace regions and those structures can be expressed in a visualization map. First, we defined manipulator pose dexterity in sub-workspace with sum of feasible arm angle in different orientations. In the second step, manipulator pose dexterity map with different pose dexterity index range is generated with the data. Right arm of HIT humanoid robot is employed as an example.

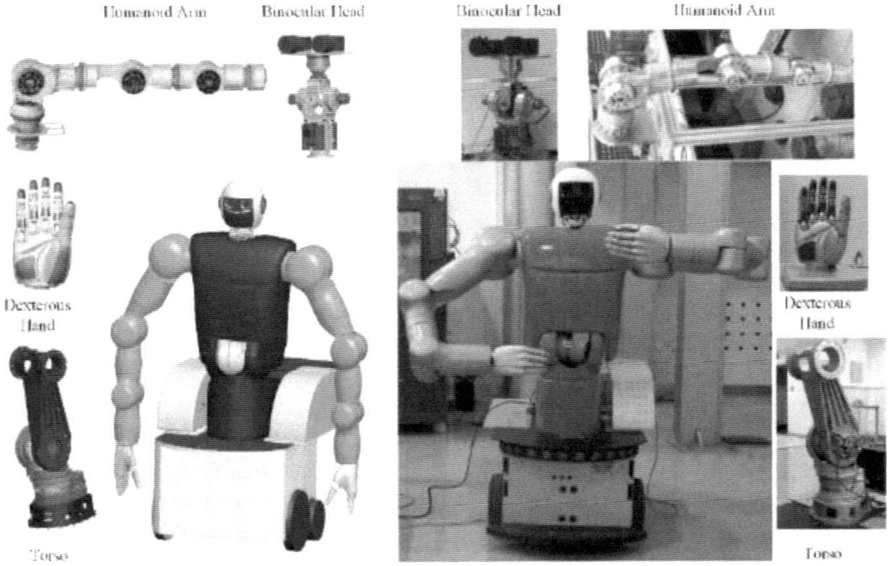

Figure 1. Overview of HIT humanoid robot system

3.1. The pose dexterity index generated

R($p_{i,j,k}$) represents quantity of reachable orientations in each sub-workspace center. In paper [16], Zacharias assign a measure called the reachability index D to characterize the workspace. But to a redundant manipulator as used in our humanoid robot, the reachability index D can not represent manipulator dexterity in region enclosed by sub-workspace. In this paper, we consider feasible arm angle range as a significant value to refer redundant manipulator dexterity,

The pose dexterity index $D_{\text{redundant}}$ in each sub-workspace is defined by,

$$D_{\text{redundant}} = \sum_{m=1}^{n_p} \sum_{n=1}^{n_{\text{ptheta}}} (\Psi \times \text{Reachable}(T_{i,j,k,m,n})) \Big/ (2\pi \times n_p \times n_{ptheta}) \quad (1)$$

Where, $\text{Reachable}(T_{i,j,k,m,n}) = \begin{cases} 1 & \text{if pose can be reached} \\ 0 & \text{otherwise} \end{cases}$.

3.2. Sub-workspace zone with different pose dexterity index range

According to the dexterity of the workspace of the robot arm, the robot workspace is divided into different dexterity workspace areas. As can be calculated with equation (1), the position of the humanoid robot workspace is from 0 to 19.41. The workspace of the robot arm is divided into 4 layers according to the dexterity, and the dexterity of each layer is gradually increased, which means that the mechanical arm has better operability in the workspace of this layer. Sub-workspace that yield different dexterity ranges for the robot arms are shown in figure 2.

In figure 2, light black part in the figure is the complete workspace of the robot arm, and the red part is the sub-workspace area where the mechanical arm is dexterity in different positions.

At this time, the sub-workspace area defining the robot arm position dexterity Dr>13 is a high-dexterity workspace of the robot arm. When the humanoid robot arm moves in the workspace region, the humanoid robot arm has better dexterity.

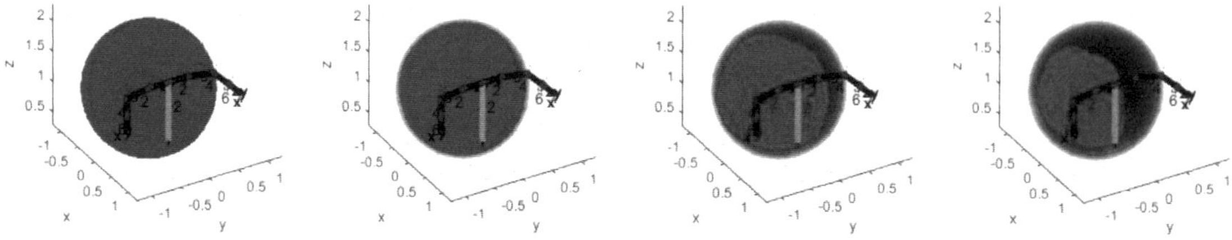

a) Dr=0 Sub-workspace zone | b) $0<Dr\leq6.5$ Sub-workspace zone | c) $6.5<Dr\leq13$ Sub-workspace zone | d) $13<D_r$ Sub-workspace zone

Figure 2. Sub-workspace zone with different pose dexterity index range

4. High pose dexterity index zone transform with torso movement

One of the main goals of the humanoid robot torso motion is to assist the robotic arm to complete the operational tasks, providing a more dexterous workspace for the robotic arm in specific operational tasks. When the operation target is outside the workspace of humanoid robot arm, the arm movement of the robot arm envelops the operation target by the movement of the humanoid torso, so that the robot arm can reach the position of operation target; When the operation target is in the low-index workspace of the robot arm, the humanoid arm cannot realize complicated operation, the movement of humanoid robot torso makes the robot arm high-dexterity workspace envelop operation target, thereby achieving dexterous operation of the operation target.

When the humanoid robot moves torso joint, the pedestal posture of the humanoid robot arm changes accordingly, resulting in a change in the dexterity of the workspace of the humanoid robot arm, as shown in figure 3. As can be seen from the illustration, the movement of the torso joint of humanoid robot has a significant influence in high-dexterity workspace distribution. For different dexterous operational task requirements, the manipulator torso joint angle should be changed to achieve the operability of humanoid arm to improve position of operation target, thereby completing the specified operation task.

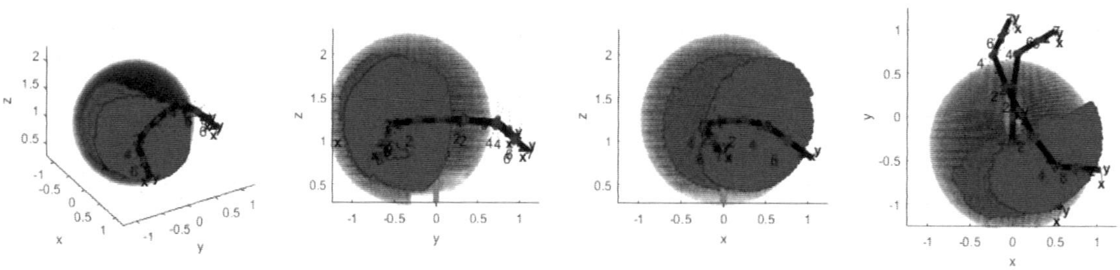

Figure 3. Example of high pose dexterity index zone transform with torso movement

5. Humanoid robot torso position optimized with manipulator pose dexterity index data

During the operation of the humanoid robot, due to the different relative positions of the base of the robot arm and operation target, the operation target in the dexterity workspace of humanoid robot arm. In order to realize dexterous operation when the robot arm is close to the operation target, this paper through changing position of humanoid robot arm base by the movement of the torso, thereby changing the relative positional relationship between the operation target and the base of humanoid robot arm, so that the dexterity workspace area of humanoid robot arm completely envelops the operation target.

In order to optimize the position of the torso joint, torso joint movement range is first divided equally into m×n group, and then torso joint angle group of the robot dexterous operation space is fully enveloped under different torso joint angle conditions. The minimum torso joint movement angle is selected as optimal torso joint position. The specific optimization flow is introducing in Table 1.

6. Experiments

6.1. Torso-arm grasp scenario for HIT humanoid robot

In this section, one application scenarios are verified with manipulator pose dexterity map. For example, in figure 4, target bottle is in lower dexterity index region of manipulator workspace, the bottle can be overlaid with the region of the pose dexterity map where the dexterity index is highest by moving humanoid robot torso. Which means manipulator could grasp the target with more alternatives. Thus torso motion could be planning with manipulator pose dexterity map to optimally reach objects.

Table 1. Torso position optimized with pose dexterity index

Algorithm 1:	Torso position optimized with pose dexterity index
Initial	Ptarget ← Vision_system
	$\theta_W(\theta_{W1},\theta_{W2}) \in R^{m\times n}$ ← Torso_Discretization
	for k = 1 to m×n do
	T_k ← Function_Trans(θ_{Wk})
	P_{tk} ← Function_Inverse(T_k)
	if $P_{tk} \in$ D_ High do
	$\theta_{opt} = \theta_{Wk}$
	end if
	end for
θ_{Wopt} ← Function_min(θ_{opt})	

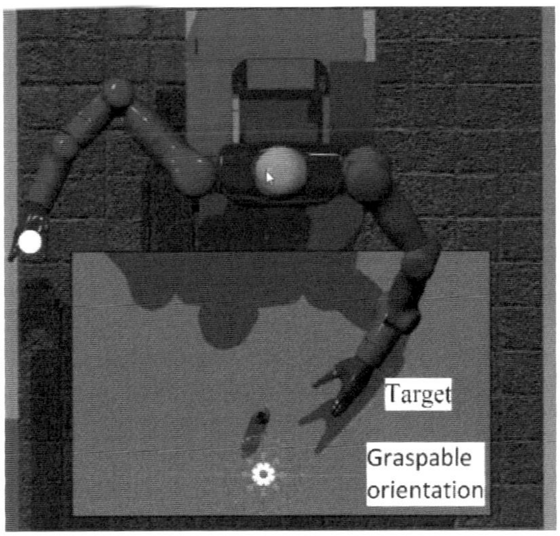

Figure 4. Target grasp scenario for HIT humanoid robot

When the operation target is in the unobtrusive/non-dexterity grasping state in the erect state of the torso joint, after the movement of the torso joint, the operation target can be in the dexterity grasping range of the mechanical arm, and the coordinated movement of torso can achieve effective grasping of the operation target. The initial position of the experimental operation target sets $p_{target} = [1.00 \quad 0 \quad 0.85]$. Through the inverse kinematics calculation, it can be seen that the operation target is out of humanoid arm workspace with rigid torso, that is, under the condition of the torso joint upright, the humanoid robot can not achieve effective operation of the target. Therefore, it is necessary to move torso joint to finish grasping task.

6.2. Torso position optimized with pose dexterity index distribution

The torso joint angle optimization was performed according to the torso joint optimization method proposed in the section. Firstly, the torso joint angle range [-90°~90°; -45°~-135°] is divided into 37×19 group joint angle with equal intervals $\Delta\theta_{\mathrm{W}} = 5^{\circ}$. Then, the equivalent position p'_{target} of the target initial position under the condition of torso joint upright with different sets of torso joint group is calculated by inverse kinematics of humanoid robot. Finally, pose dexterity index in sub-workspace of the humanoid robot can be obtained by the equivalent position. The pose dexterity index with different torso joint angles group $\boldsymbol{\theta}_{\mathrm{W}}$ are shown in figure 5.

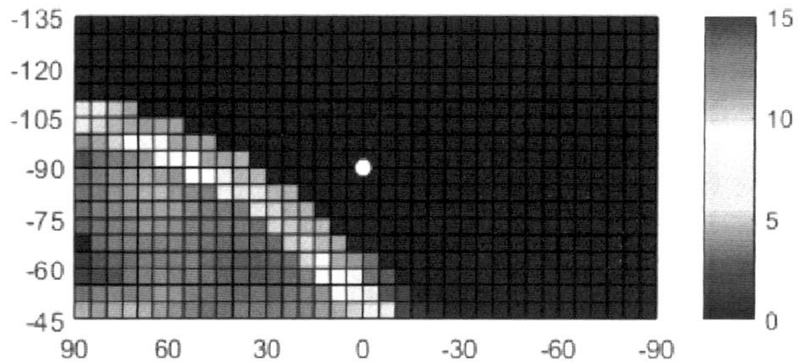

Figure 5. Pose dexterity index distribution of operation platform with dynamic torso

The white point in the figure is the pose dexterity index of arm at the target initial position p_{target} under straight up torso condition. It can be seen from the figure 6 that when the torso is upright, the end of the humanoid arm cannot reach the target initial position at any angle. The torso joint angle of the inverse kinematics of the mechanical arm is basically distributed in the direction of the left and/or forward tilt of the torso, and it can be seen that the left turn and the forward until it has ability to increase dexterity index of target initial position. The same effect. by moving torso, the pose dexterity index of humanoid arm at initial position can be increased, and the circle is centered on the white point to obtain the nearest joint angle of dexterity index of different positions. Determine torso joint angle group $\boldsymbol{\theta}_{\mathrm{W}} = [15^{\circ}; -75^{\circ}]$ for the grasp task.

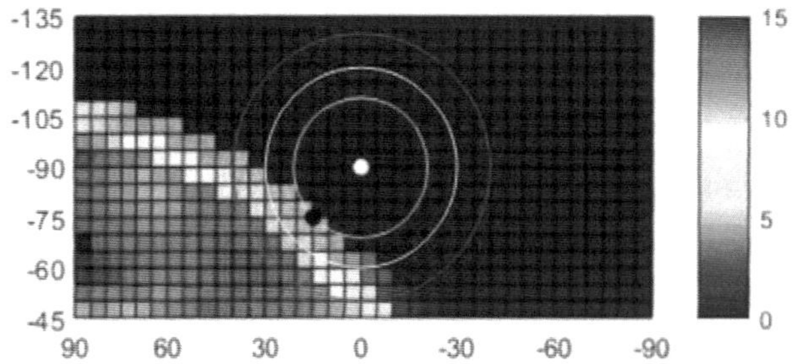

Figure 6. Torso position optimized with manipulator pose dexterity index distribution

7. Conclusion

Structure is indeed present in a manipulator's workspace and can be captured in the form of a map. We develop the representation called manipulator pose dexterity map to capture the manipulator dexterity in its workspace. The manipulator pose dexterity can be computed offline for a specific manipulator. The map is anchored manipulator base and shift position when humanoid torso moves. When target is out of manipulation workspace with rigid torso, optimizing torso position to accomplish task. And manipulator pose dexterity is used to optimize torso position with minimum energy. With the optimized torso joints angle, humanoid robots could use it to enhance humanoid arm manipulation ability. The ungraspable target with rigid torso can be reached with optimized torso joints angle. In further work, optimized torso joints combine with optimized manipulator joints to drive humanoid robot to accomplish task.

Acknowledgment

This work was supported in part by the National Key R&D program of China (Grant No. 2017YFB1300400), the National Natural Science Foundation of China (No. 61603112) and the Self-Planned Task (No.SKLRS201721A) of State Key Laboratory of Robotics and System (HIT).

References

[1]. Zacharias, F., C. Borst, and G. Hirzinger. *Capturing robot workspace structure: representing robot capabilities*. in *Intelligent Robots and Systems, 2007. IROS 2007. IEEE/RSJ International Conference on*. 2007. Ieee.

[2]. Klein, C.A. and B.E. Blaho, *Dexterity measures for the design and control of kinematically redundant manipulators.* The International Journal of Robotics Research, 1987. **6**(2): p. 72-83.

[3]. Stocco, L., S.E. Salcudean, and F. Sassani, *Fast constrained global minimax optimization of robot parameters.* Robotica, 1998. **16**(6): p. 595-605.

[4]. Yoshikawa, T., *Manipulability of robotic mechanisms.* The international journal of Robotics Research, 1985. **4**(2): p. 3-9.

[5]. Zacharias, F., C. Borst, and G. Hirzinger. *Bridging the Gap between Task Planning and Path Planning*. in *Ieee/rsj International Conference on Intelligent Robots and Systems*. 2006.

[6]. Cao, B., et al. *Workspace Analysis Based on Manipulator Pose Dexterity Map*. in *2018 3rd International Conference on Robotics and Automation Engineering, ICRAE 2018, November 17, 2018 - November 19, 2018*. 2018. Guangzhou, China: Institute of Electrical and Electronics Engineers Inc.

[7]. Zacharias, F. and C. Borst, *Knowledge representations for high-level and low-level planning*, in *ICAPS 2011*. 2011. p. 98-99.

[8]. Park, J.-Y., P.-H. Chang, and J.-Y. Yang, *Task-oriented design of robot kinematics using the grid method.* Advanced robotics, 2003. **17**(9): p. 879-907.

[9]. Konietschke, R., et al., *Optimal Design of a Medical Robot for Minimally Invasive Surgery, 2.* Jahrestagung der Deutschen Gesellschaft fuer Computer-und Roboterassistierte Chirurgie (CURAC), 2003.

[10]. Roth, B. *Performance evaluation of manipulators from a kinematic view point*. 1970.

[11]. Yang Y. *Mechanism design of robonaut head and research on its object tracing control method.* Master thesis. Harbin Institute of Technology, 2011.

[12]. Huo, X., et al. *Design and development of a 7-DOF humanoid arm.* in *Robotics and Biomimetics (ROBIO), 2012 IEEE International Conference on*. 2012. IEEE.

[13]. Liu, H., et al., *Development of a multi-DOF prosthetic hand with intrinsic actuation, intuitive control and sensory feedback.* Industrial Robot, 2014. **41**(4): p. 381-392.

[14]. Cao, B., et al. *Design and development of a two-DOF torso for humanoid robot*. in *IEEE International Conference on Advanced Intelligent Mechatronics*. 2016.

[15]. Cao, B., et al. *Development of HIT humanoid robot*. in *10th International Conference on Intelligent Robotics and Applications, ICIRA 2017, August 16, 2017 - August 18, 2017*. 2017. Wuhan, China: Springer Verlag.

[16]. Zacharias, F., *Knowledge Representations for Planning Manipulation Tasks*. 2012: Springer Berlin Heidelberg.

Gait Phase Classification Based on sEMG Signals Using Long Short-Term Memory for Lower Limb Exoskeleton Robot

Ye Yuan[1,2,3], Ziming Guo ,Can Wang*,Shengcai Duan, Lufeng Zhang, Xinyu Wu

1.Guangdong Provincial Key Lab of Robotics and Intelligent System, Shenzhen Institutes of Advanced Technology, Chinese Academy of Sciences;
2.CAS Key Laboratory of Human-Machine Intelligence-Synergic Systems, Shenzhen Institutes of Advanced Technology;
3.Chongqing Key Laboratory of Nonlinear Circuits and Intelligent Information Processing, College of Electronic and Information Engineering, Southwest University.

can.wang@siat.ac.cn.

Abstract. In this work, we present a Long Short-Term Memory Model (LSTMM) for gait phase classification based on sEMG signals to control the lower limb exoskeleton robot which can recognize 2 phases (Stand and Swing) of leg phases between the foot and ground. This model only needs four sEMG signals to control the lower limb exoskeleton robot helping the hemiplegia patient walking. Compared with the existing methods, the proposed model not only avoids the complex sensor systems but also enhances the accuracy of gait phase classification. The experimental results first verify the availability of sEMG data acquisition system on the lower limb exoskeleton robot made by the Shenzhen Institutes of Advanced Technologies (SIAT) by quantify the system with gold standard optoelectronic system Vicon, then show that the proposed LSTMM is significantly higher on prediction accuracy and has better robustness for gait phase classification to control the lower limb exoskeleton robot with different speeds. Finally, the maximum accuracy of LSTMM on the gait phase classification is 97.89%.

1. Introduction

1.1. Background

Every year nearly 800 000 people experience a recurrent or new stroke accounting for approximately 6.5 million stroke survivors in the U.S. alone [1]. About 80% of them experience significant impairments and require rehabilitation. The lower limb exoskeleton robot is a wearable human-robot interactive equipment tying to human legs and moving synchronously with the human gait. Hence, the gait phase classification is a significant study to help stroke walking easily. However, few studies classify the gait phase from sEMG signal [2]. As reviewed in [3], majority of the studies devoted to recognizing gaits for the lower limb exoskeleton robot have been performed using kinetics and kinematics data (gyroscopes, force detectors and foot inertial sensors) during locomotion. In this work, we will use the sEMG signals to classify the phase of gait cycle into two phases-Stand (ST) and Swing (SW) [4]. sEMG is acknowledged as a non-invasive approach, specifically suitable to monitor muscle activity during walking [5]. sEMG signals are typically coupled with signals able to provide the synchronization of the gait cycle, such as joint angles, to control the lower limb exoskeleton robot.

Consequently, many researchers have a growing interest in using the sEMG signals to accurately classify the gait phase to control the lower limb exoskeleton robots helping walking problem people.

1.2. Related works
In recent years, with the development of machine learning and deep learning techniques, various machine learning algorithms, such as linear discriminant analysis (LDA) [6], multilayer perceptron (MLP) [7] and SVM [8] have been used for classification problems. Specifically, Joshi et al. [6] adopt LDA to separate different phases of a gait cycle with the EMG signal data of the lower limb exoskeleton. [7] employs the MLP, which has the advantages of build non-linear fits, to classify the gait phase obtaining satisfactory results. AS for SVM, Ziegler et al. [8] use the theory of SVM to solve the classification of ST and SW phase during healthy human gait based on the muscle.

1.3. Motivation
However, all of the above methods [6]-[8] have emerged drawback on time series. Such as bigger memory spending and lower accuracy on long time series. With the development of machine learning algorithm, Long Short-Term Memory (LSTM) have demonstrated ground-breaking performance on classify, process and predict time series. For example, LSTM has been widely applied in handwriting recognition, speech recognition and image generation for gait phase detection. Due to the LSTM architecture has memory cell, it can utilize current and previous information to prediction or classification. Therefore, taking advantage of LSTM to classify the gait phase is a new direction.
Motivated by the above work, this work proposes a Long Short-Term Memory Model (LSTMM) for gait phase classification based on sEMG signals to control the lower limb exoskeleton robot which can recognize ST and SW phase of leg phases between the foot and ground. The proposed model is based on [9], where Ding et al. apply the LSTM to detect and discriminate gait phase based on single Inertial Measurement Unit (IMU). Inspired by the [9], we first attempt to apply the LSTM to solve the problem of gait phase classification based on sEMG in the lower limb exoskeleton robot. Compared with the previous studies, our specific contributions are summarized as following:
(1) We propose an LSTMM for gait phase classification based on sEMG signals to control the lower limb exoskeleton robot helping the people who have walking problem.
(2) In our model, we capture the signals from four muscles of the lower limb using sEMG signal acquisition system. Only four muscles sEMG signals can detect phases of walking in contrast to [9].
(3) The proposed LSTMM based on sEMG signals is not only higher on prediction accuracy but also has better robustness for gait phase detection to control the lower limb exoskeleton robot under different speed.

2. Methods on gait phase classification

2.1. Long Short-Term Memory
Fig. 1 shows the basic architecture of an LSTM memory block with one memory cell. Where σ denotes the gate activation function, which is generally the sigmoid function; g represents the input activation function which is usually the tanh function; h represents the output activation function that is usually the tanh function. At time t, the input gate calculates the input message of LSTM unit i_t, forget gate calculates the information f_t left by the LSTM unit, output out calculates the LSTN unit output information O_t, which are defined as:

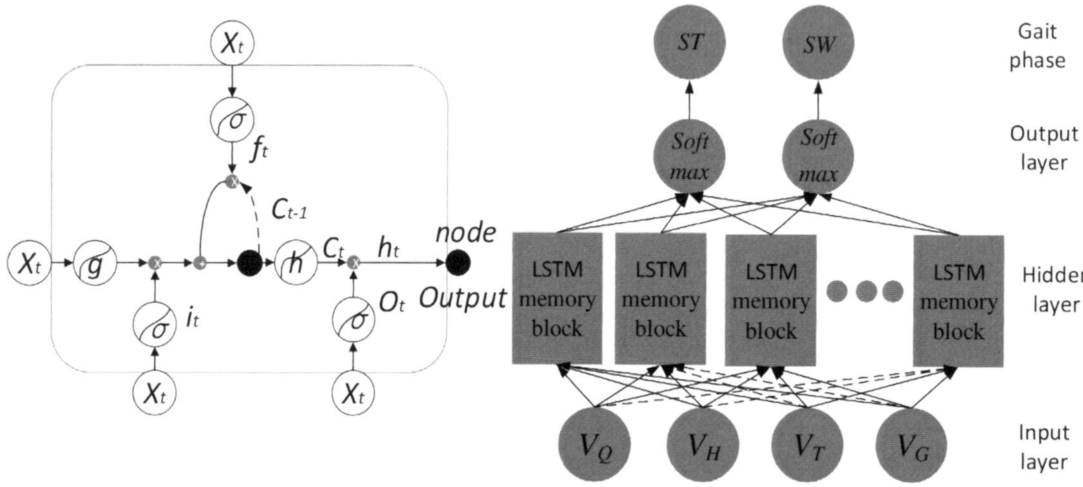

Figure 1. LSTM memory block and the Gait phase classification of LSTMM

$$i_t = \sigma\left(W_{xi}x_t + W_{hi}h_{t-1} + W_{ci}c_{t-1} + b_i\right) \tag{1}$$

$$f_t = \sigma\left(W_{xf}x_t + W_{hf}h_{t-1} + W_{cf}c_{t-1} + b_f\right) \tag{2}$$

$$o_t = \sigma\left(W_{xo}x_t + W_{ho}h_{t-1} + W_{co}c_{t-1} + b_o\right) \tag{3}$$

Where x_t denotes the input, h_t of time represents the hidden layer state of t time, {W*} represents the weight matrix of LSTM unit, and $\{b*\}$ represents the bias item of LSTM unit. The c_t updates formula and output of the whole unit h_t are as follows:

$$c_t = f_t c_{t-1} + i_t g\left(W_{xc}x_t + W_{hc}h_{t-1} + b_c\right) \tag{4}$$

$$h_t = o_t gh\left(c_t\right) \tag{5}$$

Owing to the multiple LSTM cells can be stacked for more expressive power. In the process of normal walking, the pose angle of hip and knee joint and the sEMG signals are obtained, so the gait of human body is analyzed and divided. As shown in Fig. 1, the LSTM neural network structure is divided into 1 layer input layer, 1 layer hidden layer and 1 layer output layer. The input layer is composed of 4 signals, the number of neuron nodes in hidden layer is 128, it uses relu as the activation function, the probability of signals not being set to 0.5 by dropout, the method of using dropout can make the LSTM network model more robust and avoid the problem of over-fitting. The softmax layer is selected as the output layer composing of two nodes, which are divided into two gait categories.

2.2. Support Vector Machine

An outline of the SVM classification algorithm is described as follows: let $\mathrm{x}_i = R^n$, $i = 1,\ldots,m$ be data points of a training data set in an n-dimensional feature space. Define a vector w , a scalar b and a vector of labels $y \in \mathrm{R}^m$, $y_i \in \{1,-1\}$ with $y_i = 1$ for samples and $y_i = -1$ for samples, such that

$$y_i\left(\mathrm{w}^{\mathrm{T}}\mathrm{x}_i + b\right) \geq 1 - \varepsilon_i \tag{6}$$

The slack variable $\varepsilon_i \geq 0$. Defining $y_i\left(\mathrm{w}^{\mathrm{T}}\mathrm{x}_i + b\right) - 1 = 0$ on the class borders, the margin d between the classes can be calculated as $d = 2/\|\mathrm{w}\|$. The corresponding constrained optimization problem is then

$$\min \frac{1}{2}\|\mathrm{w}\|^2 + C\sum_i \varepsilon_i \tag{7}$$

where weight factor $C > 0$.

The performance of SVM classifier relies on the choice of regularization parameter C. The value of C in this part is set to 1. During gait phase classification, the training labels are assigned to two activities as well as to the transitions between these activities.

2.3. Multi-Layer Perceptron

Multilayer perceptron (MLP) can project the input to a linear separable space with a particular set of weight values. For the lower limb exoskeleton gait phase classification based on sEMG signals, the input is the sEMG signals. Signals (V_Q, V_H, V_T, V_G) pass from the input layer through the hidden layers to the output layer. The output of MLP is used to control the movement of exoskeleton robot.

3. sEMG signals acquisition

3.1. sEMG signal acquisition process

A 16 channels wireless sEMG signals acquisition device that used in our experiments is made by Biometrics (US) shown in Fig. 2. Out of the available 16 channels, 4 channels are used for left leg with 1 channel corresponding to sEMG data from one muscle. Fig. 2 shows the activity of four muscles (Quadriceps, Hamstring, Gastrocnemius and Tibialis Anterior) of the lower limb that is captured during the gait cycle. And the electrodes are located according to the guidelines of the expert.

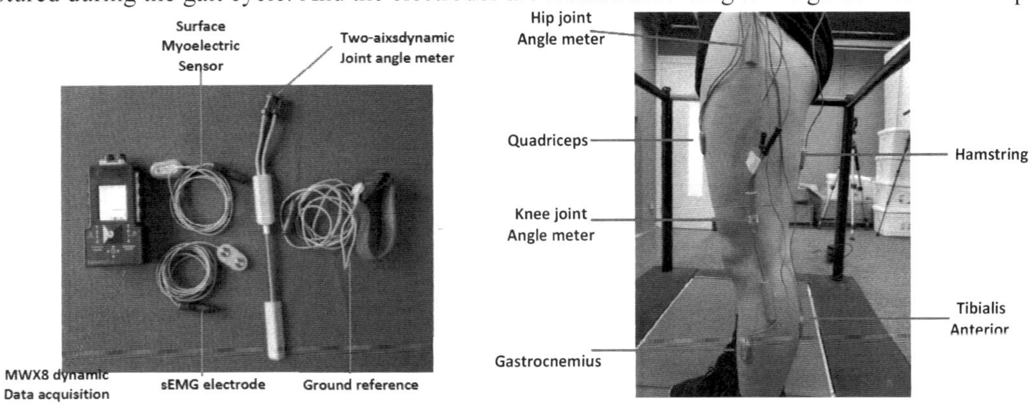

Figure 2. Biometrics device and the position of sEMG acquisition

For rehabilitation of hemiplegia patients, obtaining the real-time sEMG signals can control the exoskeleton robot exactly classifying 2 phases of leg phases between the foot and ground. The SIAT exoskeleton robot has power assist mode and zero-torque mode. The clutches are designed on each joint for switching between two modes. The power-assist mode offers enough torque to let wearers walk with the crutches to maintain balance. In zero-torque mode, the wearers walk following their own will, and the exoskeleton offers torque as need. This work is based on the zero-torque mode.

3.2. sEMG Data Source and Gait Data phase division

Six healthy and never suffered gait dysfunctions male testers are recruited in the experiments for training classifiers. The purpose and the procedure of the data collection are understood by the testers. The detail information of the testers is shown in Table 1. Each tester wears the SIAT exoskeleton robot set in zero-torque mode and walks in their comfortable way for a while on the AMTI (US) force platform under different speed to obtain the sEMG signals data. Testers are required to walk on one minute 2 times with their normal walking speed while the SIAT robot controller records the sEMG signal acquisition device data by a Bluetooth device. Finally, we get the sEMG samples data in Table1.

Table1. Model Results.

NO	*Hight(cm)*	*Weight(kg)*	*sample*	α_i
1	**175**	64	3000	0.2
2	**178**	66	3000	0.25
3	**179**	62	3000	0.23
4	**175**	69	3000	0.24
5	**176**	67	3000	0.25
6	**183**	63	3000	0.22

It's easy to find that the phases of SIAT exoskeleton robot can be classified into ST and SW phases.

3.3. Assigning labels process

The hip joint angle and knee joint angle are used to assign labels to the sEMG data. The threshold of the hip joint angle and knee joint angle to classify the gait phase is calculated as

$$\tau = \theta_{\min} + \alpha\left(\theta_{\max} - \theta_{\min}\right) \tag{8}$$

where, τ is the joint angle threshold of the walking gait, θ_{max} and θ_{min} is the max and min value of the joint angle, α is the coefficient that represents the percentage of threshold. α is from 0.2 to 0.25 in table1. While the hip angle greater than threshold and the knee angle below the threshold, the gait phase is SW. Other cases are ST.

4. Gait phase classification results

4.1. The availability of sEMG data acquisition system

To validate the output of the sEMG data acquisition system, two randomly selected healthy testers are asked to walk at least 1 minute under speed of 2.0 km/h on the AMTI force platform with retro-reflective markers placed on the SIAT lower-limb exoskeleton robot under Vicon environments shown in Fig. 3. Three Vicon cameras are used to track the trajectories of the markers while the sEMG data output signals are recorded. Based on the marker trajectories the reference gait phase signal generating. The reference signal is compared to the measured sEMG data acquisition system output in Fig. 4.
A typical example of the hip angle signal, knee angle signals, and sEMG gait data acquisition system output signal and labels recorded during the walking of tester are shown in Fig.4. Synchronized Vicon measurements of the (hip and knee) marker positions in the horizontal direction are shown in the Fig.3. Due to the sampling frequency of the sEMG is set to 1000Hz. Sampling frequency of the Vicon motion capture system is 100Hz. Thus, 1 sample of angular data corresponds to10 samples of the sEMG data according to [8]. A comparison between the reference angle signal (red line), which is generated by the hip and knee markers trajectories, and the sEMG data acquisition system output signal (blue line) is shown in the Fig. 4. The sEMG data acquisition system output correlated well with the reference gait phase signal for all trials. A gait cycles with walking speed 2.0 km/h indicate that the sEMG data acquisition system of the two gait phases relative to the reference signal is synchronization.

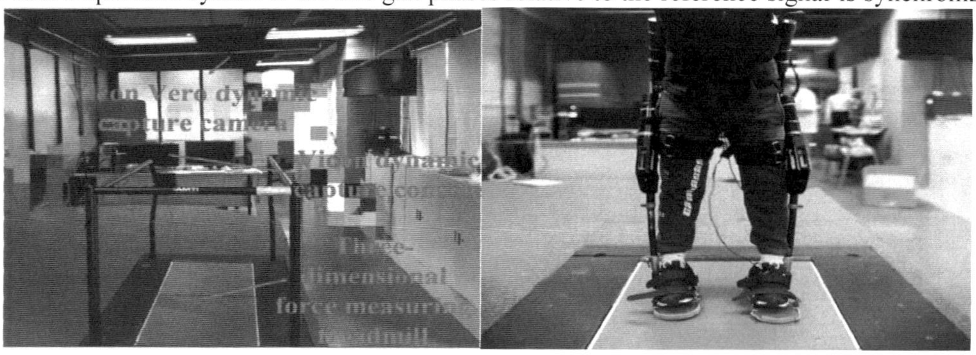

Figure 3. Vicon environments

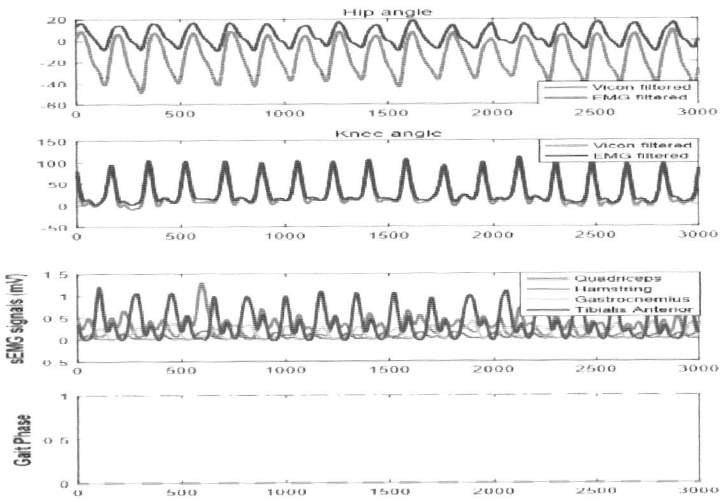

Figure 4. The comparison of the sEMG data acquisition system with Vicon

4.2. The availability of sEMG data acquisition system

Next, we begin to classify the gait phase by LSTM, SVM and MLP. Defining input x as

$$x = \left[V_Q, V_H, V_T, V_G \right] \tag{9}$$

where V_Q, V_H, V_T and V_G denote the EMG signal of left quadriceps, hamstring, gastrocnemius and tibialis anterior, respectively. To verify the robust of the above models on gait phase classification to control the SIAT exoskeleton robot, we analyse the walking speed of 1.5km/h, 2.0km/h and 2.5km/h. Each speed all selects 3000 valid samples. All of the models use the same train and test set to test the performance of the models. The train set account for 80% and the test set take up 20%.

1) LSTM: Fig.5(a) presents the gait phase classification results of LSTMM, where the gait phase prediction accuracy on 1.5km/h, 2.0km/h and 2.5km/h are 97.61%, 97.89% and 97.75%, respectively.

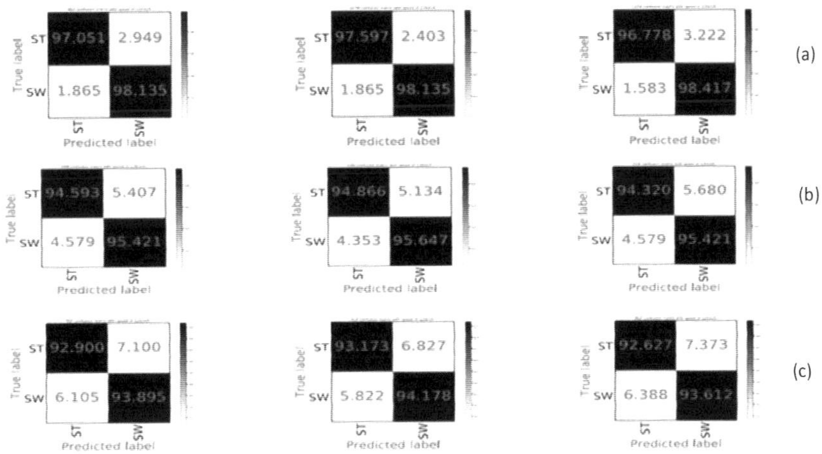

Figure 5. The gait phase classification results of LSTM, SVM and MLP

2) SVM: Fig. 5(b) shows the gait phase classification results of SVM, where the gait phase predict accuracy on 1.5km/h, 2.0km/h and 2.5km/h are 94.83%, 95.11% and 94.55%, respectively.

3) MLP: The gait phase classification results of MLP is shown in Fig. 5(c). We can see that the average accuracy walking on 1.5km/h, 2.0km/h and 2.5km/h are 93.06%, 93.33% and 92.61%.

Table2. Model Results.

accuracy	*LSTM*	*SVM*	*MLP*
1.5km/h	**97.61**	94.83	93.06
2km/h	**97.89**	95.11	93.33
2.5km/h	**97.75**	94.55	92.61

After above works, the best results of LSTM, SVM and MLP are shown in Table2. From Table2, the proposed LSTMM is shown to be robust for gait phase classification to control the lower limb exoskeleton robot under different speed in precision. The maximin accuracy of the gait phase classification is walking with 2km/h which is also the normal speed of the stroke people.

5. Conclusion

This work presents an LSTMM for gait phase classification based on sEMG signals to control the lower limb exoskeleton robot which can recognize ST and SW phases of leg phases between the foot and ground. Compared with the existing methods, our model just needs four sEMG signals to control the lower limb exoskeleton robot assisting the hemiplegia patient to walk. The experimental results show that the proposed model is not only higher on prediction accuracy but also has better robustness for gait phase classification to control the lower limb exoskeleton robot under different speed.

Acknowledgments

This work described is partially supported by the National key research and development plan (2017YFB1302303), National Natural Science Foundation-Shenzhen Joint Research Program(U1613219) and Innovation Fund Project：COMAC-SFGS-2019-262. The authors would also like to thank all subjects who participate in experiments.

References

[1] V. L. Roger, A. S. Go, D. M. Lloyd-Jones, R. J. Adams, J. D. Berry, T. M. Brown, M. R. Carnethon, S. Dai, G. de Simone, and E. S. Ford, "Executive summary: Heart disease and stroke statistics-2011 update" Circulation, vol. 123, pp. 459-463, 2011.

[2] Hermans and Anne, "Exploring the potential of using the foot sole pressure pattern to control step length in exoskeleton gait," Delft University of Technology, 2019.

[3] C. MacDonald, D. Smith, R. Brower, M. Ceberio, T. Sarkodie-Gyan, "Determination of Human Gait Phase Using Fuzzy Inference "Int. Conf. on Rehab. Robotics, 2007.

[4] Nurhazimah Nazmi, Mohd Azizi Abdul, and S. Anon Ahmad, "Generalization of ANN Model in Classifying Stance and Swing Phases of Gait using EMG Signals ," 2018 IEEE-EMBS Conference on Biomedical Engineering and Sciences , pp. 461-466, 2018.

[5] Wang,W., Stefano,A., Allen,R. "A simulation model of the surface EMG signal for analysis of muscle activity during the gait cycle". Comput. Biol. Med. 2006, 36, pp.601-618.

[6] C. D. Joshi, U. Lahiri and N. V. Thakor, "Classification of gait phases from lower limb EMG: Application to exoskeleton orthosis," 2013 IEEE Point-of-Care Healthcare Technologies (PHT), Bangalore, 2013, pp. 228231.

[7] N. Suo, X. Qian, and J Zhao. "Gait recognition based on KPCA and KNN," in Proc. Conf. Envior Sci and Inf App Tech, pp.432-435, 2010.

[8] J. Ziegler1, H. Gattringer, and A. Mueller, "Classification of Gait Phases Based on Bilateral EMG Data Using Support Vector Machines," in IEEE. Conf. Bio Rob and Bio, pp. 978-983, 2018.

[9] Z. Ding, C. Yang, K. Xing, X. Ma, K. Yang, and F. Jiang, "The Real Time Gait Phase Detection Based On Long Short-Term Memory," in IEEE. Conf. Data Science in Cyberspace, pp. 33-38, 2018.

IOP Conf. Series: Materials Science and Engineering **853** (2020) 012042 doi:10.1088/1757-899X/853/1/012042

A Mobile Robot Recognize Blurred Targets via an Improved Histogram Equalization

Longzhi Zhang, Dongmei Wu

State Key Laboratory of Robotics and System, Harbin Institute of Technology, Harbin, China

wdm@hit.edu.cn

Abstract. Target recognition is a precondition for guiding robots to complete tasks. However, if the target is blurred, recognition will be difficult, which will lead to an inefficiency operation for robots. For the sake of improving the robot operation efficiency, it is essential to improve the image quality of recognized blurred target. Since image enhancement is taken as a powerful ways for improving image quality, it has gained largely attentions from researches. In fact, most of enhance methods are proposed based on histogram equalization, and until present, focus of research is still on improvements of histogram equalization. Hence, this paper proposes an improved histogram equalization for blurred targets in mobile robot recognition. Compared with the traditional histogram equalization methods, our method has less calculation time and better performance of real-time. Furthermore, the proposed approach also has strong robustness in blurred target recognition via a mobile robot.

1. Introduction

Target recognition is always a popular topic in computer vision, which purpose is to find the correct location of the object. Thus, target recognition could be applied in robot vision to assist robots to finish specific missions. However, in practical applications, robots usually have to operations on blurred target. But, recognize blurred object is difficult for robots, thereby leading to an inefficiency operations of robots. Accordingly, in order to solve this problem, it is necessary to improve the image quality of blurred target.

As a useful ways for improving the image quality, image enhancement has attracted concerns from researches. Over past few years, numbers of enhance methods have been studied. For instance, Tarik Arici et al. proposed a general framework based on histogram equalization for image contrast enhancement [1]. Research of Wu et al. was via enhance contrast and suppress background to achieve intensify the image [2]. You put forward an enhancement method on the basis of logarithmic mapping to enhance contrast [3]. In the case of poor illumination, Bhattacharya et al. researched on an localized image enhancement method, which was via using singular value decomposition to increase local contrast for given color images [4]. Hossain et al. proposed a novel contrast enhancement approach in medical image processing, for detecting cartilage shape change of the knee joint [5]. The authors of literature [6] presented a histogram normalization enhancement method in gradient field for infrared images to identify and tracking target. Kheirkhah et al. via studied histogram-based threshold to improve the image quality of medical video image [7]. Wang et al. came up with an enhancement approach named content-adaptive histogram equalization, which was via using contrast-adaptive contrast improvement and contrast-dependent saturation adjustment to improve intensity contrast for outdoor images [8].

Apparently, among these recent years of approaches, mostly are innovations based on histogram equalization, which indicates that improvements of histogram equalization is still a focus of research on image enhancement. Hence, an improved histogram equalization approach is proposed in this paper, which is used for blurred target recognition via a mobile robot.

The overall paper is organized as follows. Section 2 briefly overview the theoretical of histogram equalization to provide a research fundamental and motivation for our method. Afterwards, Section 3 describe the proposed method. In Section 4, experiments are performed and results are exhibited. Lastly, in Section 5, the paper is summarized and future work is discussed.

2. Overview of histogram equalization

In fact, principle of histogram equalization is relative simple, and largely novel ideas are based on it. Thus, this work proposes an improved histogram equalization for blurred targets in mobile robot recognition. In this part, histogram equalization will be overviewed.

Define original image as I, the grayscale of I is L, $I(i, j)$ represents the gray value at a coordinate position (i, j), and $I(i, j) \in [0, L-1]$, $I \in I(i, j)$.

Then, the probability density function of grayscale in I could be expressed as,

$$\xi(k) = \frac{x_k}{X} \quad (k = 0,1,\ldots,L-1) \tag{1}$$

Where, X is the total number of pixels, and x_k is the number of pixels with grayscale k.

Consequently, the cumulative distribution function of grey level in I is defined as,

$$l(k) = \sum_{i=0}^{k} \xi(i) \quad (k = 0,1,\ldots,L-1) \tag{2}$$

Principle of histogram equalization is via cumulative distribution function to maps the original image into an enhanced image, with approximate uniform gray level distribution.

Thus, the corresponding mapping relation could be described as follows,

$$f(k) = (L-1) \times l(k) \tag{3}$$

Through above calculation, original image I is enhanced due to merging the gray level. Nonetheless, the brightness of enhanced image is uneven, and some details in image are easy to loss. Thus, quite a few improvement researches based on histogram equalization is proposed.

However, among these studies, five most representative research results have been widely accepted, which have been achieved successful to some extent from different aspects [9-13]. But in practical applications, compared with tradition histogram equalization, their real-time is insufficient. Thus, to solve this problem, this paper proposes an improved histogram equalization, which is used for blurred targets in mobile robot recognition.

3. Improved histogram equalization

This work is to deal with blurred targets recognition in practical applications, which means real-time is significant. Thus, our idea is the proposed algorithm could real-time enhance the image, meanwhile has relative flexibility performance. That is, we will attempt to use flexible local region to replace the global image.

3.1. Local region

In this work, entire image is replaced via local region in recognized blurred target, to achieve real-time enhance image, and finally improve the efficiency of recognition. When select a local region, we first need to convert the entire image into a gray image and display it. Afterwards, local region with suitable location and size is selected out according to the actual application needs. Next, image of selected target area is displayed, which is enhanced region. Especially noteworthy, original image always unchanged during this process.

3.2. Enhance for local region
In our approach, we first define a local region with a variable position and size to acquire enhanced image, which aims at acquiring flexible local region to achieve enhancement with different requirements in applications. Afterwards, intensify this image via histogram equalization. Lastly, replace the original image corresponds to the local region with enhanced image to form a new partially enhanced image.
Especially, in new enhanced image, only the region of local selected is enhanced, and the other regions remain the same as the original.

4. Experiments

4.1. Design of experiment
In order to confirm the proposed algorithm is effective, procedure of experiment is designed as shown in figure 1.

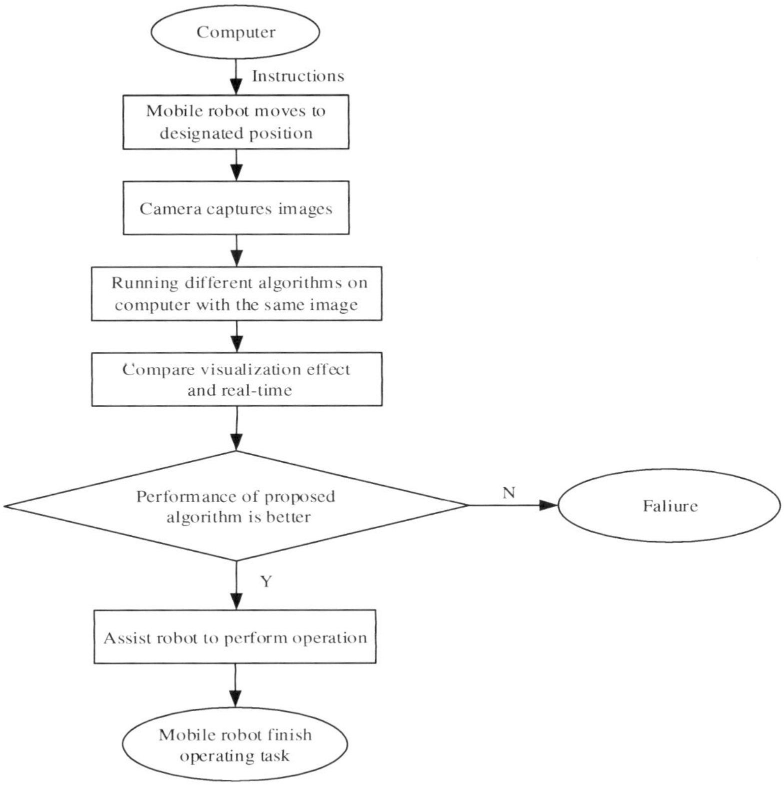

Figure 1. Procedure of experiment

4.2. Experimental results
The proposed algorithm is focus on the need for blurred target recognition via a mobile robot. Thus, a blurred image captured via the mobile visual system is shown in figure 2.
Since our method is based on histogram equalization, then in our experiments, comparison will be performed between histogram equalization and our approach.

Figure 2. Original Blurred Image

4.2.1. Visual effect. In fact, the proposed method is a local enhancement method. Hence, in our implement, region of our approach is randomly selected. The visual effect of enhancement as show in figure 3.

(a) Histogram Equalization (b) Proposed Method

Figure 3. Visual Effect of Enhancement

Obviously from figure 3, the proposed method is superiority to histogram equalization in visual effect of enhancement, which states that our method is effective and useful.

4.2.2. Real-time. In this experiment, each algorithm was tested five times with the same computation device. The computation time of each algorithm is shown in table 1.

Table1. Computation time of different algorithms (ms)

Times	Histogram Equalization	Proposed Method
1	323.425	141.450
2	311.590	139.536
3	326.614	138.589
4	331.503	136.637
5	324.433	134.562

It can be known from table 1 that the proposed method costs less computation time than histogram equalization, which computation time is shortened by nearly 1.5 times, and the real-time performance is better.

5. Conclusion and Discussion

In this paper, an improved histogram equalization method is proposed. The experimental results demonstrate that our approach has less computation time and better real-time performance than traditional histogram equalization. That is, from the aspect of applications, our method is superiority to traditional histogram equalization.

In future, we intend to perfect our work, and add more detail comparisons and analyses among our algorithm and others. Moreover, more evaluate criterions of algorithm are also could be considered and added, to further indicate the advantage of our proposed method.

Acknowledgment
This work is supported by Self-Planned Task of State Key Laboratory of Robotics and System under Grant No. SKLRS201910B.

References

[1] T. Arici, S. Dikbas, and Y. Altunbasak, "A histogram modification framework and its application for image contrast enhancement," IEEE Transactions on Image Processing, vol. 18, Sept. 2009, pp. 1921-1935.
[2] WU Zhi-guo, WANG Yan-jie, "An Image Enhancement Algorithm Based on Histogram Nonlinear Transform," Acta Photonica Sinica, vol.39, Apr. 2010, pp.755-758.
[3] You J, "Contrast enhancement using histogram equalization based on logarithmic mapping," Optical Engineering, vol. 51, pp. 7002, June 2012.
[4] Bhattacharya S, Gupta S, Subramanian V K, "Localized image enhancement," Proc. 2014 20th National Conference on Communications, Feb. 2014, pp. 1-6.
[5] Hossain M B, Lai K W, Pingguan-Murphy B, et al., "Contrast enhancement of ultrasound imaging of the knee joint cartilage for early detection of knee osteoarthritis," Biomedical Signal Processing & Control, vol. 13, Apr. 2014, pp. 157-167.
[6] Zhao W, Xu Z, Zhao J, et al., "Variational infrared image enhancement based on adaptive dual-threshold gradient field equalization," Infrared Physics & Technology, vol. 66, Sept. 2014, pp. 152-159.
[7] Kheirkhah F M, Mohammadi H R S, Shahverdi A, "Histogram non-linear transform for sperm cells image detection enhancement," Proc. 2016 Eighth International Conference on Information and Knowledge Technology, Dec. 2016, pp.25-30.
[8] Wang S, Cho W, Jang J, et al., "Contrast-dependent saturation adjustment for outdoor image enhancement," Journal of the Optical Society of America A, vol. 34, Jan. 2017, pp. 2532-2542.
[9] KIM Y T, "Contrast enhancement using brightness preserving bi-histogram equalization", IEEE Transactions on Consumer Electronics, vol. 43, Feb. 1997, pp. 1-8.
[10] Wang Y, Chen Q, Zhang B, "Image enhancement based on equal area dualistic sub-image histogram equalization method", IEEE Transactions on Consumer Electronics, vol. 45, Feb. 1999, pp. 68-75.
[11] Chen S D, Ramli A R, "Minimum mean brightness error Bi-histogram equalization in contrast enhancement", IEEE Transactions on Consumer Electronics, vol.49, Nov. 2003, pp. 1310-1319.
[12] Kim W K, You J M, JEONG J, "Contrast enhancement using histogram equalization based on logarithmic mapping", Optical Engineering, Vol. 51, June 2012, pp. 1-10.
[13] Celik Turgay, "Spatial Entropy-Based Global and Local Image Contrast Enhancement", IEEE Transactions on Image Processing, vol.23, Dec. 2014, pp. 5298-5308.

Research on gait planning of Hexapod wheel legged robot based on STM32 single chip microcomputer

Chu Zenan, Wang Wei, Caisheng Yu

Department of Computer Science and Information Engineering , Anyang Institute of Technology, Anyang, Henan, 4655000, China

chuzenan@163.com

Abstract. In this paper, the rectangular wheel legged hexapod robot is used as the experimental robot, the STM32 single-chip microcomputer is used as the main control core, the steering control board is used as the auxiliary board, the upper computer programming is used to write the action group of the hexapod robot, the command is sent to the single-chip microcomputer through the PS2 handle, the single-chip microcomputer controls the steering board to achieve the corresponding action group, and the Arduino control board is used to change the environment through the sensor in the later stage The control of action group can be realized by the detection. In addition, in order to solve the gait planning problem of hexapod robot and realize the optimal learning of robot's free gait on specific terrain, based on the discretization of the robot's single foot distance, integrating the time rhythm principle of CPG model and the space rule constraint mechanism of reflection model, a discrete gait model of Hexapod robot is constructed. Ant colony algorithm is used to study the gait path planning of the robot. MATLAB and C language technology are used to realize the simulation of the gait planning of the hexapod robot.

1. Introduction

With the continuous development and progress of society, although the scope of human involvement is increasing, there are still some special occasions that may endanger our lives in human society and nature, such as deep mountains, disaster mines[1-3], rescue and rescue, and places that human beings currently have no way to reach. These places have one common topographical feature: ruggedness and irregularity. It is not difficult to see that the wheeled robot has absolute advantages in speed and flexibility on the relatively flat road, but on the uneven road, the energy consumption of the wheeled robot is more obvious and the flexibility is greatly reduced. Compared with the crawler robot[4-6], it has more advantages in uneven ground, but it has defects in mobility and balance. Compared with the wheeled and crawler mobile robots, the walking robot on the rough road has superior performance[7-8].

The research of hexapod robot in China is relatively late compared with foreign countries, but with the development of science and technology, the development of hexapod robot technology is more and more perfect. In this paper, five conclusions are put forward for the motion planning of a six legged ant robot designed by Beijing University of technology. Huazhong University of science and technology has designed and manufactured "4 + 2" hexapod robot, which can better adapt to complex environment and work more efficiently. But at present, these hexapod robots can't really do bionics, and the influence of gait path plays an important role.

IOP Conf. Series: Materials Science and Engineering **853** (2020) 012043 doi:10.1088/1757-899X/853/1/012043

Therefore, based on the advantages of hexapod robot and wheeled mobile robot, this paper designs a wheel leg hybrid robot. Through the continuous transformation of wheel and foot, the hybrid robot can not only realize the fast driving on the flat road, but also realize the smooth passing on the rough ground. In addition, ant colony algorithm is used to study the path planning of the robot, and the gait planning simulation of the hexapod robot is carried out with STM32 single chip microcomputer.

2. The overall structure design of Hexapod Robot

2.1. Selection of mechanical structure of Hexapod Robot

There are three main ways of combination of wheel and leg: one is the combination of wheel and leg; the other is the independence of wheel and leg; the last is the combination of wheel and leg. There are many mechanical structures of Hexapod robots, most of which are mainly circular and rectangular[9]. In this project, the rectangular mechanical structure is adopted, as shown in Figure 1. The wheels and legs of the robot are relatively independent and installed on the body of the robot independently. When the legs fold in a certain direction, the wheels will contact with the ground, and then the robot can do wheeled motion; when the joints are extended and standing, the wheels will be separated from the ground, and then the robot can do leg walking. Compared with the circular mechanical structure, the parallelism of the six legs in the rectangular mechanical structure just satisfies the parallel movement of two groups of wheels when the robot changes from foot to wheel.

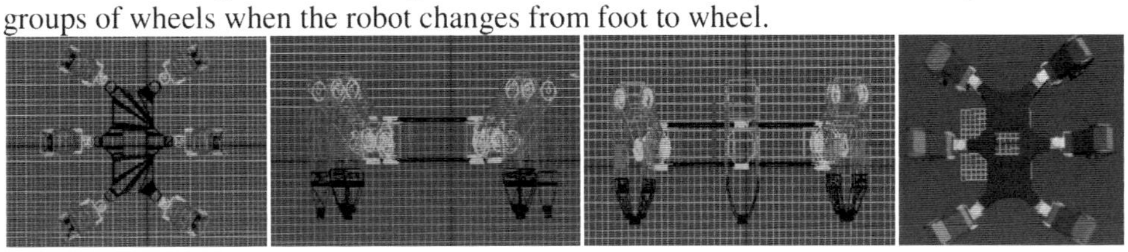

Figure 1. three dimensional mechanical structure of Hexapod Robot.

2.2. Selection of control panel

The control board is the core of the whole hexapod robot. There are many kinds of control boards, such as raspberry pie, 32 single chip microcomputer, 51 single chip microcomputer, Arduino control board and so on. Considering the influence of the whole project on intelligent control and convenient operation of hexapod robot, board area and power consumption, and the powerful PWM output efficiency, fast running speed, large memory, powerful engineering programming environment and a large number of IO ports of 32 single chip microcomputer, 32 single chip microcomputer is selected as the main control board[10].

The project adopts STM32series single chip microcomputer of STM company, 10 series single chip microcomputer has high performance, low cost, low power consumption, 144 pins, which provides more io for the project. The core adopts arm cortex ™ - m3 micro controller, 1 μ s dual 12 bit ADC, 4 megabit / s UART, 18 megabit / s SPI, 18 MHz I / O turnover speed.

3. Wheel leg conversion of Hexapod Robot

3.1. The shape of wheel leg transformation

Due to the multi joint characteristics of the hexapod robot and the rectangular mechanical structure we use, we imitate the shape of the multi wheel steering of the six wheel armored vehicle, so that the hexapod robot will take on the shape of the six wheel motor vehicle after deformation. At the same time, we use the multi joint mobility of the hexapod robot to realize the multi steering of the six wheels, making the robot more flexible and turn at the same time It is more convenient to turn to two wheels or speed than before. The requirement of motor speed is not so high, and the operation is relatively simple.

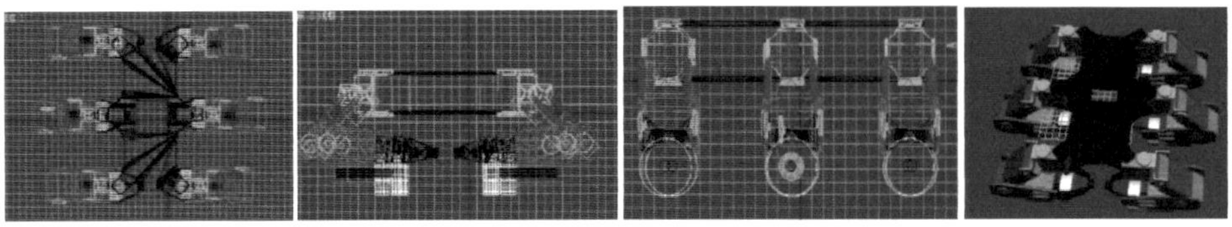

Figure 2. Wheel type three-dimensional diagram of Hexapod Robot.

3.2. The control of the deformation of six legged robot's foot wheel by STM32 single chip microcomputer

In this paper, we use 32 single chip microcomputer to control the steering gear control board, and we use PS2 handle to control the overall manual. The command is sent to 32 single chip microcomputer through PS2 handle. After receiving the command, the single chip microcomputer sends the corresponding command to the steering gear control board through serial port. After receiving the command, the steering gear control board makes the corresponding action group or action to complete the deformation task. The flow chart is shown in Figure 3:

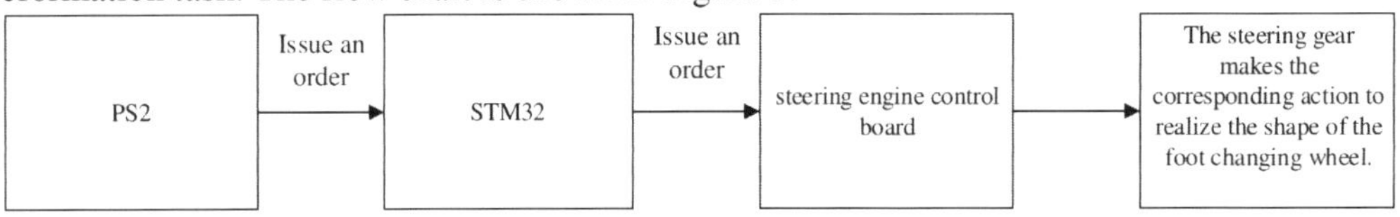

Figure 3. Single chip command flow chart.

4. The implementation of ant colony algorithm in path planning

4.1. Basic principle of ant colony algorithm

Ant colony algorithm is a kind of swarm intelligence algorithm [7-8] which is inspired by the idea of observing the real ant colony foraging. After a lot of practice by researchers, it has been proved that when ants are searching for food, they can leave a kind of substance called pheromone on their way, so that later ants can not only perceive the intensity of pheromone, but also get the direction of guidance. If the distance between ants' nest and food is relatively short, the number of ants carrying food back and forth will increase. After a long period of time, there will be a positive feedback phenomenon. The phenomenon is that when a certain path is shorter, the number of ants passing through the path will increase, and the intensity of pheromone will be higher, so the probability of ants choosing this path will increase The higher the rate. Through this positive feedback phenomenon, the optimal path will be obtained.

4.2. The preparation of ant colony algorithm

4.2.1 Environment modelling. In order to better study the path planning of robot, it is necessary to build a digital model of the random environment around the robot, and transform the environment information into matrix storage. The data 1 in the matrix indicates that there is an obstacle in front that cannot pass, and 0 indicates that there is no obstacle in front that can pass. Finally, it is expressed in the form of grid map.

4.2.2. Grid map. In this paper, the hexapod robot and obstacles are reduced in proportion, and the scale size of the hexapod robot is taken as the size of a grid, so the hexapod robot can be regarded as a particle. If the location of the map is an obstacle, the color of the grid is black, indicating that the obstacle cannot pass. If the grid is white, it means that there is no barrier. The establishment of coordinate system in this paper is closely related to grid map. Take the lower left corner of the grid map as the origin, the horizontal direction of the grid map is x axis, and the vertical direction is Y axis.

5. Ant colony algorithm for shortest path planning

5.1. Basic steps of ant colony algorithm

The basic idea of ant colony algorithm comes from the principle of the shortest path for ants in nature to find food. According to the observation of insect scientists, it is found that although the vision of ants in nature is not developed, they can find the shortest path from food source to nest without any hint, and search the new best path adaptively after the environment changes.

The specific steps of the algorithm are as follows:

Step1: Initialization of relevant parameters, including ant colony size, pheromone factor, heuristic function factor, pheromone volatility factor, pheromone constant, maximum number of iterations, etc., as well as reading data into the program, and preprocessing: for example, transforming the coordinate information of the city into the distance matrix between cities.

Step 2: Put ants at different starting points randomly, and calculate the next visiting city for each ant until there are ants visiting all cities.

Step 3: Calculate the path length LK of each ant, record the optimal solution of the current iteration times, and update the pheromone concentration on the path.

Step 4: Determine whether the maximum number of iterations is reached, if not, return to step 2; yes, end the program.

Step 5: Output the results, and output the relevant indexes in the process of optimization, such as running time, convergence iterations, etc.

Using ant colony algorithm to realize the shortest path planning of robot in the complex environment with obstacles is as follows:

Firstly, the working environment of hexapod robot is established by grid method. The environment information is represented by a matrix of only 1 and 0. Where 0 means that it can pass, 1 means that it is an obstacle and cannot pass.

The initialization map is represented by an example. The initialization grid map code is as follows:

map=[0 0 0 0 0 0 1 1 1 0 0 0 0 0 0 0 0 0 0 0 0; 0 1 1 0 0 0 1 1 1 0 0 0 0 0 0 0 0 0 0 0 0; 0 1 1 0 0 0 1 1 1 0 0 0 0 0 0 0 0 0 0 0 0; 0 0 0 0 0 0 1 1 1 0 0 0 0 0 0 0 0 0 0 0 0; 0 0 0 0 0 0 1 1 1 0 0 0 0 0 0 0 0 0 0 0 0; 0 1 1 1 0 0 1 1 1 0 0 0 0 0 0 0 0 0 0 0 0; 0 1 1 1 0 0 1 1 1 0 0 0 0 0 0 0 0 0 0 0 0;0 1 1 1 0 0 1 1 1 0 1 1 1 1 0 0 0 0 0 0 0; 0 1 1 1 0 0 0 0 0 0 1 1 1 1 0 0 0 0 0 0; 0 0 0 0 0 0 0 0 0 0 0 1 1 1 1 0 0 0 0 0 0; 0 0 0 0 0 0 0 1 1 1 1 1 1 1 0 0 0 0 0 0; 0 0 0 0 0 0 0 1 1 1 1 1 1 1 0 0 0 0 0 0 0; 0 0 0 0 0 0 0 0 0 0 0 0 0 1 1 1 0 1 1 1 1 0; 0 0 0 0 0 0 0 0 0 0 0 0 0 1 1 1 0 1 1 1 1 0; 1 1 1 1 0 0 0 0 0 0 0 0 1 1 1 0 1 1 1 1 0; 1 1 1 1 0 0 1 1 1 1 1 1 1 0 0 0 0 0 0 0 0; 0 0 0 0 0 0 1 1 1 1 1 1 0 0 0 0 0 1 1 0; 0 0 0 0 0 0 0 0 0 0 0 0 1 1 0 0 0 0 0 1 1 0; 0 0 0 0 0 0 0 0 0 0 0 0 1 1 0 0 1 0 0 0 0 0; 0 0 0 0 0 0 0 0 0 0 1 1 0 0 0 0 0 0 0 0;];

Next, we need to initialize an pheromone matrix, and some parameters, such as start position, final position and other parameters.

Then, after determining the initial position, we need to find the next accessible position. This is what we do. According to the pheromone concentration of each node, the probability of going to each node is calculated. Here, the wheel algorithm is selected to select the starting point of the next step. The formula of wheel disk method is as follows:

$$P_{ij}^{k} = \begin{cases} \dfrac{[\tau_{ij}(t)]^{\alpha} \cdot [\eta_{ij}]^{\beta}}{\sum_{k \in \{N - tabu_k\}} [\tau_{ij}(t)]^{\alpha} \cdot [\eta_{ij}]^{\beta}}, & j \in \{N - tabu_k\} \end{cases} \quad (1)$$

Table 1 shows the parameter interpretation of wheel algorithm formula

Table 1. wheel algorithm

$\tau_{ij(t)}$	The concentration of pheromones on the arc (I, j) in the graph
$\eta_{i,j(t)}$	Heuristic information associated with arc (i, j)

α	Information heuristic factor
β	Expected heuristic factor

5.2. Ant colony algorithm path planning experiment
In order to verify the effectiveness of the algorithm, a large number of simulation experiments are done in this paper, respectively considering the work space with arbitrary obstacles and different complexity. The working environment of the robot is modeled as a grid of (20 × 20). The scale of traditional ant colony algorithm is set to 30, and the maximum number of iterations of the algorithm is set to 300. Figure 4 below shows the number of iterations and the length of the shortest path of the robot.

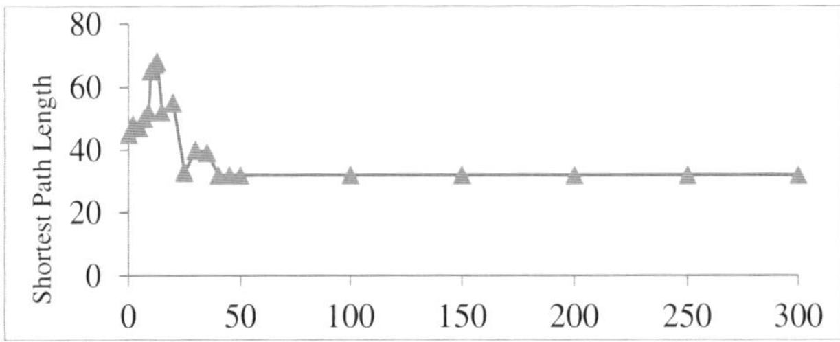

Figure 4. Variation of iteration times and shortest path length

6. Conclusion
In this project, the main control board of 32 single chip microcomputer, the wheel foot conversion controlled by PS2 handle and the corresponding action group controlled by sensor with aduino as auxiliary board are realized. Based on the flexible characteristics of the leg joints of the multi legged robot and the characteristics of being able to better adapt to the surrounding environment, this paper analyzes the gait movement of the six legged robot and the shortest path planning of the simulation robot under the ant colony algorithm in the complex environment with obstacles. The experiment proves that the ant colony algorithm can evaluate the convergence of the gait planning of the six legged robot in the complex scene And give the shortest path. How to further improve the stability and obstacles while quickly finding the shortest path is the next research goal.

References
[1]Guojie, Tian Wenkai, Lv Chengzhe, Yinlin Yin and Wang Qirui. Design of six-foot bionic robot and control system[J]. TT & C Technology,2017,36 (01):55-58

[2] Guo Zuhua, Wang Shenjiang and Zhao Shijie. Motion planning algorithm of six-legged robot based on global trajectory[J]. Journal of System Simulation,2015,27 (11):2682-2688.

[3]Zhao Huaibei, Liu Shujun, Lv. Design of six-foot search and rescue robot system based on remote control[J]. Computer measurement and control,2015,23 (07):2392-2395.

[4]Liu Yiqun, Deng Zongquan, Zhao Liang, Ding Liang, Li Zhizhong, and Gao Haibo. Hydraulic-driven six-foot robot walking leg performance[J]. Journal of Jilin University (Engineering),2015,45 (05):1512-1518.

[5]QU Jian,CHEN Hongyan and LIU Wenzhen etc. (2015)Application of Support Vector Machine Based on Adaptive Mutation Particle Swarm Optimization in Analysis of Gas Mixture,Chinese Journal of Sensors and Actuators,12:1262-1268.

[6]HU Li-kun,WANG Di and HUANG Wen-qin etc, (2014)Internal Model Control Based on GA-SVM and Inverse System of Manipulator,Control Engineering of China, 38:415-418.

[7]Kutlu M , Freeman C T , Hallewell E , et al. (2015)Goal-Orientated Upper-Limb Stroke Rehabilitation Utilising Functional Electrical Stimulation with Advanced Sensing and Control. 20:45-49.

[8]Nanotechnology-Biomimetics; New Data from Jiangxi University of Science and Technology

Illuminate Findings in Biomimetics (The Design and Building of a Hexapod Robot with Biomimetic Legs)[J]. Biotech Week,2019.

[9]Bai Long,Hu Hao,Chen Xiaohong,Sun Yuanxi,Ma Chaoyang,Zhong Yuanhong. CPG-Based Gait Generation of the Curved-Leg Hexapod Robot with Smooth Gait Transition.[J]. Sensors (Basel, Switzerland),2019,19(17).

[10]Xu Shoulin,He Bin,Hu Heming. Research on Kinematics and Stability of a Bionic Wall-Climbing Hexapod Robot.[J]. Applied bionics and biomechanics,2019,2019.

IOP Conf. Series: Materials Science and Engineering **853** (2020) 012044 doi:10.1088/1757-899X/853/1/012044

Minimum-time Trajectory Planning for Residual Vibration Suppression of Flexible Manipulator Carrying a Payload

Yong Liu, Zhe Du*, Shaonan Zhang

School of Mechanical Engineering, Hefei University of Technology, No. 193 Tunxi Road, Baohe District, Hefei 230009, China

Zincdz@163.com

Abstract. For the purpose of eliminating residual vibration of a two-link flexible manipulator as well as minimizing time consumption, a scheme of multi-objective trajectory planning in joint space is proposed in this paper. Piecewise polynomial is applied to generate a smooth joint trajectory. Lagrangian approach and the assumed modes method is used to derive motion equations of FMs. Finally, Multi-objective particle swarm optimization (MOPSO) algorithm is implemented to solve time of each polynomial. The effectiveness of proposed algorithm is demonstrated by simulation results.

1. Introduction

With the increasing demands of high speed and accuracy in manufacturing, the existing heavy rigid industrial robots turn out to be inefficient. Flexible manipulators (FMs) have several advantages over the rigid ones [1]. Whereas, residual vibration may take place for FMs because of the dynamic deflection, which causes losses in efficiency [2].

Many efforts have been done in order to suppress vibration during FMs motion. Akira Abe proposed a trajectory planning technique using PSO to minimize residual vibration considering large bending deformations [3]. Later in his work, the trajectory planning scheme was improved so that not only residual vibration but also energy consumption was optimized [4]. W. F. FARIS et.al also proposed a genetic algorithm (GA) based energy minimization scheme [5]. Kyung-Jo Park provided a method reduces residual vibration under the torque constraints considering both link and joint flexibilities [6]. Shigang Yue et.al presented a trajectory planning method to minimize amplitude and time with GA [7]. In this study, residual vibration and time consumption are optimized simultaneously using MOPSO algorithm for 2-dof FMs. Euler-Bernoulli's beam theory is applied to the flexible links, and both links are treated as simply supported beam since the heavy payload on actuator. Joint torques derived from trajectory planning of rigid manipulators are defined as input variables to calculate dynamic response of FMs. In this paper, assumed modes method is adopted because of its computational efficiency and flexible boundary conditions. Numerical simulations of the proposed optimization algorithm using piecewise polynomial trajectories are introduced and analyzed.

2. Dynamic modelling

2.1. Dynamic modelling of 2-dof flexible manipulator

Figure 1 depicts the model of rigid manipulator, whose joint motor and payload are treated as tip concentrated mass. And Figure2 is the model of flexible manipulator whose links are regarded as simply supported beams. The movement of flexible body is the combination of rigid motion and elastic deformation of the moving coordinate system. The position vector of any part on flexible body is defined as follows [8]

$$\boldsymbol{r}_P = \boldsymbol{R}_0' + \boldsymbol{A}(\boldsymbol{u}_0' + \boldsymbol{u}_f) \tag{1}$$

$\boldsymbol{r}_p$ and $\boldsymbol{R}_0'$ are position vectors point P and original point of the floating coordinate system in inertial frame, respectively. $\boldsymbol{A}$ is direction cosine matrix, $\boldsymbol{u}_0'$ is position vector of point P before flexible deformation, $\boldsymbol{u}_f$ is amount of elastic deformation. According to analysis above, key to describe deformation of manipulator is the displacement vector $\boldsymbol{u}_f$ .Modal analysis is applied. Considering both links as simply supported beam, by variables separation method the free vibration equation can be written as

$$u_f = v(x,t) = w(x)\cdot T(t) \tag{2}$$

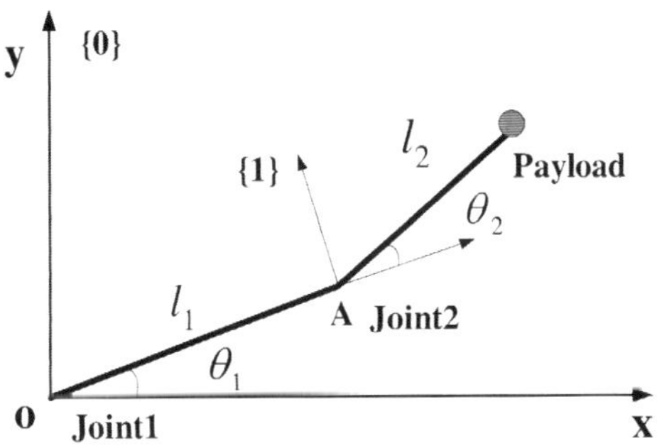

Figure 1. Model of 2-dof planar manipulator

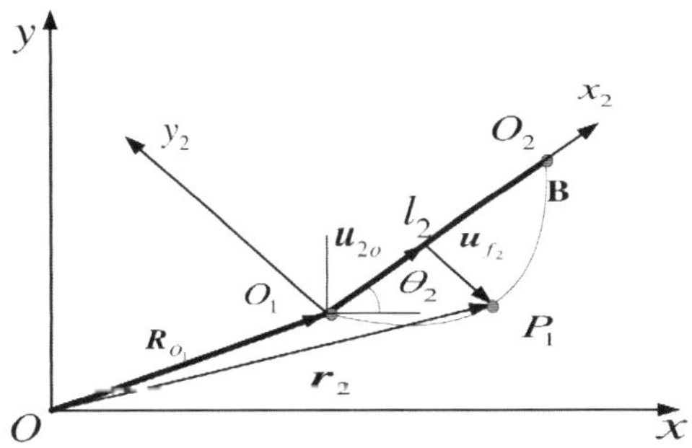

Figure 2. Model of 2-dof flexible manipulator

Modal of the beam can be described as $w_n(x) = c_n[\sin \beta_n x]$, it's frequency equation is $\sin \beta_n l = 0$ .Only first and second order of modes are taken into consideration. So that $\beta_1 l = \pi$, $\beta_2 l = 2\pi$. Define $c_n T(t)$ as a_n and b_n , Eq(2) is equivalent to

$$v_1(x_1,t) = \sin\frac{\pi}{l_1}x_1\cdot a_1 + \sin\frac{2\pi}{l_1}x_1\cdot a_2, v_2(x_2,t) = \sin\frac{\pi}{l_2}x_2\cdot b_1 + \sin\frac{2\pi}{l_2}x_2\cdot b_2 \tag{3}$$

The dynamic equation of flexible manipulators can be represented in following form when gravity is excluded. Where m_{kj} is mass coefficient, K_{kj} is stiffness coefficient, q is generalized coordinate and Q_k is generalized force.

$$\sum_{j=1}^{n} m_{kj}\ddot{q}_j + \sum_{i=1}^{n}\sum_{j=1}^{n}\left(\frac{\partial m_{kj}}{\partial q_i} - \frac{1}{2}\frac{\partial m_{kj}}{\partial q_k}\right)\dot{q}_i\dot{q}_j + \sum_{j=1}^{n} K_{kj}q_j = Q_k \quad (k = 1,2,\cdots,n) \tag{4}$$

The kinetic energy of links are as follows with ρ represents the mass of unit length

$$T = \int_l \rho\dot{\boldsymbol{r}}_p^T\dot{\boldsymbol{r}}_p \mathrm{d}l = T_1 + T_2 = \frac{1}{2}\dot{\boldsymbol{q}}^T\boldsymbol{M}\dot{\boldsymbol{q}} \tag{5}$$

Where $\boldsymbol{q}=[\theta_1,a_1,a_2,\theta_2,b_1,b_2]^T$, $\boldsymbol{M}$ is a 6 by 6 matrix whose non-zero elements are

$$\begin{cases} a_{11}=(m_1+m_2+\frac{M_1}{3}+M_2)l_1^2+\frac{M_1}{2}(q_2^2+q_3^2), a_{12}=a_{21}=\frac{M_1 l_1}{\pi}, a_{13}=a_{31}=\frac{M_1 l_1}{2\pi}, a_{22}=a_{33}=\frac{M_1}{2} \\ a_{14}=(m_2+\frac{M_2}{2})l_1 l_2\cos(\Delta\theta)+\frac{2M_2 l_1}{\pi}b_1\sin(\Delta\theta), a_{15}=a_{51}=2\frac{M_2 l_1}{\pi}\cos(\Delta\theta), a_{45}=a_{54}=\frac{M_2 l_2}{\pi} \\ a_{46}=a_{64}=-\frac{M_2 l_2}{2\pi}, a_{55}=a_{56}=\frac{M_2}{2}, a_{41}=a_{14} \end{cases}$$

The total strain energy of the system is

$$U_s=\frac{1}{2}\sum_{i=1}^{2}EI_i\int_0^{l_i}(v_i'')^2\mathrm{d}x_i=\frac{1}{2}\boldsymbol{q}^T\boldsymbol{K}\boldsymbol{q} \tag{6}$$

Where stiffness coefficient $\boldsymbol{K}$ is a 6 by 6 matrix, the non-zero elements are

$$K_{22}=\frac{\pi^4}{2l_1^3}EI_1, K_{33}=\frac{8\pi^4}{l_1^3}EI_1, K_{55}=\frac{\pi^4}{2l_2^3}EI_2, K_{66}=\frac{8\pi^4}{l_2^3}EI_2$$

Assume the driving torque of two joints are τ_1 and τ_2. The generalized force $\boldsymbol{Q}_k$ can be acquired with principle of virtual work, $\boldsymbol{Q}_k=[\tau_1-\tau_2,\frac{\pi}{l_1}(\tau_1+\tau_2),\frac{2\pi}{l_1}(\tau_1-\tau_2),\tau_2,\frac{\pi}{l_2}\tau_2,\frac{2\pi}{l_2}\tau_2]$. Thus, the dynamic equation of flexible manipulator can be obtained by substituting Eqs(5)-(6) to Eq(4).

3. Trajectory optimization with MOPSO

3.1. Trajectory parameterization

Piecewise polynomial is used to generate trajectories. The expression is defined as

$$\begin{cases} h_1(t)=a_1t^3+a_2t^2+a_3t+a_4, t\in[0,t_1] \\ h_2(t)=a_5(t-t_1)^5+a_6(t-t_1)^4+\cdots+a_{10}, t\in[t_1,t_1+t_2] \\ h_3(t)=a_{11}t^3+a_{12}t^2+a_{13}t+a_{14}, t\in[t_1+t_2,t_1+t_2+t_3] \end{cases} \tag{7}$$

t_1,t_2,t_3 are planning time of each polynomial. The constraint conditions of initial and final position, velocity, acceleration and jerk are

$$\begin{cases} h_1(0)=q_0, \dot{h}_1(0)=0, \ddot{h}_1(0)=0, h_1(t_1)=h_2(t_1), \dot{h}_1(t_1)=\dot{h}_2(t_1)\ddot{h}_1(t_1)=\ddot{h}_2(t_1), \dddot{h}_1(t_1)=\dddot{h}_2(t_1), \\ h_2(t_2)=h_3(t_2), \dot{h}_2(t_2)=\dot{h}_3(t_2)\ddot{h}_2(t_2)=\ddot{h}_3(t_2), \dddot{h}_2(t_2)=\dddot{h}_3(t_2), h_3(t_3)=q_f, \dot{h}_3(t_3)=0, \ddot{h}_3(t_3)=0 \end{cases} \tag{8}$$

Where q_0 and q_f are initial and final joint angles, respectively. Substitute constraint conditions Eq (8) to Eq(7), $\boldsymbol{a}=[a_1,a_2,\cdots,a_{14}]$ can be represented by q_0,q_f,t_1,t_2,t_3. Since q_0 and q_f are known conditions, $\boldsymbol{t}=[t_1,t_2,t_3]$ is chosen to be the decision variable.

3.2. Trajectory optimization algorithm

MOPSO applied Pareto dominance in PSO algorithm to handle problems with several cost functions [9]. In the velocity formula of particle's flight, the best position that all the particles have had (g_{best}) is replaced by a value taken from repository ($R[h]$). The velocity formula is:

$$V[k]=\chi\cdot\left[V[k]+c_1r_1\left(p_{\text{best}}[k]-P[k]\right)+c_2r_2(R[h]-P[k])\right] \tag{9}$$

$P[k]$ and $V[k]$ denote position and velocity of particle k ; $\chi = 2/(|2-\phi-\sqrt{\phi(\phi-4)}|)$ is constriction factor; $\phi = c_1 + c_2$, while c_1 and c_2 are learning factors regulates the step size of the particles flying to p_{best} (the best position a particle has explored) and g_{best} ; r_1 and r_2 are independent random numbers in range $[0,1]$. Traditional approximation method is applied to calculate the dynamic response of FMs. The cost functions are defined as follows

(1) Residual Vibration: Both links are regarded as simply supported beams, vibration is equivalent to the midpoint deflection of each link. Cost function for minimizing residual vibration is

$$f_1 = |a_1(t_f)| + |b_1(t_f)| \tag{10}$$

Where t_f end stands for the final time.

(2) Time Consumption: Time of each polynomial is chosen as decision variable.

$$f_2 = t_1 + t_2 + t_3 \tag{11}$$

4. Simulation results and analysis

A planar manipulator with two flexible links is simulated. The mass of two links is M_1=1.4kg , $M_2 = 0.62$kg and the concentrated mass at tip positions is $m_1 = 5.5$kg , $m_2 = 2.75$kg .

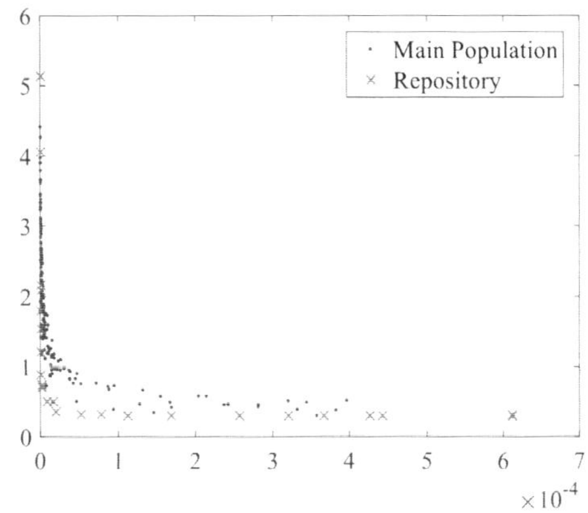

Figure 3. Pareto front of 2 objective functions

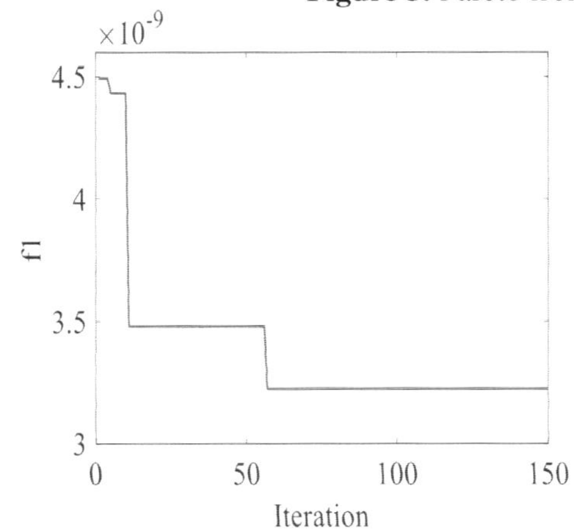

Figure 4. Change of optimal value in f_1

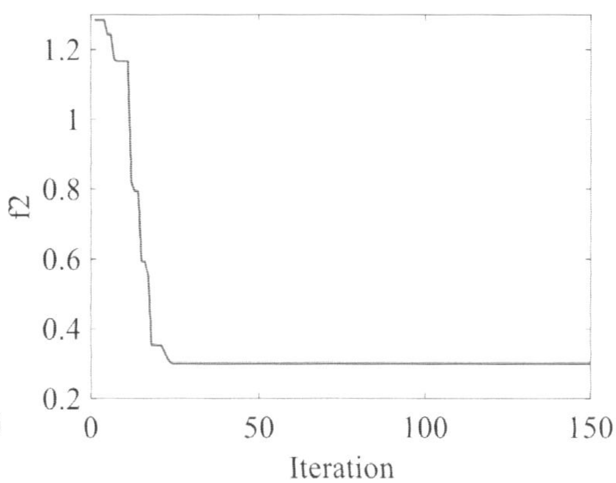

Figure 5. Change of optimal value in f_2

IOP Conf. Series: Materials Science and Engineering **853** (2020) 012044 doi:10.1088/1757-899X/853/1/012044

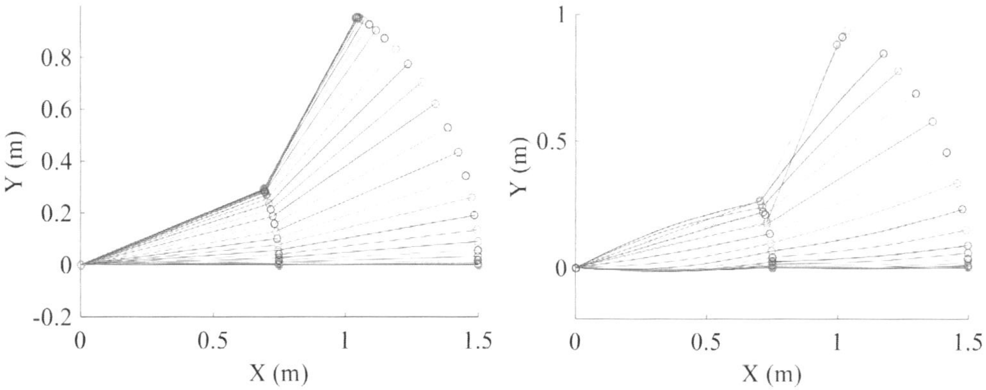

Figure 6. Configuration change for extreme solutions f_1^{min} and f_2^{min}

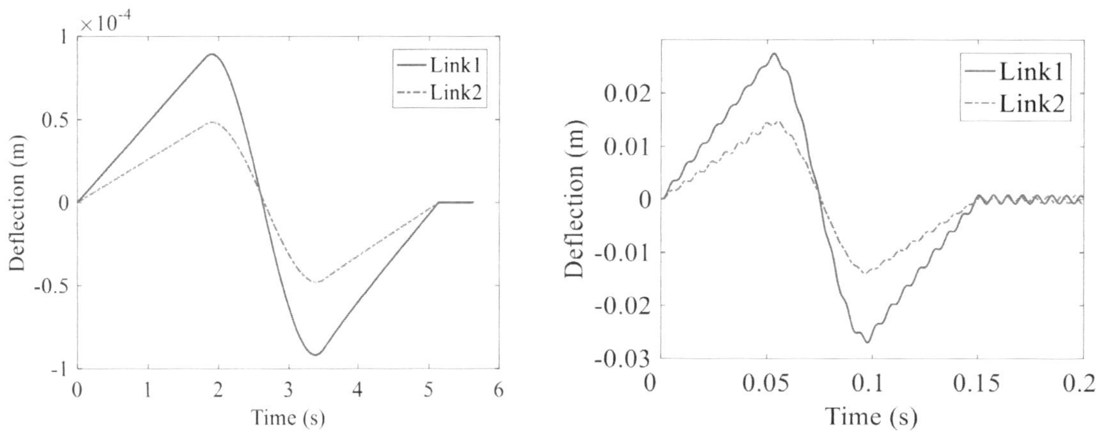

Figure 7. Midpoint deflection change for extreme solutions f_1^{min} and f_2^{min}

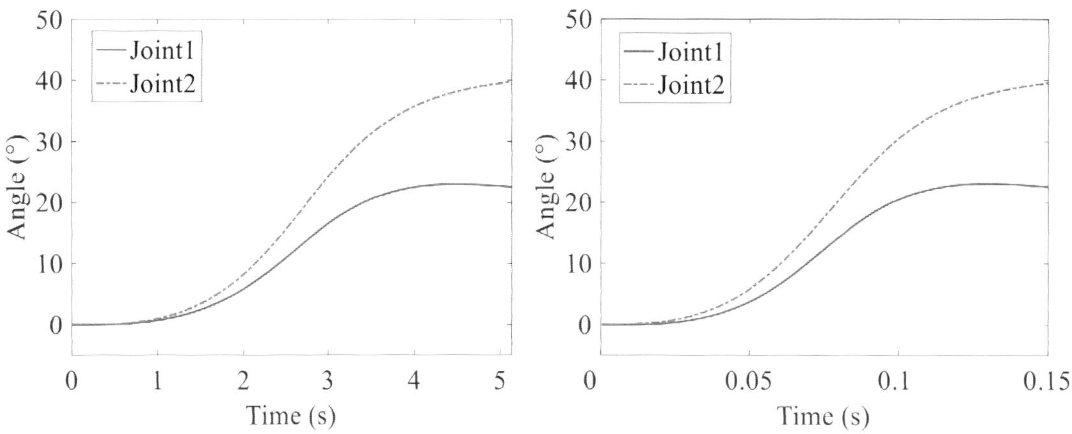

Figure 8. Joint trajectories for extreme solutions f_1^{min} and f_2^{min}

The length of each link is $l_1 = l_2 = 0.75\text{m}$ and the bending rigidities are $EI_1 = 1200\text{N/m}$, $EI_2 = 600\text{N/m}$. The initial joint angles are $\theta_1^0 = 0$, $\theta_2^0 = 0$ and the final joint angles are $\theta_1^f = 25^\circ$, $\theta_2^f = 35^\circ$. The parameters of MOPSO are set as follows: $[\lambda_{min}, \lambda_{max}] = [0.1, 3]$; $n_P = 200$; $n_R = 40$;

$c_1 = c_2 = 2.05$ .The Pareto front is shown in Figure 3. The optimal values of cost functions are shown in Figures 4-5 which testified the convergence. The minimum value of planning time is set to 0.3 second in order to limit the joint torques. As it depicts in Figure 4, as cost function f_1 converges, the optimal value of residual vibration decreased by 28.27%. The extreme solutions of f_1 and f_2 are selected and compared which are $f_1^{\min} = 3.22 \times 10^{-9}$ and $f_2^{\min} = 0.3$.

Figure 6 shows the configuration change during trajectory planning. The trajectory of solution $f_1^{\min}$ is more symmetrical and we can hardly distinguish the deformation of the links. However, in solution $f_2^{\min}$ the links deforms significantly, which indicates the non-negligible vibration. Figure 7 depicts the deflection at midpoint of link1 and link2. It is obvious that the curve of solution $f_1^{\min}$ is smoother than $f_2^{\min}$ and maximum deflection is also much smaller. Residual vibration takes place when trajectory planning ends, whereas the amplitude in solution $f_1^{\min}$ is eliminated very well. However, solution $f_1^{\min}$ takes 5.13 seconds to reach the desired joint angles which is unacceptable in manufacturing. Generally speaking, the relationship between residual vibration and time consumption is contradictory which means solution selection should connect with the practical situation.

Finally, joint trajectories are shown in Figure 8, Since approximate approach is adopted, positional error still exists.

4.1. Conclusion

In this work, a multi-objective trajectory planning scheme for two-link flexible manipulator to simultaneously reduce residual vibration and time consumption is proposed and verified.

The extreme solution to minimize residual vibration has much smaller and smoother deflection change, whereas it takes too much time to reach the desired joint angles. This phenomenon indicates the confliction between two cost functions which may lead to contradiction in solution selection for manufacturing. Since traditional approach which computes joint torques in rigid manipulator and substitute back into flexible dynamic formulations is applied, although residual vibration is well suppressed in extreme solution, tip positional deviation still takes place. Eliminating positional error would be our future work.

Acknowledgments

Authors wishing to acknowledge assistance or encouragement from colleagues, special work by technical staff or financial support from organizations should do so in an unnumbered Acknowledgments section immediately following the last numbered section of the paper.

References

[1] Lochan K , Roy B K , Subudhi B . A review on two-link flexible manipulators[J]. Annual Reviews in Control, 2016:S1367578816300736.

[2] Dwivedy S K . Dynamic Analysis of Flexible Manipulators, a Literature Review[J]. Mechanism & Machine Theory, 2006, 41(7):749-777.

[3] Abe A . Trajectory planning for residual vibration suppression of a two-link rigid-flexible manipulator considering large deformation[J]. Mechanism and Machine Theory, 2009, 44(9):1627-1639.

[4] Abe, Akira. An effective trajectory planning method for simultaneously suppressing residual vibration and energy consumption of flexible structures[J]. Case Studies in Mechanical Systems and Signal Processing, 2016, 4:19-27.

[5] Faris W F , Ata A A , Sa"Adeh M Y . Energy Minimization Approach for a Two-link Flexible Manipulator[J]. Journal of Vibration and Control, 2009, 15(4):497-526.

[6] Park K J . Flexible robot manipulator path design to reduce the endpoint residual vibration under torque constraints[J]. Journal of Sound and Vibration, 2004, 275(3-5):1051-1068.

[7] Yue S, Henrich D, Xu W L, et al. Point-to-point trajectory planning of flexible redundant robot manipulators using genetic algorithms[J]. Robotica, 2002, 20(3): 269-280.
[8] Ahmad M A, Mohamed Z, Hambali N. Dynamic modelling of a two-link flexible manipulator system incorporating payload[C]//2008 3rd IEEE Conference on Industrial Electronics and Applications. IEEE, 2008: 96-101.
[9] Coello C A C, Pulido G T, Lechuga M S. Handling multiple objectives with particle swarm optimization[J]. IEEE Transactions on evolutionary computation, 2004, 8(3): 256-279.

IOP Conf. Series: Materials Science and Engineering **853** (2020) 012045 doi:10.1088/1757-899X/853/1/012045

Design and Implementation of a Device for Reducing Road Accident at Foggy, Rainy and Drowsy Conditions

Shihab Uddin Ahmed [1], Sanzid Junaeid [2], Md. Monirul Islam [2], Bishwajit Banik Pathik [2]

[1] Hiperdyne Corporation, Japan
[2] Dept. of Electrical & Electronic Engineering, Faculty of Engineering, American International University-Bangladesh.

shihablitu09@gmail.com

Abstract. The main objective of this project is to design and implementation of microcontroller-based circuitry for vehicles to detect obstacles and hence to minimize road accident. The microcontroller is programmed in such a way that, it can give command to the other apparatus to work according to the receiving pulse from the LDR (Light Dependent Resistor). When reflection of light falls on LDR then the function of LED will start to work and after having this pulse, the microcontroller allows to turn on the dc motor instead of vibrator to aware the driver at foggy, rainy or drowsy conditions.

1. Introduction

Most of accidents occurred at night, rainy conditions, and foggy conditions and for driver's drowsiness [1]. It has been observed that most of the road accidents happened for having problem to detect opposite vehicle [2], and this device will help to detect obstacles. The main purpose of this project is to develop a system with low cost and effective at foggy, rainy weather which can be used in underdeveloped countries like Bangladesh.

2. Hardware Description

Electrical equipment was carefully selected and properly installed. The following equipment were used in this project they are: resistor, transistor (pn2222) [3], light dependent resistor (LDR), 7 segment display [4], 16x2 LCD display, laser light [5], voltage source, dc motor, 220V/12V transformer [6], microcontroller (ATMEGA16), variable resistor.

2.1. Methodology

The Project is divided into 2 steps: Simulation & Real-world development.

Simulation:

- *Step 1: The circuit diagram was implemented in PROTEUS.*
- *Step 2: The code was constructed in mikroC PRO for AVR and run into microcontroller.*
- *Step 3: The implemented circuit was simulated and run successfully.*

Real-world development:

- *Step 1: All equipment was collected for physical implementation of the circuit.*
- *Step 2: The circuit was implemented in bread board according to the design.*

IOP Conf. Series: Materials Science and Engineering **853** (2020) 012045 doi:10.1088/1757-899X/853/1/012045

- *Step 3: The code was burned into the microcontroller with the help of AVR burner.*
- *Step 4: The implemented circuit was précised to install a vehicle.*
- *Step 5: The meter box was constructed for placing inside the car.*
- *Step 6: The whole system has been checked successfully in different conditions.*

3. Design of Circuit & Working Plan

3.1. Working Principle

The module of the power source was 5V dc battery. In this circuitry common cathode 7-segment display [4] was used to show the time of fallen light over LDR and LCD display was used to show the command message. At first all the laser light was turned on. When the reflection of light was fallen on the LDR, the LDR would receive that signal and then LED [5] will glow on and the system has lost some voltage. To minimize this problem transistor Q1 was used that supply the full voltage to the circuit. When transistor Q2 got this voltage then 7 segment display [4] was started the show the time with the help of microcontroller and LCD display was also ready to display the message. When the 7-segment display was displayed 3.the motor which was used instead of vibrator was rotator and on that time LCD display was displayed the command message.

3.2. Block Diagram

The block diagram helps to demonstrate the working cycle of the component. In this block diagram all the apparatus of this circuit is assigned with working function and with this diagram the decision making of microcontroller is shown whether the vibrator either vibrate or not.

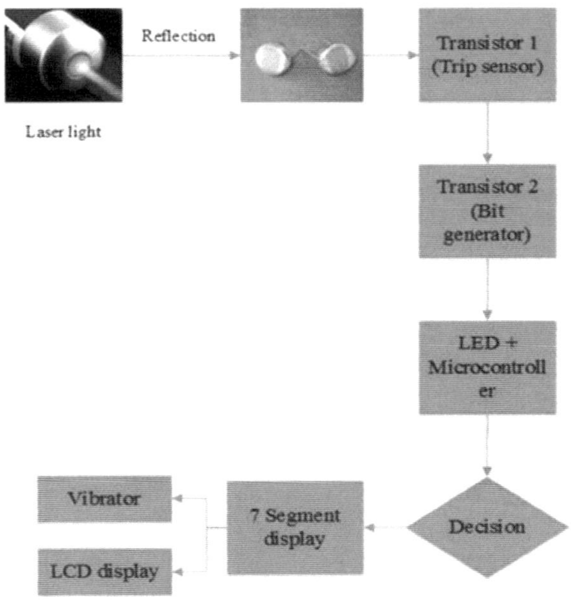

Figure 1. Block Diagram of Experimented Circuit.

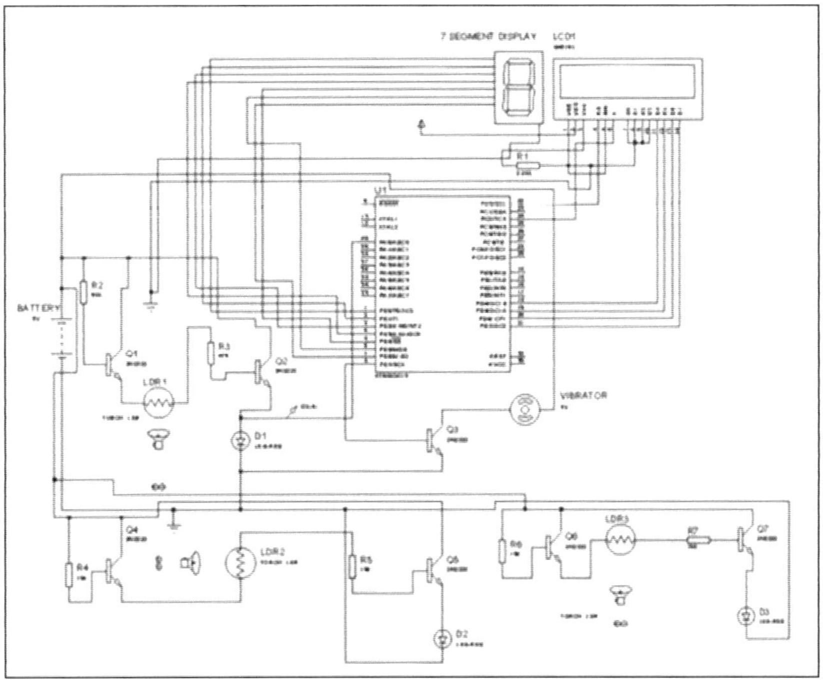

Figure 2. Circuit Diagram of the Designed System

3.3. Circuit Diagram

The circuit was connected as shown in the diagram. Microcontroller ATMEGA16 was used to make circuit work easier. PORT40 in the microcontroller was used to get the input pulse to show the output result. PORTB, PORTC, PORTD was used as output and connected with various components. The higher level of voltage was set to 5V.

4. Principle and Simulation

4.1. Software Simulation

The code was simulated in mikroC PRO for AVR to estimate the design.

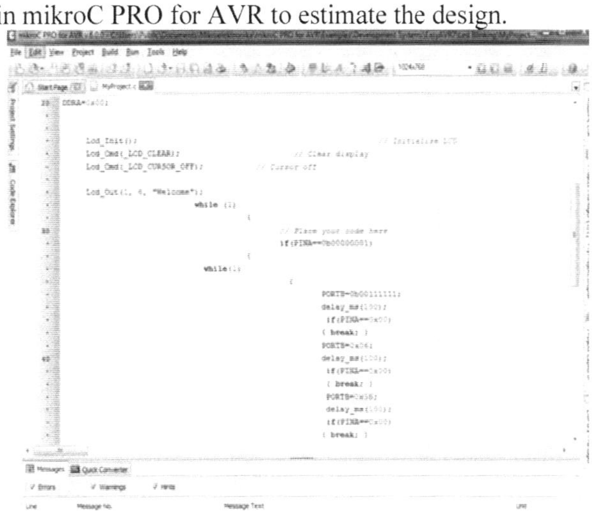

Figure 3. Software Simulation of Experimented Circuit.

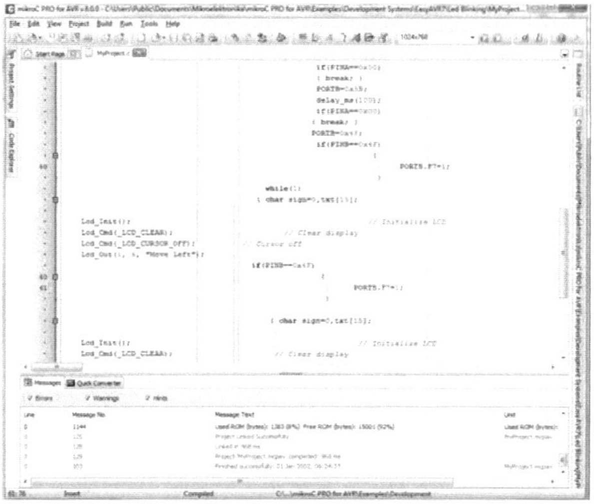

Figure 4. Software Simulation of Experimented Circuit.

4.2. Software Simulation with Time

7-segment display [4] and LCD display will be on upon passing through microcontroller. The LCD display will show "Welcome" when seven segment display shows time 0. There will be no change in LCD display upon seven segment display showing time 1 and time 2. Whenever 7-segment display shows 3 vibrator will turn on and there will be change in LCD display displaying 'Move Left'.

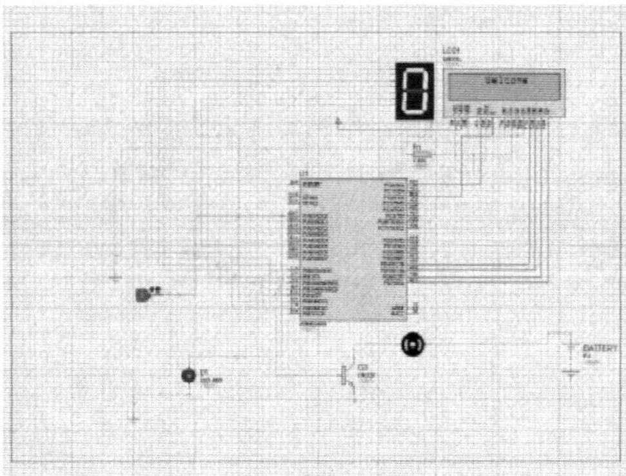

Figure 5. Initial condition of simulation

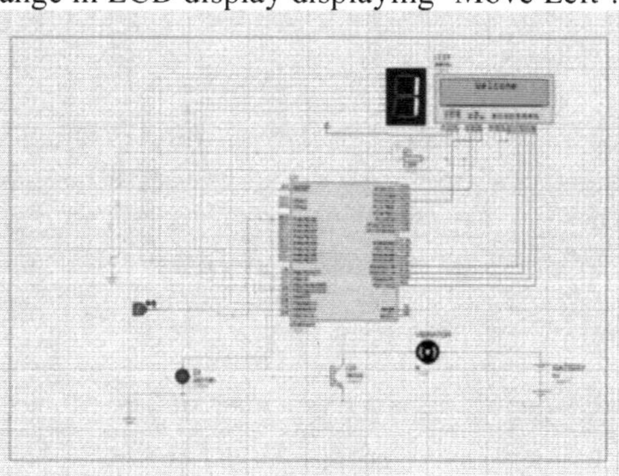

Figure 6. Simulation condition after passing 1 second

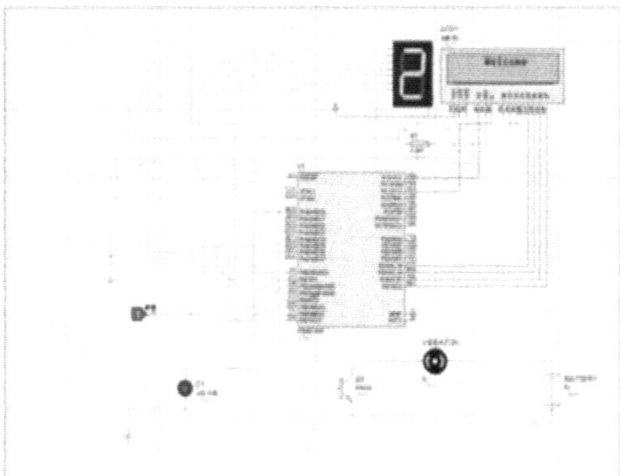

Figure 7. Simulation condition after passing 2 seconds

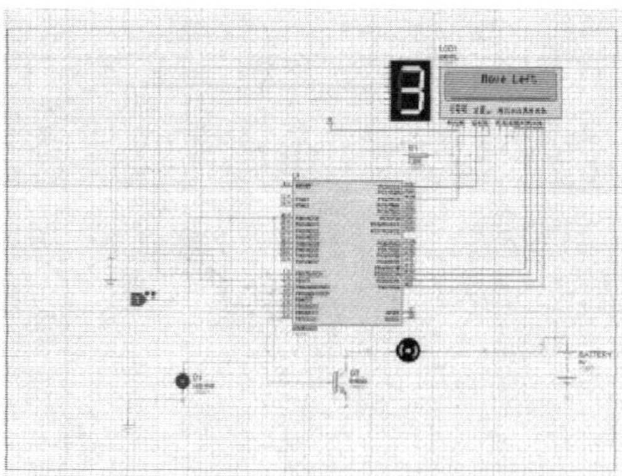

Figure 8. Simulation condition after passing 3 seconds

IOP Conf. Series: Materials Science and Engineering **853** (2020) 012045 doi:10.1088/1757-899X/853/1/012045

5. Experiment and Result

In this section implementation was evaluated by final result. To obtain accuracy of the model, trial the system and the controller test is needed in different angle.

5.1. Practical implementation

In practical simulation the result was observed under distinct 6 cases. Case 1 is for device's attachment in vehicle whereas case 2 is for LED's function upon face to face position of opposite vehicle. Case 3 define LED's function when two opposite vehicle passes two thirds portion of each other. Case 5 for foggy condition and case 6 for rainy conditions.

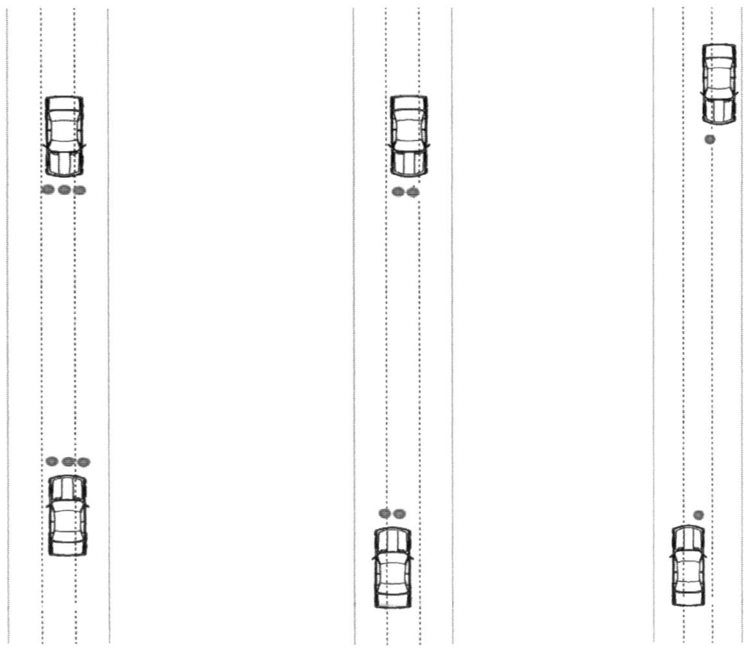

(a) Vehicles are head to head condition (b) 2/3 portion of the vehicles are at face to face (c) 1/3 portion of the vehicles are at face to face

Figure 9. Alignment of car and proposed warning system for avoiding accidents

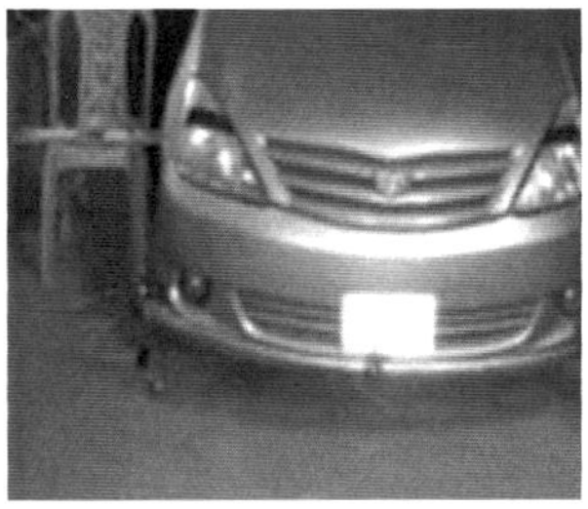

a. Attachment of Device in front of car.

b. LED's function upon face to face position of opposite vehicle

c. LED's function when opposite vehicle passes 2/3 portion of each other

d) LED's function when opposite vehicle passes 1/3 portion of each other.

e. LED's function for foggy condition.

f. LED's function for rainy condition.

Figure 10. Practical Implementation in different conditions

Conclusion

The project was intended to implement and design a simple and low-cost device for declining the road accidents in foggy, rainy and drowsy conditions. The Atmega16 microcontroller was taken and it roles the main part of this exciting project. The main objective for making this device will help to save human life and reducing road accidents as well.

References

[1] Road Accident in Bangladesh. IATTSS Research, Quium, A.S.M.A (1995)
[2] Strategic Transport Plan, Loiuis Burger, Group-Bc 1 Dhaka Transport Coordination Board, Ministry of Communication, govt. of Bangladesh(2005).
[3] V. Ryan (2009) transistor, Available at http://www.technologystudent.com/ (Accessed: 16th Apr. 2018).
[4] (2018) Common cathode 7-Segment Display, Available at: http://jameco.com/ (Accessed: 12th Mar, 2018.)
[5] (2013) 532-655 nm Red Beam Laser Pointer HLM1845, Available at: http:// www.roithner-laser.com/ (Accessed: 14th Apr. 2018).
[6] B.L THEREJA (2005) TRANSFORMER, First Multi color radiation edn..., New Delhi: S.CHAND &COMPANY LIMITED. [Transformer]

New Dcsign of All-terrain Smart Car Based on Arduino

ZhengYang He, Ying Zhang, HongMing Zheng, Yi Zhang, HongTao Yu, MingNan Luo

Sichuan University Jinjiang College, Meishan, China

cquxiaomage@qq.com

Abstract. Due to bad working environment, the flexibility and steering stability of all-terrain vehicles have become the main factors affecting the performance. Designing a new all-terrain electric vehicle is aimed at improving the stability of driving on low-attach roads and maneuverability at low speeds. Compared with traditional internal combustion engines, four-wheel-drive all-terrain vehicles have obvious advantages in vehicle motion control and optimal allocation of driving force, thus making it play an important role in all-terrain vehicle technology in the future. Within the scope of the requirements of the "Explorer" modular component equipment, by means of prototype design, assembly, commissioning, operation, the smart vehicles that can automatically and smoothly pass through the obstacles of ten different characteristics and different difficulty without any intervention are designed and completed.

1. Introduction

Smart cars have great advantages over manned driving. They can go to places where people can't survive, small places and places with high risk factors. Off-road is a necessary feature for most smart cars. The purpose of this car is to consider the environment faced by the car as much as possible, and constantly innovate and improve the off-road and obstacle avoidance methods, while increasing the function of the car.

2. Smart car structure design

2.1. Chassis structure design

The all-terrain smart chassis body is composed of "Explorer" modular smart components. Because the smart car needs a variety of terrains, it must ensure that the overall quality is light, and the components are all made of aluminum-magnesium alloy material, which is light in weight and high in strength. The stability is effective to ensure that the smart has good off-road performance. The chassis of the smart car is higher enough to ensure that the phenomenon on the chassis of grounding doesn't happen when climbing the ladder and passing the pit.

2.2. actuating device

The drives adopt four A171:10 model tires with four M06 dual-axis DC motors for forwarding, reversing, turning left and right upon the all-terrain smart car. The four-wheel drive car is the most stable form of driving, which greatly avoids the "pushing head" of the front wheel drive and the "fishtailing" of the rear wheel drive, thus reducing the impact of hardware problems on the obstacle crossing system.

3. Mechanical structure innovation

The smart car as shown in figure 1 uses a simple structure to complete complex work. According to the functional requirements, in order to smoothly cross various terrains including steps, potholes, slopes, etc., the smart car adopts the front wheel as the rear wheel of the big wheel as the structure of the centre of gravity of the small wheel, which is more conducive to the function of driving wheel and the increase to the power of smart cars[1].

Figure 1. The overall structure of the smart car

The tire as shown in figure 2 is equipped with a track to make the smart car more gripped to prevent slipping; screwing the bolt on the track makes it easier for the smart car to climb the ladder through the reaction force of the step on the bolt when the smart car is going uphill.

Figure 2. The overall structure of the tire

Figure 3. The Installation location of four guide wheels

Four guide wheels as shown in figure 3 are installed in front of and behind the smart car to avoid obstacles. The auxiliary steering module is formed by the upper guide wheel of the leading phoenix tail, which not only reduces the forward resistance, but also helps to increase the speed of the vehicle, and also protects the front end. The guide wheel allows the car to quickly pass through the tunnel without slowing down.

4. Each part hardware circuit design

4.1. Main control board module

The software design and entry of the car design use the software "Arduino", which greatly reduces the difficulty of software development and is suitable for beginners. Arduino is a convenient, flexible and easy-to-use open source hardware product with rich interfaces, digital I/O ports, analogue I/O ports, and support for SPI, IIC, and UART serial communication. It senses the environment through a variety of sensors by controlling the wheels to turn over obstacles and reach the end. It has no complicated underlying code of the microcontroller, no incomprehensible assembly, just a simple and practical function [2-3]. And Arduino has a simple programming environment IDE, great freedom, and high scalability. The standardized interface model provides a solid foundation for its sustainability. The pin definition for Arduino is shown in Figure 4.

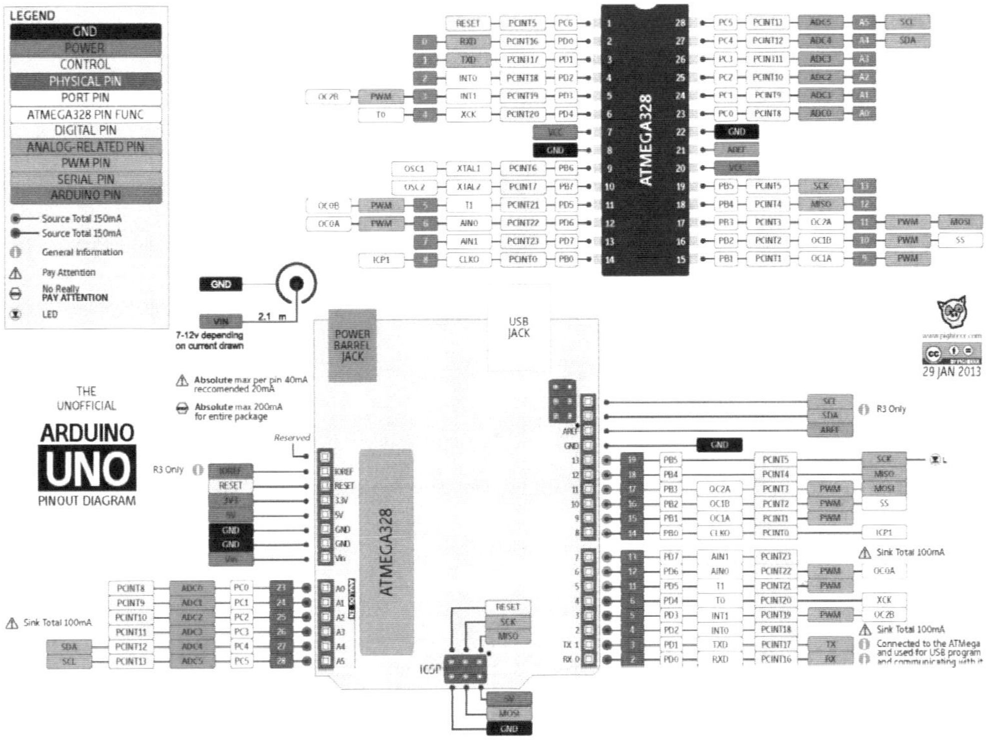

Figure 4. The pin definition for Arduino.

4.2. Path detection module

The grayscale sensor is an analogue sensor. The gradation sensor uses different colours to detect the difference in the degree of reflection of light, and the photoresistor performs colour depth detection on the principle that the resistance of the light which is returned from different detection surfaces is different. The grayscale sensor is used to distinguish black from the others when ambient light interference is not severe. It also has a wide operating voltage range and can still work normally when the power supply voltage fluctuates greatly. It outputs a continuous analogue signal, so it can be easily judged by the A/D converter or a simple comparator to determine the degree of deviation of the car. It is a practical intelligent line sensor. At the same time, we can add different path detection modules in accordance to different needs [4].

The ultrasonic ranging module(HC-SR04) provides 2cm-400cm non-contact distance sensing function with a range accuracy of up to 3mm. The module includes an ultrasonic transmitter, receiver and control circuit. Car avoids obstacle through distance measurement. The Schematic diagram of ultrasonic transmission modulet is shown in figure 5.

IOP Conf. Series: Materials Science and Engineering **853** (2020) 012046 doi:10.1088/1757-899X/853/1/012046

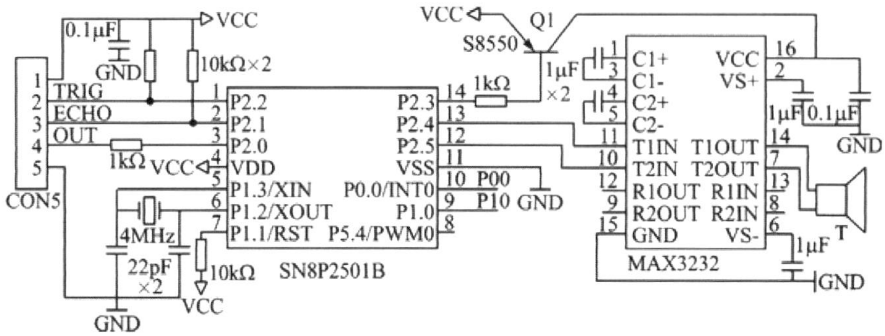

Figure 5. The Schematic diagram of ultrasonic transmission module.

4.3. PSCM

The steering module of the all-terrain smart car uses the differential wheel of the rear wheel of the car, so that the distance from the braking point to the path detecting module is far enough, and the car has sufficient response time to achieve precise control of the car. Compared with the direction of steering engine control, the differential steering will not cause the deviation of the steering gear of the car in the stepped and pit section.

5. Host programming language

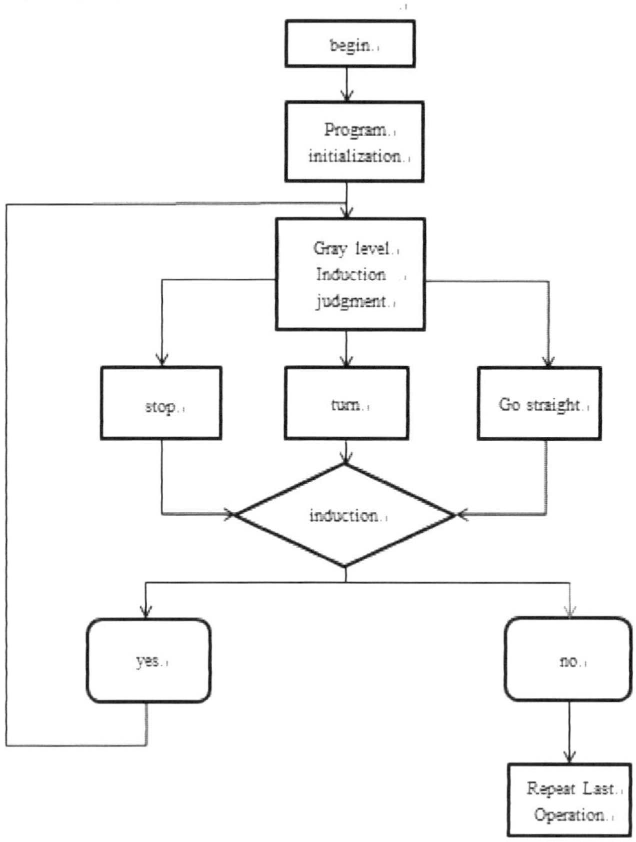

Figure 6. The logic schematic of smart car working.

IOP Conf. Series: Materials Science and Engineering **853** (2020) 012046 doi:10.1088/1757-899X/853/1/012046

This all-terrain trolley uses BASRA as the main control chip, the power is provided by four DC machines, and the tracking sensor is 3 gray scales. The installation method is that the middle gray level is horizontally placed, and the gray scales are placed vertically on both sides, and the middle gray level is used to judge the state of the vehicle body by the gray level of the two sides of the track black line, and whether the black line is guided away from the centre. When the sensor detects that the vehicle body is off centre black line, the DC motor is controlled by the program to form a differential speed while slowly adjusting the car to restore the vehicle body to the straight state. When entering the tunnel, the track is a 180° curve, and the car also makes a turn through the differential. The degree of modification of the steering can be increased according to the number of added gradations. The main program flow chart is: The program is initialized, the gray level sensor receives the signal, and the car goes straight or turns and stops. The gray level sensor continues to sense. If the black line is not sensed, the car repeats the last operation, and at the same time avoids the possibility that the car moves too fast and does not sense the black line to rush out of the specified route. The car continuously executes the program in a cyclical manner to achieve the purpose of automatic tracking [5]. The logic schematic of smart car working is shown in Figure 6.

Part of the car driving control code: The moving path of the smart car is to adjust the forward and reverse rotation of the four DC motors through the main control board connection sensor and motor drive to complete the forward and reverse, and different degrees of steering and direction.

5.1. DC motor control

```
void Forwards ()
{
digital Write (5, HIGH);
digital Write (6, LOW);
digital Write (9, HIGH);
digital Write (10, LOW);
analogue Write (9,200);
analogue Write (10,0);
analogue Write (5,200);
analogue Write (6,0);
}
```

5.2. Smart car steering control code

The smart car decomposes complex control logic into a finite number of stable states and determines events in each state by applying a finite state machine. Since the finite state machine has a limited number of states, it can be implemented in practice. The finite state machine can be widely applied to the judgment of the combined state of multiple sensors of the all-terrain car, which greatly improves the detection efficiency. The following are some of the control procedures:

```
{
value = 0;
for(int i=0; i<3; i++)
{
value |= (digital Read(pin[i]) << i);
}
if(value == 0x07){ //Default to last value when no sensors are triggered
value = value his;//Record the last sensor value
}
```

5.3. Smart car obstacle avoidance controlling code

The smart car uses the ultrasonic ranging module to judge the distance of the obstacle and then avoids the obstacle through the steering module. Some codes are as follows:

```
{
digital Write(TRIGPIN, LOW);
delay Microseconds(2);
digital Write(TRIGPIN, HIGH);
delay Microseconds(10);
digital Write(TRIGPIN, LOW);
float distance = pulse In(ECHOPIN, HIGH);
distance= distance/58;
Serial.println(distance);
delay(500);
}
```

6. System debugging

In order to solve the bug, the functions of the car should be inspected: It can smoothly pass the terrain uphill, downhill, steps, grille, etc. identify the ditch that cannot pass; The Smart car test site schematic is shown in figure 7. The test procedure is:

(1) Testing the car to cross three obstacles by arranging the site and placing the grille, steps and slope in order.

(2) Using a square to form a platform and leaving a gap in the middle of the platform. For the first time, the gap distance should be controlled less than 10cm to see if the car can pass smoothly.

(3) Placing the car on the open space and making it drive automatically without black line, seeing if the car can stop immediately or execute the last instruction.

(4) Simulating grassland and snow slide test.

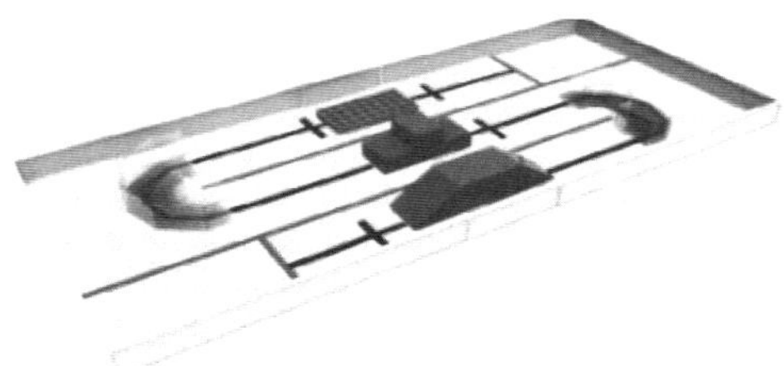

Figure 7. The smart car test site schematic.

(5)Multiple obstacle avoidance tests.

1) For the problem that the car front is not against downhill when the car is going downhill, the solution is to accurate the direction of uphill, and the steering module is quickly debugged.

2) When the car is slipping in a curve, the reaction speed is slow and does not meet the expectation. The solution is to reduce the speed of turning at an acute angle.

3) For the problem that the car encounters multiple obstacles without optimal route, the solution is to measure the distance of the car and determine whether the distance meets the obstacle avoidance condition (the obstacle avoidance distance can be determined according to the actual situation), When it is necessary to turn to avoid obstacles, the car stops. The ultrasonic wave swings left and right with the rotation of the steering gear while detecting car to the obstacle distance, thereby judging the optimal route.

7. Summary

The smart car has a simple hardware structure, a high degree of modularity, convenient debugging, and a smooth system response. The car uses the BASRA control board as the control core, the sensor feedbacks the signal of the environment in which the car is located, and the BASRA controls the car to perform the corresponding action. The operator only needs to open the switch to allow the all-terrain smart car to drive on the pre-prepared all-terrain track. From the final test results, the design proposal of the all-terrain smart car is correct, and the indicators are stable and achieve the intended purpose.

Acknowledgments
I sincerely appreciate for the assistance or encouragement in my scientific research from Sichuan University Jinjiang College.

References
[1] Zhao Wang, Eng Gee Lim, "Design of An Arduino-based Smart Car", 2014 International SoC Design Conference (ISOCC), November 2014, DOI: 10.1109/ISOCC.2014.7087683
[2] BOLUN ZENG 1, JIANBO ZHANG (Corresponding Author) 2, LANG CHEN 1 ,YAO WANG,"Self-balancing car based on ARDUINO UNO R3", 2018 IEEE 3rd Advanced Information Technology, Electronic and Automation Control Conference(IAEAC 2018).
[3] Ruslan T. a, Yerden K. a, and Md. Hazrat Ali a, "Development of Arduino Glove-Based utonomous Car", 2017 21st International Conference on System Theory, Control and Computing (ICSTCC).
[4] Mohammad O. Khan and Gary Parker, "Learning Live Autonomous Navigation: A Model Car with Hardware Arduino Neurons" , 2016 IEEE International Conference on Systems, Man, and Cybernetics • SMC 2016 | October 9-12, 2016 • Budapest, Hungary.
[5] Hemant Chaudhary, Prateek Bansal, Dr. B. Valarmathi, "Advanced CAR Parking System using Arduino",2017 International Conference on Advanced Computing and Communication Systems (ICACCS -2017), Jan. 06 – 07, 2017, Coimbatore, INDIA.

IOP Conf. Series: Materials Science and Engineering **853** (2020) 012047 doi:10.1088/1757-899X/853/1/012047

On Computation of Fractional-Order Differentiations

Soichiro Ebihara* **and Noriyuki Hori+**

Intelligent Interaction Technologies, Graduate School of Systems and Information Engineering, University of Tsukuba, 1-1-1 Tennoudai, Tsukuba, 305-8573 Japan

* ebihara@digicon-lab.esys.tsukuba.ac.jp
+ hori@iit.tsukuba.ac.jp

Abstract. It is shown in this paper that the conventional approach for computing fractional-order differentiation using gamma-functions can lead to numerical difficulties and result in large errors. To overcome this issue, a new algorithm based on matrix computations is proposed. Typical simulation results are presented to explain the concepts and show that, while the gamma-function-based derivative diverges away from the analytical derivative, the proposed technique does not at least for the range of discrete-time periods tested. However, the computation time required with the proposed method is longer, at present, than the conventional method, and further research is on-going.

1. Introduction

The history of fractional calculus is said to have begun when Leibniz referred to its existence in 1695 and the theoretical framework set in place in the 1800s [1, 2]. It is now recognized by many that laws of fractional calculus can be found in a variety of phenomena, such as properties of viscoelastic bodies, characteristics of anomalous diffusion models, and control of undulation motions [3, 4, 5]. Fractional-order devices can have characteristics that are not prevalent in integer-order devices. For instance, it is said that the CRONE suspension system [6], which uses fractional-order derivative, is capable of making the vertical overshoot of the vehicle height constant regardless of the vehicle weight, which is not possible using integer-order derivatives. When the control law is realized using analog devices, the suspension can be bulky and expensive, while it can be compact and inexpensive when it is implemented digitally. Computations of fractional derivatives are important for simulations and controls, but have been difficult to implement and further studies are desired to improve efficiency and accuracy.

There are several definitions for fractional calculus, such as those proposed by Riemann-Liouville, Grünwald-Letnikov, and Caputo [7,8]. The class of functions to which these definitions can be applied differ so that depending on what function is considered, a different definition may need to be considered. In the discrete-time domain, there are two approaches; the first is to carry out fractional calculus on the given function and then discretize this result. Since the calculus must be carried out first, this is not suitable for computation using the past data only. The second is to discretize the given function and carry out discrete-time version of fractional calculus. While this has conventionally been achieved using gamma functions, a large number of memories is required for computations [6, 7, 9]. In

the present study, a new method is proposed for the second approach using matrix calculations, such that only the past data are required for computation.
The paper is organized as follows: After Introduction in Section 1, a review of discrete calculus from the discrete-time model perspective is given in Section 2. The description of the conventional method based on gamma-functions is given using Examples in Section 3, followed by the proposed method based on matrix computations in Section 4. Conclusions are drawn in Section 5.

2. Discrete Calculus as Discrete-Time Model

2.1. Fractional Derivatives

Let $\mathbb{C}$ be the set of all complex numbers, $\mathbb{R}$ that of all real numbers, and $\mathbb{Z}$ that of all integers. Denoting the continuous-time variable as $t \in \mathbb{R}$, let $\bar{f}(t)$ represent a continuous-time signal. Similarly, denoting the number of steps as $k \in \mathbb{Z}$ and the sampling period as $T \in \mathbb{R}^{+}$, let $f(k,T) = f_k$ represent a discrete-time signal defined for $t=kT$. Discrete-time signals are assumed to be zero for negative time, while continuous-time signals are not in general. It is assumed that T is a constant, while $r \in \mathbb{R}$ and $n \in \mathbb{Z}$ are variables. To consider fractional α-th derivative, let an integer β and a fraction γ be such that

$$\alpha < \beta < \alpha + 1, \quad \gamma = \beta - \alpha . \tag{1}$$

The Riemann-Liouville (RL) derivative is defined as [7]

$${}_{c}^{RL}\bar{D}_t^{\alpha} \bar{f}(t) = \frac{1}{\Gamma(\gamma)} \frac{d^{\beta}}{dt^{\beta}} \int_c^t (t-\tau)^{\gamma-1} \bar{f}(\tau) d\tau , \tag{2}$$

while the Caputo derivative as

$${}_{c}^{C}\bar{D}_t^{\alpha} \bar{f}(t) = \frac{1}{\Gamma(\gamma)} \int_c^t (t-\tau)^{\gamma-1} \left\{ \frac{d^{\beta}}{dt^{\beta}} \bar{f}(\tau) \right\} d\tau . \tag{3}$$

In these definitions, the fractional portion is taken care of by the integral, with the value of c chosen accordingly to function $\bar{f}(t)$ as needed. The Grünwald–Letnikov (GL) derivative is defined by [7]

$${}_{c}^{GL}\bar{D}_t^{\alpha} \bar{f}(t) = \lim_{h \to 0} \frac{1}{h^{\alpha}} \sum_{j=0}^{\left\lfloor \frac{t-c}{h} \right\rfloor} (-1)^j \binom{\alpha}{j} \bar{f}(t - jh) , \tag{4}$$

where c is a constant, $\lfloor r \rfloor$ represents the largest integer not exceeding r,

$$\binom{\alpha}{j} = \frac{\Gamma(\alpha+1)}{\Gamma(j+1)\Gamma(\alpha-j+1)} , \tag{5}$$

and Γ is the gamma function defined by [7],

$$\Gamma(\alpha) = \int_0^{\infty} e^{-t} t^{\alpha-1} dt . \tag{6}$$

When α is integer, eq. (5) is the usual binomial coefficient. In this definition, the fractional derivative is based on the sequence obtained by sampling the continuous-time signal.

2.2. Discrete-Time Equivalent of Fractional Derivatives

A time derivative is a rate of change of a continuous-time function during an infinitesimal time interval. Since there is no discrete-time counterpart for such an interval, a finite and nonzero time interval has to be chosen as a discrete-time counterpart. This is usually chosen as the discrete-time period T.
Let the discrete-time signal f_k be a sequence of numbers with integer k indicating the step number and T the discrete-time interval. The backward-difference operator ∇_T is defined as

$$\nabla_T f_k := f_k - f_{k-1} . \tag{7}$$

where f_{k-1} is the shifted signal of f_k by one sampling period or, using the shift operator [10] q_T,

$$q_T f_k := f_{k+1} . \tag{8}$$

The backward-quotient operator D_∇ is a discrete operator defined by

$$D_\nabla f_k := \frac{\nabla_T f_k}{T}. \tag{9}$$

The delta operator [10,11] is defined as

$$\delta_T := \frac{q_T - 1}{T}. \tag{10}$$

Using q_T and δ_T operators, D_∇ and its n-th order operations can be written as

$$D_\nabla^n f_k = \left(\delta_T q_T^{-1}\right)^n f_k. \tag{11}$$

where n is any integer. Using the binomial theorem given by

$$\left(1 - q_T^{-1}\right)^n = \sum_{j=0}^{n} \binom{n}{j} (-q_T^{-1})^j, \tag{12}$$

eq. (11) can be written as

$$D_\nabla^n f_k = \frac{1}{T^n} \sum_{j=0}^{n} \binom{n}{j} (-1)^j f_{k-j}. \tag{13}$$

Replacing the binomial theorem by Newton's generalized binomial theorem, taking the causality of the final result into account, and eliminating the use of inter-sample values, the α-th order backward-quotient operator is defined as

$$D_\nabla^\alpha f_k = \frac{1}{T^\alpha} \sum_{j=0}^{\infty} (-1)^j \binom{\alpha}{j} f_{k-j}. \tag{14}$$

This is one of possible discrete-time models for the fractional differentiation, eq. (4), under the framework of discrete-time models of a continuous-time system. In eq. (4), the infinitesimal time interval h goes to zero, so that the number of sums approaches infinity. While the number of nonzero binomial coefficients are finite for integer values of α, making the sum finite, it is infinite for fractional α. Basically the same occurs in discrete-time equivalent eq. (14), where h is replaced with a finite T and discrete-time functions are defined on the sampling instants $t=kT$ from the present instant back into the negative time at infinity.

3. Conventional Model: Gamma-Function-Based Method

The conventional discrete-time model of a fractional-differentiator is given as the combination of eqs. (14) and (5). The numerator of eq. (5) is finite for a given fraction α, while the denominator is the product of two gamma functions, each with an increasing and an decreasing argument as j increases. Let us see this issue by an example.

Example 1: To examine the discrete-time model given by eqs. (14) and (5), simulation studies are conducted for

$$\bar{f}(t) = t \tag{15}$$

with α=1/2. The analytical half-derivative is known to be [12]

$${}_0^{RL}\bar{D}_t^{1/2} \bar{f}(t) = \frac{2}{\sqrt{\pi}} t^{1/2}. \tag{16}$$

The conventional discrete-time model of this half-derivative is given [7,8] by

$$D_\nabla^{1/2} f_k = \frac{1}{\sqrt{T}} \sum_{j=0}^{k} \frac{(-1)^j \Gamma(3/2)}{\Gamma(j+1)\Gamma(3/2-j)} f_{k-j}. \tag{17}$$

Looking at eq. (17), it might be noticed that the product of two gamma functions, $\Gamma(j+1)\Gamma(3/2-j)$, in the denominator of the infinite sum can be a source of numerical problem as integer j increases. Mathematically, $\Gamma(j+1)$ grows unbounded as j increases, while it becomes computationally infinite so that $1/\Gamma(j+1)$ is numerically zero after j reaches some integer value. MATLAB used in this study returns

$$\Gamma(171) = 7.2574\text{e}+306, \quad \Gamma(172) = \text{Inf}. \tag{18}$$

This suggest that the terms to be added in eq. (17) reaches zero after a fixed number of iterations (discrete steps), independently of T, causing the discrete-time result to depart from the true values.

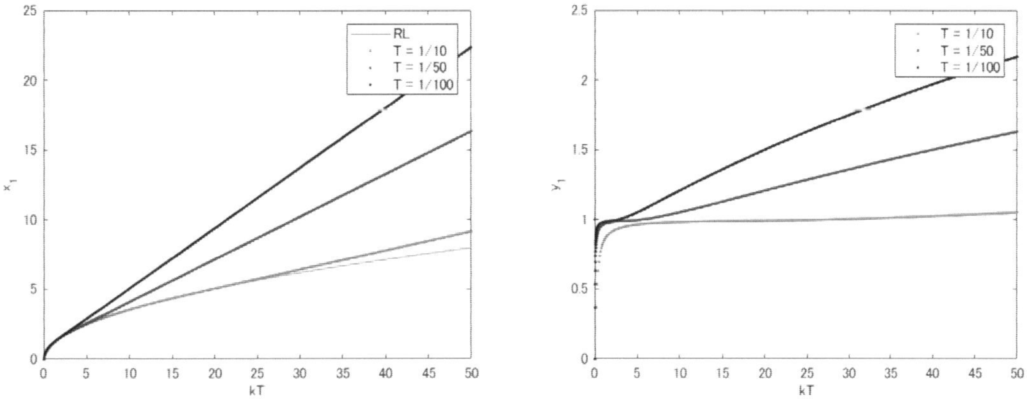

Figure 1. (Left) The gamma-based half-derivatives of t. (Right) The ratios to the analytical derivative.

Fig. 1 shows the result. In the figure on the left, the vertical axis x_1 represents either ${}^{RL}_{0}D_t^{1/2} f(t)\big|_{t=kT}$ or $D_\nabla^{1/2} f_k$, where the line in pink is the true analytical half-derivative given by eq. (16). The red, blue, and black lines are the discrete-time models, eq. (17), based on the gamma-functions evaluated at T=1/10, 1/50, and 1/100, respectively. The figure on the right shows the ratio of these models to analytical model; i.e., y_1 represents $\sigma[D_\nabla^{1/2} f_k]/\sigma\left[{}^{RL}_{0}D_t^{1/2} f(t)\big|_{t=kT}\right]$, where σ denotes $\sigma[g_k] = \sum g_k$. The value of y_1 being unity implies that the discrete-time half-derivative gives values that are identical to the true values at the sampling instants. The discrete-time equivalents, eq. (17), of the half-derivative give values smaller than the true values in the beginning and seem to converge to the true value for a brief while. Although the convergence to unity becomes faster as the discrete-time interval T becomes smaller, their value tends to diverge away from the true value sooner, against expectations. This happens soon after the discrete-time step reaches a certain number of steps, which depends on the CPU and the software used, when the gamma function becomes numerically infinite, although it is not mathematically.

Example 2: For the function given by

$$\bar{f}(t) = \sin t, \tag{19}$$

the half-derivative is known to be [12]

$${}^{RL}_{0}\bar{D}_t^{1/2} \sin t = \sqrt{2}\left\{C\left(\sqrt{\frac{2}{\pi}t}\right)\cdot\cos t + S\left(\sqrt{\frac{2}{\pi}t}\right)\cdot\sin t\right\}. \tag{20}$$

where $C(u) = \int_0^u \cos\left(\frac{\pi}{2}x^2\right)dx, S(u) = \int_0^u \sin\left(\frac{\pi}{2}x^2\right)dx$. Fig. 2 shows the results for the sinusoidal function under consideration, as in Fig. 1 for t. In this example, the discrete-time model for T=1/50 gives better results than others. The error is larger with T=1/100s than with T=1/10, against expectations.

4. Proposed Model: Matrix-Based Method

To rearrange the sum in eq. (14) in a matrix form, such that no gamma function, eq. (5), is necessary, the concept of signal lifting is used [13]. Let $\mathbf{f}_{\ell,k}$ be the lifted (k+1)-dimensional vector written as

$$\mathbf{f}_{\ell,k} = \begin{bmatrix} f_\ell & f_{\ell+1} & \cdots & f_{\ell+k} \end{bmatrix}^T, \tag{21}$$

where ℓ is integer, and $\mathbf{D}_k$ be the (k+1) × (k+1) matrix (k>0) defined as

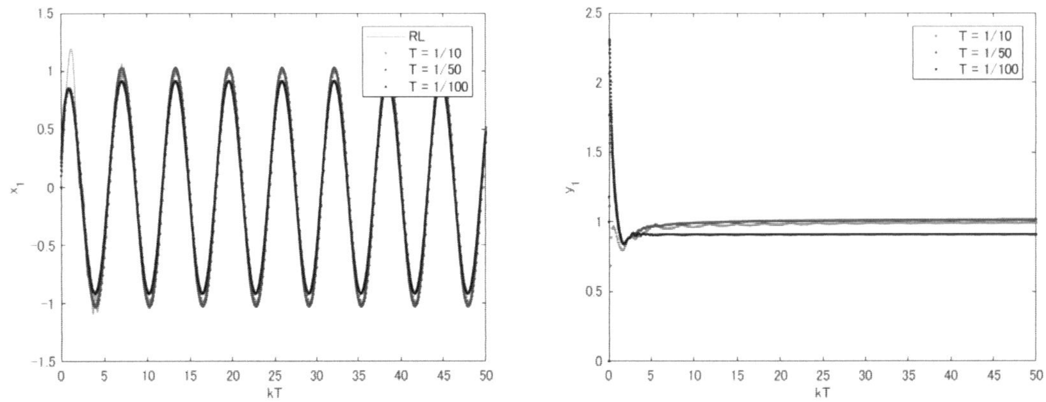

Figure 2. (Left) The gamma-based half-derivatives of sin$t$. (Right) The ratios to the analytical derivative.

$$\mathbf{D}_k = \frac{1}{T}\begin{bmatrix} 1 & 0 & \cdots & 0 & 0 \\ -1 & 1 & \ddots & \vdots & \vdots \\ 0 & -1 & \ddots & 0 & 0 \\ \vdots & \vdots & \ddots & 1 & 0 \\ 0 & 0 & \cdots & -1 & 1 \end{bmatrix}, \tag{22}$$

whose eigenvalues are all unity. Furthermore, let ϕ be the function that returns the last element of its input vector, as

$$f_{\ell+k} = \phi[\mathbf{f}_{\ell,k}]\,. \tag{23}$$

Using these notations, it can be seen that

$$\phi\left[\mathbf{D}_k \mathbf{f}_{0,k}\right] = \frac{f_k - f_{k-1}}{T} = D_{\nabla}^{1} f_k\,. \tag{24}$$

The operation that is repeated α times should equal $\mathbf{D}_k$, so that

$$\phi\left[\mathbf{D}_k^{\alpha} \mathbf{f}_{0,k}\right] = D_{\nabla}^{\alpha} f_k\,. \tag{25}$$

For instance, for the half-derivative, α=1/2, one has

$$\mathbf{D}_k^{1/2} = \frac{1}{\sqrt{T}}\begin{bmatrix} 1 & 0 & \cdots & 0 & 0 \\ -\frac{1}{2} & 1 & \ddots & \vdots & \vdots \\ -\frac{1}{8} & -\frac{1}{2} & \ddots & 0 & 0 \\ \vdots & \vdots & \ddots & 1 & 0 \\ & & \cdots & -\frac{1}{2} & 1 \end{bmatrix}. \tag{26}$$

Some fractional-order matrix computations are possible, the one-third power of matrix (22) is not. In such cases, MATLAB's sqrt command can be used to compute matrix powers of $m/2^n$, where m and n are integers. Using series expansions, this can practically cover most rational fractions as shown in Example 4 for α=1/3.

Example 3: Examples 1 and 2 for the half-derivative are repeated here using the proposed matrix method.

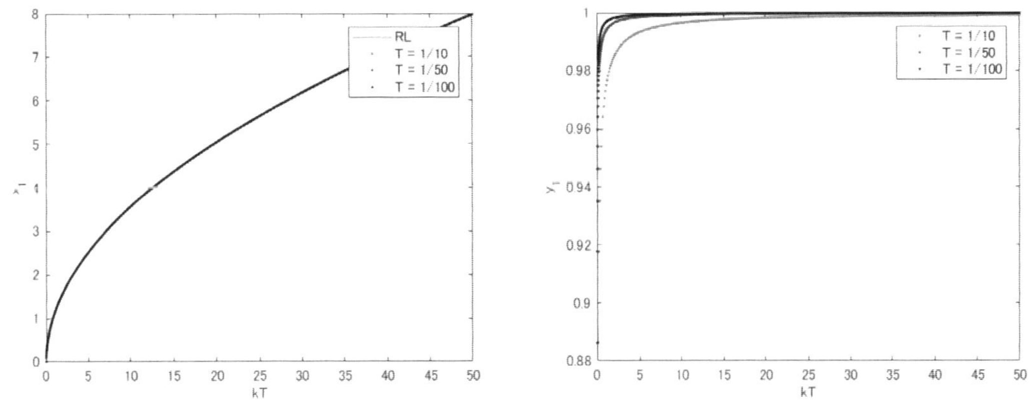

Figure 3. (Left) The matrix-based half-derivatives of t. (Right) The ratios to the analytical derivative.

The proposed discrete-time model of a half-differentiator for t is given as the combination of eqs. (14) and (5). Comparing Fig. 3 to Fig. 1, one can see that the proposed method is much more accurate than the conventional discrete-time model based on the gamma-function for the sampling period simulated. Fig. 4 is the result of the proposed model for sin$t$ and should be compared to Fig. 2. There is an error at the start, which may be attributed to the fact that sin$t$ is actually zero for negative t in the simulation. Otherwise, it is seen that the proposed method performs consistently better than the conventional model for all Ts tested.

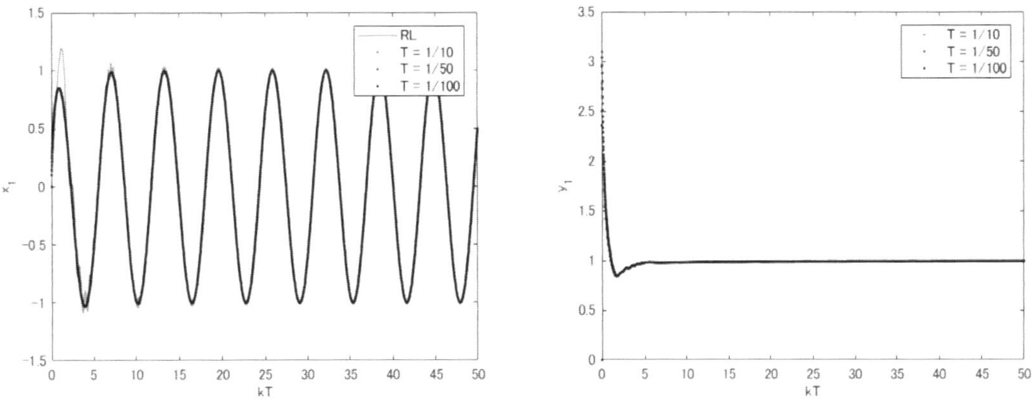

Figure 4. (Left) The matrix-based half-derivatives of sin$t$. (Right) The ratios to analytical derivative.

Example 4: The result of one-third derivative of t^2 [12] is shown in Fig. 5 for the conventional model and in Fig. 6 for the proposed model, where α=1/3,

$$\overline{f} = t^2, \quad {}_0^{RL}\overline{D}_t^{1/3}\overline{f}(t) = \frac{2}{\Gamma(8/3)} t^{5/3}. \tag{27}$$

One-third derivative is computed using $1/3 \approx \sum_{n=1}^{N} 4^{-n}$ with N=5. Fig. 5 is for the conventional model and Fig. 6 for the proposed model, where the vertical axis x_2 on the left shows ${}_0^{RL}\overline{D}_t^{1/3} f(t)\big|_{t=kT}, D_{\nabla}^{1/3} f_k$, while that y_2 on the right $\sigma[D_{\nabla}^{1/3} f_k]/\sigma\left[{}_0^{RL}\overline{D}_t^{1/3} f(t)\big|_{t=kT}\right]$. It is clear by comparing these figures that the proposed method is superior to the conventional model even though one-third is approximated using 5 terms in the proposed model. Actually N=3 may suffice for approximate computations.

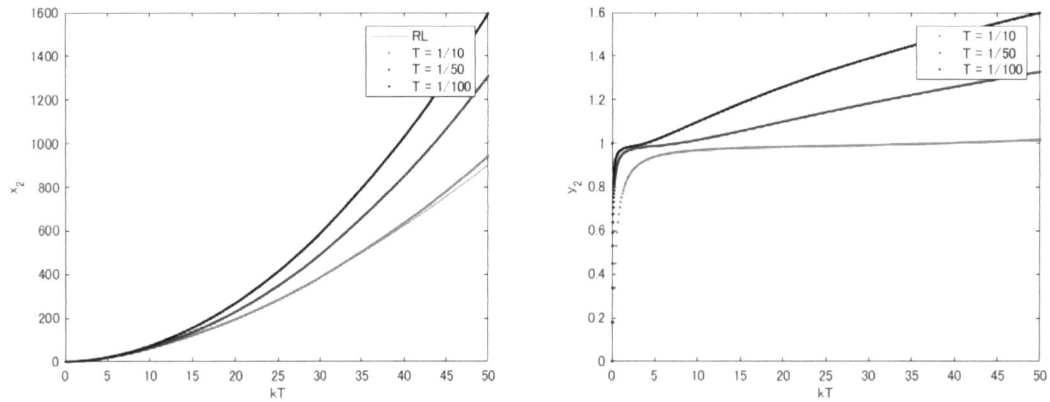

Figure 5. (Left) The gamma-based one-third-derivatives of t^2. (Right) The ratios to the analytical derivative.

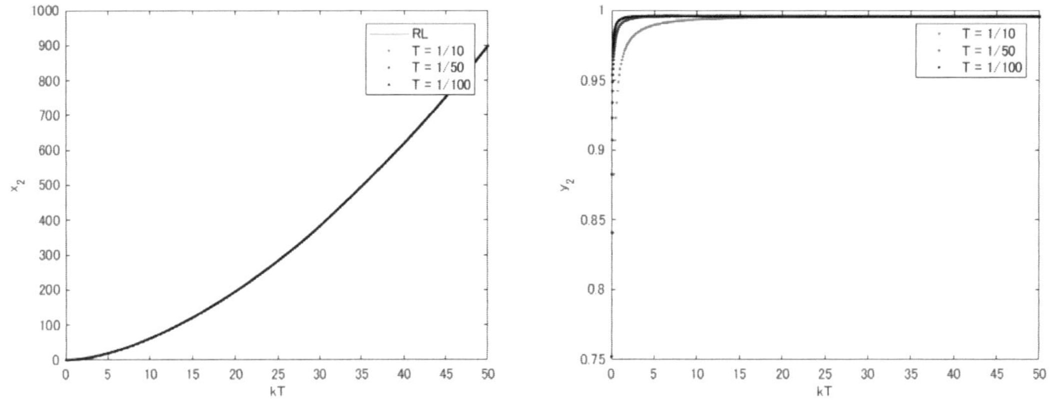

Figure 6. (Left) The matrix-based one-third-derivatives of t^2 with N=5. (Right) The ratios to the analytical derivative.

5. Conclusions

It has been shown that a conventional method of computing a fractional derivative using gamma-functions has numerical problems due to numerical overflows of gamma functions. To overcome this issue, a matrix-based computation method has been proposed in the present paper. Through some simulations, it has been shown that the proposed model performs better than the conventional model in terms of accuracy. However, it has also been found that it takes longer with the proposed matrix method to compute the fractional derivatives (see Table 1). Further studies are desired to arrange the matrix in a recursive form so that they can be used for digital implementation, for instance, of a control algorithm. It is noted also that simulations for other fractions, such as α=3/4, lead to similar results and conclusions.

Table 1. Conventional model vs Proposed model

	Gamma-Based	Matrix-Based
Advantage	Shorter computation time (Fig. 1: 31.4s, Fig. 2: 32.6s, Fig. 5: 33.6s)	High accuracy
Disadvantage	Poor accuracy	Longer computation time (Fig. 3: 214.8s, Fig. 4: 216.6s, Fig. 6: 1711.8s)

Processor: Intel Core i5-7500T, MATLAB: R2017b

Acknowledgments
A part of the work reported in this paper has been supported financially by JSPS research grant under #19K04451.

References

[1] C. Goodrich and A. C. Peterson, *Discrete Fractional Calculus*, Springer, Berlin, Germany 2016
[2] N. Sugimoto, "On Differential and Integral Calculus of Non-Integer Order," Autumn Synthesis Subcommittee Meeting Public Lecture of the Mathematical Society of Japan, 2016.
[3] H. Nasuno, N. Shimizu, and T. Yasuno, "Geometrical Nonlinear Statical and Dynamical Models of Fractional Derivative Viscoelastic Body," Transactions of the Japan Society of Mechanical Engineers, Series C, Vol. 72, No. 716, pp. 27-34, 2006.
[4] N. Shimamoto, "Modeling of Anomalous Diffusion based on Time Fractional Differential," Research Institute for Mathematical Sciences Kokyuroku, Vol. 1847, pp. 116-128, 2013.
[5] M. Kuroda, "Control Undulation Motions using Fractional Differential," Dynamics, Measurement and Control Division News, No. 59, 2017.
[6] Alain Oustaloup, "Diversity and Non-integer Differentiation for System Dynamics," John Wiley & Sons, Inc., Hoboken, NJ, USA 2014.
[7] Igor Podlubny, "Fractional Differential Equations," Academic Press, San Diego, CA, USA, 1999.
[8] Ivo Petráš, "Fractional-Order Nonlinear Systems : Modeling, Analysis and Simulation," Springer, Berlin, Germany 2011.
[9] Duarte Valério and José Sá da Costa, *An Introduction to Fractional Control*, The Institution of Engineering and Technology, UK 2012.
[10] R. H. Middleton and G. C. Goodwin, Digital Control and Estimation: A Unified Approach, Prentice-Hall, Englewood Cliffs, N. J., 1990.
[11] N. Hori, T. Mori and P. N. Nikiforuk, "A New Perspective for Discrete-Time Models of a Continuous-Time System," IEEE Trans. on Automatic Control, Vol. 37, No. 7, pp. 1013-1017, 1992.
[12] K. B. Oldham and J. Spanier, *The Fractional Calculus – Theory and Applications of Differentiation and Integration to Arbitrary Order*, Dover Publications, Inc., Mineola, New York, 1974.
[13] T. Chen and B. Francis, *Optimal Smapled-Data Control Systems*, Springer-Verlag, London, 1995.

Complex Network Community Detection based on Genetic Algorithm using K-cliques

Jian Ma[1,2], **Jianping Fan**[1,3]

[1]School of Computer and Information Technology, Beijing Jiaotong University, Beijing 100044, China
[2]Research Center for High-Speed Railway Network Management of Ministry of Education, Beijing Jiaotong University, Beijing 100044, China
[3]Shenzhen Institutes of Advanced Technology , Shenzhen 518055, China

13112083@bjtu.edu.cn

Abstract. Inspired by the genetic algorithm, this paper proposes a complex network community detection algorithm, which searches for complete sub-graphs in a network, the numbers of nodes are greater than or equal to k, also referred to maximum k-cliques. K-cliques are the most connected communities in a network. Using the maximum k-clique initializes the population, which can improve the accuracy and efficiency of population initialization. K-clique-based population initialization, crossover, mutation, $\mu+\lambda$ selection strategy, and evaluation function (Q function) are adopted to select the next generation of population, the superior traits formed by parents during the process of evolution cannot be destroyed and can also be effectively inherited by offspring individuals. Finally, community partition is optimized using the fast Newman algorithm. This step performs further clustering on the communities. The algorithm can reduce the search space of community partition and improve the search efficiency of the algorithm. It is tested on benchmark networks and real-world networks. The algorithm has an acceptable time complexity. The experimental results show that the algorithm can effectively divide the communities.

1. Introduction

In a sense, complex networks are made up of different communities. The problem of complex network clustering (community mining) is how to reveal the real communities in which the internal nodes are densely connected, and the connections between different community nodes are sparse. In recent years, many researchers have proposed many classic community structure detection algorithms from different perspectives.

The genetic algorithm is a method for simulating natural evolution process and searching for optimal solution. As an adaptive global optimization probability search algorithm, the genetic algorithm is applied to community detection problems and is an effective means for solving complex optimization problems. Therefore, a complex network community detection method based on the genetic algorithm has attracted the attention of scholars. Tasgin et al. [1] first introduced the genetic algorithm for solving complex network community detection problems. The advantage was that the algorithm did not require prior knowledge, such as the number of communities. The disadvantage was that the algorithm searched entire solution space, rather than the neighboring nodes, when searching for community labels, it would lead to a meaningless search space with slow convergence. Yang et al. [2]

proposed an improved Genetic Algorithm based on local modularity (IGALM). Hafez et al. [3] proposed a novel multi-objective optimization problem for the genetic algorithm in solving community detection problems. Li et al. [4] proposed MAGA-Net, which was based on the genetic algorithm, each agent represented a community partition, and different operators for crossover and mutation were proposed, the purpose was to increase community partition evaluation values. The k-path algorithm was used to obtain the topology of a network, applying the k-path algorithm in initialization phase could speed up the convergence of the algorithm and increase the modularity function for community structure detection [5]. Guerrero et al. [6] proposed a novel genetic algorithm called GGA+. The algorithm analyzed a network from different perspectives, optimized the operator of the genetic algorithm based on modularity function, and obtained relatively good experimental results compared with those by other algorithms. Due to the limitation of the algorithm itself. Liu et al. [7] proposed a genetic method using node-to-community function mutation.

Inspired by the genetic algorithm, this paper proposes a complex network community detection algorithm based on maximum k-cliques (GAC algorithm). The algorithm, referred to the GAC algorithm, uses maximum k-cliques to improve the genetic representation of the genetic algorithm and continues to use the maximum k-cliques during crossover and mutation phases. Finally, the test is carried out on benchmark networks and real-world networks. The experimental results show that the algorithm can effectively divide communities.

2. Genetic algorithm

2.1. Gene expression

A genetic algorithm (GA) evolves to the optimal solution by simulating selection, crossover, and mutation of biological inheritance. The genetic algorithm begins with a population that represents a possible solution set of problems, and the population consists of a certain number of genetically encoded individuals. By encoding, the solution to the problem can be transformed into a chromosome or an individual.

The encoding methods currently applied to GA-based community mining algorithms include string-based encoding and locus-based adjacency representation. This paper uses the former.

Let the two-tuple $G=(N,E)$ denote a complex network, where $V=\{v_1,v_2,\ldots,v_m\}$ represents a node set, and $E=\{e_1,e_2,\ldots,e_n\}$ represents an edge set. $A=\left(a_{ij}\right)_{n\times n}$ represents the adjacency matrix of the network; if node v_i and v_j are adjacent, $a_{ij}=1$; otherwise, $a_{ij}=0$.

$$\kappa(t)=[\kappa_1,\kappa_2,\kappa_3,\ldots,\kappa_n] \quad (1)$$

where κ_i indicates the community label to which node i belongs; if $\kappa_i=\kappa_j$, node i and node j are in the same community; otherwise, node i and node j belong to different communities.

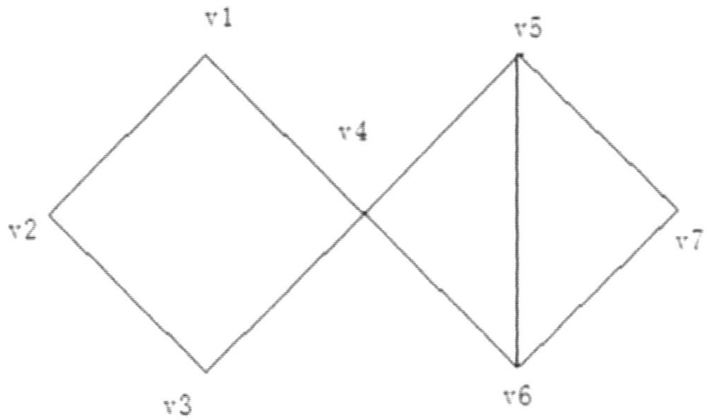

Figure 1. An example of community structures

This paper adopts string-based encoding. Each individual (chromosome) in the population contains n gene positions, $1 \ldots n$, where n is the number of nodes in the network. In Figure 1, the community label of each node is represented as in Equation (1). Let there be 7 nodes in the network, and the chromosome is initialized as shown in Table 1.

Table 1. Gene expression

v	individual (chromosome)
v_1	1
v_2	2
v_3	3
v_4	4
v_5	5
v_6	6
v_7	7

The k-clique is the most closely connected community in a network. The algorithm searches for all complete sub-graphs which the numbers of all nodes are greater than or equal to k, that is, maximal k-cliques. In this paper, the maximal k-cliques generated by the above steps are to initialize the population. The algorithm generates a chromosome using the following steps: Initialize the community label of each node in the network as a unique label. When the algorithm generates an individual, we randomly select a node and seek any maximal clique containing this node, If all the nodes in this selected clique are not assigned to any community, that is, the community labels of the nodes are still at their initial labels, then update the community labels corresponding to all nodes in the clique with the minimum initialized community label among all the nodes in this clique. The reason for this is that nodes in the same clique are more likely to be in the same community. If the node does not belong to any maximal clique, the algorithm does nothing. We use the same operation $\alpha \times n$ times. The process forms many small local communities, but it is still far from optimal solution.

The maximal k-clique initializes the population, which can improve the accuracy and efficiency of population initialization, nodes belonging to the same community are assigned together. For example, the network in Table 2 has 7 nodes. A maximal 3-clique is a complete sub-graph which the number of nodes is greater than or equal to 3. The nodes are randomly arranged, for example, in the order v_5, v_2, v_4, v_6, v_1, v_3, v_7. We select the node v_5, there are three cliques containing node v_5, which are $\{v_4, v_5, v_6\}$ $\{v_5, v_6, v_7\}$ and $\{v_4, v_5, v_6, v_7\}$, the algorithm arbitrarily selects one, for example, $\{v_4, v_5, v_6\}$, then update the community labels of nodes v_4, v_5, and v_6 to the minimum label of the nodes in clique $\{v_4, v_5, v_6\}$, which is 4.

Table 2. Population initialization

v	maximal 3-clique
v_1	$\{v_1,v_2,v_3,v_4\}$
v_2	$\{v_1,v_2,v_3,v_4\}$
v_3	$\{v_1,v_2,v_3,v_4\}$
v_4	$\{v_1,v_2,v_3,v_4\}, \{v_4,v_5,v_6\}, \{v_4,v_5,v_6,,v_7\}$
v_5	$\{v_4,v_5,v_6\}, \{v_5,v_6,v_7\}, \{v_4,v_5,v_6,,v_7\}$
v_6	$\{v_4,v_5,v_6\}, \{v_5,v_6,v_7\}, \{v_4,v_5,v_6,,v_7\}$
v_7	$\{v_5,v_6,v_7\}, \{v_4,v_5,v_6,,v_7\}$

2.2. Selection operator

The algorithm adopts a traversal sampling method to implement the survival of individuals in the population. In this paper, the selection strategy $\mu+\lambda$ is adopted, which the size of the parent population is μ. After crossover and mutation, the sub-population size is $\mu+\lambda$, from which μ individuals with the greatest fitness function (Q function) [8] are selected to be the next-generation parents.

2.3. Crossover operation

The crossover operation of the genetic algorithm in this paper does not simply exchange some gene positions of the parents, this operation could not only produce excellent individuals but destroy the good community structures that has been formed. Therefore, in the process of crossover, the algorithm adopts the single-way crossover strategy. Firstly, select two individuals, one as the source individual called κ_{src}, and the other as the target individual named κ_{dest}, randomly select node v_i belonging to two or more maximal k-cliques, find the nodes that are in the same community as v_i in the source individual κ_{src}, then update the corresponding node community labels in the target individual κ_{dest} to the community label of node v_i, thereby generating an individual κ_{new}. If the above node v_i does not exist, the algorithm does nothing.

For example, Table 3 shows the single-way cross operation, node v_4 which belongs to three maximal k-cliques is selected. In the source individual, nodes v_1, v_4 are in the same community. After one crossover operation, the community labels of the corresponding nodes in the new individual are all updated to 1.

Table 3. Single-way crossing over

v	$\kappa\, src$	$\kappa\, dest$	$\kappa\, new$
v_1	**1**	1	**1**
v_2	3	2	2
v_3	3	2	2
v_4	**1**	4	**1**
v_5	7	5	5
v_6	2	5	5
v_7	3	2	2

2.4. Mutation operator

The mutation operation adopts basic position mutation. The source individual κ_{src} randomly assigns a certain gene position or a certain number of genes to the mutation operation. Firstly, find the nodes that do not belong to any maximum k-clique, randomly pick node v_i from these nodes, obtain the collection of nodes adjacent to this node, then randomly select a neighboring node, and update the community label of node v_i to the label of this node, the rest remain unchanged. If the above node v_i does not exist, the algorithm does nothing. The purpose is to solve the isolated node that does not belong to any maximal k-cliques.

2.5. Optimization operation

This paper combines the communities formed by the above steps and finds the maximum Q value to increase. Starting with an origin network which each community is formed by the above steps as a sole member and no edge, then repeatedly join communities together in pairs, at each step choose the greatest increase value in Q. "dendrogram", tree-like diagram that shows the order of the nodes join. Cut through this dendrogram at different levels to give divisions of the network, the network will be split into several communities and the partition corresponding to the largest Q value is the most accurate partition. The optimization operation is the further clustering operation of the communities.

2.6. Algorithm Description

In summary, the algorithm is described as follows:

Algorithm 1: Initiation	
Input: maximum k-cliques	/* maximum k-cliques is a set of all maximum k-cliques*/
Output: P	/* population P*/
1. for i=1: $\alpha \times n$	
2. randomly select a node v_j;	

3. if exists maximal k-clique *mc* contains this node v_j and all the nodes in *mc* are their initial labels;
4. update the gene positions corresponding to all nodes in *mc* with the minimum community label of all nodes in *mc*;
5. end if
6. end for

Algorithm 2: crossover

Input: *P*
Output: *P*
1. randomly select a chromosome g_1 from *P*;
2. randomly select a node v_i which belongs to more than one maximal k-cliques, find the node set v_s in the same community as node v_i in g_1;
3. randomly select a chromosome g_2 from *P*;
4. update v_s's community labels in g_2 to the community label of node v_i;

Algorithm 3: mutation

Input: *P*
Output: *P*
1. randomly select a chromosome g_1 from *P*;
2. randomly select a node v_i which do not belong to any maximal k-cliques, obtain the community labels of node set v_s linking to v_i;
3. update v_i's labels to any of the community label of node set v_s;

Algorithm 4:

1. *P*= Initiation(maximum k-cliques) /*population initialization, the number is μ*/
2. for *i*=1:*NG* /**NG* is the number of iterations*/
3 *Pnew*= $\varnothing$
4. for *j*=1:λ
5. g=choose two individuals from *P*; /*crossover*/
6. g'=choose one individual from *P*; /*mutation*/
7. *Pnew*=*Pnew* $\cup$ g $\cup$ g';
8. end for
9. *P'*=*P* $\cup$ *Pnew*;
10. the greatest fitness μ are selected from *P'* to *P* , the next-generation is *P*; /*μ+λ*/
11. end for
12. *bestv* = max(*P*); /**bestv* is the maximal fitness individual*/
13.*cluster*=fast newman(*bestv*); /*optimization operation, *cluster* is the community with maximum Q value */

Figure 2 is a flow chart of the GAC algorithm, it reveals the whole process of the GAC algorithm and outlines Algorithm 1 to Algorithm 4. This gives an intuitive idea of our algorithm.
The GAC algorithm starts from population initialization, then calculate Q value, selection operation, crossover operation, mutation operation, finally, if it is greater than the number of iterations, the optimization operation is performed, otherwise the iteration continues.

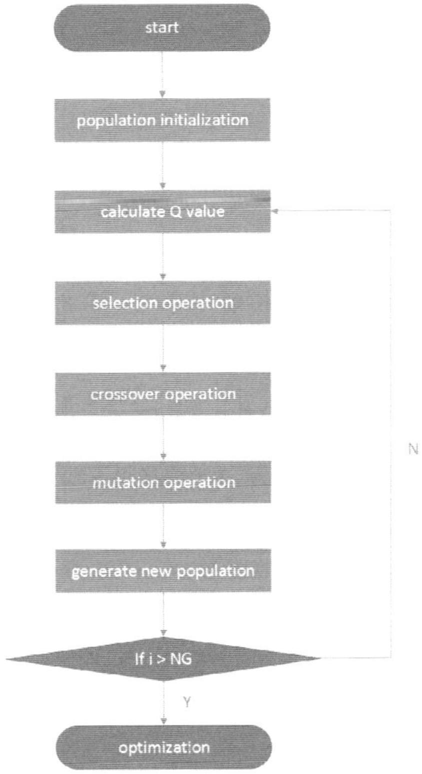

Figure 2. A flow chart of the GAC algorithm

2.7. Algorithm complexity analysis

The recursive algorithm is adopted to solve maximum k-cliques. Let t is the number of maximum k-cliques in a network, the network has n nodes and m edges, and the average degree of all nodes in the network is d. The time complexity of algorithm is $O(n*d)$.

The time complexity of population initialization is $O(\alpha*n*\mu)$, n is the number of nodes. The number of operations in the genetic algorithm is λ and the number of iterations is NG, the time complexity of the whole iterative process is $O(NG*\lambda)$. If there are a total of c communities after decoding, the time complexity of the optimization operation scales as $O(c^2)$.

3. Experiment and analysis

3.1. GN benchmark network

The GN artificial network was proposed by Girven and Newman. According to the artificial data sets generated by computer program, there are 128 nodes in the network; each community has 32 nodes, and there are 4 communities in total. Each node in the network is connected to z_{in} nodes in the same community and to z_{out} nodes of other communities. The sum of z_{in} and z_{out} is 16. The larger z_{in}, the more obvious the community structures. Otherwise, there will be more obstacles in the community structures.

Figure 3 shows a comparison of GAC with IGALM [2], Ref. [7] and MAGA-Net [4] algorithms; the x-axis represents z_{out}, the y-axis represents Normalized Mutual Information (NMI), and the experimental data are the average of 20 runs of the algorithm. The experimental results show that good experimental results can be achieved. Figure 3 shows that the algorithm has good clustering accuracy

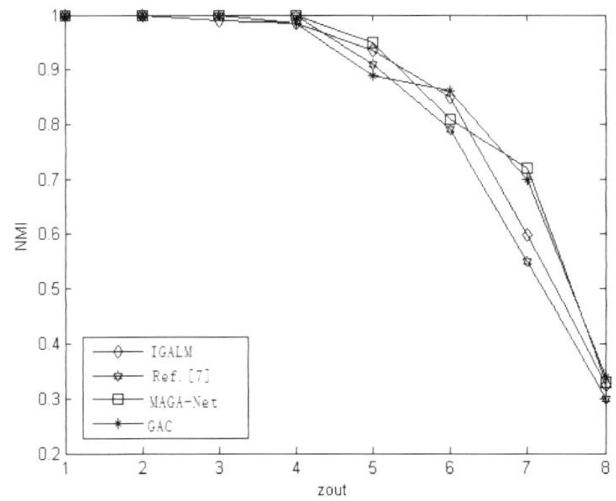

Figure 3. Comparison of clustering results of GN networks

Figure 3 shows that when z_{out} is less than or equal to 0.4, the algorithm can correctly detect the real community structures in the network. When z_{out} is greater than 0.4, the community structures of the network becomes an obstacle, and the GAC algorithm still has a higher community detection accuracy than those of other algorithms, which shows that the algorithm is feasible and can well detect community structures.

3.2. LFR benchmark network

The LFR benchmark network model [9] was proposed by Lancichinetti et al. It can simulate the statistical characteristics of real complex networks and has been widely used in testing network community detection algorithms. The parameter settings of the benchmark network are: LFR(N, k, $k_{\max}$, c_{min}, c_{max},t_1,t_2,mu,on,om). Table 4 shows the meanings of the parameters.

Table 4. The meaning of the parameters

parameters	meaining
N	the number of nodes
k	the average degree of the nodes
$k_{\max}$	nodes' maximum degree
c_{min}	nodes' minimum cluster size
c_{max}	nodes' maximum cluster size
t_1	the exponent of the nodes degree distribution
t_2	the exponent of the cluster distribution
mu	the mixed parameter
om	the number of overlapping nodes' communities
on	the number of multiple communities' nodes

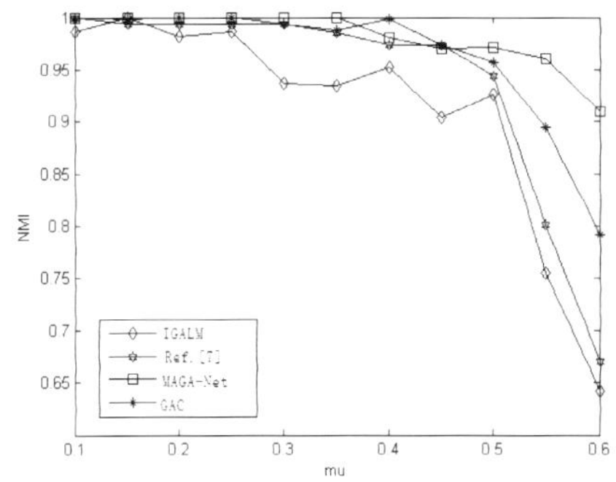

Figure 4. Comparison of clustering results of LFR benchmark networks

In the experiment, we used the network parameters LFR (n = 1000, k = 15 $kmax$ = 50, $cmin$ = 20, $cmax$ = 50, t_1 = 2, t_2 = 1, mu = 0.1–0.6).

Figure 4 shows the experimental results of the comparison of GAC and IGALM, Ref. [7] and MAGA-Net algorithms. The x-axis represents mu, the y-axis represents NMI, and the experimental data are the average of 20 runs of the algorithm. It can be seen that the algorithm has good clustering accuracy.

It can be seen from Fig. 4 that when mu is less than or equal to 0.5, the algorithm can correctly detect the real community structures in the network, and when mu is greater than 0.5, the community structures of the network become blurred, but the GAC algorithm can still correctly divide 95.69% of the nodes.

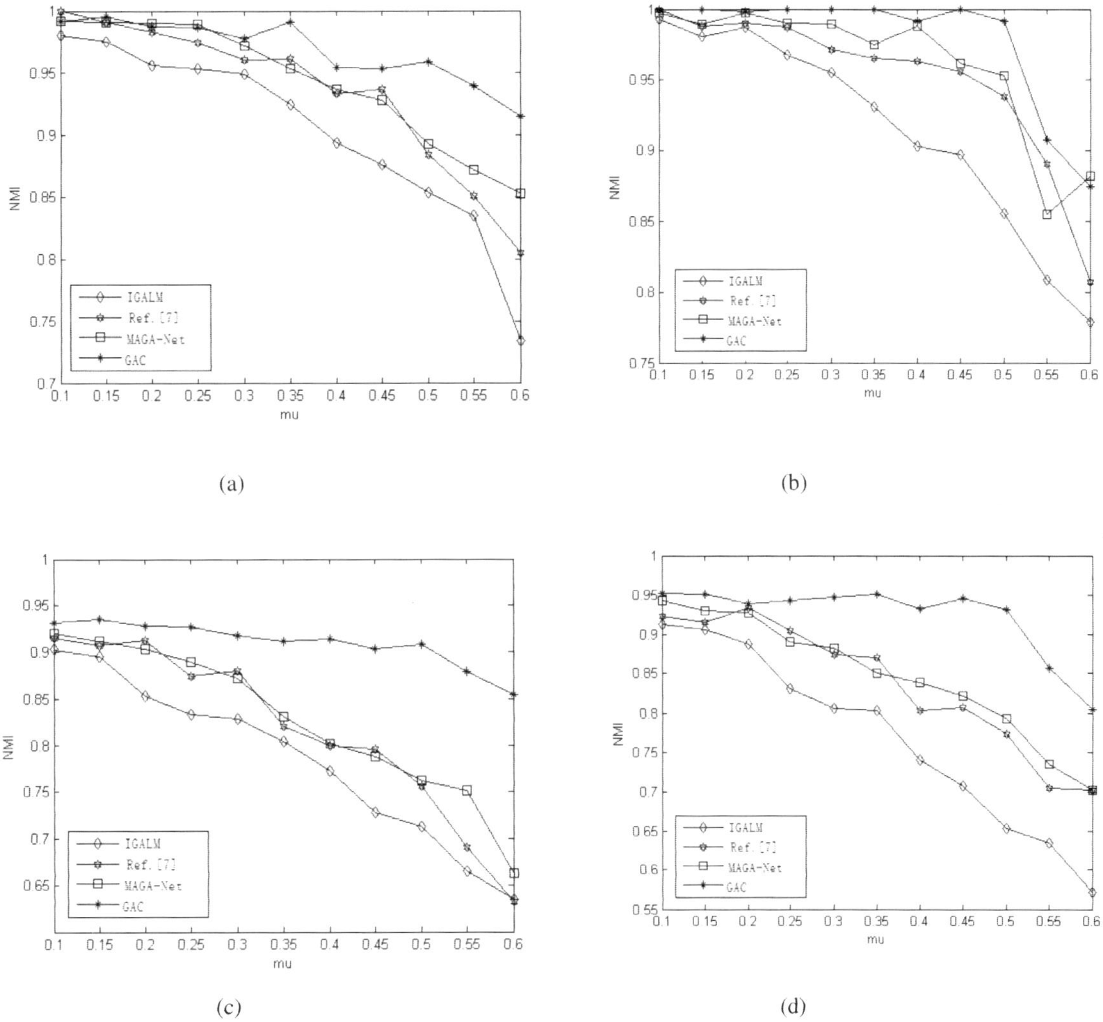

Figure 5. Comparison of NMI of LFR benchmark networks

Figure 5 (a) and (b) are the changes of NMI of four overlapping community detection algorithms with the increase of mixed parameter mu in the LFR (N=1000/5000, k = 20, k_{max} = 50, t_1 = 2, t_2 =1, on= 0, om = 1, c_{min}=10, c_{max}=50). The network size N and the distribution interval of community [c_{min}, c_{max}] also affect the community discovery algorithm results, Figure 5 (c) and (d) are the experimental results of parameters LFR (N=1000/5000, k = 40, k_{max} = 100, t_1 = 2, t_2 =1, on= 0, om = 1, $cmin$=20, c_{max}=100). It can be concluded from the experimental results that GAC achieves higher detection accuracy than the other three algorithms with different network sizes and distribution intervals.

3.3. *Running time*

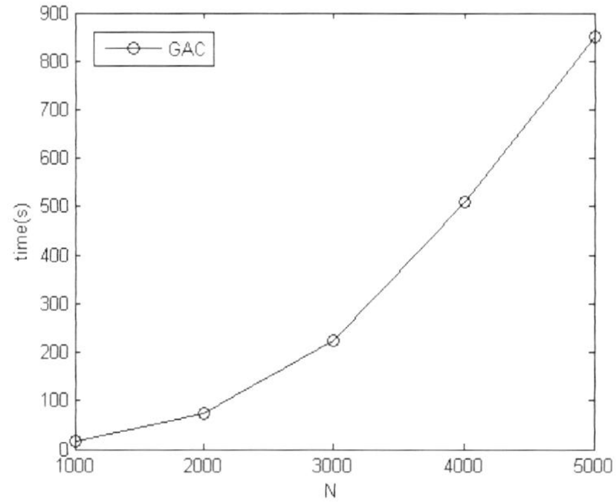

Figure 6. GAC algorithm runtime

This test uses the network parameter LFR (n=1000-5000, k=15, k_{max}=50, c_{min}=20, c_{max}=50, t_1=2, t_2=1, mu=0.1 on= 0, om = 1). Figure 6 shows the algorithm runtime as the number of nodes increases, and the running time of the GAC algorithm is within an acceptable range.

3.4. *Parameter analysis*

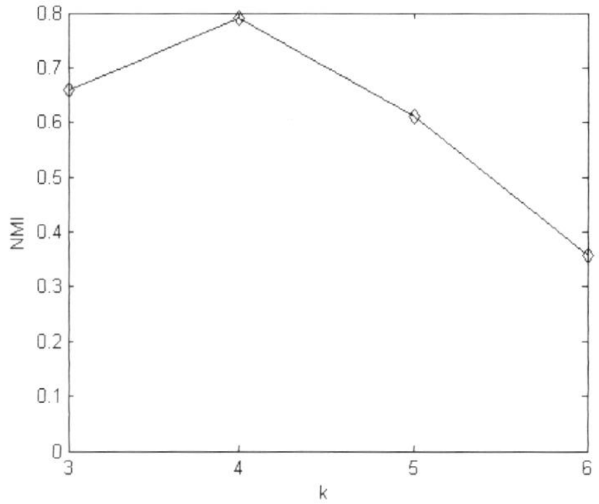

Figure 7. Comparison of clustering results of LFR network with different k values

The parameters of the algorithm are α , the population size μ, λ, the number of iterations NG, and the k in k-clique. The first four parameters are easy to set. In the text, α is 0.2, NG is taken as 20, μ and λ are 100. For the parameter k, we use the networks LFR (n=1000, k=15, k_{max}=50, c_{min} =20, c_{max}=50, t_1=2, t_2=1, mu=0.6, on= 0, om = 1). Figure 7 shows that when the value of k is 3 or 4, you can get better results.

3.5. *Real world network*

We test the algorithm on the real-world networks. The data used in the experiment are as follows: the Karate club network provided by Newman [10], Dolphin social network [11], American College football, Word adjacencies, Books about US politics, Jazz musicians network, and Email network of human interactions. Figures 8 and 9 show the clustering results for the real networks Karate club and Dolphins, respectively, with Fig. 10 listing the different shapes and representing different community divisions. Table 5 shows seven real network datasets.

Table 5. Real networks used in the experiment

NO.	networks	node	edge	communities
1	karate	34	78	2
2	Dolphins	62	159	2
3	Football	115	613	12
4	adjnoun	112	425	2
5	Political books	105	441	3
6	jazz	198	5484	4
7	email	1133	5451	11

Table 6. Comparison of Q value of real world networks

Q-value	MAGA-Net	Ref.[7]	IGALM	GAC
Zachary's club	0.4194	0.4183	0.4179	0.4174
Dolphins	0.5286	0.5264	0.5163	0.5268
Football	0.6046	0.6046	0.5986	0.6046
adjnoun	0.2940	0.2851	0.2851	0.2945
Political books	0.5273	0.5264	0.5231	0.5268
jazz	0.4448	0.4431	0.4438	0.4433
email	0.5115	0.5246	0.5168	0.5210

Table 6 shows the Q function values for different real-world networks. With the Karate club, the Dolphin social network, and the Jazz musician network, the GAC algorithm is higher-performing than the Ref [7]. With American College football and Word adjacencies, the GAC algorithm has the highest Q value. Through analysis and comparison, the GAC algorithm can find small communities that exist in the network and can effectively divide the network.

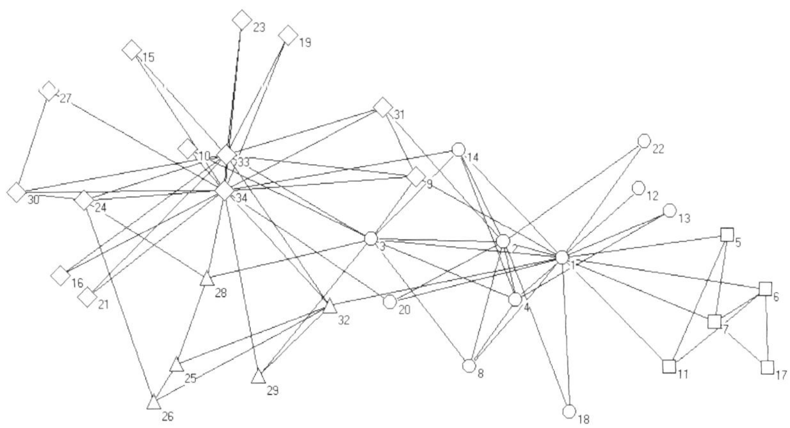

Figure 8. Communities structure of Karate obtained by the GAC

Figure 8 shows the network partition obtained by the GAC algorithm applied to the Karate club network. The algorithm divides the network into four communities. The circular nodes and square nodes is one part, and the triangle nodes and diamond nodes is the other parts. The combination of the circular nodes and square nodes, and the combination of the triangle nodes and diamond nodes are the correct division of this network.

The phenomenon suggests that the GAC algorithm can uncover smaller, more compact communities in communities.

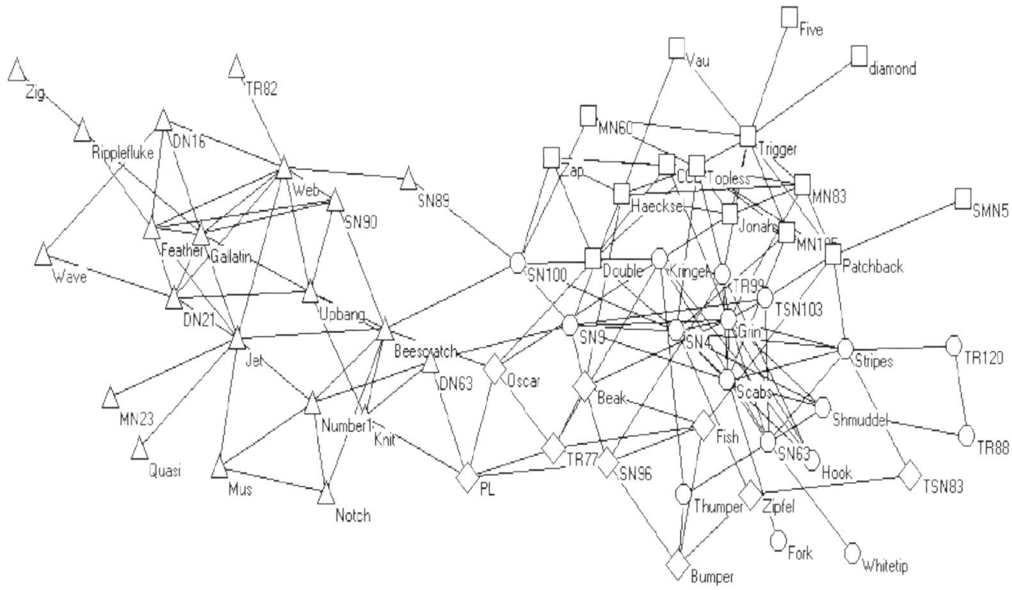

Figure 9. Communities structure of Dolphins obtained by the GAC

Figure 9 shows the network partition obtained by applying the GAC algorithm to the Dolphins network. The algorithm divides the network into four communities. The nodes represented by the triangles are consistent with the correct division of the network, and the combination of the rest corresponds to the proper partition of the network. The results of the partitioning also show that the algorithm can find a tightly connected community structures.

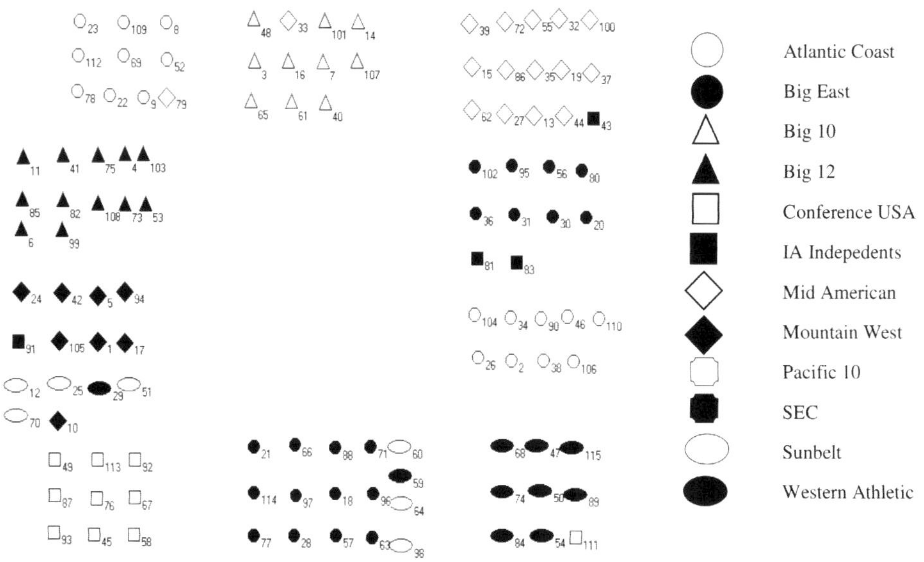

Figure 10. Communities structure of football obtained by the GAC

Figure 10 shows the network partition obtained by applying the GAC algorithm to the football network. The algorithm divides the network into 10 communities. Different communities are represented by different node markers, and the GAC algorithm correctly divides 87% of the community structures.

4. Conclusions

This paper proposes the GAC algorithm, a k-clique complex network community algorithm based on the genetic algorithm which uses maximum k-cliques for genetic evolution selection, crossover, and mutation operations. It does not require prior knowledge and reduces the search space for community partition. The algorithm improves the search efficiency and is tested on benchmark networks and real-world networks. The experimental structure shows that the algorithm can effectively divide the community. This algorithm can also be applied broadly to clustering problems in many others domains.

Acknowledgments

The authors would like to acknowledge the funding support of the National Key R&D Program of China under Grant 2016YFB1200100.

References

[1] M TASGIN, A HERDAGDELEN, H BINGOL. Community Detection in Complex Networks Using Genetic Algorithms. eprint, arXiv:0711.0491,2007.
[2] X Yang, R Li. An Improved Genetic Algorithm Based on Local Modularity for Community Detection in Complex Network, International Conference on Information Science & Cloud Computing Companion. 2014.
[3] A I HAFEZ, E T AlSHAMMARI, A E HASSANIEN, et al. Genetic Algorithms for Multi-Objective Community Detection in Complex Networks, In Social Networks: A Framework of Computational Intelligence:Springer,2014:145-171.
[4] Z LI, J LIU. A Multi-agent Genetic Algorithm for Community Detection in Complex Networks. Physica A Statistical Mechanics & Its Applications, vol. 449, no. 5, pp. 336-347, 2016.
[5] X MENG, L DONG, Y LI, et al. A Genetic Algorithm Using K-path Initialization for Community Detection in Complex networks. Cluster Computing, vol. 20, no. 1, pp. 1-10, 2016.
[6] M GUERRERO, F G MONTOYA, et al. Adaptive Community Detection in Complex Networks Using Genetic Algorithms. Neurocomputing, vol. 226, no. Suppl. C, pp. 101-113, 2017.
[7] H LIU, F YANG, D LIU. Genetic Algorithm Optimizing Modularity for Community Detection in Complex Networks.// Proceedings of the 35th Chinese Control Conference. July 27-29, 2016, Chengdu, China.
[8] M E J NEWMAN. Fast Algorithm for Detecting Community Structure in Networks. Physical Review E Statistical Nonlinear & Soft Matter Physics, vol. 69, no. 6, pp. 066133, 2004.
[9] A. Lancichinetti, S. Fortunato, János Kertész, Detecting the overlapping and hierarchical community structure in complex networks, New J. Phys. vol. 11, no. 3, 033015, 2009.
[10] W W Zachary. An information flow model for conflict and fission in small groups. Journal of Anthropological Research, vol. 33, no. 4, pp. 452-473, 1997.
[11] D Lusseau, K Schneider, O J Boisseau, P Haase, E Slooten, S M Dawson. The bottlenose dolphin community of doubtful sound features a large proportion of long-lasting associations. Behavioral Ecology and Sociobiology, vol. 54, no. 4, pp. 396-405, 2003.

IOP Conf. Series: Materials Science and Engineering **853** (2020) 012049 doi:10.1088/1757-899X/853/1/012049

R-MUCH: A Clustering Routing Algorithm Using Fuzzy Logic for WSNs

Alia Ghaddar[1], Mohammad Houssein Ghosn[1,2], Nathalie Mitton[2]

[1] International University of Beirut, BIU/LIU, Lebanon
[2] Centre de recherche Inria, Lille - Nord Europe, France

mohammad-hussein.ghosn@inria.fr

Abstract. An adequate usage of energy on nodes with restricted energy source is one of the major challenges in WSNs. Since data transmission is the main task that shortens node's lifetime, it is very necessary to balance the transmission of data among the network paths. Cluster-based architecture in WSNs is one of the keys to improve energy efficiency and extend network lifetime. It reduces the number of messages transmitted towards the sink or Base Station. This is due to restricting the communication with the sink to few nodes, called Cluster Head nodes (CHs). One of the major concerns is how to choose cluster heads and route data through energy-efficient paths towards destination. In this paper, we propose R-MUCH a clustering routing algorithm. It is a multi-hop version of MUCH algorithm (Multi-Criteria Cluster Head Delegation Based on Fuzzy Logic). CHs send data in a multi-hop fashion to the sink by choosing the path that has the lowest cost in terms of energy consumption. R-MUCH chooses for each CH its next-hop. It uses fuzzy logic and relies on three factors: the distance, the node's remaining energy and the number of times the node has served as next hop. Simulation shows that network lifetime is increased over existing approaches in the state of the art.

1. Introduction

Wireless Sensor Network (WSN) technology is described as very promising for the future [1][2]. It becomes an active research area, especially with the great progress of miniaturization technology and wireless autonomous systems. A WSN is composed of spatially distributed tiny nodes, which sense the environment and work cooperatively to deliver information in a multi-hop fashion to a sink or Base Station (denoted BS). Nodes are equipped with limited battery ($\leq$0.5 Ah, 1.2 V) and have limited transmission range, processing and storage capabilities [3]. In many application domains, battery cannot be recharged (under-sea surveillance, tracking, disaster scenarios, etc.). Consequently, the power consumption should be reduced and carefully managed in order to keep the network alive for the longest period of time. Different components in a wireless sensor node contribute in energy consumption: the power supply, the sensing unit, the processing and transmission units. Many researches showed that data communication is the most energy consuming part, while processing consumes much less [4]. Clustering architectures [5] are widely adopted to tackle challenges related to data aggregation and routing. It groups nodes into clusters, each with cluster head. A CH is in-charge of different tasks, mainly receiving data from other nodes in the same cluster, aggregating and forwarding it [6][7]. The other nodes are Cluster Members (CM) that communicate with the sink (or

base station) through CHs. This structure is an effective way to reduce communication distance as well as the amount of data transmitted or replicated in the network [8]. On the other hand, energy depletion is faster on CHs than other nodes, since they are loaded with different duties in processing, aggregation and communication. Consequently, the selection of CHs is a critical issue in designing clustering routing protocols. In this work, we propose a distributed clustering routing algorithm that uses both single hop communication (between CH and CM), and multi-hop communication (between CHs to route data to the sink). We use fuzzy logic for: 1 Giving CMs weight and elect the convenient one for the CH mission, 2- path Cost calculation for the routing part and choosing the energy-efficient next-hop node.
The rest of the paper is organized as follows: Section 2 presents preliminaries on fuzzy logic and clustering. Section 3 discusses related works. Section 4 presents our contribution. Section 5 shows the simulation results. Section 6 concludes this paper.

2. Preliminaries

2.1. Basics of clustering
The main goal of clustering is to reduce the overall energy consumption [7]. Since CMs communicate with CHs over a shorter distance, their energy dissipation decreases. There are two types of communications: One between CMs and their CHs (called Intra-cluster) and the other is between the CH and the sink (called Inter-cluster). How to choose CHs is a great challenge. A bad selection of CHs may cause a fast energy depletion and declines the network performance. A clustering algorithm consists of two main phases: (a) The setup phase that organizes the clusters and selects the CHs; (b) the steady-state phase where member nodes in each cluster start sensing data and communicating with their CHs, which perform data collection and fusion and send the aggregated data to a sink. This process of setup and communication is repeated in each round. Here the term 'round' refers to the time elapsed before a network re-clustering event occurs. During cluster formation, the decision of a sensor node to become CH and join an existing cluster is typically based on different metrics. In probabilistic approaches, a random CH election is performed (as in Low Energy Adaptive Clustering Hierarchy algorithm-LEACH [9]). Other approaches use different criteria (such as the remaining energy, the distance from the sink, the centrality, the density in the network, etc.)[10]. These approaches are called attribute-based. We present an attribute-based algorithm that uses three inputs to a fuzzy system. The inputs are the node's energy cost, its distance and the number of times it has served as CH in the previous rounds.

2.2. Fuzzy Logic System (FLS)
2.2.1. Overview. Fuzzy logic (FL) is a method of reasoning that resembles human reasoning by making decisions based on rules. It is also defined as the multi-valued logic. It differs from the boolean logic (single-valued logic) by the fact that every event can take two or more values with a degree of membership (called degree of truth) to each one of these. Let us consider the following question: "Are you sure of what you've heard?" In boolean logic, the answer would be either yes or no (0 or 1). However, in fuzzy logic the answer would be: "I am sure for a degree of 60 % and unsure for a degree of 40 %". Knowledge in FL is interpreted as a collection of fuzzy rules on a collection of non-numeric values. These rules operate using a series of if-then statements [11]. In order to exemplify, the temperature is qualified using non-numeric terms such as "hot", "warm", etc. These terms are called fuzzy sets. However, the real value for the temperature, called the crisp value, is a numeric value (example 27°C). Each fuzzy set covers a range of numerical values. Values in [15°C, - 25°C[can be thought of as warm, [25°C, - 30°C[can be thought of as hot and so on, depending on the defined membership functions. In Figure 1, for example, 13°C can be considered 40% "COOL" and 60% "NORMAL".
2.2.2. FLS structure. A FLS consists of three main parts (Figure 2) - A fuzzifier (using membership functions), rules evaluation (rules & inference engine), and defuzzifier. Assume an air conditioner

system is controlled by a FLS. The system adjusts the temperature of the room according to the current room temperature. The Fuzzifier block converts the given crisp input (ex: 13°C) into the appropriate fuzzy linguistic variable (ex: cold or warm). The inference is made based on a set of rules. An example of rule can be: IF (temperature is cold OR too-cold) THEN command is heat. Finally, the defuzzifier block converts the fuzzy output into the crisp output using a suitable defuzzification method.

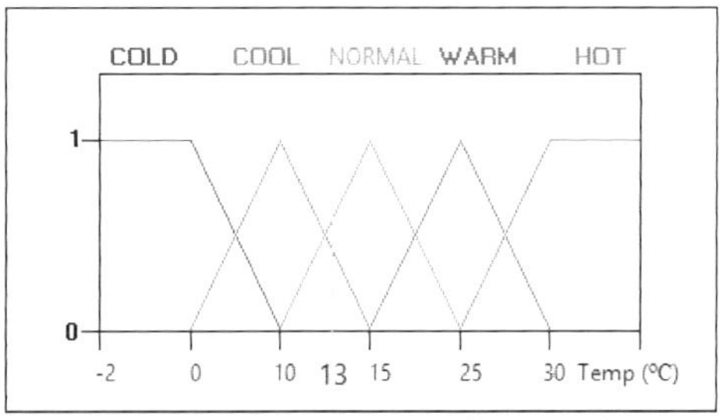

Figure 1. Membership Function

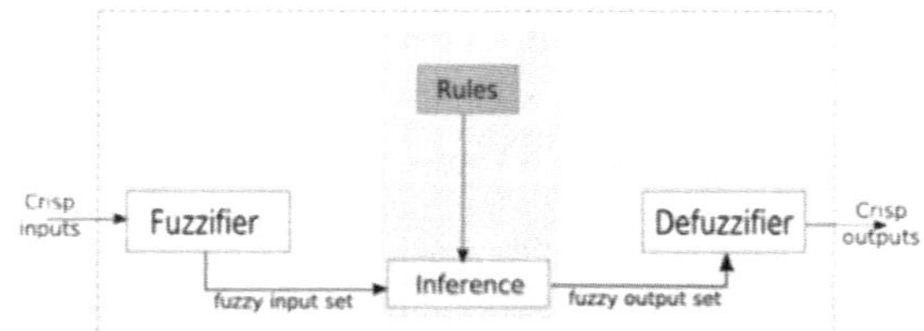

Figure 2. Fuzzy Logic System

2.2.3. Fuzzy Logic in our work. We use fuzzy logic to determine the cost of routing from a specific CH toward the sink. We aim to decide on the convenient path for routing (i.e the path that has the lowest overall cost in terms of energy). We use Gaussian membership function (see Table 1) for fuzzification as it is common because of its smoothness and non-zero values at all point [12]. We use five fuzzy sets (denoted A): VL (Very Low), L (Low), M (Medium), H (High) and VH (Very High). For the defuzzication function, we used Center of Gravity (see Table 2). It is one of the most popular defuzzification technique used as an assessment model [13]. This method avoids the discontinuities that could appear using other defuzzication functions such as the Mean of Maxima function [14].

Table 1. Gaussian Membership Function caption.

Gaussian Function	$\mu_{A^k} = \exp(-\frac{(c_k - x)^2}{2\sigma_k^2})$
Variable	**Meaning**
c_k and σ_k	center of the kth fuzzy set A[k]
σ_k	width of the kth fuzzy set A[k]

Table 2. Center of gravity function.

Center of gravity	$U = \frac{\int_{min}^{max} \upsilon\, \mu(\upsilon)\, d\upsilon}{\int_{min}^{max} \mu(\upsilon)\, d\upsilon}$
Variable	**Meaning**
U	result of defuzzification
υ	output variable
μ	membership function after accumulation
min	lower limit for defuzzification
max	upper limit for defuzzification

3. Related works

This section presents a list of clustering routing algorithms that are either single-hop or multi-hop. LEACH [9] and MUCH [15] are considered single-hop. Elected CHs send the aggregated data to the sink through direct communication hop. One of the main concerns in single-hop communication is that the distance between CHs and the sink could be large. Consequently, CHs located far from the sink are subjected to death earlier than the nodes located near the sink [16][17]. To overcome this, data is transmitted to the destination through large number of smaller hops. CHs send their data to intermediate CHs along the path to the sink. This multi-hop routing is considered to be more efficient [18]. Few nodes will be transmitting directly their data to the sink, while most of CHs transmit to their next-hop selected by the routing protocol. Since the transmission energy is proportional to distance squared, multi-hop routing consumes less energy than single-hop communication. LEACH [9] is based on the idea of randomized rotation of cluster-heads. It distributes the energy load equally among all the sensor nodes. In the setup phase, every node generates a random number between 0 and 1. This number is compared to a threshold T(n). If the number is less than T(n), that node will be selected as CH for that round. The threshold is as follows:

$$T(n) = \begin{cases} \frac{p}{1-p*(r \bmod \frac{1}{p})} & n \in G \\ 0 & Otherwise \end{cases} \tag{1}$$

where p is the ratio of cluster-head nodes in the total number of nodes. r is the current round number. G is the set of nodes that have not been selected as CH nodes in the former $\frac{1}{p}$ rounds. Sensors elect themselves to be CHs at a given time with a certain probability, and broadcast their status to other sensor nodes. Each node chooses the CH with the highest signal. LEACH carries out energy optimization and also reduces the amount of data transmitted. However, it has some drawbacks. It may happens that the same node is selected as CH again in other rounds. The CHs are selected randomly, so if the node with low energy is elected, it will be heavily loaded and more energy will be consumed. This results in early death of these nodes and consequently reduce the network lifetime. In FLCMN [19] (Fuzzy Logic-Based Clustering Algorithm for Multi-hop Wireless Sensor Networks), every node has its own satisfaction factor. It is calculated using fuzzy inference system that takes three inputs into consideration: node residual energy, number of neighbor nodes and the average residual energy of neighbor nodes. In every cluster the node that has the lowest factor will be CH. In this algorithm, a multi-hop transmission scheme based on Fibonacci is used. The goal is to avoid the hot spot problem between clusters and to distribute data load among the neighboring CHs. In EMHR [20], the node that has the highest remaining energy is elected as CH. Then every cluster head (say CH_i) selects its Next-Hop (CH_j) based on a weighting function. This function takes into consideration: the distance between the two neighboring CHs (CH_i and CH_j), the distance between CH_j and the sink, and the remaining energy of CH_i.

EMHR-FL [21] (Energy-Efficient Multi-hop Hierarchical Routing Protocol using Fuzzy Logic) is an extension of EMHR[20]. Authors consider the parameters proposed in EMHR not sufficient for energy efficient routing. They highlighted the problem of load occurring on the next-hop CH. For this purpose, they introduced a new weighting parameter which is the node's density. It reflects the number of neighboring CHs that may use this node as Next-Hop and thus consume more of its energy. Unlike EMHR, the weighting parameters are inputs to a fuzzy system instead of a static function and the output is the cost of adopting CH_j as a next-hop for CH_i. In MUCH [15], every CM computes its weight in a cluster and sends it to its CH. The latter chooses the node that has the lowest weight to be its delegate for the next round. The weight of a CM_i is computed based on three parameters: its distance to sink and to other CMs, its energy cost, and the number of times it has been chosen as CH in the previous rounds. MUCH decreases the energy dissipation during intra-cluster communication. The communication between CHs and the sink is single-hop. These nodes are subjected to early death when transmitting to far sink [18]. The integration of multi-Hop ("short" paths) communication in MUCH could be a solution to increase the nodes' lifetime. Many metrics were suggested in the literature for the selection of energy-efficient paths. In [19][21, 22], the selection is based on the least energy dissipation path. Authors in [17][23], choose the path that has the lowest number of hops to sink. Authors in [19–21] give each neighboring CH a cost and the one that has the lowest cost will be the next-hop. In this work, we use three parameters to determine for each CH its next-hop that decreases energy depletion during inter-cluster communication.

4. Our Contribution

The energy dissipated on a node is directly related to the transmission distance [18]. We suppose that every CH is connected to each of its neighbouring CHs via single-hop link and has a direct path to the sink. The goal is to avoid routing to next-hop nodes that are low energy, located far from the sink or have heavy load to transmit. We assign a cost to every single path link between two CH nodes. It is the cost in energy for sending data via this link. The value ranges from zero to one. As the cost decreases, the priority for a CH to transmit via this link increases.

4.1. Setting the cost

We compute the cost of each single-hop (between CHs) using FL system (Figure 3). The system has three inputs:

- The remaining energy in the source CH denoted "ReSrc".
- The remaining energy on the candidate next-hop node denoted "ReDst".
- The distance between the two nodes composing the single-hop link, denoted Distance".
- The number of times a candidate CH has been next hop before, denoted " NextHopNb". If this number increases, the cost increases.

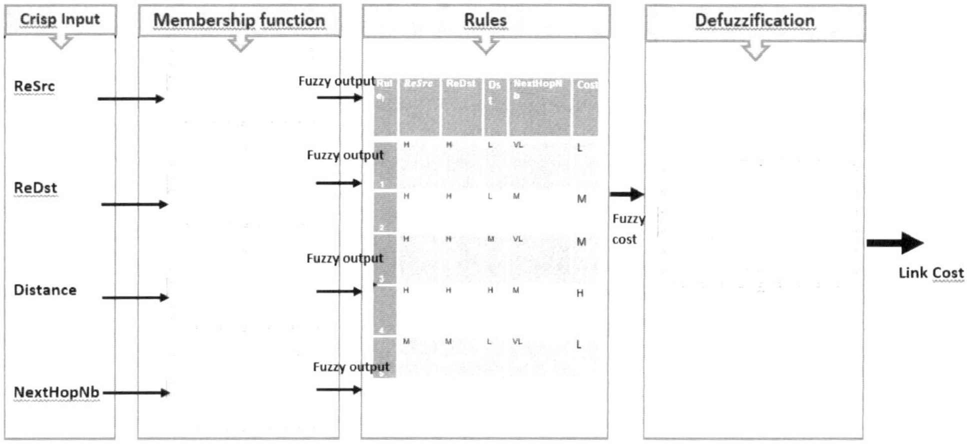

Figure 3. Fuzzy system

The output is the cost denoted LinkCost. A link (or hop) between two consecutive CHs has low cost when the remaining energy of the two nodes is high. It is also low when the distance between nodes is low. The relation between inputs and the cost is represented by if-then rules (Fuzzy Rules), illustrated in Table 3. It has three fuzzy values {L, M, H}. NextHopNb is expressed by five fuzzy values {VL, L, M, H, VH}.

ReSrc	ReDst	Distance	Next Hop Nb	Link Cost
H	H	L	VL	L
H	H	L	M	M
H	H	M	VL	M
H	H	H	M	H
M	M	L	VL	L

Table 3. Fuzzy Rules.

Table 5. Possible paths from CH_1 to sink.

Path	Total Cost
Direct (P_1)	0.4
via CH_2 (P_{12})	>0.75
via CH_7 (P_{71})	>0.7

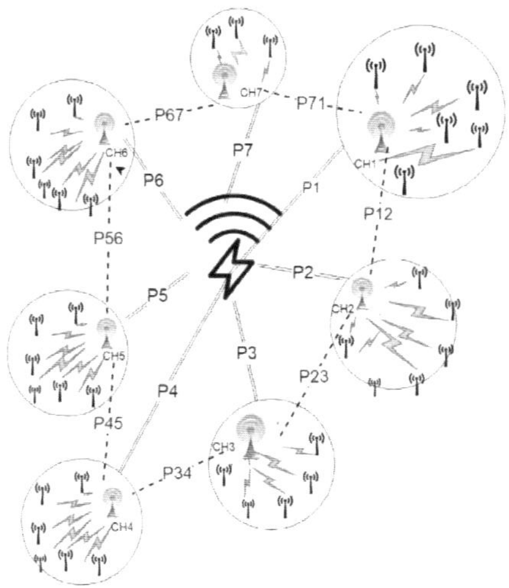

Figure 4. Sample Topology

Path	Link Cost	Path	Link Cost
P_1	0.4	**P_{12}**	0.75
P_2	0.3	**P_{23}**	0.50
P_3	0.15	**P_{34}**	0.25
P_4	0.05	**P_{45}**	0.45
P_5	0.8	**P_{56}**	0.55
P_6	0.2	**P_{67}**	0.60
P_7	0.8	**P_{71}**	0.70

Table 4. Link Costs

Table 6. Possible paths from CH_5 to sink.

Path	Total Cost
Direct (P_5)	0.8
via CH_4 (P_{45}) to P_4	0.45+0.05=0.5
via CH_6 (P_{56}) to P_6	0.55+0.20=0.75

4.2. Selection of the paths for CHs

R-MUCH runs after the selection of CHs. At each round, every CH sends its remaining energy to the sink (or BS) via the old path. The sink runs the fuzzy system to get the cost on each hop between CHs.

Then it finds for every CH the best path in terms of energy efficiency. It selects the path that has the lowest total cost. To exemplify, given the topology in Figure 4, let P_i be the direct hop from CH_i to the sink, and P_{ij} be the hop between two neighbouring CHs, where CH_i is the source and CH_j is the destination. Every path is composed of consecutive hops to the sink. Table 4 presents the link costs obtained after running the fuzzy system. The sink will have then the information about paths. Table 5 and Table 6 show the path costs from CH_1 and CH_5 respectively. The sink will choose the link P_1 because it has the lowest total cost and informs CH_1 that his next hop is the sink. On the other hand, for CH_5 the sink chooses the path $P_{45}P_4$ because it has the lowest total cost. The sink then informs CH_5 that the next hop is CH_4. The path selection for a specific CH at each round is described in Figure 5(a).

4.3. Description of R-MUCH

R-MUCH aims to reduce energy consumption during intra-/inter-cluster communication. In the setup phase, clusters are organized and CHs are selected using MUCH algorithm. Each CH selects the node that has the lowest weight among the cluster members, to be its delegate for the coming round. In the steady state phase, CHs adopt the path that has the lowest cost (as described in the previous section). The whole process continues until all nodes die (Figure 5(b)).

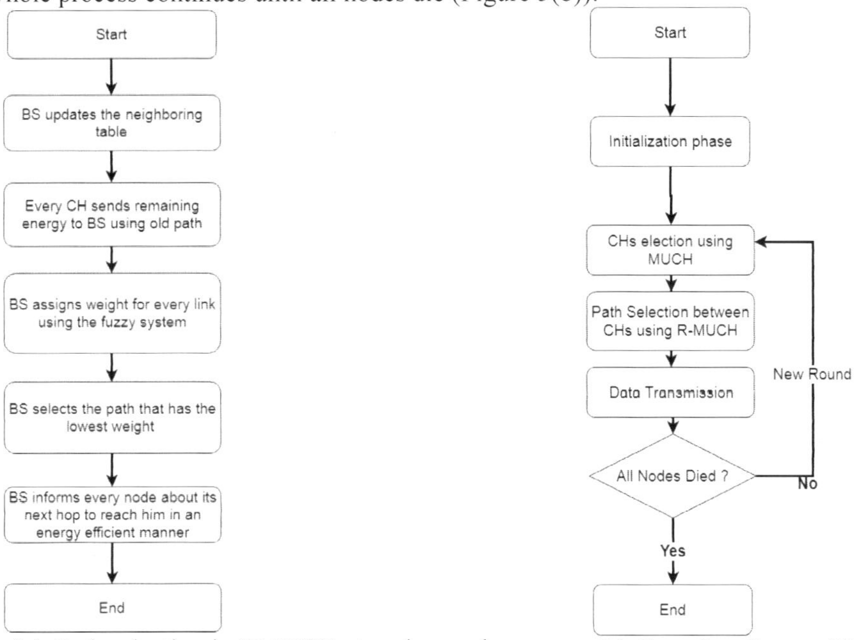

(a) Path selection in RMUCH at each round **(b)** Complete System Flow Chart

Figure 5. Complete System Flow Chart

5. Simulation

In our simulation, we have a rectangular area, which consists of one static sink and several sensor nodes. We made the following assumptions: nodes are homogeneous, location-aware and energy constrained. Each sensor node has a unique identifier (ID), and the same initial energy. Each CM node attempts to transmit a q bit message at its allocated time-division multiple-access (TDMA) schedule and the received messages from CMs can be aggregated in a single message by the CH. Note that the TDMA schedule [24] makes certain that there are no data collisions during the network activities. In such a way, the TDMA schedule saves energy by allowing nodes to sleep all the time except during the node's transmission time. As all the data collected by the sensor nodes are forwarded to a sink node, the placement of the sink has a great impact on the energy consumption and network lifetime [25–29]. Therefore, we took two scenarios.

In the first scenario, the sink is located at the center of the network. In the second scenario, the sink is located outside the network rectangular area. We used MATLAB to evaluate the performance of our algorithm. We compared our work to FLCMN[19] and EMHR-FL[21]. We refer to the following performance metrics: the nodes' remaining energy, the number of dead nodes and the network lifetime. Two definitions are used for the network lifetime: First Node Dies (FND) and Last Node Dies (LND). FND is the time when the first node dies in the network and LND is the time when the last node dies.

5.1. Scenario 1

The network consists of 250 nodes randomly distributed over an area of 100mx100m. Simulation parameters are presented in Table 7. Figure 6 shows the remaining energy on each node in the network at round 200. We notice that the energy saving is increased using our approach compared to other algorithms. Nodes' remaining energy in R-MUCH is higher than in FLCMN and EMHR-FL by 51.2% and 89.6% respectively. This is because in FLCMN and EMHR-FL there is no consideration of the distance from the CH to sink, and the distance to CMs. This causes the selection of far CHs possible, hence, they will be subject to die at early rounds, and CM nodes to consume too much energy.

Table 7. Simulation parameters

Parameters	Value	Parameters	Value
Network Size	100m*100m	Initial energy	0.5 J
Number of nodes	250	E_{elec}	50 nJ/bit
Location of sink	Center	ε_{mp}	0.0013 pJ/bit/m^4
Location of nodes	Random	ε_{fs}	10 pJ/bit/m^2
Aggregation energy E_{DA}	5 nJ/bit	Number of rounds	200

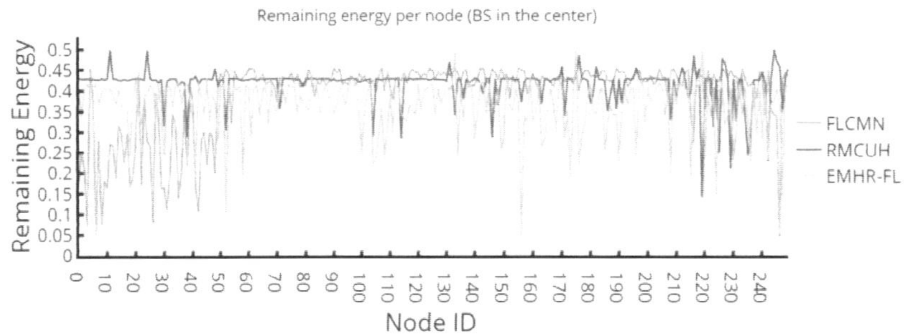

Figure 6. Remaining energy on each node at round 200

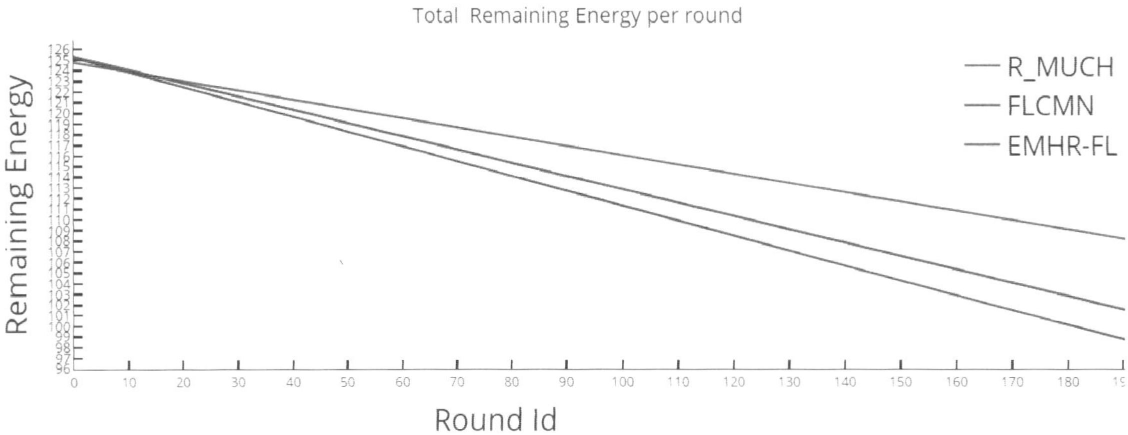

Figure 7. Total remaining energy per round

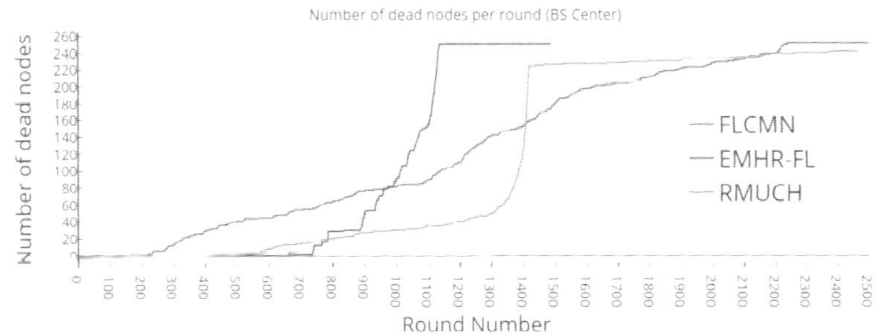

Figure 8. Number of dead nodes per round

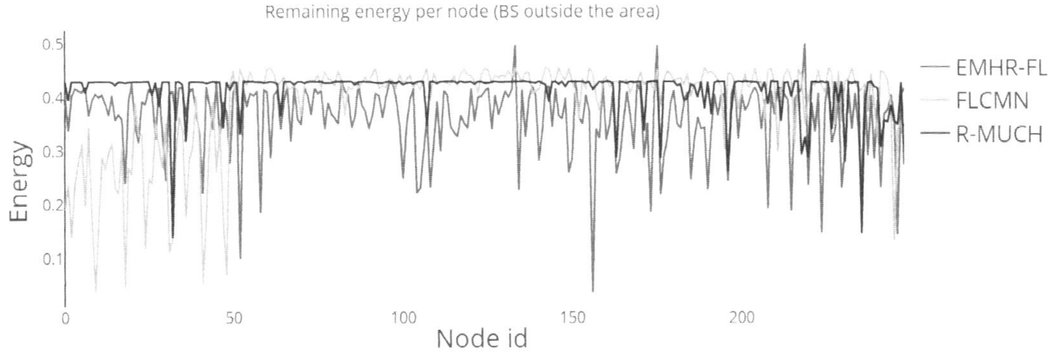

Figure 9. Remaining energy per node at round 200

In RMUCH we avoided the selection of these CHs. Figure 7 shows the total remaining energy in the network per round until round 200. We can notice that R-MUCH gives the higher remaining energy. Figure 8 illustrates the death of nodes. We evaluated the lifetime of the network by three metrics: FND, HND (half node dies) and LND (as shown in Figure 10). We can notice that the earliest death in nodes occurs in FLCMN at round 218. However, First node dies in R-MUCH occurs at round 609. The whole network goes down in EMHR-FL at round 1138, earlier to the other algorithms. RMUCH doubles the network lifetime compared to EMHR-FL. The LND in our work was earlier than FLCMN that is because of two reasons:

- In FLCMN, authors distribute the traffic between CHs and their neighbouring CHs according to Fibonacci sequence. However, they didn't take into consideration the lifetime of CMs while selecting CHs. There is no consideration of the distance between member nodes and the elected CH. This causes member nodes far from the CHs to die at early rounds. This is obvious in the FND in Figure 8. FND occurred at round 168 earlier than RMUCH.
- As result of the above point, during late rounds many of the cluster members far from their CHs in FLCMN were subjected to earlier death. This decreased the energy consumption load on the new selected CHs and made them last longer because of the absence of almost 80% of CM neighbours.

Late death of nodes isn't alone an indication of the network lifetime because the energy consumption could be unequally distributed between nodes. This unbalanced power consumption may cause network partition, blocking the transmissions from sensors to CH nodes [30]. To measure the energy balancing we used the standard deviation. It measures the energy variation between nodes according to the mean. The lower this value is, the closer the energy level to the mean (in energy balancing the lower the deviation is, the more is the equilibrium). Figure 11 shows the energy variation in FLCMN and R-MUCH. It indicates that R-MUCH maintains better energy balancing.

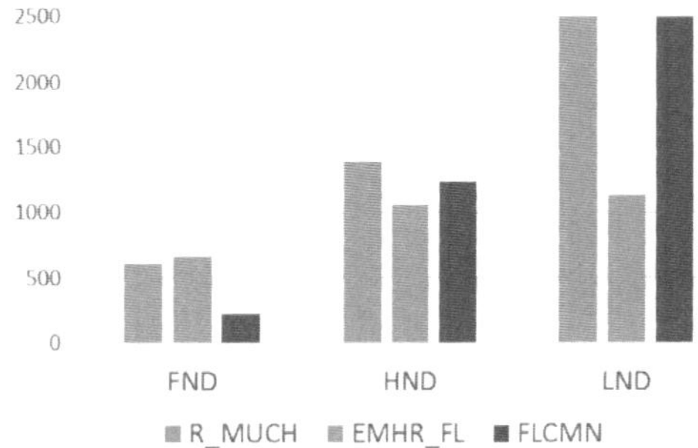

Figure 10. FND, HND and LND for each algorithm

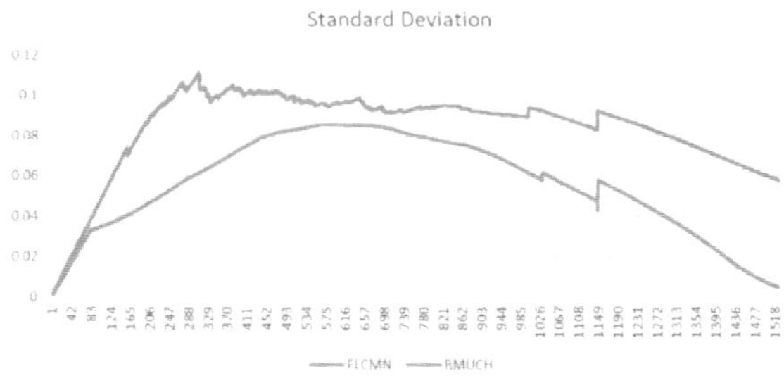

Figure 11. Standard Deviation Variation between RMUCH And FLCMN

5.2. Scenario 2

In this scenario, the sink is located outside the network area. Figure 9 shows the remaining energy at each node in round 200. It shows that R-MUCH maintains higher remaining energy compared to other algorithms. 48% of the nodes in RMUCH have higher energy level than the nodes in FLCMN. Moreover, 90% of the nodes have higher energy level than the nodes in EMHR-FL. The total remaining energy per round is presented in Figure 12. We can notice that R-MUCH shows the highest total remaining energy per round, when the sink is located outside the network area. This delays the death of nodes and increases the network lifetime.

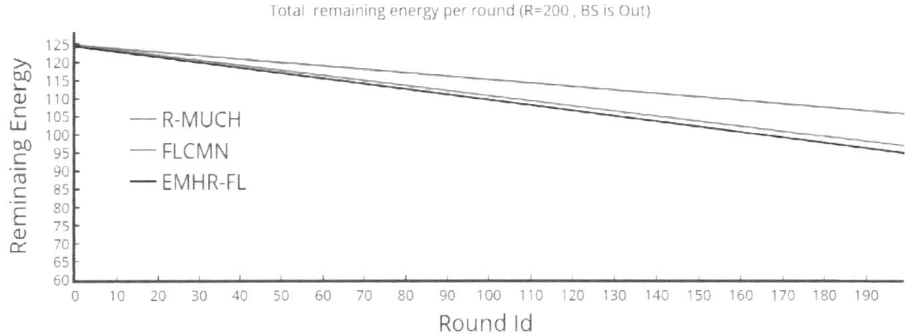

Figure 12. Total remaining energy per round

5.3. RMUCH versus MUCH

Our simulation results show that RMUCH outperforms MUCH for multihop communication scenario where the sink is located far from the sensor nodes (Figure 13(a)). 95% of the nodes in RMUCH have higher remaining energy. However, for single-hop scenarios where the sink resides close to the nodes, a slight difference in the number of dead nodes per round exists between the two algorithms (see Figure 13(b)). The reason behind this is the energy that RMUCH consumes for finding the convenient next-hop node on the path. The next-hop could be the sink itself or a neighbour CH. However in MUCH, CHs send data directly to the sink. This explains why FND in RMUCH occurs earlier than MUCH. On the other hand, the remaining energy per node in R-MUCH is higher than MUCH. 65 % of the total nodes in R-MUCH have higher remaining energy. Figure 14(a) shows the variation in terms of LND, HND and LND between RMUCH and MUCH. The network lifetime (LND-FND) shows an enhancement of 4% in RMUCH over MUCH algorithm. Figure 14(b) shows the variation in terms of LND, HND and LND when the sink is located outside the area.

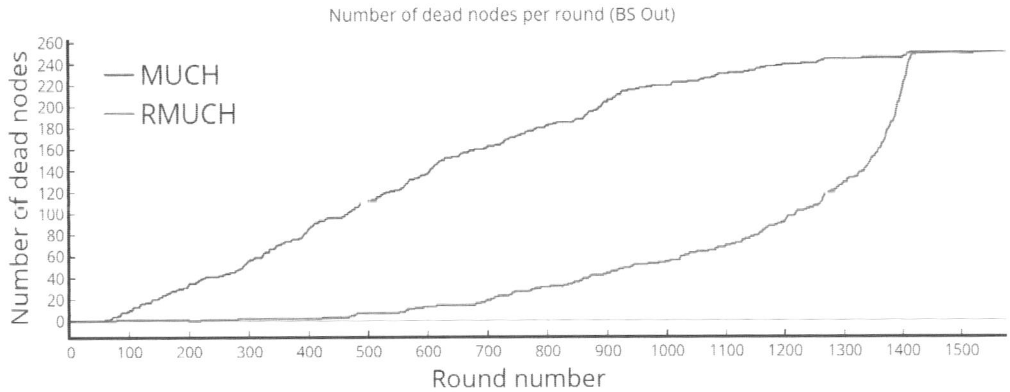

(a) Number of dead nodes per round (sink outside)

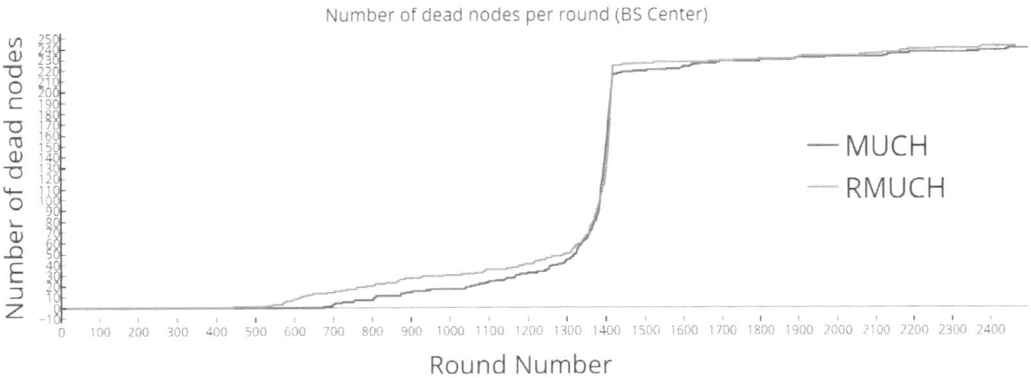

(b) Number of dead nodes per round (sink at the center)

Figure 13. MUCH versus RMUCH

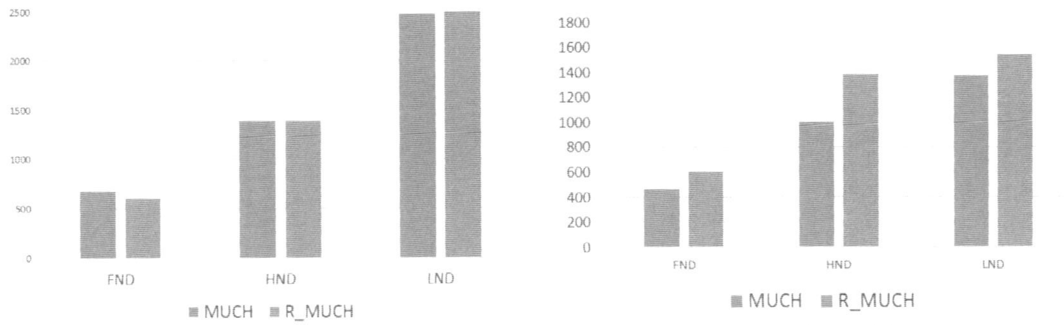

(a) Sink at the center (b) Sink outside

Figure 14. Number of dead nodes per round

RMUCH achieves 31% gain in FND over MUCH. 37% gain in HND and 12% gain in LND were achieved by RMUCH over MUCH. The reason behind this gain is due to the multihop mode in RMUCH that can ease the nodes far away from the cluster head by reducing the burden of long distance communication with the sink. Single-hop mode in MUCH needs to go over a longer distance to reach a far sink with more energy transmission.

6. Conclusion

In this paper, we proposed a new clustering routing algorithm (R-MUCH) for multi-hop communication scenarios. The algorithm uses fuzzy logic for the CHs election and routing. Our simulation results show that R-MUCH reduces the energy consumption in the network. It achieves a balanced energy usage and prolongs the network lifetime compared to the other presented algorithms. This achievement is due to the distribution of CH mission among the nodes and the selection of nexthop nodes on the least energy consuming path. In the future works, we aim to explore more complex fuzzy functions for routing in heterogeneous sensor network.

Acknowledgments
The authors would like to thank the editors and the anonymous reviewers, for their valuable comments which improved the clarity and quality of the paper.

References
[1] A. Ali, Y. Ming, S. Chakraborty, and S. Iram, "A comprehensive survey on real-time applications of wsn," Future Internet, vol. 9, no. 4, p. 77, nov 2017.
[2] D. M. SRJ Ramson, "Applications of wireless sensor networks-a survey," International Conference on Innovations in Electrical, Electronics, Instrumentation and Media Technology (ICEEIMT), pp. 25–329, November 2017.
[3] P. Sallis, Ed., Wireless Sensor Networks - Insights and Innovations. InTech, oct 2017.
[4] R. A. S. G. M.Sheik Dawood, P.Jayalakshmi, "A survey on energy efficient clustering protocols for wireless sensor network," International Journal of Computer Science and Mobile Computing, vol. 3, no. 3, pp. 1158–1163, 5 2014.
[5] A. Zeb, A. K. M. M. Islam, M. Zareei, I. A. Mamoon, N. Mansoor, S. Baharun, Y. Katayama, and S. Komaki, "Clustering analysis in wireless sensor networks: The ambit of performance metrics and schemes taxonomy," International Journal of Distributed Sensor Networks, vol. 12, no. 7, 2016.
[6] D. U. Maheswari, M. Meenalochani, and S. Sudha, "Influence of the cluster head position on the lifetime of wireless sensor network a case study," in 2nd International Conference on Next Generation Computing Technologies (NGCT), Oct 2016, pp. 378–381.
[7] Y. Singh and U. Chugh, "Clustering, information fusion and event detection in wireless sensor networks: A review," in 2015 2nd International Conference on Computing for Sustainable Global Development (INDIACom), March 2015, pp. 566–571.
[8] S. R. K. Anita Chavan, "Survey on efficient clustering with energy aware routing in wireless sensor networks," International Journal of Science and Research (IJSR), vol. 5, no. 2, pp. 2115–2118, February 2016.
[9] Reeta and A. Meenu, Mrs., "A study on leach protocols and its variants in wireless sensor networks," 2 2016.
[10] P. C. Srinivasa Rao and H. Banka, "Energy efficient clustering algorithms for wireless sensor networks: Novel chemical reaction optimization approach," Wirel. Netw., vol. 23, no. 2, pp. 433–452, February 2017. [Online]. Available: https://doi.org/10.1007/s11276-015-1156-0
[11] V. M. Mirjana Maksimovic, Vladimir Vujovic, "Fuzzy logic and wireless sensor networks - a survey," pp. 877–890, 2014.
[12] S. Singh, S. Chand, and B. Kumar, "Multilevel heterogeneous network model for wireless sensor networks," Telecommunication Systems, vol. 64, pp. 259–277, February 2017.
[13] F. Dernoncourt, "Introduction to fuzzy logic," January 2013.
[14] ——, "Fuzzy logic: introducing human reasoning within decision support systems," February 2011.
[15] M. H. Ghosn and A. Ghaddar, "Much: Multi-criteria cluster head delegation based on fuzzy logic," International Journal of Advanced Research in Computer and Communication Engineering (IJARCCE), vol. 6, no. 6, pp. 415–421, 2017.
[16] H. A. s. A. S. M. B. Ali Asghar Rahmani Hosseinabadi, Ali Asghar Rahmani, "Survey on clustering in heterogeneous and homogeneous wireless sensor networks," Journal of SuperComputing, 2017.

[17] R. A. Mekkaoui Kheireddine, "Analysis of hops length in wireless sensor networks," Wireless Sensor Network, 2014.
[18] G. Khanna and S. Chaturvedi, "A comprehensive survey on multi-hop wireless networks: Milestones, changing trends and concomitant challenges," in Wireless Personal Communications, vol. 101, April 2018.
[19] S. Zhang, Y. Tao, J. Dai, and L. Yang, "A multi-hop wsns clustering algorithm based on fuzzy logic," Chinese Journal of Sensors and Actuators, vol. 31, pp. 1085–1090, 07 2018.
[20] W. Huang, Y. Peng, J. Wen, and M. Yu, "Energy-efficient multi-hop hierarchical routing protocol for wireless sensor networks," in 2009 International Conference on Networks Security, Wireless Communications and Trusted Computing, vol. 2, April 2009, pp. 469–472.
[21] S. Amri, M. L. Kaddachi, and A. Trad, "Energy-efficient multi-hop hierarchical routing protocol using fuzzy logic (emhr-fl) for wireless sensor networks," in 2014 World Congress on Computer Applications and Information Systems (WCCAIS), January 2014, pp. 1–6.
[22] S. Chelbi, M. Abdouli, R. Bouaziz, and C. Duvallet, "Multi-hop energy efficient routing protocol based on data controlling for wireless sensor networks," proceedings of IEEE ACS International Conference on Computer Systems and Applications, AICCSA, pp. 1–6, 05 2013.
[23] W. L. Zuo Chen, Min He and K. Chen, "Trust-aware and low energy consumption security topology protocol of wireless sensor network," Journal of Sensors, 2015. [Online]. Available: https://doi.org/10.1155/2015/716468
[24] L. P. C. G. J. P. J. R. Agre, "Self-organizing distributed sensor networks." Unattended Ground Sensor Technologies and Applications, vol. 3713, p. 229–237, 1999.
[25] K. A. I. G.Sateesh, M.V.S.Sastry, "Placement of sink in wireless sensor networks," International Journal of Computer Science and Information Technology Research, vol. 3, no. 2, pp. 340–346, January 2015.
[26] Y. V. A. S. Shaik Beejan, "Multiple optimal sink placements in wireless sensor network using network lifetime efficient pso routing method," International Journal of Advancement in Engineering Technology, Management and Applied Science (IJAETMAS), vol. 5, pp. 13–19, April 2018.
[27] S. P. Tirani and A. Avokh, "On the performance of sink placement in WSNs considering energy-balanced compressive sensing-based data aggregation," Journal of Network and Computer Applications, vol. 107, pp. 38–55, April 2018.
[28] P. Pardesi and J. Grover, "Improved multiple sink placement strategy in wireless sensor networks," in International Conference on Futuristic Trends on Computational Analysis and Knowledge Management (ABLAZE). IEEE, February 2015. [Online]. Available: https://doi.org/10.1109%2Fablaze.2015.7155032
[29] Z. Rehena, D. Das, S. Roy, and N. Mukherjee, "Multiple sink placement in partitioned wireless sensor networks." International Journal of Next-Generation Computing, vol. 6, no. 2, 2015.
[30] O. O. Ogundile and A. S. Alfa, "A survey on an energy-efficient and energy-balanced routing protocol for wireless sensor networks," Computer networks, vol. 17, May 2017.

UFSP-Net: a neural network with spatio-temporal information fusion for urban fire situation prediction

Guangyin Jin[1], **Cunchao Zhu**[1], **Xiaoxuan Chen**[1], **Hengyu Sha**[1], **Xingchen Hu**[1], **Jincai Huang**[1]

[1]College of System Engineering, National University of Defense Technology, Changsha, China

Guangyin Jin: jinguangyin18@nudt.edu.cn

Abstract. Capturing the dynamics of urban fire situation is a basic but challenging task, which takes an indispensable role in the field of urban security and fire emergency decision. Traditional methods approach the urban fire prediction via stochastic process based on physics or statistics, which may be interpretable but less practical in real applications. Recently, some data-driven models, Convolutional Neural Network (CNN), Recurrent Neural Network (RNN) and Graph Convolutional Neural Network (GCN), seem to be fruitful in capturing spatio-temporal dynamics with massive high-dimension data. In this paper, we process some regional urban fire dataset of recent six years in the fire situation awareness images (FSAIs) and extract pixel-level latent representations with CNNs while GCNs are applied to process some spatial graph structure auxiliary information to obtain graph-level latent representations. And then Urban Fire Situation Prediction Neural Network (UFSP-Net) is formulated as a novel urban fire prediction model, integrating these two different kind of spatial latent representations and RNN structure. In comparison to other traditional algorithms, such as Conv-RNN, UFSP-Net demonstrates its superior prediction performance for multi type urban fire in spatio-temporal scale.

Key Words: Fire Situation Prediction, Graph Convolutional Neural Network, Convolutional Neural Network, Spatio-Temporal Analysis

1. Introduction

With the advancement of urbanization, the prevention and control of urban fires has become a tricky task in the field of urban security. According to the official fire statistical report of America in 2018 [1], a total of 237,000 fires were reported nationwide, 1,407 people were killed and 798 people were injured. The direct property loss has been calculated to exceed 500 million dollars. Among them, urban fires still account for the largest proportion and nearly 80% of the deaths are concentrated in urban residential region. Compared with wildfires, urban fires manifest more complex dynamics and lower predictability due to more impact factors.

According to the types of urban fire impact factors, it can be divided into two categories: spatial impact factors and temporal impact factors. In the space perspective, a modern city can be divided into many different functional regions, for instance, residential regions, commercial regions, industrial regions, tourist regions, etc. Empirically speaking, the incentives of urban fires vary with spatial regions. For instance, fire in residential regions are often caused by fires in boilers and appliances while fires in industrial regions are often caused by chemical and electrical failures. In addition to the functional region, the regional fire situation is also inseparable from the fire equipment investment and deployment in the particular region [2]. In the time perspective, fire incidents also have strong temporal correlation and we can argue from two aspects [3]. For one thing, the occurrence of fires varies with the seasonal climate. According to previous investigation researches, the hot and dry seasons are more prone to fire than usual. For another, the occurrence of fires varies with the time of one day. Empirically speaking, working hours in industrial regions during the day may have a higher probability of fire than non-working hours at night due to equipment's operation or other reasons. In summary, space and time are the two most crucial dimensions of urban fire prediction tasks. However, the most serious challenge of urban fires prediction tasks is modeling spatio–temporal dynamics to achieve real-time prediction [4]. In some previous literatures, the theoretical model of urban fire prevention and control has been developed. In the past decade, as some data-driven approaches [5, 6] were gradually proposed, these models have also been transferred to urban fire prediction tasks. Although some statistical learning models have achieved state-of-art performance, these models still have difficulties to capture complex spatio–temporal dynamics in urban fire prediction. In recent year, development of deep learning is considered promising to capture complex spatial and temporal dynamics [7]. However, the application of deep learning in the field of fire prediction has not been fully developed.

In this paper, we proposed a deep learning approach, Urban Fire Situation Prediction Neural Network (UFSP-Net), which integrates CNN, GCN and RNN model to cope with urban fire prediction. We initially arrange spatio-temporal fire record data and region boundary data into Fire Situation Awareness Images (FSAIs) and region graphs respectively. Then, convolutional neural networks (CNN) [8] and graph neural network (GCN) [9] are applied to embed the spatial information from the sequences of FSAIs and region graphs into latent space to obtain joint spatial representation. Meanwhile, extraction of temporal information can be tackled by Gated Recurrent Unit Networks (GRUN) [10], which can capture relevance of former and future time slots. Subsequently, the correlation of spatial and temporal dynamics has been established. To summarize, we make the following main contributions:

- We first proposed the real-time deep learning model to capture the spatial temporal dynamics from joint latent representation of Urban Fire Situation Awareness Images and Region Boundary Distance Graph graphs.
- We conduct experiments on real world dataset, and the proposed approach achieves remarkable performance in comparison to state-of-art baseline methods for urban fire prediction.

In the remainder of the paper, we begin with a review of the related work on the traditional urban fire prediction methods in section 2. Then we review background of CNN, GCN and GRU in section 3. The research data process and methodology adopted in this study is presented in section 4, followed by the experimental results and analysis in section 5. The final section concludes the achievements of this research and proposes future work.

2. Related work

There exists a large body of work on fire prediction and early methods for addressing this task were mainly based on some mathematical modeling and stochastic process simulation technology. Himoto et al investigated the physics-based computational model for urban fire spread [11], which proposed a mathematical physics model that interpreted urban fire as a collection of multiple building fires and simulated the spread of a fire by predicting the behavior of a single building fire under the thermal influence of a neighboring building fire. To better simulate the stochastic process of urban fire spread, the Cellular Automata (CA) model was introduced into the field of fire control and prevention in the work [12]. Based on the traditional cellular automata model, the work [13] presented a novel model integrating the Extreme Learning Machine (ELM) with the CA framework to improve simulation performance. However, these simulation methods are difficult to capture the spatio-temporal

correlations. To solve the problem, some improved mathematical models have been developed. For instance, Markov chain technology was involved to establish a spatio-temporal dynamic model of urban residential fire risk in a recent research [14] ,which has been well correlated with time and space dimensions. However, the lack of real-time prediction is the most serious shortcoming of this model.
Another intriguing line of former researches is application of data-driven models for fire prediction, which has achieved spectacular performance on many scenarios. Some early studies [15] and [16] first utilized K-means clustering model to analyze the urban fire risk. There are also some studies that have made substantial progress in urban fire prediction by regression model, for instance, the work [17] based on logistic regression with spline functions to accommodate nonlinear relationships between fire-danger predictors and probability of large fire events. Some other statistical learning models, Support Vector Machines (SVM) and Random Forests were also utilized in fire prediction tasks, which were applied in the work [5]. Based on some traditional statistical learning models, a unified pipeline framework for spatio-temporal fire risk prediction was proposed by [6], which demonstrated XGBoost and has superior performance to other algorithms.
More recently, some deep neural networks models have gradually demonstrated their potential in urban fire prediction tasks. To the best of our knowledge, recurrent neural network approach was first proposed for forest fire prediction in the work [18]. Neural network method was also developed by Mathibela et al to predict forest fire risk with vegetation, climate and location features of Cape Town, the most fire-prone city in South Africa [19]. Based on previous work, the application of fire detection systems also have achieved some breakthroughs that an energy-friendly, and computationally efficient CNN architecture was applied for complex semantic understanding of urban fire scene [20]. However, previous methods focus on the capture of spatial information while ignoring the simultaneous modeling of spatio-temporal relationships. Although some achievements have been made, deep learning approaches have not yet been explored in fire prediction, especially the methods that combine spatiotemporal relationships efficiently.
Some methods applied in other spatio-temporal prediction tasks can be worth learning from. The state-of-art deep learning model Conv-LSTM for precipitation nowcasting was proposed by Xingjian et al [21], which captures spatial and temporal dynamics by CNN and RNN respectively. Graph Neural Networks also have been widely applied in some spatio-temporal prediction tasks and the work [22] proposed a GCN-LSTM approach by graph modeling. Many data of spatio-temporal prediction tasks are non-European, so graph modeling is a fruitful solution. Motived by the works [21] and [22], we make an attempt to combine CNNs, GCNs and RNNs to capture spatial–temporal dynamics for addressing urban fire prediction task, which have advantages on real time prediction as well as full integration of geographic information and time series information.

3. Background

In this section, we briefly call the related methods of this research. In particular, the basic principles of Convolutional Neural Network (CNN), Graph Convolutional Neural Network (GCN) and Recurrent Neural Network (RNN).

3.1. *Convolutional neural network*

Convolutional Neural Network has achieved remarkable breakthroughs in image processing, speech recognition, etc. This model is designed to automatically and adaptively learn spatial hierarchies of features by using multiple building blocks, such as convolution layers, pooling layers, and fully connected layers. Among them, convolution layer is the most crucial part of CNN and convolution layer operation is formulated as:

$$Y^l(m,n) = X^k(m,n) * H^{kl}(m,n) = \sum_{k=0}^{K-1}\sum_{i=0}^{I-1}\sum_{j=0}^{J-1} X^k(m+i,n+j)H^{kl}(i,j) \tag{1}$$

It is assumed that the convolutional layer has *L* output channels and *K* input channels, so that KL convolution kernels are needed to realize the conversion of the number of channels. X^k is the two-dimensional feature map of the *k-th* input channel, Y^l is the two-dimensional feature map of the *l-th* output channel, and H^{kl} represents the *k-th* row and the *l-th* column of the two-dimensional convolution kernel. Assuming that the convolution kernel size is $I*J$ and the feature map size of each output channel is $M*N$.
The pooling layer can reduce the dimension of the feature map while maintaining the most important information. It has different types such as max pooling, average pooling, sum pooling and so on, but max pooling is default in this paper. The fully connected layer is a traditional multi-layer perceptron whose main function is to combine the features extracted by the previous convolutional layer and pooling layer into one-dimensional vectors.

3.2. *Graph convolutional neural network*
GCN is defined on graph structure data G= (V, E) where V is the set of graph nodes and E is edges between two different nodes. In a defined graph, the set of connected edges weights between different nodes can be represented by an adjacency matrix $A \in R^{|V|*|V|}$and a feature matrix $F\in R^{|V|*|N|}$. The notation N represents the feature dimensions of each node. GCN is able to obtain local features with different reception fields from translation variant non-Euclidean structures [9]. Let $\mathrm{L} = \mathrm{I} - D^{-1/2}AD^{-1/2}$denotes the graph Laplacian matrix, where D is the degree matrix, a graph convolution operation is defined as:

$$\mathrm{f}(F, A) = \sigma\left(L^{(1)}FW^{(1)}\right), l = 1 \tag{2}$$

$$\mathrm{f}\left(H^{(l)}, A\right) = \sigma\left(L^{(l)}H^{(l)}W^{(l)}\right), l \geq 2 \tag{3}$$

Where $H^{(l)}$ denotes the hidden layer features in the l_th layer, $W^{(l)}$ is the weight of each layer, $L^{(l)}$ is the graph Laplacian matrix of l_th layer, σ represents an arbitrary activation function. And the input feature of the first layer equals the original feature matrix *F*.

3.3. *Recurrent neural network*
RNN has been widely applied to capture the dynamics of sequence data in recent year. However, some disadvantages of RNN, for instance, gradient explosion, gradient disappears and long-term dependency problem, have hindered its promotion and application [23]. Hence, many variants of RNN were proposed such as long and short term memory network (LSTM) and gated recurrent units network (GRUN). GRUN is easier to train but can achieve performance comparable LSTM, so it is applied in our research to capture temporal dynamics. GRU is the basic component of GRUN and two gate operators are included in GRU as reset gate $\mathrm{R_t}$ and update gate $\mathrm{U_t}$:

$$\mathrm{R_t} = \sigma(w_{xr}x_t + w_{hr}h_{t-1}) \tag{4}$$

$$\mathrm{U_t} = \sigma(w_{xu}x_t + w_{hu}h_{t-1}) \tag{5}$$

Meanwhile, the candidate hidden information $\mathrm{h}_t^{\sim}$ and merged hidden information $\mathrm{h_t}$ can be calculated as:

$$\mathrm{h}_t^{\sim} = \tanh\left(w_{xh}x_t + w_{hh}(R_t \odot h_{t-1})\right) \tag{6}$$

$$\mathrm{h_t} = U_t \odot h_{t-1} + (1 - U_t) \odot h_t^{\sim} \tag{7}$$

4. Data processing and methodology
In this section, we provide details of data processing approach and our proposed Urban Fire Situation Prediction Neural Network (UFSP-Net).

4.1. Data processing

The urban fire dataset in this research originates from a public safety database[1] managed by San Francisco government. And the interest region is defined in San Francisco city, which cover the square region in the map with Latitude Interval [37.71, 37.8] and Longitude Interval [-122.50, 122.38]. Taking into account the integrity and time-effectiveness of the information, fire incidents in the period of last six years until June 2019 are processed, and detailed column attributes mainly include some temporal information (incident date, incident time, incident year), some spatial information (latitude, longitude, location description and region boundary), fire types (Potentiality Life-threatening, Non-Life-threatening and Alarm), battalion, neighborhoods and police district information. **Fig 1** illustrates the general situation of fire incidents data.

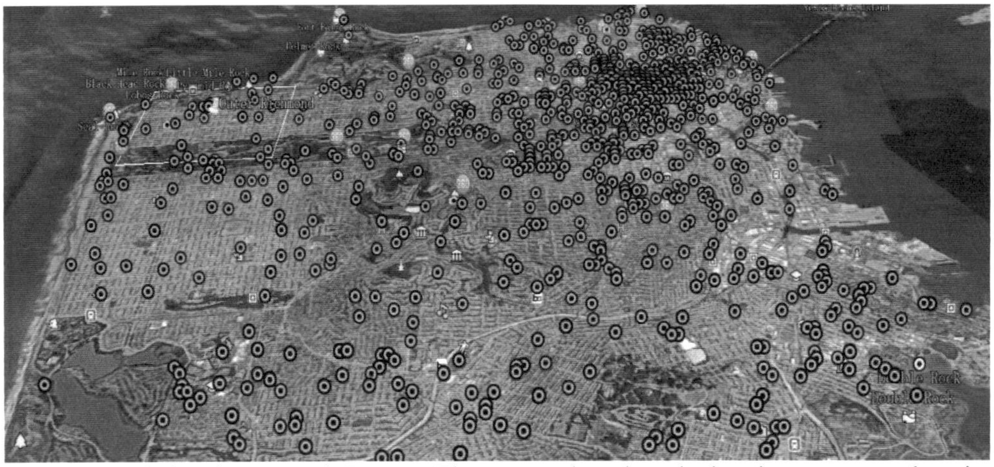

Figure 1. Spots of Fires in Research Dataset. The orange dotted marked regions are some locations fire incidents happened.

With this dataset, we focus on the spatio-temporal distribution of different types of fire, and the frequency of some fire types in some region lattice at some time slot is a direct statistics revealing the spatio-temporal situation of urban fire. Hence, we apply one channel fire situation awareness image to represent spatio-temporal distribution of each fire type.

Definition 1. Fire Situation Awareness Image of specific fire type $\mathbf{X}^{(t)}$. For the region lattice indexed (i, j), the time slot indexed t and one specific fire type, we define the element $x_{i,j}^{(t)}$ as the occurrence count of the specific fire type in the particular time slot and region. And FSAIs of one specific fire type are composed of basic elements$\mathbf{X}^{(t)} = \left[x_{i,j}^{(t)}\right]_{I*J}$, where $I * J$ represents geographical indexes.

Target area are uniformly partitioned into 10*10 lattices in this research and the FSAIs of three different fire types respectively as Potentiality Life-Threatening (Fire Type 1), Non-Life-Threatening (Fire Type 2) and Alarm (Fire Type 3).

Definition 2. Region Boundary Distance Graph. In urban fire prediction task, for two region boundaries near each other, they may share similar fire patterns. Following this idea, we use the distance to construct the region-based graphs. More specifically, reciprocal of the distances are used to mark the weight between two region boundaries so that closer regions will be linked with higher weights. The adjacency matrix of Region Boundary Distance Graph is defined as:

[1] https://data.sfgov.org/Public-Safety/Fire-Department-Calls-for-Service/nuek-vuh3

$$G(V,E)\quad Weight = region\ distance^{-1}$$

$$A_t = \begin{bmatrix} 0 & \frac{1}{d_{0,1}} & \cdots & \frac{1}{d_{0,n-1}} & \frac{1}{d_{0,n}} \\ \frac{1}{d_{1,0}} & 0 & \cdots & \frac{1}{d_{1,n-1}} & \frac{1}{d_{1,n}} \\ \vdots & \vdots & \ddots & \vdots & \vdots \\ \frac{1}{d_{n-1,0}} & \frac{1}{d_{n-1,1}} & \cdots & 0 & \frac{1}{d_{n-1,n}} \\ \frac{1}{d_{n,0}} & \frac{1}{d_{n,1}} & \cdots & \frac{1}{d_{n,n-1}} & 0 \end{bmatrix} \quad (8)$$

In this dataset, region of San Francisco has been divided into 41 region boundaries and we can mark the latitude and longitude of each region boundary by GIS.

4.2. Urban fire situation prediction neural network

To address the challenge of capturing spatio-temporal dynamics from non-linear events, we propose a deep neural network framework, referred to as Urban Fire Situation Prediction Neural Network (UFSP-Net), to understand fire situation and perceiving potential dynamics. The overview of UFSP-Net is revealed in **Fig 2**.

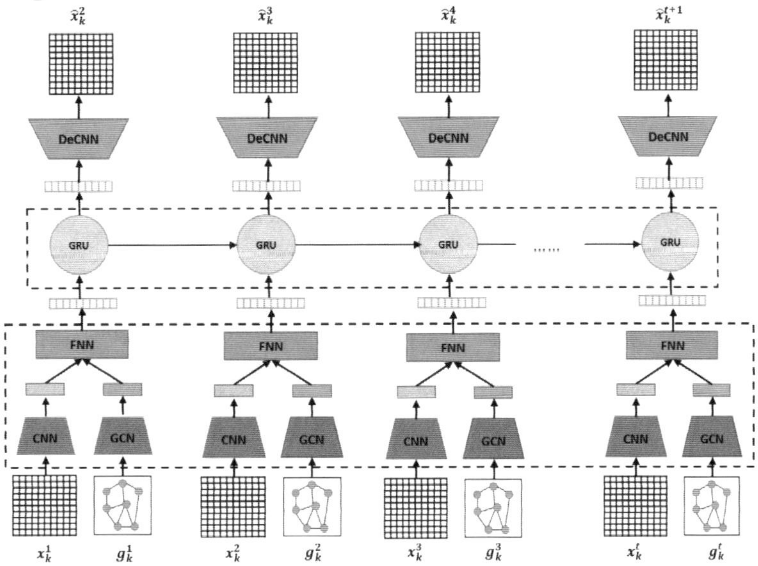

Figure 2. Overview of the Urban Fire Situation Prediction Neural Network

As illustrated in **Fig 2**, UFSP-Net is mainly composed of three components, the spatial latent representation extraction model, temporal dynamics capture model and decoder model respectively. The first model consists of CNN, GCN and information fusion feedforward neural network (FNN), and attempts to map the high dimensional FSAIs and Region Boundary Distance Graphs into low dimensional joint latent representation. The second model is actually GRUN model to capture temporal dynamics while the third model is decoder CNN model to decode FSAIs from predicted latent representation. The goal of UFSP-Net is to predict the next time slot FSAI by the last time slot FSAI and Region Boundary Distance Graph.

In UFSP-Net model, we applied two convolution layers and one fully connected layer in CNN. The parameters of the first convolution layer are [1, 16, 3, 1, 1], which represent input channel, output channel, kernel size, stride and padding value respectively. The parameters of the second convolution layer are [16, 32, 3, 1, 1], which represent the same definition as the first convolution layer. After each convolution layer, there will be a batch normalization layer and an activation function LeakyReLU. Following the second convolutional layer is a fully connected layer to map the FSAIs to one-dimensional latent variables with size 1*10. The process is formulated as:

$$z_{pk}^{t} = Linear(LeakyReLU\left(BN2\left(Conv2\left(LeakyReLU\left(BN1\left(Conv1(x_{k}^{t})\right)\right)\right)\right)\right)) \quad (9)$$

Where $Conv$ and BN represent convolution operation and batch normalize operation respectively. x_k^t is input FSAI and z_{pk}^t is pixel-level latent representation of fire type k.

In GCN model, we applied only one graph convolution layer and the parameters of it are [1, 10], which represent the input dimension of feature matrixes and output dimension of hidden representation. In this paper, feature matrixes are one-dimensional vectors which represent number of fire incidents in some regions for particular time slot. Following the graph convolution layer is a fully connected layer to map the Region Boundary Distance Graphs to one-dimensional latent variables with size 1*10. The process is formulated as:

$$z_{gk}^{t} = Linear\left(LeakyReLU\left(GC(A_t, f_k^t)\right)\right) \quad (10)$$

Where A_t, f_k^t are adjacency matrix and feature matrix respectively. GC represents the graph convolution operation and z_{gk}^t is graph-level latent representation of fire type k.

The pixel-level latent representation and graph-level latent representation are extracted by CNN and GCN respectively while the FNN acts as a spatial information fusion network to obtain a joint latent representation with size 1*10. And the process is defined as:

$$\mathrm{z}_{k}^{t} = FNN\left(z_{pk}^{t} \oplus z_{gk}^{t}\right) \quad (11)$$

The operator $\oplus$ represents concentration operation and z_k^t is the joint latent representation converged by z_p^t andz_g^t.

Through the process of the spatial latent representation extraction model, we get a series of joint latent representation $[\mathrm{z}_k^1, \mathrm{z}_k^2, \ldots, \mathrm{z}_k^{t-1}, \mathrm{z}_k^t]$. For GRUN, the training process of the model is also a process that maximizes the conditional probability as:

$$\arg\max_{\theta} \prod_{k=1}^{3} \prod_{t=1}^{T} P_{\theta}\left(z_k^{t+1} \middle| z_k^t\right) \quad (12)$$

The mathematical form of GRUN is defined as:

$$z_k^{t+1} = \mathrm{GRU}\left(z_k^t, h_k^{t+1}\right) \quad (13)$$

Where h_k^{t+1} represents the hidden representation in the time slot $t+1$.

For the last part of UFSP-Net, decoder CNN model consists of one fully connected layer and two deconvolution layers. The parameters of the first deconvolution layer are [32, 16, 3, 1, 1], which represent input channel, output channel, kernel size, stride and padding value respectively. The parameters of the second deconvolution layer are [16, 1, 3, 1, 1], which represent the same definition as the first deconvolution layer. The computational process of decode is defined as:

$$\hat{x}_k^t = Deconv2\left(LeakyReLU\left(BN\left(Deconv1(z_k^t)\right)\right)\right) \quad (14)$$

The loss function of UFSP-Net model is formulated as:

$$L_2(x_k^t, \hat{x}_k^t) = ||x_k^t - \hat{x}_k^t||_2^2 \quad (15)$$

Where $x_k^{(t)}, \hat{x}_k^{(t)}$ are respectively the original FSAIs and predicted FSAIs.

5. Experiments and Analysis

In this section, the training details are described, and some numerical analysis are presented to show the performance of our model.

5.1. Training details and comparison algorithms

Training Details: The urban fire dataset employed in this research ranges from Jun. 1st, 2014 to June. 12th, 2019, and the time series can be roughly partitioned into 1810 continuous disjoint time slots with one day as unit to formulate FSAIs and Region Boundary Distance Graphs. Taking variations in three types of fire incidents in spatio-temporal patterns, we respectively train three CNNs and GCNs, then three series of joint latent representation at initial time slot t are obtained by information fusion as$[z_1^t, z_2^t, z_3^t]$. For three types of FSAIs and region-based graphs, 90% of dataset are for model training with the rest for the performance testing. For UFSP-Net model, the Adam optimizer is used in training process and learning rate is defined as 0.001.

Comparison Algorithms: In the experiments, four algorithms are included in comparison to Urban Fire Situation Prediction Neural Network (UFSP-Net), including Random Forest, XGBoost, Gated Recurrent Unit Neural Network (GRUN) and Convolutional Gated Recurrent Unit Neural Network (Conv-GRUN). And the prediction process for comparison algorithms is to automatically generate the next time slot FSAIs by the last time slot FSAIs.

5.2. Result analysis

We evaluate the effectiveness of proposed framework UFSP-Net in this section. No matter for CNN, GCN or GRUN training, the quality of prediction of FSAIs is the most concerned. Three metrics are applied to evaluate the prediction performance of each algorithm, respectively as the Root Mean Square Error (RMSE), Mean Square Percentage Error (MSPE) and Jessen-Shannon Divergence (JS), whose definitions are as follows:

$$\mathrm{RMSE}_k = \left(\frac{1}{T}\Sigma_{t=1}^{T} \left\|x_k^{(t)} - \hat{x}_k^{(t)}\right\|_2^2\right)^{\frac{1}{2}} \tag{16}$$

$$\mathrm{MAPE}_k = \frac{1}{T}\Sigma_{t=1}^{T} \frac{|x_k^{(t)} - \hat{x}_k^{(t)}|}{|x_k^{(t)}|} \tag{17}$$

For the normalized spatial frequencies:

$$p_{i,j,k}^{(t)} = \frac{x_{i,j,k}^{(t)}}{\Sigma_{i=1}^{M}\Sigma_{j=1}^{N} x_{i,j,k}^{(t)}} \tag{18}$$

$$p_{i,j,k}^{(t)\sim} = \frac{\hat{x}_{i,j,k}^{(t)}}{\Sigma_{i=1}^{M}\Sigma_{j=1}^{N} \hat{x}_{i,j,k}^{(t)}} \tag{19}$$

The JS-Divergence is computed as:

$$D_{JS}(P, P^{\sim}) = \frac{1}{2}D_{KL}(P||1/2(P + P^{\sim})) + \frac{1}{2}D_{KL}(P^{\sim}||1/2(P + P^{\sim})) \tag{20}$$

Where $x_k^{(t)}, \hat{x}_k^{(t)}$ are respectively the original FSAIs and predicted FSAIs. In this case, MSPE can be viewed as a squared individualized weighted RMSE, measuring the ratio of approximation regardless of scales. And JS divergence measures the spatial distribution similarity between two FSAIs.

In **Table 1**, the comparison results on Random Forest, XGBoost, GRUN, Conv-GRUN and UFSP-Net are displayed. We find UFSP-Net is significantly superior to others in testing dataset and performances fluctuate for three types of fire. On the datasets of these three fire types, the effect of these four algorithms are successively decreasing and this suggests that deep learning model could be harder to perform well on sparse datasets. We also find that combination of pixel-level latent representation and graph-level latent representation can improve the evaluation metrics on all of fire type dataset.

Table 1. Prediction Results on Five State-of-art Models. We conducted 10-fold cross validation and the results by the best performer in each column are in boldface. We assume that ten results are consistent with a normal distribution and statistical significance of pairwise differences of the structure UFSP-Net vs. the other state-of-art Models is determined by a t-test (▲/▼ for $\alpha = 0.1$)

Algorithms	Metrics	Fire Type 1	Fire Type 2	Fire Type 3
Random Forest	RMSE	2.6391±0.0312	1.8320±0.0291	1.0371±0.0268
	MAPE	0.1803±0.0035	0.2201±0.0042	0.4788±0.0139
	JS	0.1059±0.0031	0.1281±0.0029	0.2286±0.0038
XGBoost	RMSE	2.5478±0.0277	1.7679±0.0167	0.9901±0.0125
	MAPE	0.1712±0.0017	0.2107±0.0019	0.4611±0.0073
	JS	0.1048±0.0015	0.1275±0.0008	0.2264±0.0019
GRUN	RMSE	2.5762±0.0260	1.7710±0.0145	0.9939±0.0119
	MSPE	0.1719±0.0014	0.2115±0.0016	0.4656±0.0058
	JS	0.1047±0.0013	0.1273±0.0004	0.2265±0.0012
Conv-GRUN	RMSE	2.4612±0.0256	1.7012±0.0141	0.9815±0.0118
	MAPE	0.1683±0.0017	0.2078±0.0018	0.4589±0.0054
	JS	0.1045±0.0011	0.1272±0.0006	0.2223±0.0017
UFSP-Net	RMSE	2.2503±0.0229▲	1.6860±0.0132▲	0.9747±0.0148▲
	MAPE	0.1642±0.0012▲	0.2023±0.0024▲	0.4512±0.0057▲
	JS	0.1045±0.0008	0.1271±0.0009	0.2217±0.0011▲

As the instantiation, the predicted FSAIs at one randomly selected time slots from UFSP-Net in **Fig 3** are satisfying to some extent. Notice that, to more easily outline the spatial distribution, the normalized FSAIs form are applied in this case. For Fire type1 and Fire type2 with more adequate spatial distribution statistics, details in images can be well predicted, accompanied with lower MSPE and JS divergence. And for Fire type3, rough hotspots of situation can be earlier perceived though regional details need further polished. The predicted situation images through UFSP-Net enable us to roughly master the spatial fire dynamics of different types.

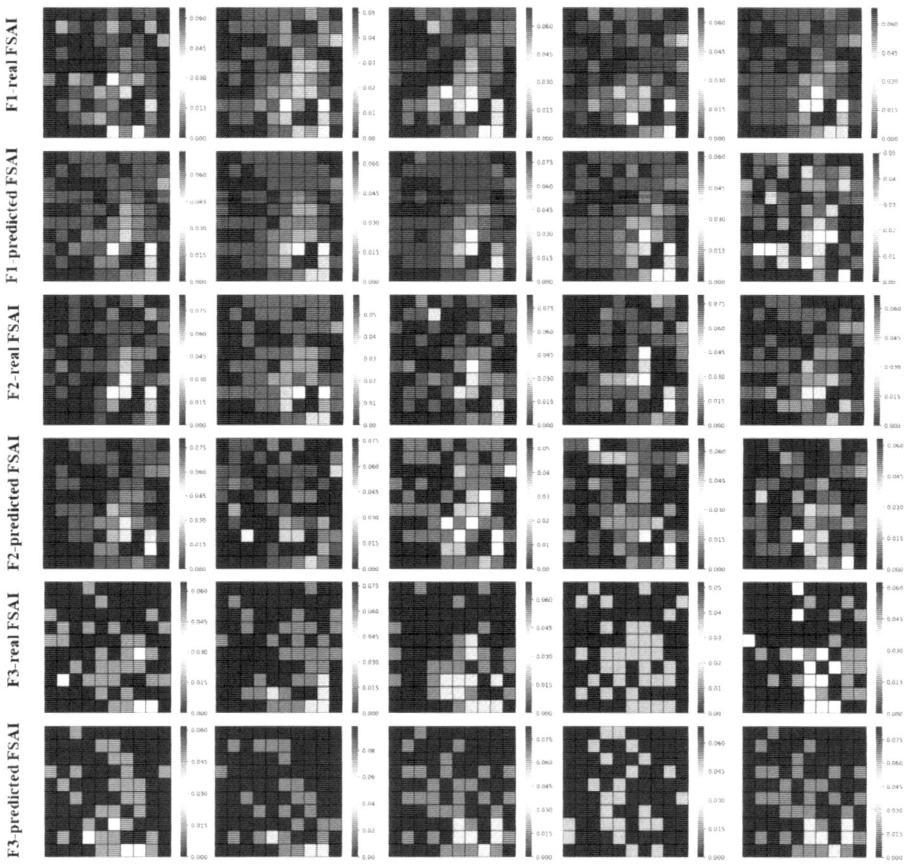

Figure 3. Predicted Performance Visualization. One initialized time slot is randomly selected from testing dataset on UFSP-Net. The previous first column is original and predicted FSAIs initialized from Jan 10th, 2017.

6. Conclusion

In this paper, we make attempts to utilize deep learning approach for urban fire situation modelling and a novel deep learning framework UFSP-Net is introduced in this domain. Though it is still tough to precisely predict urban fire situation in both scales and spatio-temporal distribution, the experimental results of UFSP-Net is remarkable in comparison to other state-of-art traditional models in this domain and suggests that the combination of pixel-level information and graph structure information makes the awareness about urban fire situation more foreseeable.

In the future, it is potential to make further improvements on the structure of UFSP-Net by fusing more domain graph structure information that can be associated to real word statistical indexes and involving attention mechanism for capturing long-term dependence. Meanwhile, some possible extensions could be made to predict other social events for urban safety, for instance, urban drought and flood forecast and the urban air pollution monitoring with improved deep learning frameworks.

References

[1] "Fire department operations, management and safety," https://www.usfa.fema.gov/operations/.

[2] J. Corcoran, and G. Higgs, "Special issue on spatial analytical approaches in urban fire management," *Fire Safety Journal*, no. 62, pp. 1-2, 2013.

[3] K. Wuschke, J. Clare, and L. Garis, "Temporal and geographic clustering of residential structure fires: A theoretical platform for targeted fire prevention," *Fire Safety Journal*, vol. 62, pp. 3-12, 2013.

[4] M. Angelidou, A. Psaltoglou, N. Komninos, C. Kakderi, P. Tsarchopoulos, and A. Panori, "Enhancing sustainable urban development through smart city applications," *Journal of Science and Technology Policy Management,* vol. 9, no. 2, pp. 146-169, 2018.
[5] P. Cortez, and A. d. J. R. Morais, "A data mining approach to predict forest fires using meteorological data," 2007.
[6] B. Singh Walia, Q. Hu, J. Chen, F. Chen, J. Lee, N. Kuo, P. Narang, J. Batts, G. Arnold, and M. Madaio, "A Dynamic Pipeline for Spatio-Temporal Fire Risk Prediction." pp. 764-773.
[7] R. M. Cichy, A. Khosla, D. Pantazis, A. Torralba, and A. Oliva, "Deep neural networks predict hierarchical spatio-temporal cortical dynamics of human visual object recognition," *arXiv preprint arXiv:1601.02970*, 2016.
[8] Y. LeCun, and Y. Bengio, "Convolutional networks for images, speech, and time series," *The handbook of brain theory and neural networks,* vol. 3361, no. 10, pp. 1995, 1995.
[9] T. N. Kipf, and M. Welling, "Semi-supervised classification with graph convolutional networks," *arXiv preprint arXiv:1609.02907*, 2016.
[10] T. Mikolov, M. Karafiát, L. Burget, J. Černocký, and S. Khudanpur, "Recurrent neural network based language model."
[11] K. Himoto, and T. Tanaka, "Development and validation of a physics-based urban fire spread model," *Fire Safety Journal,* vol. 43, no. 7, pp. 477-494, 2008.
[12] J. Quartieri, N. E. Mastorakis, G. Iannone, and C. Guarnaccia, "A cellular automata model for fire spreading prediction," *Latest Trends on Urban Planning and Transportation*, pp. 173-178, 2010.
[13] Z. Zheng, W. Huang, S. Li, and Y. Zeng, "Forest fire spread simulating model using cellular automaton with extreme learning machine," *Ecological Modelling,* vol. 348, pp. 33-43, 2017.
[14] R. Ardianto, and P. Chhetri, "Modeling Spatial–Temporal Dynamics of Urban Residential Fire Risk Using a Markov Chain Technique," *International Journal of Disaster Risk Science,* vol. 10, no. 1, pp. 57-73, 2019.
[15] L. Wu, and A. Ren, "Urban fire risk clustering method based on fire statistics," *Tsinghua Science and Technology,* vol. 13, no. S1, pp. 418-422, 2008.
[16] J. Clare, L. Garis, D. Plecas, and C. Jennings, "Reduced frequency and severity of residential fires following delivery of fire prevention education by on-duty fire fighters: Cluster randomized controlled study," *Journal of safety research,* vol. 43, no. 2, pp. 123-128, 2012.
[17] H. K. Preisler, and A. L. Westerling, "Statistical model for forecasting monthly large wildfire events in western United States," *Journal of Applied Meteorology and Climatology,* vol. 46, no. 7, pp. 1020-1030, 2007.
[18] T. Cheng, and J. Wang, "Integrated spatio - temporal data mining for forest fire prediction," *Transactions in GIS,* vol. 12, no. 5, pp. 591-611, 2008.
[19] S. Lall, and B. Mathibela, "The application of artificial neural networks for wildfire risk prediction." pp. 1-6.
[20] K. Muhammad, J. Ahmad, Z. Lv, P. Bellavista, P. Yang, and S. W. Baik, "Efficient deep CNN-based fire detection and localization in video surveillance applications," *IEEE Transactions on Systems, Man, and Cybernetics: Systems*, no. 99, pp. 1-16, 2018.
[21] S. Xingjian, Z. Chen, H. Wang, D.-Y. Yeung, W.-K. Wong, and W.-c. Woo, "Convolutional LSTM network: A machine learning approach for precipitation nowcasting." pp. 802-810.
[22] D. Chai, L. Wang, and Q. Yang, "Bike flow prediction with multi-graph convolutional networks." pp. 397-400.
[23] T. , M. Karafiát, L. Burget, J. Černocký, and S. Khudanpur, "Recurrent neural network based language model."

IOP Conf. Series: Materials Science and Engineering **853** (2020) 012051 doi:10.1088/1757-899X/853/1/012051

Cattle identification using muzzle print images based on feature fusion

Cong Sian[1], Wang Jiye[1], Zhang Ru[2], Zhao Lizhi[1]

[1]College of Information Engineering, Minzu University of China, Beijing, 100081, China
[2]College of Science, Minzu University of China, Beijing, 100081, China

a1691806320@163.com, jiye_wang@163.com,847132978@qq.com, lizhi3285@126.com

Abstract. Individual identification of animals is an important means to modernize the livestock industry. In recent years, the research on individual identification of cattle has also received more and more attention. Individual cattle identification is necessary for many important reasons including registration, traceability, production management and animal disease control. Biometric features are unique, which often do not change over time. In this paper, muzzle print is used as biometric feature. The fusion of texture features extracted from Weber Local Descriptor(WLD) and local binary pattern was used to represent individual cattle. Some improvements were made to WLD algorithm. Finally, support vector machine was employed to identify head of cattle from their fusion feature. The proposed method achieved 96.5% identification accuracy.

1. Introduction

Animal individual identification has received a great research attention as an important way to achieve modern management of the livestock industry. Identification of cattle can apply to the variety of applications, such as registration, traceability, production management and animal disease control.

To identify cattle, different traditional methods were used such as tattoos, ear notching, hot iron branding, plastic ear tags and Radio Frequency Identification(RFID) tags. Tattoos, ear notching and hot iron branding can cause great pain to cattle and are not permanent methods as they can be altered or removed over time. Ear tags have been found to be vulnerable to damage, duplication, loss and replacement [1]. RFID tags have certain reliability and security, but RFID systems also have security flaws such as tag content changes and system fraud possibility.

Therefore, biometric identification has become a new choice. Biometric recognition not only ensures accuracy, but also provides high security. Nowadays, human fingerprint recognition has been widely used, and has a very high accuracy. Muzzle print is similar to a human's fingerprint and muzzle print of different animals of the same species are mostly unique [2]. Cattle muzzle print has rich texture characteristics. The oval, rounded or irregular structures are called "beads", and slender straight lines or curves are known as "ridges" [3].

Initially, people used ink to collect muzzle print images. This method has disadvantage such as inconvenient and time inefficient process, special skills to control the animal. It is difficult to obtain good-quality muzzle print. Nowadays, the muzzle photos are used as input data for cattle identification.

In 2012, Noviyanto et al. [4] used the Speeded Up Robust Features (SURF) to extract the features of the muzzle print images. Experiments on the data set of eight head of cattle showed that the accuracy was higher than 90% with sufficient training data. In 2013, Awad et al. [5] used Scale Invariant Feature Transform (SIFT) to detect the key points of muzzle print texture, and Random Sampling Consistency (RANSAC) to remove outliers. They achieved 93.3% accuracy of cattle identification. The approach also improved the reliability and robustness of the algorithm. In 2014, Tharwat et al. [6] applied local binary pattern(LBP) to describe local features. The linear discriminant analysis (LDA) was used to address high dimensionality problem. The authors used different methods including nearest neighbors, naive Bayes, support vector machine(SVM), and K nearest neighbors to identify cattle. Experimental results showed that support vector machine achieved the best classification effect. In 2016, Gaber et al. [7] proposed a method for identifying cattle based on Weber Local Descriptor (WLD). They used WLD method to extract the characteristics of muzzle print texture and the Adaboost classifier to classify the features. The experimental results showed that the identification accuracy was 99.5%. In 2018, Kusakunniran et al. [8] proposed a cattle individual identification method based on feature fusion. The author extracted Gabor features in different scales and specific frequency directions, extracted LBP histogram for each local sub-image and fused Gabor features and LBP features. Experimental results showed that this method achieved high accuracy.

In this paper, two kinds of descriptors, local binary pattern [9] and improved WLD method are used to extract texture features. Then, The two features are fused and classified by SVM. The gray information of image can be obtained by LBP operator, while differential excitation and gradient direction value can be provided by WLD.

The rest of the paper is organized as follows. Section 2 describes our proposed approach in detail. Experimental results and discussion are introduced in Section 3. Finally, conclusions are summarized in Section 4.

2. Proposed method

The flow chart of the individual identification algorithm of cattle is shown in Figure 1, which mainly includes image preprocessing phase, feature extraction phase, feature fusion phase and recognition phase. After building the database, the images are processed into grayscale images. The next step is to extract features from each muzzle print image, based on the fusion of LBP and improved WLD. Finally, support vector machine is used as the classifier for identifying the ID of individual cattle.

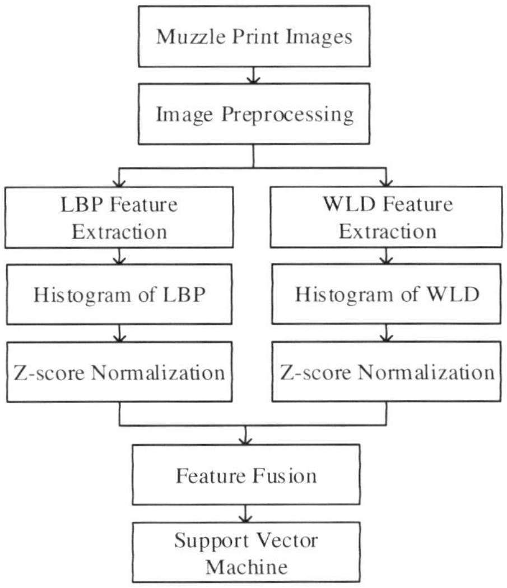

Figure 1. The flow chart of cattle identification algorithm

2.1. Image acquisition

The experimental data were collected from the practice base of the school in Fenghuang County, and photographed in the cattle shed under natural light. The cattle breed is the Wu Ling cattle. The data set contains 900 images. It consists of 45 head of cattle. The beads and ridges of the muzzle texture are shown in Figure 2. All images require grayscale processing.

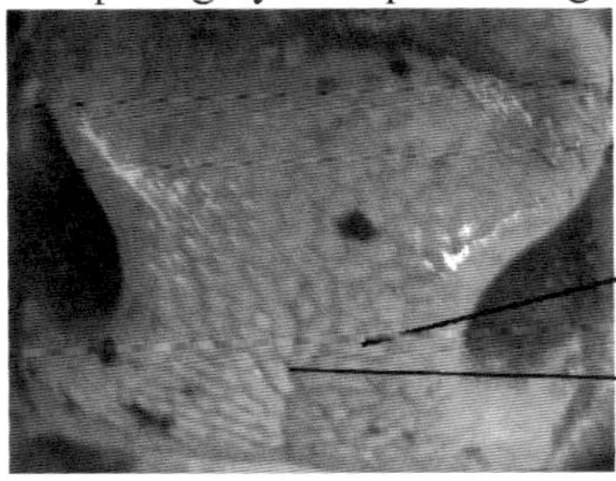

Figure 2. Texture features of muzzle image

2.2. Feature extraction
There are two kinds of feature to extract, including local binary pattern and Weber Local Descriptor.

2.2.1. Local binary pattern(LBP)
LBP is an operator used to describe local features of an image. It has illumination invariance and robustness to image rotation. The idea of LBP technology is to assign a label to every pixel of an image by thresholding the 3×3 neighborhood of each pixel with the center pixel value and considering the result as a binary number [10]. If the intensity of the center pixel is greater than (or equal to) the intensity of the neighboring pixel, the label is set to 1 at the corresponding position; otherwise, it is set to 0. The basic formula of LBP is:

$$LBP(x_c, y_c) = \sum_{i=0}^{7} s(g_i - g_c) \times 2^i \tag{1}$$

$$s(x) = \begin{cases} 1, x \geq 0 \\ 0, x < 0 \end{cases} \tag{2}$$

g_c is the gray value of center pixel (x_c, y_c) and g_i is the value of neighbor pixel.

2.2.2. Weber local descriptor
Weber local descriptor is a local image descriptor proposed by Chen jie et al. [11] in 2009.It is inspired by Weber's law. The difference threshold of the sense changes with the change of original stimulus, and it shows a certain regularity. The ratio of the stimulus increment (ΔI) to the original stimulus value (I) is a constant. Weber local descriptor consists of two parts, differential excitation and gradient direction.

2.2.2.1 Differential excitation
Differential excitation compares the difference between the center pixel and the neighbor pixel, which reflects the intensity information of local gray variation.
Firstly, filter f_{00} is used to calculate the sum of the differences between the current pixel and its adjacent pixel gray values, and the calculation formula is as follows:

$$v_s^{00} = \sum_{i=0}^{p-1} (\Delta x_i) = \sum_{i=0}^{p-1} (x_i - x_c) \tag{3}$$

Where $x_i\,(i = 0,1,...p-1)$ is the i-th adjacent pixel of the center pixel x_c, and p is the total number of adjacent pixels around it.

Next, according to Weber's law, the ratio of v_s^{00} and its current pixel x_c is calculated by two filters f_{00} and f_{01}, and the inverse tangent transformation is performed to obtain the difference excitation $\xi(x_c)$ of the current pixel x_c. The calculation formula is as follows:

$$\xi(x_c) = \arctan\left[\frac{v_s^{00}}{v_s^{01}}\right] = \arctan\left[\sum_{i=0}^{p-1}\left(\frac{x_i - x_c}{x_c}\right)\right] \tag{4}$$

v_s^{01} is calculated by the filter f_{01}, it's actually the value of the current pixel x_c.
Figure 3 shows the calculation process of differential excitation.

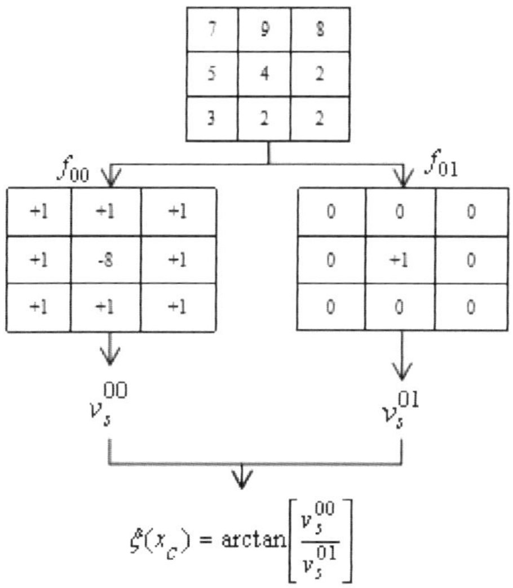

Figure 3. The computation of difference excitation

2.2.2.2 Gradient direction
The calculation formula of gradient direction is as follows:

$$\theta(x_c) = \arctan\left(\frac{v_s^{11}}{v_s^{10}}\right) \tag{5}$$

Where v_s^{11} and v_s^{10} respectively represent the gray scale difference between neighboring pixels on the two sides of the central pixel in the horizontal direction and the vertical direction. v_s^{10} and v_s^{11} are calculated from the filter f_{10} and f_{11}.

0	-1	0
0	0	0
0	1	0

0	0	0
1	0	-1
0	0	0

Figure 4. The filter f_{10} (left) and f_{11} (right)

2.2.3 Improved weber local descriptor
In the WLD feature, the method only considers the current 4 neighboring pixels when calculating the gradient direction.It will lose some discriminative details and are also susceptible to noise interference. In this paper, the Scharr operator is used to calculate the gradient direction. It takes into account all

eight domain pixels of the current pixel, which can better extract the direction information and effectively suppress the noise. The Scharr operator is shown in Figure 5.

-3	0	3
-10	0	10
-3	0	3

a

-3	-10	-3
0	0	0
3	10	3

b

Figure 5. Scharr kernels: (a) Horizontal kernel (b)Vertical kernel

The image processing effects of filter f_{10}, f_{11} and Scharr operator are shown in figure 6 and figure 7 respectively:

Figure 6. The image processed by filter f_{10} and f_{11}

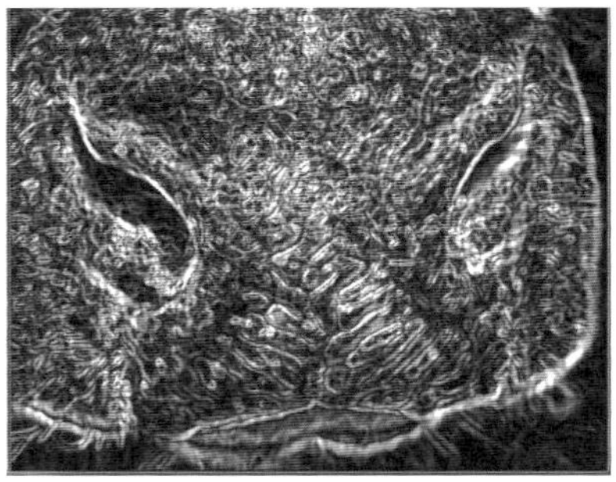

Figure 7. The image processed by Scharr operator

After the differential excitation and gradient direction are calculated, two-dimensional WLD histogram can be obtained. In order to make the feature discrimination of WLD stronger, the two-dimensional histogram is further transformed into the one-dimensional histogram.
In order to express the spatial structure characteristics of the image, the image is divided into m(R_1,R_2..., R_m) blocks. Then, the histogram information of each block is calculated to obtain the histogram features and connect them to form the characteristic of the muzzle print image.

2.3. Data normalization and feature fusion
The LBP method acquires the grayscale information of the image, and the WLD method obtains the differential excitation and gradient information of the image. A single feature does not fully reflect the texture information, so the LBP histogram feature is combined with the WLD histogram feature. After feature extraction, z-score is used for standardized processing to make the data within the interval of [0,1],and the features are concatenated to obtain the feature vectors.

2.4. Classification
The final critical step in cattle individual identification is classification. Support vector machine (SVM) is a new generation of learning algorithm developed on the basis of statistical learning theory, which has a better generalization ability for test samples that have not been seen . Since most problems are nonlinear and separable, kernel functions of SVM can map low-dimensional space to high-dimensional space. Therefore, kernel functions turn SVM into nonlinear classifier, and SVM is selected for classification in this paper.

3. Results and discussion

3.1. Experimental environment
The experiment in this paper is completed under the Windows 7 operating system, the host memory is 4.0 GB, the CPU is a 4-core i5-3470, and the main frequency is 3.2 GHz.

3.2. Experimental results

In this paper, there are 900 pictures in the data set. In order to eliminate the interference of random results, the following data are the average results obtained by repeating 10 experiments under the same conditions.

The performance of both WLD histogram and LBP histogram is affected by the number of sub-regions. As shown in Table 1, the recognition rates of the LBP descriptor, the WLD method and the fusion method in the case where the image is divided into 4×3, 5×5, 6×8,10×10 sub-regions are shown respectively .The support vector machine is used for identification.

Table 1. Recognition rate of four models(%)

Feature descriptor	Number of regions			
	4×3	5×5	6×8	10×10
LBP	92.9	93.5	92.6	92.5
WLD	91.7	92.9	91.6	90.8
Improved WLD	93.3	94.4	93.5	93.2
LBP+ Improved WLD	95.7	96.5	95.9	95.3

As can be seen from the table, the recognition rate of the algorithm is improved by integrating the features extracted from LBP descriptor and WLD method. The number of sub-regions of the image has a great influence on the performance of the algorithm. For LBP algorithm and WLD algorithm, the recognition rate is the highest when the image is divided into 5×5.The recognition rate increases first and then decreases as the number of sub-images increases. As the number of blocks increases, the statistical information of each small histogram becomes smaller, which affects the recognition effect.

In order to select an appropriate cattle individual recognition algorithm, the performance of SVM was compared with that of decision tree and KNN , and the feature fusion of LBP feature and WLD feature was selected for identification. The specific experimental results are shown in table 2.

Table 2. Performance comparison of different classifiers(Recognition rate%)

Classifiers	Number of regions			
	4×3	5×5	6×8	10
KNN	80.5	81.2	82.8	83.4
Decision Trees	41.9	42.4	41.5	40.8
SVM	95.7	96.5	95.9	95.3

All three algorithms were repeated for 10 times under the same conditions to get the average of the experimental results.

In KNN algorithm, the value of k is set as 3, Euclidean distance is used for similarity measurement, and the highest recognition rate is 83.4. The algorithm recognition rate of decision tree is low, which is 42.4 and can not meet the requirements of individual recognition accuracy of cattle. SVM has the best recognition effect, up to 96.5.

In order to verify the robustness of the proposed method in the case of image rotation, the following experiment is designed. Experiment rotated the pictures of the test set by 0°, 45°, 90°, 135°, 180°, 225°, 270° and 315°respectively. The experimental results are shown in table 3.

Table 3. Accuracy in (%) of our approach when rotated images are used

Algorithm	Angle of rotation							
	0°	45°	90°	135°	180°	225°	270°	315°
LBP+ Improved WLD	96.5	96.3	97.2	95.8	96.4	95.6	96.8	96.2

From table 3, it can be seen that when the test images are rotated with different angles, the recognition accuracy is similar to that of the non-rotated images. This proves that our method is robust against any rotations in the image.

4. Conclusion

In this paper, improved Weber Local Descriptor and local binary pattern are applied to extract the feature, and the SVM, decision tree and KNN algorithm are used for recognition. A single feature cannot fully reflect the texture information, so the LBP feature is integrated with the WLD feature. Experiments show that the feature fusion algorithm has a good effect on accuracy. Compared with the decision tree and KNN algorithm, SVM has the best recognition effect and can be used to study the individual identification of cattle.

Although the current algorithm recognition accuracy has reached more than 96%, it still needs to be improved. In the future, it is necessary to increase the number of pictures in the database and evaluate whether the algorithm proposed in this paper will get good results.

Acknowledgments

This work was supported by the National Natural Science Foundation of China under Grant 61701554

References

[1] A. I. Awad, "From classical methods to animal biometrics: A review on cattle identification and tracking," Computers and Electronics in Agriculture, vol. 123, pp. 423 - 435, 2016.

[2] A. S. Baranov, R. Graml, F. Pirchner, and D. O. Schmid, "Breed differences and intra-breed genetic variability of dermatoglyphic pattern of cattle," Journal of Animal Breeding and Genetics, vol. 110, no. 1-6, pp. 385 - 392, Dec. 1993.

[3] B. Barry, U. A. Gonzales-Barron, K. Mcdonnell, F. Butler, and S. Ward, "Using Muzzle Pattern Recognition as a Biometric Approach for Cattle Identification," Transactions of the ASABE, vol. 50, no. 3, pp. 1073 - 1080, 2007.

[4] A. Noviyanto, and A. M. Arymurthy, "Automatic cattle identification based on muzzle photo using speed-up robust features approach," Proceedings of the 3rd European conference of computer science (ECCS),vol. 110, pp. 114, 2012.

[5] A. I. Awad, H. M. Zawbaa, H. A. Mahmoud, E. H. H. A. Nabi, R. H. Fayed and A. E. Hassanien, "A robust cattle identification scheme using muzzle print images," 2013 Federated Conference on Computer Science and Information Systems, Krako¿w, 2013, pp. 529-534.

[6] A. Tharwat, T. Gaber, A. E. Hassanien, H. A. Hassanien, and M. F. Tolba, "Cattle Identification Using Muzzle Print Images Based on Texture Features Approach," Advances in Intelligent Systems and Computing Proceedings of the Fifth International Conference on Innovations in Bio-Inspired Computing and Applications IBICA 2014, pp. 217 - 227, 2014.

[7] T. Gaber, A. Tharwat, A. E. Hassanien, and V. Snasel, "Biometric cattle identification approach based on Weber's Local Descriptor and AdaBoost classifier," Computers and Electronics in Agriculture, vol. 122, pp. 55 - 66, 2016.

[8] W. Kusakunniran, A. Wiratsudakul, U. Chuachan, S. Kanchanapreechakorn and T. Imaromkul, "Automatic cattle identification based on fusion of texture features extracted from muzzle images," 2018 IEEE International Conference on Industrial Technology (ICIT), Lyon, 2018, pp. 1484-1489.

[9] T. Ojala, M. Pietikainen and T. Maenpaa, "Multiresolution gray-scale and rotation invariant texture classification with local binary patterns," in IEEE Transactions on Pattern Analysis and Machine Intelligence, vol. 24, no. 7, pp. 971-987, July 2002.

[10] C. Cai and J. Li, "Cattle face recognition using local binary pattern descriptor," 2013 Asia-Pacific Signal and Information Processing Association Annual Summit and Conference, Kaohsiung, 2013, pp. 1-4

[11] J. Chen et al., "WLD: A Robust Local Image Descriptor," in IEEE Transactions on Pattern Analysis and Machine Intelligence, vol. 32, no. 9, pp. 1705-1720, Sept. 2010.

IOP Conf. Series: Materials Science and Engineering **853** (2020) 012052 doi:10.1088/1757-899X/853/1/012052

An enhanced stochastic algorithm for optimal contract capacity of furniture stores

Chia-Sheng Tu[a], Ming-Tang Tsai[b,*], Chun-Chieh Huang[b], Kuei-Ching Su[b]

[a]College of Intelligence Robot, Fuzhou Polytechnic, Fujian, China

[b]Department of Electrical Engineering, Cheng-Shiu University, Kaohsiung, Taiwan, R.O.C.

*E-mail address: k0217@gcloud.csu.edu.tw

Abstract. This paper integrated the Grey Theory and Particle Swarm Optimization to propose an Enhanced Stochastic Algorithm (ESA) for dealing with the optimal contact capacity of the furniture store. ESA is used to find the minimal total annual electricity bill and simultaneously find the optimal contract capacity under the operational and system's constraints. The historical demand for a furniture store is first collected in this paper. Grey theory is used to predict the next year demand of every month. Based on the Time-Of-Use (TOU) rates employed by the Taiwan Power Company (TPC), the annual electric bill including basic charge and penalty charge is derived to formulate an objective function. Particle Swarm Optimization (PSO) was adopted to minimize the annual electricity bill and simultaneously find the optimal contract capacity. Practical load consumption data were used to prove the validity of the ESA. Users can able to estimate the contract capacity by themselves and curtail the electricity cost.

Keywords: Contract Capacity, Grey Theory, Particle Swarm Optimization, Time-Of-Use

1. Introduction

The power apparatus in a utility must invest tremendous money to consistently supply sufficient electric energy. The electric energy will immediately be supplied by the generating plants to meet the requirements of consumers. The load demand is widely violated in a power system, it is difficult to precisely predict load demand growth [1-3]. For the users, they need to obtain an economical tariff for curtailing their electricity charge. The appropriated contract capacity must be carefully planned to avoid the penalty charge or overestimated contract [4].

For the utility, contract capacity provides customer load information and render correct development and investment in the future. Customers worrying about electric bill, lead to pay extra money due to overestimate the contract capacity or penalty charge. How to achieve a proper contract amount? Many studies were investigate for specific users. [5] use fuzzy theory to analyse and simulate the peak electricity consumption by using the average temperature and electricity fee structure of each month over the years, and then use the Fuzzy Genetic Algorithm (FGA) to find the best contract capacity value. In [6], Regression Analysis is used to calculate contract capacity. [7] considered both_the planning and dispatch of power scheduling to find the optimal contract capacity. [8] used Evolutionary Programming (EP) to develop a software package for solving the optimal contract capacity of industrial users based on two-section and three-section time-of-use(TOU) rate schedule [9]. [10] uses statistical and optimization tools to take advantage of the information obtained from the historical behaviour of the customer to search an appropriate contract capacity of its future needs. [11] proposed linear program to determination of electricity contract capacity while using much less computation time. However, this problem is an important issues both the utility and the customers. It is worth of deeper discussion.

In this paper, the Grey Theory and Particle Swarm Optimization is integrated to propose an Enhanced Stochastic Algorithm (ESA) for solving this problem according to individual load data and the rate structure of Taiwan Power Company (TPC). The historical demand for a furniture store is first collected. Grey theory [12] is used to predict the load demand of every month. Based on the Time-Of-Use (TOU) rates employed by the TPC, the annual electric bill including basic charge and penalty charge is derived to formulate an objective function. Particle Swarm Optimization (PSO) [13] is adopted to minimize the annual electricity bill and simultaneously find the optimal contract capacity. Practical load consumption data were used to prove the validity of the novel search. Users can able to estimate the contract capacity by themselves and curtail the electricity cost

2. The electricity rate structure

The total electricity bill is the summation of the contract capacity charge, the energy charge, and the penalty charge. The contract capacity charge and penalty charge is basically composed three levels.

1. If the monthly demand is less than or equal to the contract capacity, the customers must pay the bill based on contract capacity. This is called the Demand Charge.
2. If the monthly demand exceeds the contract capacity less than 10% of contract capacity, the exceeding amount will be billed with double Demand Charge Rate.
3. If the monthly demand exceeds the contract capacity over 10% of contract capacity, the exceeding amount will be billed with triple Demand Charge Rate.

In our study, the proposed algorithm is tested on a furniture store. In TOU structure, this furniture store is suitable for two-section rate, which consists only of peak load contract capacity. The objective function of contract capacity can be formulated as follows.

$$\text{Min.} \sum_{i=1}^{24} TDC_i \tag{1}$$

$$TDC_i = \begin{cases} Ar_i \times Xc & when \quad Pc_i \le Xc \\ Ar_i \times Xc + (Pc_i - Xc) \times 2Ar_i & when \quad (Pc_i - Xc)/Xc \le 10\% \\ Ar_i \times Xc + (0.1 \times Xc) \times 2Ar_i + (Pc_i - 1.1 \times Xc) \times 3Ar_i & when \quad (Pc_i - Xc)/Xc \ge 10\% \end{cases} \tag{2}$$

Ar_i :the charge of contract capacity during the i-th month(NT$/KW).

Xc : the contract capacity during i-th month(KW)

Pc_i : the demand during the i-th month (KW)

The summer-season in the TOU tariff represents the season between June 1 to September 30, and the non-summer-season represents the rest of the year. In our study, the TOU rate of contract for summer-season and non-summer-season is 236.2 NT$/KW and 173.2 NT$/KW, respectively.

3. The proposed methodology

The purpose of the optimal contract capacity is to minimize the annual bill of the customers. The grey theory is used to predict the load demand of every month. PSO is adopted to fulfil the optimal contract capacity. The searching procedure is described as follows.

3.1 Grey theory

The Grey Model G(1,1) is developed to the load demand forecast. With historical demand data, a grey differential equation is setup through the data processing and accumulating generation. The procedure of GM(1,1) is describe as follows.

Step 1. In put the initial data

$$x^{(0)} = [x^{(0)}(1), x(2)^{(0)}, x(3)^{(0)}, ..., x(i)^{(0)}] \tag{3}$$

Step 2. Formulated the GM(1,1) model

Step 3. Calculated the accumulating series data $x^{(1)}(k)$

$$x^{(1)}(k) = \sum_{i=1}^{k} x^{(0)}(i) \ , k = 1, 2, ..., n \tag{4}$$

Step 4. Set the grey differential equation through the data process and accumulating generation.

$$\frac{dx^{(1)}(k)}{dk} + ax^{(1)}(k) = u \tag{5}$$

a and u, which are the coefficient value, can be described as matrix $[a,u]^T$. $[a,u]^T$ is obtained by least square method as Eq.(6)

$$[a,u]^T = (B^T B)^{-1} B^T Y_N \tag{6}$$

$$B = \begin{bmatrix} -\frac{1}{2}(x^{(1)} + x^{(2)}) & 1 \\ -\frac{1}{2}(x^{(2)} + x^{(3)}) & 1 \\ \vdots & \\ -\frac{1}{2}(x^{(k-1)} + x^{(k)}) & 1 \end{bmatrix}$$

Step 5. Solve the grey differential equation

$$\hat{x}^{(1)}(k) = [x^{(0)}(1) - \frac{u}{a}]e^{-a(k-1)} + \frac{u}{a} \tag{7}$$

Step 6.The original serial value is calculated by Inverse accumulated generating operation in Eq.(8).

$$\hat{x}^{(0)}(k) = \hat{x}^{(1)}(k) - \hat{x}^{(1)}(k-1) \tag{8}$$

Step 7. Used the Q-test and C-test Variance(C_e) to calculate the residual between the real data and predicting data

$$Qe = \frac{1}{n}\sum_{k=1}^{n}\left|\frac{x^{(0)}(k) - \hat{x}^{(0)}(k)}{x^{(0)}(k)}\right| \tag{9}$$

$$Ce = \frac{1}{n-1}\sum_{k=1}^{n}[x^{(0)}(k) - \hat{x}^{(0)}(k)]^2 \tag{10}$$

3.2 particle swarm optimization

In a PSO system, Birds (particles) flocking optimizes a certain objective function. Each agent knows its best value so far (*Xbest*) and its position. This information is analogy of personal experiences of each agent. Moreover, each agent knows the best value so far in the group ($Gbest$) among other agents around them have performed. The proposed methodology can be summarized in the following steps.

(a) Set the value of input monthly load demand from GM(1,1).
(b) Randomly initialized 30 particle with feasible contract capacity.
(c) Calculate the value of objective function(annual bill) for each particle.
(d) Compare each particle's objective value with the $Xbest$. If the objective value is smaller than $Xbest$, set the value as the current $Xbest$.
(e) Determine the best particle associated with minimal $Xbest$ of all particles, and set the value of this $Xbest$ as current $Gbest$.
(f) Update velocity and position vectors according to (11) and (12) for each particle.

$$V_{i,d}^{j+1} = K \times \begin{bmatrix} V_{i,d}^{j} + c_1 \times rand(0,1) \times (Xbest_{i,d}^{j} - X_{i,d}^{j}) \\ + c_2 \times rand(0,1)(Gbest^{j} - X_{i,d}^{j}) \end{bmatrix} \tag{11}$$

$$X_{i,d}^{j+1} = X_{i,d}^{j} + V_{i,d}^{j} \tag{12}$$

Where,

$$K = \frac{2}{\left|2 - c - \sqrt{c^2 - 4c}\right|}, \quad c = c_1 + c_2, \quad c > 4$$

c_1, c_2: acceleration constant, In this paper, $c_1 = c_2 = 2.05$
$rand(0,1)$: uniform random value with a range of [0,1]
$X_{i,d}^{j}$:dimension d of the position of particle i at iteration j
$V_{i,d}^{j}$: dimension d of the velocity of particle i at iteration j
$Xbest_{i,d}^{j}$: dimension d of the own best position of particle i at iteration j
$Gbest^{j}$: dimension d of the best particle in the swarm at iteration j

(g) The terminating condition is maximal number of iterations. If the preset target is not yet attained, then go back to Step (d) and repeat operation. In this paper, 20 generations is set to stop condition.

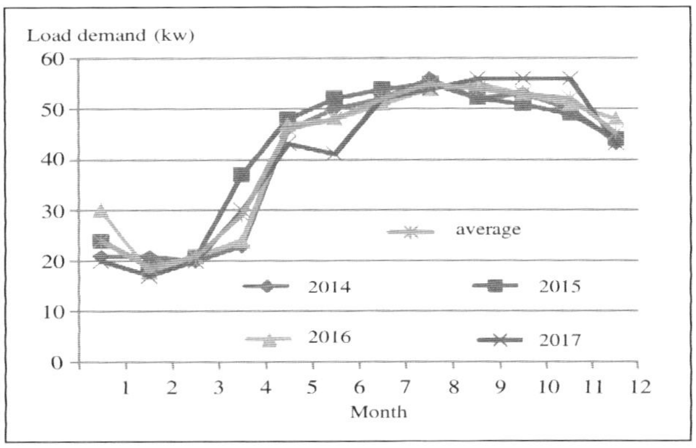

Figure 1. The monthly demand from 2014 to 2017

4. Case study

The proposed algorithm was applied to solve the optimal contract capacity. The associated data is adapted from a furniture store. The monthly demand from 2014 to 2017 is shown in Figure 1. The

numerical computations were performed using the Matlab language on an Intel Core i5-34705, 2.90GHZ CPU computer with 8GB RAM.
To evaluate the accuracy for GM(1,1) in forecasting, the Q-test and C-test all used in this paper. Least-Square(LS), GM Verhulst and the GM(1,1) were all developed and compared to check performance. Table 1 is the average accuracy of the various methods. With the same training data, the GM(1,1) shows an overall better performance than LS and GM Verhulst. Figure 2 shows the error curves of monthly in the various methods. From the Figure 2, GM(1,1) is also better accuracy than other methods.

Table 1. The average accuracy of the various methods

Method		LS	GM(1,1)	GMVerhulst
one-year average	Q-test	0.026	0.027	0.141
	C-test	0.319	0.408	3.007

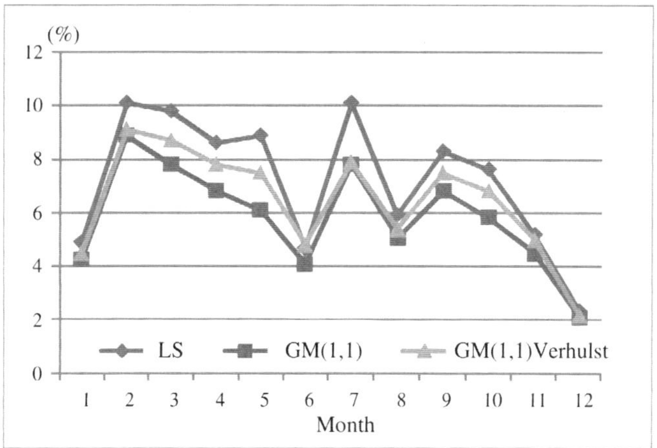

Figure 2. The error curves of monthly

PSO is used to solve the optimal contract capacity problem of the furniture store. The original contract capacity of furniture store is 49kw. Two cases including with TOU rate and non-TOU rate have been studied based on the proposed algorithm. In our study, the optimal contract capacity is 54kw. Table 2 displays the comparison between with TOU rate and non-TOU rate. PSO approach can lead to significant saving of electricity bill no matter what types of contract are used. It can be obviously seen that the electricity bill can be tremendously curtailed through the calculations of our program.

Table 2. The comparison between with TOU rate and non-TOU rate

	Total cost with optimal contract (NT$)	Total cost without optimal contract (NT$)
Non-TOU	446703.43	448829.23
TOU	491304.08	493429.88

*Original contract capacity=49kw
*Optimal contract capacity=54kw

Figure 3 is the convergent characteristics of the optimal contract capacity and Figure 4 is the convergent characteristics of the annual bill. The convergent generation is about 4-th generation.

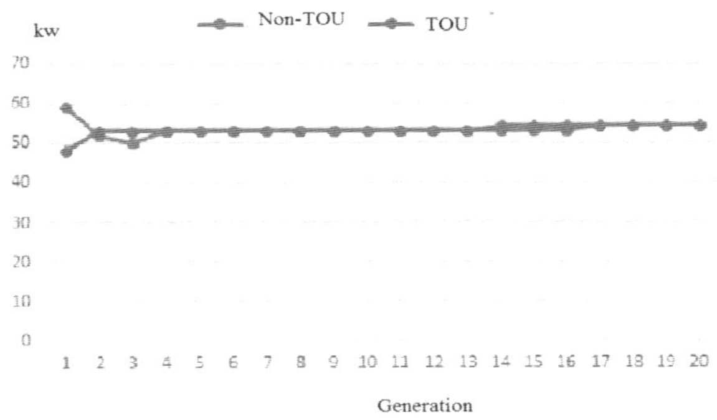

Figure 3. The convergent characteristics of the optimal contract capacity

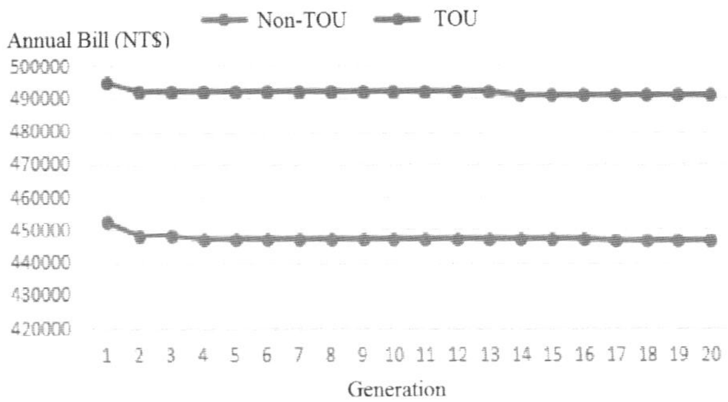

Figure 4. The convergent characteristics of the annual bill

5. Conclusions

This paper proposes an ESA to predict the load demand calculate the optimal contract capacity for furniture store. Accurately constructing the monthly demand by grey theory is a very useful tool to predict the monthly demand. PSO calculates the optimal contract capacity of low-voltage power tariffs suitable for furniture stores including two modes: non-TOU and TOU. Field demand data of customers are examined to prove the validity of this software. From the execution, users can view the billing for various contract capacities, and also find the optimal one.

References

[1] N. Hassan, et al., Customer Engagement Plans for Peak Load Reduction in Residential Smart Grids, IEEE Transactions on Smart Grid, 2015, pp.3029-3041.

[2] Ahmed S. A.et al., Optimal ESS Allocation for Load Management Application, IEEE Transactions on Power Systems, 2015, pp.327-336.

[3] Ali K. Ferdavani and H. B. Gooi, The very fast method for contracted capacity optimization problem in Singapore. 2016 IEEE Region 10 Conference (TENCON), 2016, pp.2100-2103.

[4] M.R. Lee et al., The maximum power demand forecasting with fuzzy theory, 2010 International Symposium on Computer, Communication, Control and Automation (3CA), 2010, pp419-422.

[5] Feng Liang, Cycle time estimation approach based on mathematical programming and statistical regression. 2010 IEEE 17Th International Conference on Industrial Engineering and Engineering Management, 2010, pp.742-745.

[6] S.H. Liao rt al., Optimal Energy Storage System planning for microgrids with contract capacity constraint. 2014 International Power Electronics Conference (IPEC-Hiroshima 2014 - ECCE ASIA), 2014, pp.1452-1457.

[7] M.T. Tsay et al., Optimal Contracts Decision of Industrial Customers," International Journal of Electrical Power & Energy Systems, 2001, pp.795-803.
[8] M. A. Fernández et al., Cost optimization of electrical contracted capacity for large customers, International Journal of Electrical Power & Energy Systems, 2013, pp.123-131.
[9] C.Y. Chen and C.J. Liao, A linear programming approach to the electricity contract capacity problem. Applied Mathematical Modelling, 2011, pp.4077-4082,
[10] Q. Li at al., Fundamental definitions and calculation rules of grey mathematics: A review work, Journal of Systems Engineering and Electronics, 2015, pp.1254-1267.
[11] Time-of-Use Rate Structure of Taiwan Power Company, May, 2018.
[12] Y. Shi and R.C. Eberhart, A modified particle swarm optimizer. Proceedings of the IEEE International Conference on Evolutionary Computation Anchorage 1998, pp.69-73.

Complex morphological filtering algorithm based on multi-structure elements

WANG Yiyan[1,2] LUO Jiang[1]

[1]School of Intelligent manufacturing, Sichuan University of Arts and Science, Dazhou 635000, Sichuan, China;
[2]Dazhou Industrial Technology Institute of Intelligent Manufacturing, Dazhou 635000, Sichuan, China

E-mail: wang_yiyan8207@163.com

Abstract: Morphological filter is a kind of nonlinear filter. Based on the erosion and dilation operations a complex morphological filter was proposed with multi-structure elements. In order to protect the image detail effectively, a noise detect method was introduced firstly. The noise pixels are removed using the proposed filter, while signal pixels hold their pixel value and are left unprocessed. The experimental results demonstrate that the proposed algorithm has good filtering performance, which can preserve edge details effectively during denoising. For both gray image and binary image with salt & pepper noise pollution, the proposed algorithm is superior to the conventional morphological methods in the aspect of objective performance evaluation and subjective visual effect. Meanwhile, the proposed method can also be extended to the color image processing.

Keywords: nonlinear filter; noise detection; multi-structural elements; morphological filtering

1. Introduction

In engineering applications, images are often affected by various noises during acquisition and transmission, degrading the final image quality. In order to facilitate the subsequent processing of images (such as edge detection, image segmentation, image recognition, etc.) [1], image filtering has become a key part of image preprocessing. The challenge is to remove noise and effectively protect the edge details of the image. Salt and pepper noise is the most common kind of noise, which usually appears as the gray-scale extremum of the image neighborhood, and the gray value of the noise pixel is independent of the gray value of the adjacent uncontaminated pixels. The median filtering method is often used for the removal of salt and pepper noise [2], but the traditional median filtering is affected by the filtering window, and it cannot be considered at the same time in terms of detail protection and noise removal. The morphological filter created by G. Matheron and J. Serra [3] in the early 1980s is a new type of nonlinear filter developed from mathematical morphology. Due to its fast parallel implementation, It has received extensive attention in the field of signal processing and image analysis [4-5]. The morphological filter is based on the set structure characteristics of the image, and uses a predefined structural element (corresponding to a filter window) to match or partially correct the signal to achieve the purpose of extracting the signal and suppressing noise[6-7]. Based on the traditional morphological transformation, a complex morphological filter based on multi-structural elements is constructed. At the same time, in order to protect the image details more effectively, the

IOP Conf. Series: Materials Science and Engineering **853** (2020) 012053 doi:10.1088/1757-899X/853/1/012053

noise is first detected before filtering, and only the noise points are processed during filtering, and the pixel values are kept unchanged for non-noise points. Simulation results show that the proposed method has better filtering performance. Compared with the traditional filtering method, this method can achieve better visual effects and objective evaluation indicators for both grayscale images and binary images. At the same time, the method can be applied to color image processing.

2. The proposed method

2.1 Noise model

First, we give a theoretical model of salt and pepper noise, whose expression is as follows [8]:

$$y_{i,j} = \begin{cases} S & \text{with probalibity } p \\ x_{i,j} & \text{with probalibity } 1-p \end{cases} \qquad S \in [0, 255] \tag{1}$$

In the above formula, $x_{i,j}$ is a real image, and $y_{i,j}$ is a noisy image. The noise density $p \in [0,1]$, Figure 1 below shows the experimental results of adding a noise to a constant image. It is obvious from its histogram that the salt and pepper noise exhibits gray-scale extremum.

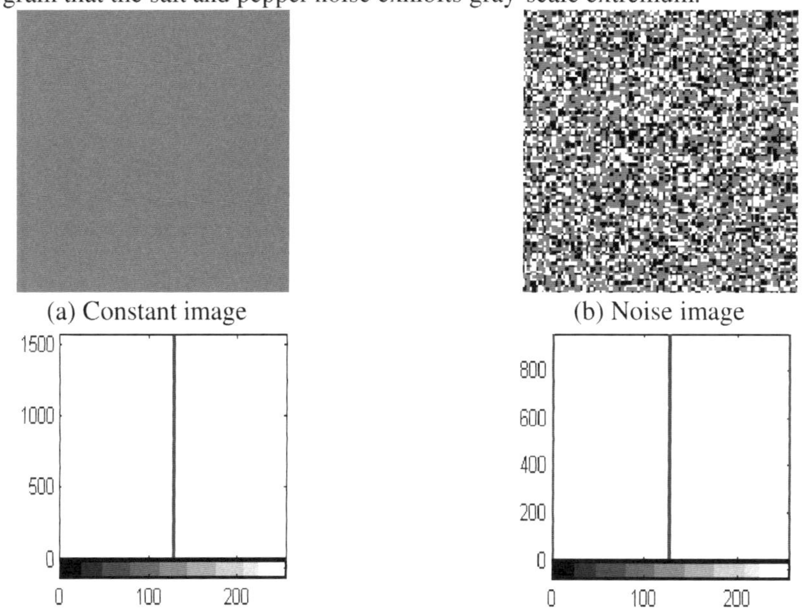

(a) Constant image (b) Noise image

(c) Histogram of (a) (d) Histogram of (b)

Figure1. Adding salt and pepper noise (50%) to the constant image

2.2 Noise detection

Let matrix $u_{M_1 M_2}$ denote the noise gray image to be detected of size $M_1 \times M_2$, Use A to represent the collection of all pixels inside, $A = \left\{(i,j) \mid 1 \le i \le M_1, 1 \le j \le M_2; M_1, M_2 \in Z^+\right\}$. $\Psi_{M_1 M_2}$ is a noise identification matrix whose elements are represented by "1" and "0".

2.2.1 The first noise detection:

Let $S_{i,j}^{w}$ denote the size $w \times w$ (w is odd), The center point is the window of $u(i,j)$. The initial detection is performed for each pixel point $u(i,j)$ in the image according to the adaptive median method [9], and the result is identified by "1" or "0", namely:

$$\Psi_{i,j}=\begin{cases}1 & u(i,j) \text{ is the candidate noise point}\\ 0 & \text{otherwise}\end{cases} \tag{2}$$

Estimate the noise density τ of the image as follows:

$$\tau=\frac{1}{M_1M_2}\sum_{i=1}^{M_1}\sum_{j=1}^{M_2}\Psi_{i,j} \tag{3}$$

If τ exceeds a given threshold τ_0, the candidate noise point is treated as a true noise point; otherwise, it is secondarily detected.

2.2.2 The second noise detection

For all candidate noise points that satisfy $\Psi_{i,j}=1$, define the mean of all signal points in window $w\times w$:

$$\bar{u}(i,j)=\frac{1}{W}\sum_{r=-(w-1)/2}^{(w-1)/2}u(i+r,j+r) \tag{4}$$

Where W represents the number of all signal points in window $w\times w$, and $u(i+r,j+r)$ represents the signal point in the window.

If the absolute difference between the gray values $u(i,j)$ and $\bar{u}(i,j)$ of the current candidate noise point exceeds a certain threshold, it is regarded as a true noise point, namely:

$$\Psi_{i,j}=\begin{cases}1 & |u(i,j)-\bar{u}(i,j)|>T \quad \text{and} \quad \Psi_{i,j}=1\\ 0 & \text{otherwise}\end{cases} \tag{5}$$

In this paper, T is defined by the mean square error of all signal points in window $w\times w$:

$$T=\sqrt{\frac{1}{W}\sum_{r=-(w-1)/2}^{(w-1)/2}\left[u(i+r,j+r)-\bar{u}(i,j)\right]^2} \tag{6}$$

2.3 Multi-structural element composite morphological filter

The traditional morphological filters mainly use morphological open operations, closed-loop, and combinations of them [10-11].

2.3.1 Gray-scale corrosion

Let $f(x,y)$ denote the input image, $B(i,j)$ represents a structural element (function), gray-scale corrosion can be defined as:

$$(f\Theta B)(x,y)=\min\left\{f(x+i,y+j)-B(i,j)\,|\,(x+i,y+j)\in D_f;(i,j)\in D_B\right\} \tag{7}$$

In above formula, D_f and D_B are the domains of functions f and B, respectively, The displacement parameter $(x+i,y+j)$ must be included in the domain of the function f. Corrosion of gray-scale images can yield two results: if all structural elements are positive, the output image will tend to be darker than the input image; the effect will be diminished after brightly etched in areas smaller than the structural elements The degree of attenuation depends on the gray value surrounding the luminance region and the shape and amplitude of the structural element itself[12].

Corrosion operation is performed point by point. The operation result of a point is the difference between the gray value of the point in a local range and the corresponding point in the structural element, and the minimum value is selected[13]. It can be seen that after the etching operation, the gray value of the point where the gray value of the edge portion is relatively large is lowered, so that the edge shrinks to the area where the gray value is higher than the adjacent area.

2.3.2 Gray-scale expansion

Gray-scale expansion is defined as:

$$(f \oplus B)(x, y) = \max\left\{f(x-i, y-j) + B(i, j) \mid (x-i, y-j) \in D_f; (i, j) \in D_B\right\} \tag{8}$$

In above formula, D_f and D_B are the domains of functions f and B, respectively. The displacement parameter $(x-i, y-j)$ must be included in the domain of the function f. Since the expansion operation is the maximum value of $f+B$ selected in the neighborhood defined by the shape of the structural element. Thus, generally, the expansion processing method for gray-scale images can obtain two results: if all structural elements are positive, the output image will tend to be brighter than the input image; black detail reduction or removal depends on the value and shape of structural element correlation in the expansion operation.

The expansion operation is performed point by point. The calculation involves the gray value of the point around it and the value of the structural element. It is actually the sum of the gray value of the point in the local range and the corresponding point in the structural element value[14]. Therefore, after the expansion operation, the edges are extended.

2.3.3 Gray-scale open operation

The opening operation of the image is to first perform the etching operation on the gray-scale image with the structural element, and then perform the expansion operation on the obtained result, namely:

$$f \circ B = (f \Theta B) \oplus B) \tag{9}$$

2.3.4 Gray-scale closed operation

The closing operation of the image is to first expand the gray-scale image with the structural element, and then perform the etching operation on the obtained result, namely:

$$f \bullet B = (f \oplus B) \Theta B) \tag{10}$$

Since the corrosion removes less bright detail and there is no recovery in subsequent expansion, in practical applications, the gray-scale opening operation is often used to remove small bright spots relative to structural elements while retaining all grays. Degrees and large bright area characteristics are unchanged. Similarly, the expansion operation removes less dark detail, and these details are not recovered in the erosion operation[15]. Therefore, the closed operation of the gray-scale image is often used to remove smaller dark details while maintaining the overall gray level and Large, bright areas are unaffected.

2.3.5 Morphological filter

Combining open and closed operations eliminates noise. If you use a small structural element to open an operation and then close an image, it is possible to remove a similar noise structure in the image that is smaller than the structural element[16]. In order to simultaneously remove peak (positive pulse) and bottom valley (negative pulse) noise in the image, open-closed (OC) and closed-close (CO) filters can be constructed in open and closed cascades, which are defined as follows:

$$OC(f(i, j)) = (f \circ B \bullet B)(x, y) \tag{11}$$

$$CO(f(i, j)) = (f \bullet B \circ B)(x, y) \tag{12}$$

2.3.6 Composite morphological filter

In morphological filtering, the selection of structural elements plays a key role in the filtering results and performance. Using a fixed structural element to construct a filter operator, the noise immunity is limited, and it is difficult for a single structural element to detect the edges of various geometric shapes [17-18]. Based on this, this paper constructs a new composite morphological filter $\psi(f(x,y))$ whose expression is:

$$\psi(f(x,y)) = \frac{1}{2}((f \circ B_1 \bullet B_2)(x,y) + (f \bullet B_1 \circ B_2)(x,y)) \tag{13}$$

among them , $B_1 = \begin{bmatrix} 1 & 0 & 1 \\ 0 & 1 & 0 \\ 1 & 0 & 1 \end{bmatrix}$, $B_2 = \begin{bmatrix} 1 & 0 & 1 & 0 & 1 \\ 0 & 1 & 1 & 1 & 0 \\ 1 & 1 & 1 & 1 & 1 \\ 0 & 1 & 1 & 1 & 0 \\ 1 & 0 & 1 & 0 & 1 \end{bmatrix}$

In order to protect the image details more effectively, the noise is first detected by the method of Section 1.2 before filtering. Only the noise points are processed during the filtering, and the pixel values are kept unchanged for the non-noise points.

At the same time, the method can be applied to color images. Based on the RGB space, the three components of R, G and B in the color image to be processed are filtered by the equation (13), and then the filtering results of the three components can be superimposed.

3. Experimental results and analysis

In order to verify the filtering performance of the proposed method, this section deals with the image of salt and pepper noise pollution. In Experiment 1, a grayscale image "rice" and a binary image "imagetech" were used as test objects. The morphological filters such as open, closed, open-close, and closed-open are used for processing, and the peak signal-to-noise ratio (PSNR) is used as an objective criterion for evaluating the filtering performance. The experimental results are shown in Figure 2 and Figure 3. Table 1 gives the corresponding results of the objective performance evaluation indicators.

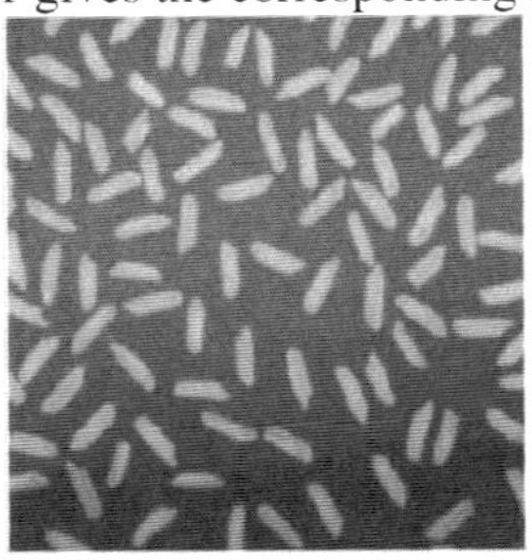

(a) Original image

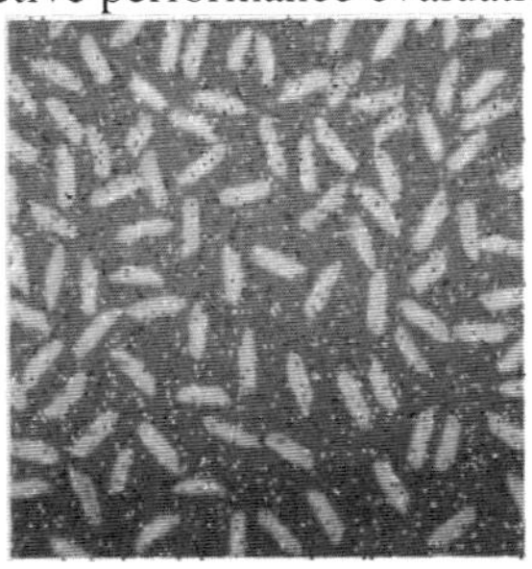

(b) Noise image

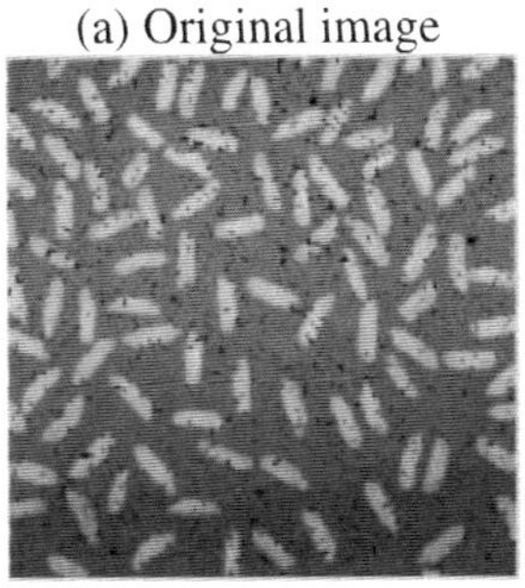

(c) Open operation

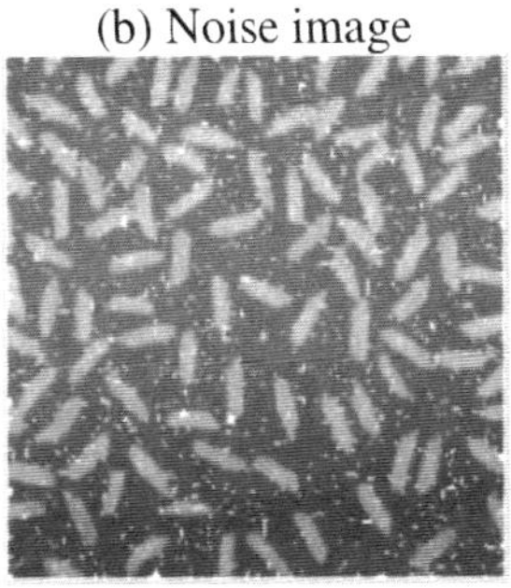

(d) Closed operation

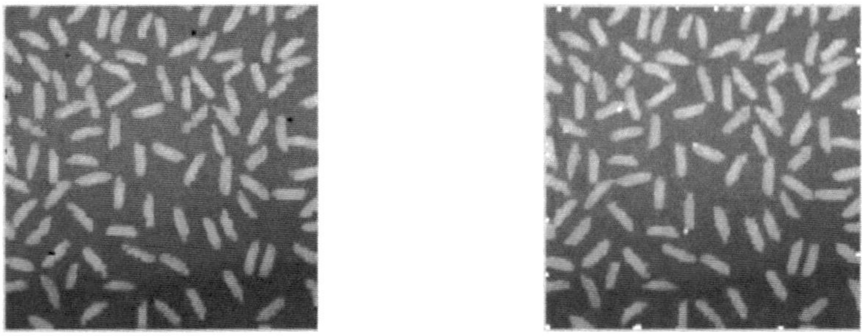

(e) Open-close filter (f) Close-open filter

Figure 2. Gray-scale image experiment results

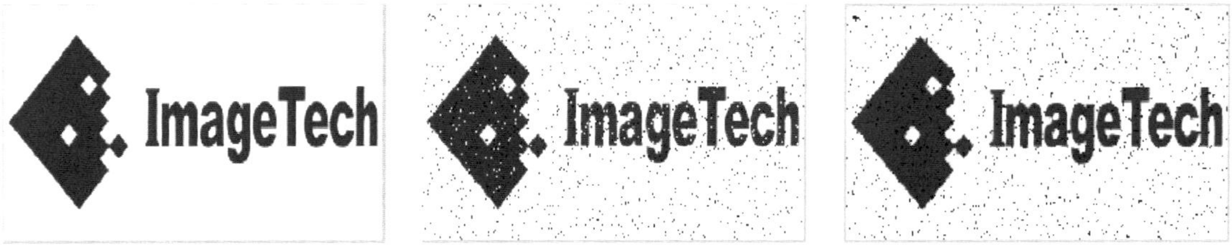

(a) original image (b) noise image (c) open operation

(d) Closed operation (e) Open-close filter (f) Close-open filter

Figure 3. Experiment results of Binary image

Table 1. Comparison of objective performance evaluation indicators (PSNR) of different morphological filters

testing object	open operation	Closed operation	Open-close filter	Close-open filter
Gray-scale image "rice"	18.6607	19.1454	21.3279	21.6610
Binary image "imagetech"	15.9032	22.7749	27.7963	27.8318

As can be seen from the above experimental results in Figure 2 - 3, the open operation removes the small bright spots (salt noise) relative to the structural elements while retaining all gray levels and large bright area features. The closed operation removes less dark detail (pepper noise) while the relative gray level and larger bright areas remain unaffected. The open-close and close-open filters simultaneously filter out salt and pepper noise. As can be seen from Table 1, the closed-close filter can achieve better results than the other three operations.

Experiment 2, using the grayscale image "lena" as the test object. The open-closed, closed-open filters and the methods herein are used for processing. The experimental results are shown in Figureure 4 below.

(a) Lena original image (b) Noise image

(c) Open-close filter (d) Close-open filter (e)Proposed method

Figure 4. Experiment results of Lena image

It can be seen from the filtering results in Figure. 4 that all three morphological filters can effectively remove the salt and pepper noise, and the method of this method has some improvement over the visual effect of the open-close or close-open filter alone. The original image is moderately bright and dark.

In Experiment 3, the color image "peppers" was selected as the test object. Under the condition of different noise density pollution, the method is filtered by the method of this paper. The experimental results are shown in Figure 5.

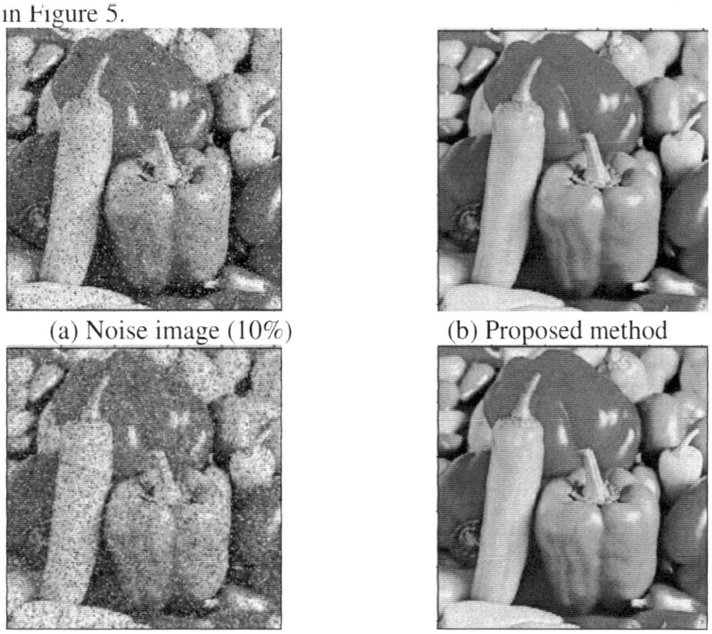

(a) Noise image (10%) (b) Proposed method

(c) Noise image (20%) (d) Proposed method

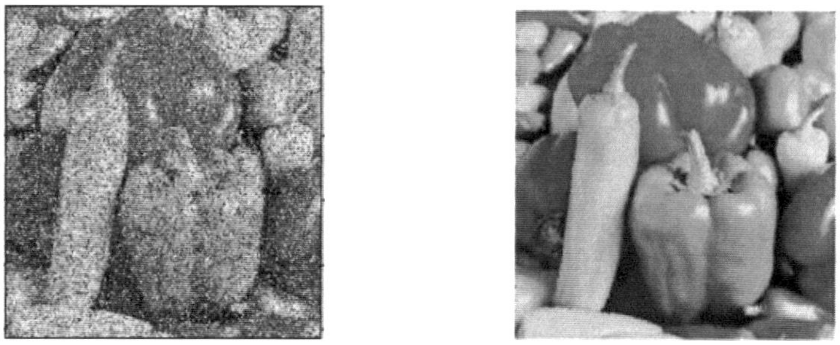

(e) Noise image (30%) (f) Proposed method

Figure 5. Experiment results of peppers color image

It can be seen from Figure 5 that for the color images with different density noise pollution, the method can achieve good filtering effect.

4. Conclusion

A complex morphological filter is constructed by using multiple structural elements based on the traditional morphological transformation. At the same time, before the filtering, the noise detection process is added, which effectively avoids the diffusion of noise points in the filtering process. The experimental results demonstrate the effectiveness of the proposed method. Compared with the traditional morphological filtering method, this method can achieve better visual effects and can be further applied to color image filtering.

Acknowledgments

This work was supported in part by Scientific Research Fund of Sichuan Provincial Education Department, China under Grant No.18ZB0509, by Science and Technology Program of Dazhou, China under Grant No.KJJ2015001, and by a Grant from the Key Programs of Sichuan University of Arts and Science, China under Grant No.2017KZ003Z, which are greatfully acknowledged.

References

[1] Yao Min. Digital Image Processing [M]. Beijing: Mechanical Industry Press, 2006.

[2] Omer A A , Hassan O I , Ahmed A I , et al. Denoising CT Images using Median based Filters: a Review[C]// 2018 International Conference on Computer, Control, Electrical, and Electronics Engineering (ICCCEEE). 2018.

[3] Chen P Y , Lien C Y . An Efficient Edge-Preserving Algorithm for Removal of Salt-and-Pepper Noise[J]. IEEE Signal Processing Letters, 2008, 15:833-836.

[4] Lu C T , Chou T C . Denoising of salt-and-pepper noise corrupted image using modified directional-weighted-median filter[J]. Pattern Recognition Letters, 2012, 33(10):1287-1295.

[5] Bai T , Tan J , Hu M , et al. A novel algorithm for removal of salt and pepper noise using continued fractions interpolation[J]. Signal Processing, 2014, 102:247-255.

[6] EVANS A N, LIU X U. A morphological gradient approach to color edge detection[J]. IEEE Trans on Image Processing, 2006, 15(6): 1454-1463.

[7] SONG Chunyan, Zhaomeng. MI-Cored: A Mathematical Morphological Filtering Method by Using Structural Element of “M” Glyph Based on Machine Vision[J]. Computer Science and Application, 2017,7(05): 438-443.

[8] Sree P S J , Pattnaik P K , Ghrera S P . A Novel Algorithm for Suppression of Salt and Pepper Impulse Noise in Fingerprint Images Using B-Spline Interpolation[J]. Advances in Intelligent Systems and Computing, 2015, 328:521-528.

[9] WANG Chao, YE Zhong-fu. Salt-and-pepper noise removal by adaptive median filter and TVinpainting[J]. Journal Of university of science and technology of China, 2008(3): 282-287.

[10] YU Wen, MAO Qiqi, WEN Hong. Morphological Filtering of UV Images in Smart Grid [J].

Smart Grid, 2014,4(04): 145-153.
[11] Zhang Jianmin, Hu Yingsa, Zeng Zhaofa, Du Wei, Zhang Ling, Huai Nan, Hu Zhipeng. Edge Enhancement Method Based on Mathematical Morphology for Potential Field Data[J]. Journal of Jilin University(Earth Science Edition), 2019, 49(3): 845-856.
[12] Li Jie, Peng Yueying, Yuan Chang'an, et al. Edge Thinning Based on Mathematical Morphology Thinning Algorithm[J]. Journal of Computer Applications, 2012, 32(2):514-516.
[13] Chen Hui, Guo Ke, Hu Ying. A Study on Application of Mathematical Morphology to Seismic Signal Processing[J]. Progress in Geophysics, 2009, 24(6):1995-2002.
[14] Huang W L, Wang R Q, Zhang D, et al.Mathematical Morphological Filtering for Linear Noise Attenuation of Seismic Data[J]. Geophysics, 2017, 82(6):V369-V384.
[15] Huang W L, Wang R Q, Zu S H, et al.Low-Frequency Noise Attenuation in Seismic and Microseismic Data Using Mathematical Morphological Filtering[J]. Geophysical Journal International, 2017, 211(3):1296-1318.
[16] Li L L, Ma G Q, Du X J. Edge Detection in Potential-Field Data by Enhanced Mathematical Morphology Filter[J]. Pure and Applied Geophysics, 2013, 170(4):645-653.
[17] Deng Caixia, Wang Guibin. Improved Algorithm of Morphology in Edge Detection for Noise Resistance[J]. Journal of Data Acquisition and Processing, 2013, 28(6):739-745.
[18] Tong Xiaoling. The Research of Image Edge Detection Algorithm Based on Mathematical Morphology[D]. Xi'an:Chang'an University, 2013.

A decision-making technique for solving order allocation problem using a genetic algorithm

Ijaz Ahmad[1], Yan Liu*[1], Danish Javeed[1], Shahab Ahmad[2]

[1]School of Computer Science and Technology, Changchun University of Science and Technology, Changchun 130022, China.
[2]School of Economics & Management, Chongqing University of Post & Telecommunication, Chongqing, 400065, China.

*e-mail:yanl@cust.edu.cn, engeerijaz04@gmail.com

Abstract. The selection of proper suppliers is one of the most complicated works of the purchasing department. Today, supplier selection includes different conflicting objectives. Because of contradictory multi-objective supplier selection is solving by using the decision-making technique. This paper is presented a modified genetic algorithm by using a combination of crossover operators, Order crossover (OX), Simulated binary crossover (SBX) to assign the optimal order quantities to each supplier, with criteria of transportation cost, product quality, and delivery time with a quantity discount. The result shows that the modified genetic algorithm is an allocated optimal order for multi vendors with improves quality as well as less computational times.

Keywords. Genetic Algorithm; Multi-objective; Order crossover, Simulated Binary Crossover

1. Introduction

Today fierce competitive environment, many industries the product cost is related to the cost of raw material and parts because, for example, the industries of automotive/textile machinery have more than 70% revenues due to purchasing components and parts from external sources that's why the literature of supplier selection is focused from the previous three decades. In Supplier selection, the appropriate supplier is selected to supply the sources and allocate optimal order for each vendor. Additionally, the multi-objective nature of vendor selection, select more suppliers based on different criteria is an issue, because due to various contradictory multi-objective factors [22][23][24].
Supplier selection can be categorized into four sections. (1) Acquiring the desired aspects through supplier selection (2) description of product criteria(3) appropriate supplier (4)select profer supplier based on supplier criteria[21]. The criteria of modern supplier selection are transportation cost, product quality, and timely delivery. Under these criteria, the vendors have a different level of performance. For example, a vendor that has less price may have less quality or poor delivery given by industries. Besides this, due to multi-objective supplier selection, the appearance of a pricing discount is also a difficult task for purchasing managers to make the best policy of purchasing. At present, there are different models for various product discount models that represent the different range of price breaks for each vendor. Quantity discounts depend on the sale of volume. The issue of purchasing is solved by using the MP(Mathematical Programming) techniques. It represents various order quantities for each vendor with different discounts provided by the supplier. In this research, a mathematical model is evaluated to various order quantities, a quantity discount, with constraints of each vendor for

the selection of appropriate suppliers. By solving such complex multi-object optimization problems is needed meta-heuristic technique (Genetic algorithm).In this paper, a Multi-object genetic algorithm is used to select proper suppliers based on various criteria. This research has the flowing sections; section 3 contains the Mathematical model to represents objectives function with constrains, section 5 methodology part, which have an explanation of the multi-object genetic algorithm, section 6 described experimental results, section 7 represents the conclusion and future work.

2. Literature review

Regarding the decision-making technique for order allocation problem with price discounts in supplier selection, there exists a wide literature. In the literature of [1], [2], [3], highlighted supplier selection problem with decision-making technique based on multi_objective optimization problems (MOPs). These articles provided systematic approaches for the purchasing department to estimate and achieve the suppliers with optimal order allocations base on multiple various criteria.

In the literature of Nourmohamadi Shalke, Paydar [4], have developed the idea of a multi-objective model to address supplier selection, order allocation in a multi-period, multi-item, and multi-supplier problem. In another study, Tavana, Yazdani, and Di Caprio [5] explained an integrated multi-criteria decision making(MCDM) approach for the selection of proper vendors with optimal order. A review of decision-making techniques in order allocation is as follows [6-17].

Table 1. The literature of order allocation problem with DM techniques

Literature	DM techniques	Order allocation
Arunkumar, N., L. Karunamoorthy.[6]	Genetic algorithm(GA)	✓
Hamdan, Sadeque.[7]	analytic hierarchy process (AHP)	✓
Scott, James, et al. [8]	analytic hierarchy process (AHP)	✓
Lee et al. ,[10].	Multi-objective decision analysis	✓
Tsai & Hung.[11]	Multi-objective programming	✓
Yeh and Chuang. [12]	Multi-objective genetic algorithm	✓
Kannan et al. [13]	FAHP, FTOPSIS, multi-objective programming	✓
Nourmohamadi , Parvin. [14]	multi-choice goal programming	✓
Azadnia et al. [15]	analytic hierarchy process (AHP)	✓
Zhang, Ju-liang. [16]	branch-bound algorithm	✓
Hamdan, Sadeque. [17]	analytic hierarchy process (AHP)	✓
Jadidi, O. M. I. D. [18]	normalized goal programming	✓
Govindan, Kannan. [19]	Fuzzy TOPSIS	✓

Recently, quantity discount schemes have received substantial attention from researchers. When a purchaser places a large order, quantity discounts are considered, and the unit purchase price is decreased by the supplier under a predetermined scheme. Hence, studies have been done that consider the problems of supplier selection, order allocation, and lot sizing, each affecting supply chain activities (Azadnia et al.[13], Mohave, Emadikhiav, & Parsa, [18], Rezaei, Davoodi, Tavasszy, & Davarynejad,[19]).

From quantity discount strategies,[6] under different business volume discounts schedules a mathematical model to represents the mathematical formulation of a single item. Most realistic optimization problems, like supplier selection, have more than one objective need for simultaneous optimization. Frequently, the conflict in the objectives. For example, the lowest per-unit price of the vendor would not have the optimal delivery. Hence the company must analyze in the making decisions the tradeoffs among the relevant criteria. The multi-objective approach allows different criteria in the evaluated process with their natural units instead of using a common unit for measurement [20].In order allocation problems to select proper vendors needs decision-making techniques. This paper presents a multi_objective programming model for select vendors.

3. MOP: formulation Objectives, Various constraints

The problems of supplier selection have appropriate constraints, multiobjective with various settings. This paper proposes a multiobjective programming model to demonstrate the supplier data of textile machinery, which consists of various components, such as a heater, feed. Each component represents by there mathematical notation are as follows [6],

N Entire number of vendors.

n Entire number of components.

γij Percentage of defective of the ith component from vendor j.

βij Percentage of late delivery for the ith component from vendor j.

λij Cost of ith component per uni from vendor j.

Xij Number of i purchased components from vendor j.

Cij Capacity of j vendor for the ith component.

Wij Vendor j minimum order supply to ith component.

Bij Minimum business for the vendor j for the ith component.

Mij Vendor j maximum trade for the ith component.

η'ij ith component of cost per unit from vendor j with no discount.

η"ij ith component of cost per from vendor j with no discount.

Mij ith component of middle-order quantity of vendor j.

Φi Upper boundary desired from the purchase manager for the number of defective

items for the ith component.

Ωi An Upper boundary desired from the purchase manager for the number of late delivery for the ith component.

ξij selection of vendor j for supplying the ith component shows by A binary variable.

3.1 Objectives function

In this problem of appropriate selection of vendors the objective function is to minimize multi factors, total cost, late deliveries, and defective items, represented by mathematical notation Y1, Y2, Y3 as follows [31],

Objective 1: Minimize no of defective items(Quality)

$$Min\,Y1 = \sum_{j=1} \gamma ijXij \quad j = 1, 2 \ldots.. n \tag{1}$$

Objective 2: Minimize time delivery

$$Min\,Y2 = \sum_{j=1} \beta ijXij \quad j = 1, 2 \ldots.. n \tag{2}$$

Objective 3: Minimze product cost

$$Min\,Y3 = \sum_{j=1} \lambda ijXij \quad j = 1, 2 \ldots\ldots n \tag{3}$$

3.2 Vendor constraints

At the point of view of the capacity of vendors, it has some business constraints, by which the capacity of vendors would not be exceeded during order allocation. As a business policy, the number of components should meet the requirements of manufacturers, as min order and max order.

Constraint 1:(Vendors Capacity)

$$X\,ij \leq (Cij \times \xi ij)\ \forall\ i=1,2,\ldots,n \quad j=1,2,\ldots N. \tag{4}$$

Constraint 2: (Min Quantity):

$$X\,ij \geq (Qij \times \xi ij)\ \forall\ i=1,2,\ldots,nj=1,2,\ldots,N. \tag{5}$$

4. Methodology

A multi-criteria supplier selection problem with considers to objectives is solved by using a step-by-step procedure one by one. By using real coded multi_object genetic algorithms to get an optimal order quantity for each supplier. The solution of MOGA is compared with other techniques. It examined that GA provides an optimal solution. This will be helpful for the purchase department manager to find the alternate set of suppliers. The problems of combinatorial optimization are involved with the optimal allocation of restricted resources to satisfy the desired objectives. The possible alternatives constraints on basic resources restrict all the various possible solutions that considered being feasible. These supplier selection problems the combinations of the vendor are input to the Multi-object GA optimize the optimal order allocation based on constraints, objectives, and business values.

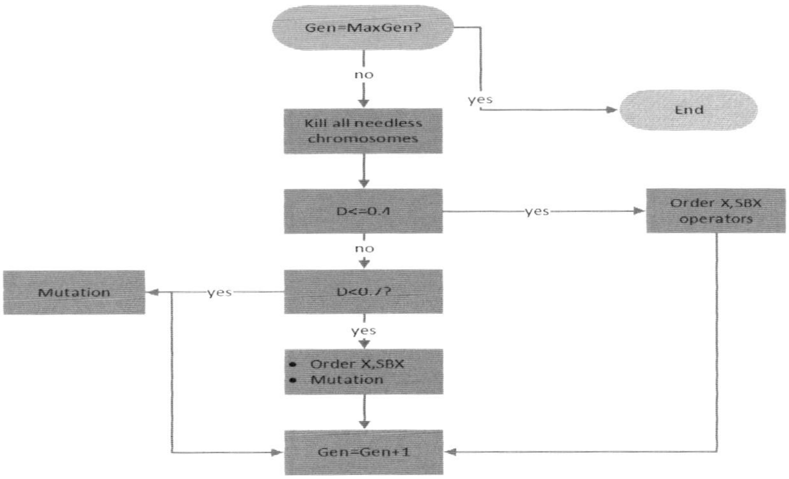

Figure 1. Evolution of multi-object genetic algorithm.

4.1 The evolution process of the Multi_object Genetic algorithm

The evolution process of the multi_object genetic algorithm is just like the guidance system. In Figure 2.shows, the evaluation process of MOGA.In the given below the figure, use the ratio of several needless chromosomes in each generation to the population size. The values of D>0.7, the number of needles chromosomes will be high. So the population will show a very less the desired solution, then for a satisfactory solution mutation operator is used, and this way, GA randomly searches the space to get the optimal solution, if the values of D<0.4, then the number of needless chromosomes is very low, the chance is more to select a satisfactory solution. If the value of D is very small, say <= 0.3, its shows that ratio of the optimal is high in the population, and combination of order crossover with simulated binary crossover (OX, SBX) operator apply to produce different solutions. The value of D found out mined in(trial /error approach, for the suitability of the problem.

4.2 Combination of order & simulated Binary crossovers (OX, SBX)

Crossover operators in one of the most important processes of genetic algorithms. The process of crossover is convergence operation by which the population of the given problem leads towards local minima or maxima. This paper presents the combination of two crosses over operators (simulated binary cross over & order crossover. In figure 2. represents the combination of (OX, SBX).

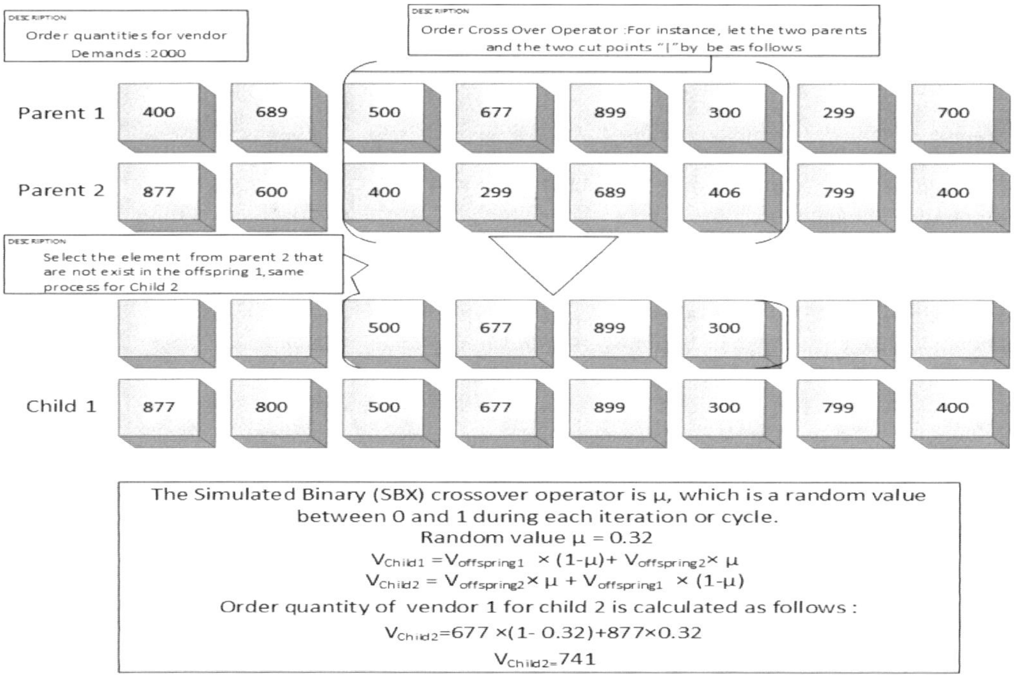

Figure 2. Combination of(SBX,OX) operators.

4.5 Mutation operator

A random way is thereby eliminating the local minima, which the problem being stuck. A general way to represents the mutation operator would change one or more bit or genes values. Mutation operators perform a vital role in genetic algorithm, by changes or eliminating the genes in the population to provide divergence or premature convergence. The mutation is the opposite process of crossover. For this problem, the rate mutation is 0.07; if the total order exceeds the demand, then the mutation takes place.

4.6 Fitness function

This section represents the fitness value of each chromosome by using the mathematical model. The fitness function shows how much the solution is fit or the approaches to compare the given solution with the optimal solution in the given problem. In this problem, the first fitness function of multi_objectives converts into a single objective of the combination of the objective function, e.g., F=f1+f2. We can obtain objective function values quality, time, and cost and also with their weights. Therefore, the objective function is the total weighted sum of these three objectives.

5. Case study: Experimental results

In this section, the models evaluate the data of the textile industry of India, having seven vendors and demands of 2000 components, as given in Table 2.

5.1 Manufacturer details

The demand for Textile machinery-manufacturing is a maximum of 2000 components. The minimum and maximum range with considering to business values of order quantities are between 100 and 1200, respectively [23].

Table 2. Vendor's details

Vendors	Min Order	Max order	Defective %	Cost1 Rs. B/discount	Late Delivery%	Cost2 after discount
V1	100	600	2.5	2.5	3.25	9
V2	200	750	4.5	4.5	5.25	10
V3	250	800	5	5	6.25	11
V4	350	750	3.5	3.5	15	9
V5	100	700	1.5	1.5	0.2	10
V6	300	950	6	6	2.5	11.5

5.2 Selection of vendors with quantity discounts

In this problem in Figure 3, explain various combinations set of vendors besides their multi-objectives, defectives/Late delivery & cost with include of various rang of price discounts.The set (1,2,3,5,6) represents vendors group,V1,V2,V3,V5 and V6.The combination set of V4 and V7 is not mentioned in the given graph because of there defective percentage is very high as compared to the product cost. In this graph, the combination set of vendors (1,2,3,5,6) represents the values of defective., time delivery with product cost which is minimum; the values of the given set are (69,47,22386), so from this graph, it is evident that the combined set of vendors (1,2,3,5,6) is best and has an optimal solution of multicriteria defectives/Late delivery & costs, price discount in supplier selection problem.

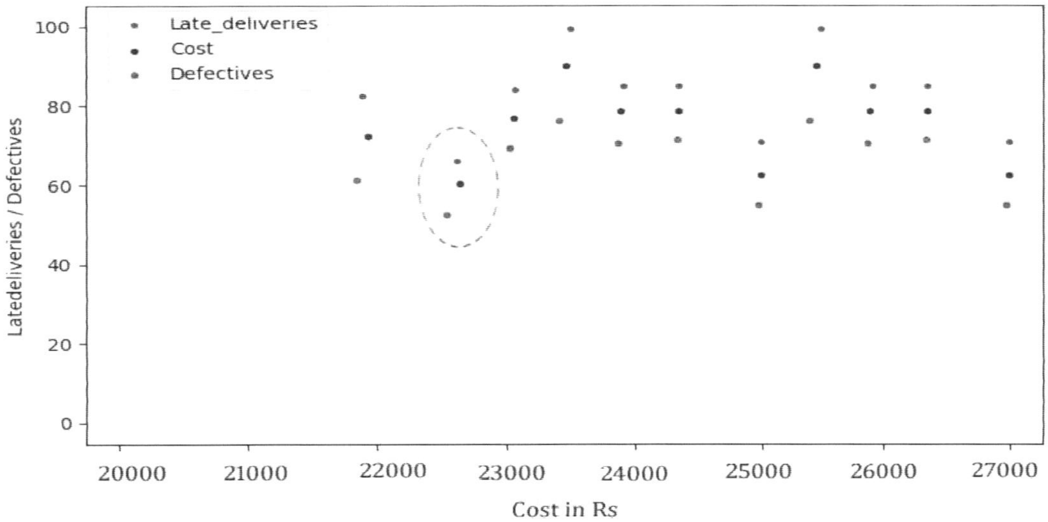

Figure 3. Combination set of vendors with their defectives/Late delivery & cost

Figure 4 showed the fitness values of the vendor in this problem. It is originated that when increases the no of generations, the feasible solutions goes towards an optimal. Figure 4 already that the optimal values to be found (151, 99.8154) with generation number is 151, against their fitness values 99.8154. This program was implemented and executed on Intel Pentium P6, 2.80 GHz of PC.from feasible solutions towards an optimal solution, the program took 30 seconds.

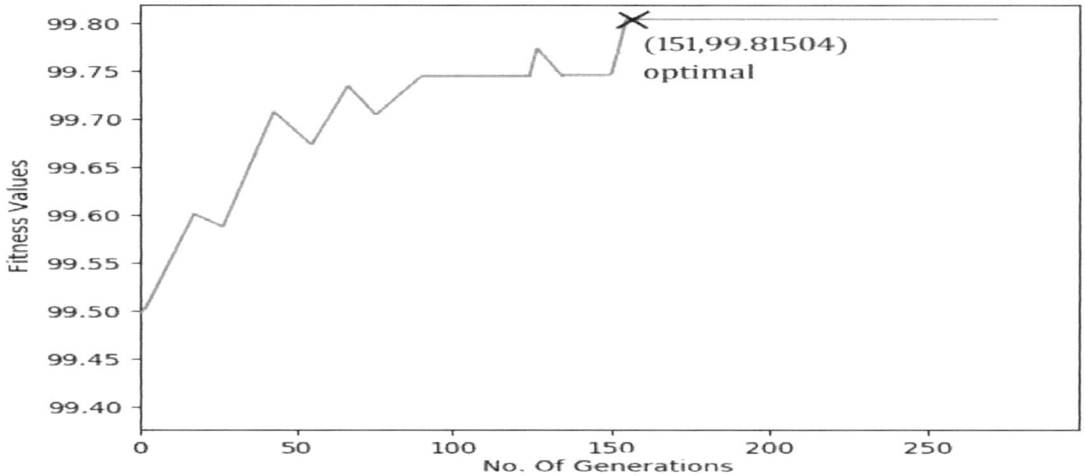

Figure 4. Fitness values Against generations.

When GA compared to Exact, the solution of GA can be more effective and optimal for incremental quantity discounts. As given in figure 5.

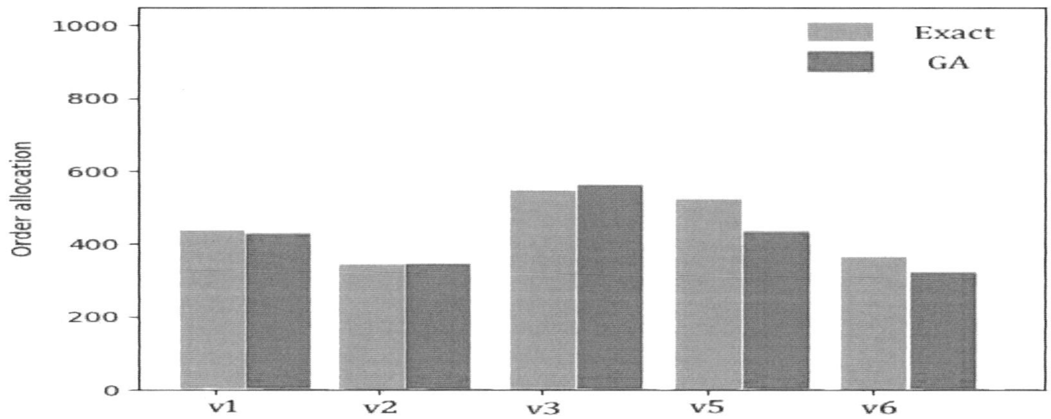

Figure 5. Comparison of genetic algorithm vs. Exact methods contains order allocation

As order allocation quantity of (V1,V2,V3,V5,V6) with respect to genetic algorithm the values is (454,324,539,448,304) and demand is 2000 so ,and the defective are (10,22,12,10,15) which total is 69.and vendors order are 2069.

Total Vendor Order=V1+V2+V3+V5+V6 ,Order – Defectives = 2069 – 69 = 2000 = Demand

6. Conclusion

This paper examined the use of MOGA selection of vendors with various quantity discounts, which is a combinatorial and multi-objective problem in nature. The previous methodology, do not define the defectives present in the supply selection while allocating order quantity to each vendor when satisfying the firm demand. One of the novel approaches planned in this model is that it define defectives during order al-location so that the accurate demand will always be fulfilled. In the supplier selection problem. The product demand given by industries would always be equal to the total order of the vendors. Besides this, the benefit of the given model is to find out the best solution. The benefit of applying MOGA as a design tool of their capability to get solutions, which is free of preconceptions that what is best or not. The results obtained by applying MOGA were found to make a competitive conclusion to that using conventional techniques. This approach will be very helpful for the purchase manager to find an alternate set of suppliers. For the future, various extensions possibility of this research follows,(1)improve the computational time, (2)improve the quality of the solution, especially when implementing big data of suppliers given by enterprises.

Acknowledgments
This research was supported by the Natural Science Foundation of Jilin province of China (Project no. 20180101035JC).

References

[1] Nourmohamadi Shalke, Parvin, Mohammad Mahdi Paydar, and Mostafa Hajiaghaei-Keshteli. "Sustainable supplier selection and order allocation through quantity discounts." International Journal of Management Science and Engineering Management 13.1 (2018): 20-32.

[2] Erdem, Aslı Sencer, and Emir Göçen. "Development of a decision support system for supplier evaluation and order allocation." Expert Systems with Applications 39.5 (2012): 4927-4937.

[3] Chen, Xiao-Fang, Li-Hui Cen, and Yi-Gang Cen. "A raw material purchasing decision-making model of order allocation with multiple suppliers and its optimization." 2009 First International Conference on Information Science and Engineering. IEEE, 2009.

[4] Tavana, Madjid, Morteza Yazdani, and Debora Di Caprio. "An application of an integrated ANP–QFD framework for sustainable supplier selection." International Journal of Logistics Research and Applications 20.3 (2017): 254-275.

[5] An Arunkumar, N., L. Karunamoorthy, and MAKESHWARAA N. UMA. "An optimization technique for vendor selection with quantity discounts using a genetic algorithm." (2007): 1-13.

[6] Hamdan, Sadeque, and Ali Cheaitou. "Supplier selection and order allocation with green criteria: An MCDM and multi-objective optimization approach." Computers & Operations Research 81 (2017): 282-304.

[7] Scott, James, et al. "A decision support system for supplier selection and order allocation in stochastic, multi-stakeholder and multi-criteria environments." International Journal of Production Economics 166 (2015): 226-237.

[8] Lee, Amy HI, et al. "A green supplier selection model for the high-tech industry." Expert systems with applications 36.4 (2009): 7917-7927.

[9] Tsai, W. H., & Hung, Shih-Jieh (2009). A fuzzy goal programming approach for green supply chain optimization under activity-based costing and performance evaluation with a value-chain structure. International Journal of Production Research, 47, 4991–501

[10] Kannan, Devika, et al. "Integrated fuzzy multi-criteria decision-making method and multi-objective programming approach for supplier selection and order allocation in a green supply chain." *Journal of Cleaner Production* 47 (2013): 355-367.

[11] Azania, Amir Hossein, Muhamad Zameri Mat Saman, and Kuan Yew Wong. "Sustainable supplier selection and order lot-sizing: an integrated multi-objective decision-making process." *International Journal of Production Research* 53.2 (2015): 383-408.

[12] Zhang, Ju-Liang, and Ming-yu Zhang. "Supplier selection and purchase problem with a fixed cost and constrained order quantities under stochastic demand." *International Journal of Production Economics* 129.1 (2011): 1-7.

[13] Hamdan, Sadeque, and Ali Cheaitou. "Green supplier selection and order allocation using an integrated fuzzy TOPSIS, AHP, and IP approach." *2015 International Conference on Industrial Engineering and Operations Management (IEOM)*. IEEE, 2015.

[14] Jadidi, O. M. I. D., S. Zolfaghari, and Sergio Cavalieri. "A new normalized goal programming model for multi-objective problems: A case of supplier selection and order allocation." *International Journal of Production Economics* 148 (2014): 158-165.

[15] Govindan, Kannan, and R34637471334 Sivakumar. "Green supplier selection and order

allocation in low-carbon paper industry: integrated multi-criteria heterogeneous decision-making and multi-objective linear programming approaches." *Annals of Operations Research* 238.1-2 (2016): 243-276.

[16] Mazda, Mohammad Mahdavi, Mohsen Emadikhiav, and Iman Parsa. "A heuristic to solve the dynamic lot-sizing problem with supplier selection and quantity discounts." *Computers & Industrial Engineering* 85 (2015): 33-43.

[17] Rezaei, Jafar, et al. "A multi-objective model for lot-sizing with supplier selection for an assembly system." *International Journal of Logistics Research and Applications* 19.2 (2016): 125-142.

[18] Fonseca, Carlos M., and Peter J. Fleming. "Genetic Algorithms for Multiobjective Optimization: FormulationDiscussion and Generalization." *Inga*. Vol. 93. No. July. 1993.

[19] Cao, Yan, et al. "Supplier pre-selection for platform-based products: a multi-objective approach." *International Journal of production research* 52.1 (2014): 1-19.

[20] Agha Mohammad Ali Kermani, Mehrdad, et al. "Supplier selection in a single-echelon supply chain with the horizontal competition using Imperialist competitive algorithm." *International Journal of Computer Integrated Manufacturing* 28.6 (2015): 628-638.

[21] Luo, X. G., et al. "Optimal product positioning with consideration of negative utility effect on consumer choice rule." *Decision Support Systems* 54.1 (2012): 402-413.

[22] Che, Z. H. "A genetic algorithm-based model for solving multi-period supplier selection problem with assembly sequence." *International Journal of Production Research* 48.15 (2010): 4355-4377.

[23] Che, Z. H., and Tzu-An Chiang. "Designing a collaborative supply-chain plan using the analytic hierarchy process and genetic algorithm with cycle-time estimation." *International Journal of Production Research* 50.16 (2012): 4426-4443.

[24] Kumar, Sri Krishna, M. K. Tiwari, and Radu F. Babiceanu. "Minimisation of supply chain cost with embedded risk using computational intelligence approaches." *International Journal of Production Research* 48.13 (2010): 3717-3739.

A review of artificial intelligence techniques for selection & evaluation

Ijaz Ahmad[1], Yan Liu*[1], Danish Javeed[1], Nadia Shamshad[1], Danish Sarwr[1], Shahab Ahmad[2]

[1]School of Computer Science and Technology, Changchun University of Science and Technology, Changchun 130022, China.
[2]School of Economics &Management, Chongqing University of Post & Telecommunication, Chongqing,400065, China.

engineerijaz04@gmail.com, *e-mail:yanl@cust.edu.cn

Abstract. Today, supplier selection &evaluation has been an important issue of strategic planning for a long time. Various artificial intelligence (AI) methods have been applied in the methodology of supplier section & evaluation from different researchers. This paper provides a review of past various artificial intelligence methods (AI) from 2010 to 2018 in the area supplier selection & evaluation problems. The review report is based on the literature of fifty articles include six review papers. Our survey presents and tries to find out the most useful and prominent AI method recommendable for the future in the problem of supplier selection &evalution.

1. Introduction

In every industry, the selection of proper suppliers and evolution is vital for the development and progress of the enterprise. In the competitive market, industries are more emphasis and focus on quality improvement to the wide use of their products. Mangers try to implement improvement policies base on budget and product. Supplier selection and evaluation need proper decision-making approaches such as artificial intelligence techniques, which lead to the revenues of industries as compared to other competitors. Because the industries need suppliers based on various multicriteria, such as product quality, delivery performance, cost, reliability. In some situations, the suppliers supply the components with low prices, but the low quality, high quality at a higher price. Different researchers have presented the idea of artificial intelligence techniques applied in supplier selection &evalution problem to provide an optimal and feasible solution. Which have Evalionaray genetic computing(EGC), artificial neural network(ANN), swarm intelligence, and knowledge-based optimization? This paper presents different artificial intelligence techniques (AI) of the past literature, tries to find out the most useful and prominent AI method recommendable for the future in the problem of supplier selection &evalution. The rest of this paper is structured as follows; Section 2 contains the methodology of various artificial intelligence technique include tables which distribute AI methods using in the application of supplier selection & evaluation. Section 3 represents the conclusion and future work.

2. Methodology

2.1 Distribution of the number of articles under various AI approaches

With regards to supplier selection and evaluation problem using AI methods, a wide literature work can be found; this paper address 50 articles from 2010 to 2018 in table .1

Table 1. Artificial Intelligence Method in supplier selection

Approaches	literature	Approaches	literature
	Genetic Evalionrary Computing		
Cao et al. [1]	NSGA-11		
Molamohamadi et al. [2]	Pareto genetic algorithm	Chen, Chwen-Tzeng.[18]	genetic algorithm (GA)
Che, Tzu-An Chiang. [3]	genetic algorithm (GA)	Mohammad et al. [19]	Psychoclonal algorithm
Kumar, Babiceanu.[4]	genetic algorithm (GA)	Aleadelat, Waleed.[20]	genetic algorithm (GA)
Che, Cheng-Jui.[5]	genetic algorithm (GA)	Molamohamadi, Zore,[21]	Pareto genetic algorithm
Bandyopadhyay,Ranjan[6]	(NSGA-II)	Hamdan, Sadeque ([22]	Genetic algorithm
Pasandideh, Reza. [7]	Pareto genetic algorithm	Scott, James, et al. [23]	hybrid-spanning
Nourmohamadi,parvin[8]	Pareto genetic algorithm	Kannan et al. [24]	(hst-GA)
Nepal, Famuyiwa.[9]	non-dominated ranking	Azania et al. [25]	genetic algorithm (GA)
Tsai, Shiu-Hung.[10]	NSGA-11	Zhang, Ju-liang. [26]	genetic algorithm (GA)
Li, Jizi, et al.[11]	genetic algorithm (GA)	Jadidi, O. M. I. D.[27]	Pareto genetic algorithm
Wang, Zi Jun.[12]	genetic algorithm (GA)	Govindan, Kannan. [28]	steady-state (GA)
Jamshidi, Karimi.[13]	genetic algorithm (GA)	Mazda,Mahdavi. [29]	NSGA-11
Hajiaghaei-Keshteli.[14]	steady-state(GA)	Kumar, Radu.[30]	MOGA
Brunet et al. [15]	parallel (hst-GA)	Carr, Amelia S., et al . [31]	multi-objective(GA)
Min, Hokey[16]	genetic algorithm (GA)	Chen et al.[32]	genetic algorithm (GA)
Chang, Ying-Hua .[17]	NSGA-11	Li, , Guang et al.[33]	Hybrid (GA)
	Swarm intelligence		
Park, Gyouhyung. [34]	PSO	Chen, Kuei-Hsien[37]	PSO
Panicker, Vinay[35]	Ant colony optimization	Hsiung, Chao Agnes.[38]	PSO
Carr, Amelia [36]	PSO		
	Artificial neural network		
Nepal, Bimal[39]	Bayesian belief networks	Golmohammadi, Davood.[42]	backpropagation Neural network
Tsai, Juin-Ming[40]	artificial neural network	Xu, Linen[43]	artificial neural network
Bahadori et al.[41]	artificial neural network	Kuo, Renjieh J.[44]	artificial neural network
Lee, Young, Hae.[45]	backpropagation NN		
	Knowledge-based		
Pitchipoo,Venkumar[46]	Expert System	Wu, Chong.[49]	Expert System
Shahroudi, Kambiz[47]	Expert System	Sanayei, Amir[50]	Expert System
Schalke, Parvin[48]	Expert System		

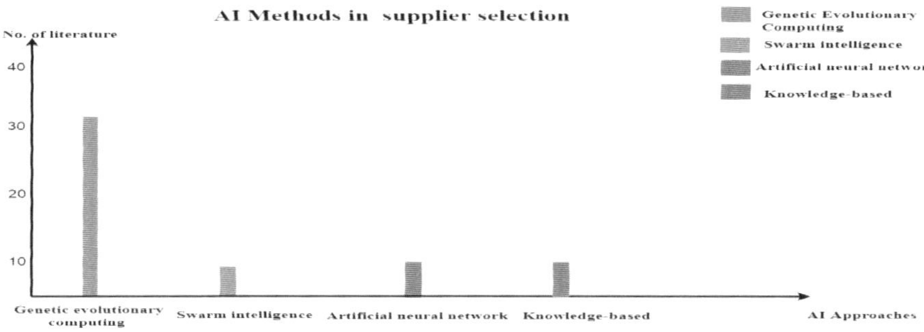

Figure 1. Distribution of AI methods in supplier selection

In the given literature presents that most of the articles of supplier selection & evaluation using AI methods, Evalionaray genetic computing have addressed more as compared to other AI methods as decision-making technique, due to better self-learning, and better convergence.as given the figure 1, Evalionaray genetic computing has 33 articles, Artificial neural networks 8, swarm intelligence 6, and knowledge-based optimization is 4 articles

3. Conclusions

In this paper, fifty articles have been studied from different journals, conferences to find out the most useful and prominent AI methods addressed by various researchers and propose to review various AI Methods in the supplier selection &evaluation problems. The distribution of different articles on supplier selection & evaluation problems based on various AI methods from 2010 to 2018 is shown in Fig. 1. quantities factor of AI articles in the supplier selection, which has 50% Evalionaray genetic computing, artificial neural networks 25 %, swarm intelligence 10 %, knowledge-based optimization is 10 % articles. Evalionaray genetic computing (EGC) can be capable of assisting the researchers and decision-makers in using EGA for the difficult task of the supplier selection & evaluation problem efficiently in the future.

Acknowledgements
This research was supported by the Natural Science Foundation of Jilin province of China (Project no. 20180101035JC).

References

[1] Cao, Yan et al."Supplier pre-selection for platform-based products: a multi-objective approach." International Journal of Production Research 52.1 (2014): 1-19.
[2] Molamohamadi, Zohreh, et al. "Supplier selection in a sustainable supply chain." Journal of Advanced Management Science 1.3 (2013).
[3] Che, Z. H., and Tzu-An Chiang. "Designing a collaborative supply-chain plan using the analytic hierarchy process and genetic algorithm with cycle-time estimation." International Journal of Production Research 50.16 (2012): 4426-4443.
[4] Kumar, Sri Krishna, M. K. Tiwari, and Radu F. Babiceanu. "Minimisation of supply chain cost with embedded risk using computational intelligence approaches." International Journal of Production Research 48.13 (2010): 3717-3739.
[5] Che, Z. H., and Cheng-Jui Chiang. "A modified Pareto genetic algorithm for multi-objective build-to-order supply chain planning with product assembly." Advances in Engineering Software 41.7-8 (2010): 1011-1022.
[6] Bandyopadhyay, Susmita, and Ranjan Bhattacharya. "Solving a tri-objective supply chain problem with modified NSGA-II algorithm." Journal of Manufacturing Systems 33.1 (2014)
[7] Pasandideh, S. H. R., Niaki, S. T. A., & Asadi, K. (2015). Bi-objective optimization of a multi-product multi-period three-echelon supply chain problem under uncertain environments: NSGA-II and NRGA. *Information Sciences*, *292*, 57-74.
[8] Nourmohamadi Shalke, Parvin, Mohammad Mahdi Paydar, and Mostafa Hajiaghaei-Keshteli. "Sustainable supplier selection and order allocation through quantity discounts." *International Journal of Management Science and Engineering Management* 13.1 (2018)
[9] Nepal, Bimal, Leslie Monplaisir, and Oluwafemi Famuyiwa. "A multi-objective supply chain configuration model for new products." *International Journal of Production Research* 49.23 (2011): 7107-7134.
[10] Tsai, Juin-Ming, and Shiu-Wan Hung. "Supply chain relationship quality and performance in technological turbulence: an artificial neural network approach." *International Journal of Production Research* 54.9 (2016): 2757-2770.
[11] Li, Jizi, et al. "A parallel Lagrange algorithm for order acceptance and scheduling in cluster supply chains." *Knowledge-Based Systems* 143 (2018): 271-283.
[12] Wang, Zi Jun, et al. "Highly porous conjugated polymers for selective oxidation of organic sulfides under visible light." *Chemical Communications* 50.60 (2014): 8177-8180.

[13] Jamshidi, R., SMT Fatemi Ghomi, and B. Karimi. "Multi-objective green supply chain optimization with a new hybrid memetic algorithm using the Taguchi method." *Scientia Iranica* 19.6 (2012): 1876-1886.
[14] Hajiaghaei-Keshteli, M. "The allocation of customers to potential distribution centers in supply chain networks: GA and AIA approaches." *Applied Soft Computing* 11.2 (2011): 2069-2078.
[15] Brunet, Gilbert, et al. "Collaboration of the weather and climate communities to advance sub-seasonal-to-seasonal prediction." *Bulletin of the American Meteorological Society* 91.10 (2010): 1397-1406.
[16] Min, Hokey. "Artificial intelligence in supply chain management: theory and applications." *International Journal of Logistics: Research and Applications* 13.1 (2010): 13-39.
[17] Chang, Ying-Hua. "Adopting co-evolution and constraint-satisfaction concept on genetic algorithms to solve supply chain network design problems." *Expert Systems with Applications* 37.10 (2010): 6919-6930.
[18] Chen, Kuei-Hsien, and Chwen-Tzeng Su. "Activity assigning of fourth-party logistics by particle swarm optimization-based preemptive fuzzy integer goal programming." *Expert Systems with Applications* 37.5 (2010): 3630-3637.
[19] Agha Mohammad Ali Kermani, Mehrdad, et al. "Supplier selection in a single-echelon supply chain with the horizontal competition using Imperialist competitive algorithm." *International Journal of Computer Integrated Manufacturing* 28.6 (2015): 628-638.
[20] Aleadelat, Waleed, Promotes Saha, and Khaled Kiribati. "Development of serviceability prediction model for county paved roads." *International Journal of Pavement Engineering* 19.6 (2018): 526-533.
[21] Molamohamadi, Zohreh, and Napsiah Ismail. "Developing a new scheme for sustainable manufacturing." *International Journal of Materials, Mechanics, and Manufacturing* 1.1 (2013): 1-5.
[22] Hamdan, Sadeque, and Ali Cheaitou. "Supplier selection and order allocation with green criteria: An MCDM and multi-objective optimization approach." Computers & Operations Research 81 (2017): 282-304.
[23] Scott, James, et al. "A decision support system for supplier selection and order allocation in stochastic, multi-stakeholder and multi-criteria environments." International Journal of Production Economics 166 (2015): 226-237.
[24] Kannan, D., Khodaverdi, R., Olfat, L., Jafarian, A., & Diabat, A. (2013). Integrated fuzzy multi-criteria decision-making method and multiobjective programming approach for supplier selection and order allocation in a green supply chain. Journal of Cleaner Production, 47, 355–367.
[25] Azania, Amir Hossein, Muhamad Zameri Mat Saman, and Kuan Yew Wong. "Sustainable supplier selection and order lot-sizing: an integrated multi-objective decision-making process." International Journal of Production Research 53.2 (2015): 383-408.
[26] Zhang, Ju-Liang, and Ming-yu Zhang. "Supplier selection and purchase problem with a fixed cost and constrained order quantities under stochastic demand." International Journal of Production Economics 129.1 (2011): 1-7.
[27] Jadidi, O. M. I. D., S. Zolfaghari, and Sergio Cavalieri. "A new normalized goal programming model for multi-objective problems: A case of supplier selection and order allocation." International Journal of Production Economics 148 (2014): 158-165.
[28] Govindan, Kannan, and R34637471334 Sivakumar. "Green supplier selection and order allocation in the low-carbon paper industry: integrated multi-criteria heterogeneous decision-making and multi-objective linear programming approaches." Annals of Operations Research 238.1-2 (2016): 243-276.
[29] Mazda, Mohammad Mahdavi, Mohsen Emadikhiav, and Iman Parsa. "A heuristic to solve the dynamic lot-sizing problem with supplier selection and quantity discounts." Computers & Industrial Engineering 85 (2015): 33-43.

[30] Kumar, Sri Krishna, M. K. Tiwari, and Radu F. Babiceanu. "Minimisation of supply chain cost with embedded risk using computational intelligence approaches." International Journal of Production Research 48.13 (2010): 3717-3739.
[31] Carr, Amelia S., et al. "RFID technology: Implications for healthcare organizations." American journal of business 25.2 (2010): 25-40.
[32] Chen, Kuei-Hsien, and Chwen-Tzeng Su. "Activity assigning of fourth-party logistics by particle swarm optimization-based preemptive fuzzy integer goal programming." Expert Systems with Applications 37.5 (2010): 3630-3637.
[33] Liu, Guang, et al. "Encapsulation of Ni/Fe 3 O 4 heterostructures inside onion-like N-doped carbon nanorods enables synergistic electrocatalysis for water oxidation." *Nanoscale* 10.8 (2018): 3997-4003.
[34] Park, KyoungJong, and Gyouhyung Kyung. "Optimization of total inventory cost and order fill rate in a supply chain using PSO." The International Journal of Advanced Manufacturing Technology 70.9-12 (2014): 1533-1541.
[35] Panicker, Vinay V., Ratnaji Vanga, and R. Sridharan. "Ant colony optimization algorithm for distribution-allocation problem in a two-stage supply chain with a fixed transportation charge." International Journal of Production Research 51.3 (2013): 698-717.
[36] Carr, Amelia S., et al. "RFID technology: Implications for healthcare organizations." American journal of business 25.2 (2010): 25-40.
[37] Chen, Kuei-Hsien, and Chwen-Tzeng Su. "Activity assigning of fourth-party logistics by particle swarm optimization-based preemptive fuzzy integer goal programming." Expert Systems with Applications 37.5 (2010): 3630-3637.
[38] Hsiung, Chao Agnes, et al. "The 5p15. 33 locus is associated with risk of lung adenocarcinoma in never-smoking females in Asia." *PLoS genetics* 6.8 (2010): e1001051.
[39] Nepal, Bimal, and Om Prakash Yadav. "Bayesian belief network-based framework for sourcing risk analysis during supplier selection." *International Journal of Production Research* 53.20 (2015): 6114-6135.
[40] Tsai, Juin-Ming, and Shiu-Wan Hung. "Supply chain relationship quality and performance in technological turbulence: an artificial neural network approach." *International Journal of Production Research* 54.9 (2016): 2757-2770.
[41] Baghban, Alireza, et al. "Prediction of CO2 loading capacities of aqueous solutions of absorbents using different computational schemes." *International Journal of Greenhouse Gas Control* 57 (2017): 143-161.
[42] Lee, Young Hae, Jung Woo Jung, and Kyong Min Lee. "Vehicle routing scheduling for cross-docking in the supply chain." *Computers & Industrial Engineering* 51.2 (2010): 247-256.
[43] Golmohammadi, Davood, et al. "Supplier selection based on a neural network model using a genetic algorithm." *IEEE Transactions on Neural Networks* 20.9 (2010): 1504-1519.
[44] Xu, Linen, and Feng Qian. "Supplier selection using the ANN-based predictive model." *2010 Sixth International Conference on Natural Computation*. Vol. 4. IEEE, 2010.
[45] Kuo, Renjieh J., Yu C. Wang, and Fangchih C. Tien. "Integration of artificial neural network and MADA methods for green supplier selection." *Journal of cleaner production* 18.12 (2010): 1161-1170.
[46] Pitchipoo, P., P. Venkumar, and S. Rajakarunakaran. "Development of a fuzzy expert system for supplier evaluation and selection." *IEEE-International Conference On Advances In Engineering, Science, And Management (ICAESM-2012)*. IEEE, 2012.
[47] Shahroudi, Kambiz, and S. Maryam Shafaei Tonekaboni. "Application of TOPSIS method to supplier selection in Iran auto supply chain." *Journal of Global Strategic Management* 12 (2012): 123-131.
[48] Nourmohamadi Shalke, Parvin, Mohammad Mahdi Paydar, and Mostafa Hajiaghaei-Keshteli. "Sustainable supplier selection and order allocation through quantity discounts." International Journal of Management Science and Engineering Management 13.1(2018): 20

[49] Wu, Chong, and David Barnes. "Formulating partner selection criteria for agile supply chains: A Dempster–Shafer belief acceptability optimization approach." International Journal of Production Economics 125.2 (2010): 284-293.
[50] Sanayei, Amir, S. Farid Mousavi, and Ahmad Yazdankhah. "Group decision-making process for supplier selection with VIKOR under fuzzy environment." Expert Systems with Applications 37.1 (2010): 24-30.

IOP Conf. Series: Materials Science and Engineering **853** (2020) 012056 doi:10.1088/1757-899X/853/1/012056

Fatigue detection based on multi-feature fusion of fatigue behavior

Chen Xing[1], Su Lumei[1,2], Qin Meixin[1]

[1]School of Electrical Engineering and Automation, Xiamen University of Technology, Xiamen430070, China

[2]Xiaolong Institute of Engineering and Technology, Second floor, Building A, Longzhou, Industrial Park, Longyan364000, China

Corresponding author: sulumei@163.com

Abstract: The research of fatigue driving detection is of great significance due to the contribution in fatigue driving prevention. However, most of current fatigue detection study are limited to unavailable real-time performance, unsatisfied accuracy rate, rough fatigue level and complicated implementation. In this paper, a novel method based on multi fatigue behavior features are proposed to improve the accuracy of fatigue detection. First of all, multi facial behavior features including eyes, mouth and head features are extracted. And then a new multi-feature fusion fatigue detection approach based on specific fatigue behavior feature anaylsis is proposed to recognize the previously extracted features. For better accuracy and applicability, a more detailed fatigue levels are newly used to subdivide fatigue behavior into three degree of fatigue. The experimental results based Matlab simulation and Android App indicate that the proposed method could accurately and real-time detect whether the tester is in a state of fatigue driving, and also determine the fatigue level.

1. Introduction

Most fatigue driving detection studies have shown that when a driver is in a fatigue state, it can easily cause a traffic accident [1]. The University of Pennsylvania uses the PERCLOS rule to measure the degree of psychophysiological fatigue [2]. Professor Nikolaos P. Papani kolo-poulos designed a system for binocular tracking and positioning of drivers [3]. Jie et al[4] proposed that facial touch can be used as a clue to extract new yawns to extract geometric and appearance features of mouth and eye areas. Ruiz et al [5] proposed a concise and robust method for determining head posture.

This paper proposes a new fatigue detection method based on multi-feature fusion. Recognizing fatigue behavior to a more detailed level will greatly improve the accuracy of fatigue detection.

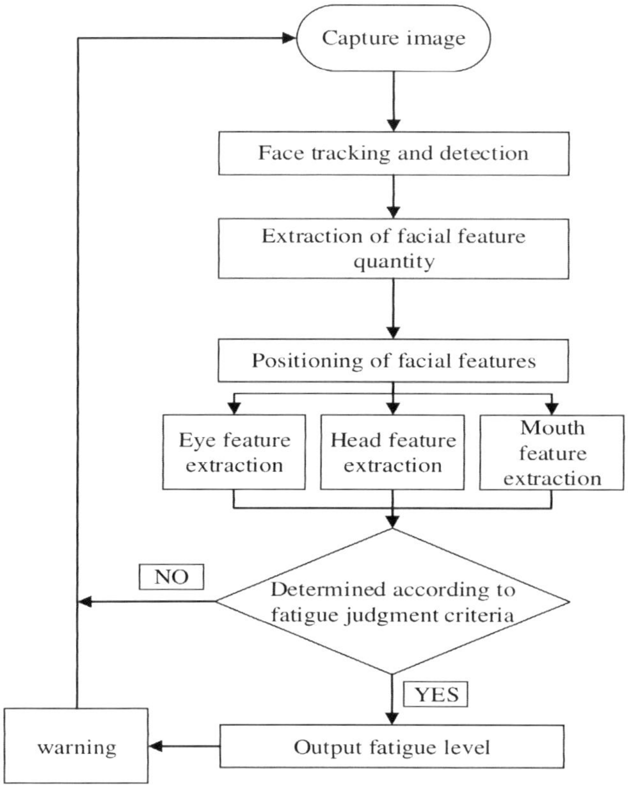

Figure 1. Fatigue detection system structure.

Table1. Detection and analysis of pupil.

Number	Pupil center coordinates	Pupil center coordinates	Pupil coordinate	Pupil longitudinal axis length	Eye state
1	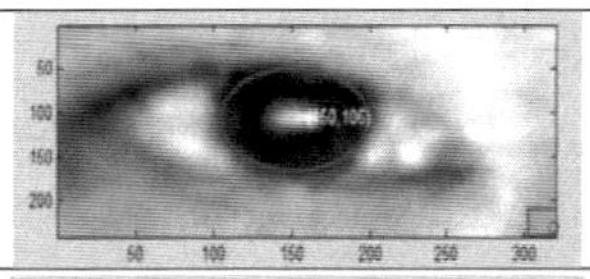	(150,106)	(104.5,49.5)	117	Open
2	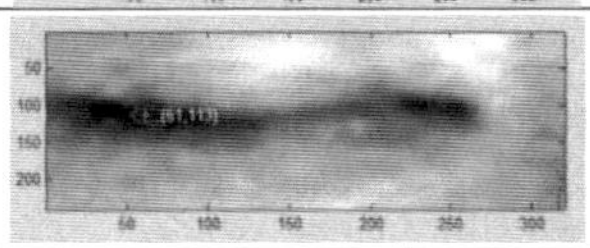	(61,113)	(53.5,103.5)	20	Close

2. Fatigue driving detection method

This paper will focus on the detection methods of fatigue driving. which including fatigue driving detection system construction, fatigue feature extraction, and fatigue level determination based on multi-feature fusion. The fatigue level determination based on multi-feature fusion is more accurate than the traditional PERCLOS rule.

2.1 Architecture of fatigue driving detection system

The overall architecture of the fatigue driving detection system is shown in Figure 1. It mainly consists of 6 parts:

1) Image acquisition

2) Face tracking and detection
3) Extraction of facial features
4) Location of face facial features
5) Extraction of human eyes, head and mouth features
6) Fatigue identification

2.2 Fatigue feature extraction

Feature extraction will be introduced below.

(1) Calculation of closed eye time

Eye feature extraction is shown in Table 1.

(2) Extraction of mouth feature quantity

Mouth feature extraction is shown in Table 2.

Table 2. Analysis of detected mouth features.

Number	Mouth grayscale picture	Mouth binarized image	Mouth area	Mouth state
1	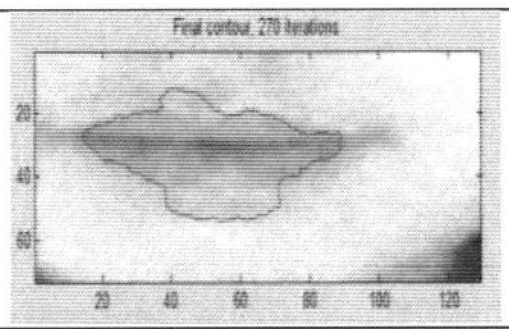	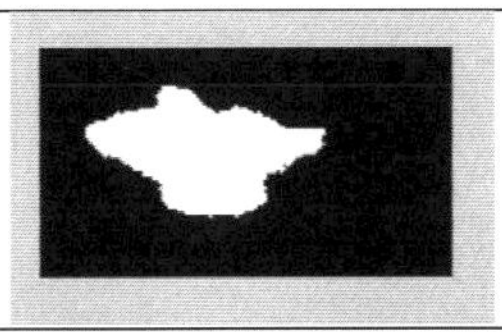	2018	Normal
2	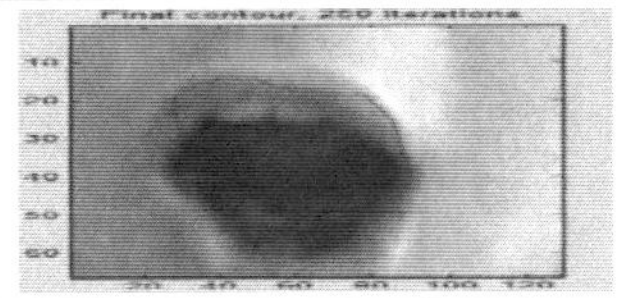	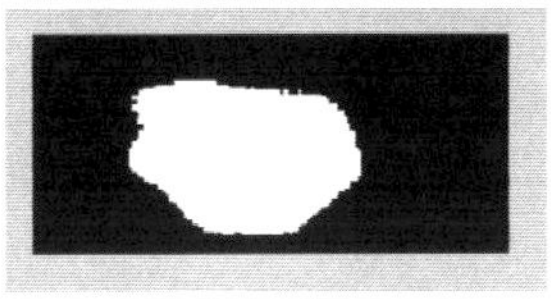	2237	Tired

(3) Extraction of head feature values

Head feature extraction is shown in Table 3

.

Table 3. Analysis of detected head features.

Number	Figure	Face coordinates	Face ordinate	Face state
1	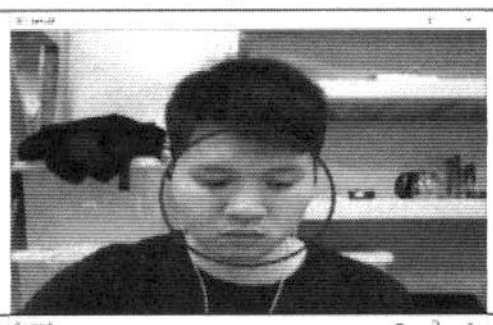	(248,147)	147	Bow down
2	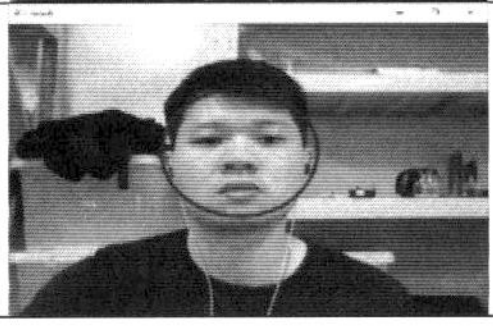	(239,94)	94	Gain ground

2.3 Fatigue rating of multi-feature fusion

In this paper, the fatigue level determination is based on the PERCLOS rule. The improved fatigue judgment criterion for merging the nodding frequency and the mouth characteristics is referred to herein as the fatigue determination criterion of multi-feature fusion. As shown in Table 3.

The PERCLOS rule determines whether the most critical amount of fatigue is the ratio of closed eye time. Since the detection time is fixed, it is assumed that the detection time is I and I>3s. Note that the detection time must be greater than 3s. The time calculation formula can calculate the closed eye time T, so Equation 1, which can obtain the closed eye time ratio f, is as follows.

$$f = \frac{T}{I} \cdot 100\% \tag{1}$$

Table4. The proposed criteria of fatigue level identifiation.

Status	Monitoring site	parameter	Indicator (T is the closed eye time)	Performance
normal status	head	Nodding frequency	f <=1 andT>5s	Nominal frequency is less than or equal to one time within the specified time
	eyes	PERCLOS	f<40%andT>3s	Closed eye time less than 40% within the specified time
	mouth	Mouth shape	L<15(L is the length of the mouth)	The height of the mouth is less than 15 pixels
First fatigue	head	Nodding frequency	f =1 andT>5s	The nod frequency is equal to one time within specified time
	eyes	PERCLOS	40%<=f<=50% andT>3s	Closed eye time between 40% and 50% within specified time
	mouth	Mouth shape	15<L<25	The height of the mouth is greater than 15 pixels and less than 25 pixels
Secondary fatigue	head	Nodding frequency	f >2 andT>5s	The nod frequency is equal to twice within the specified time
	eyes	PERCLOS	50%<f<=70%andT>3s	Closed eye time between 50% and 70% within specified time
	mouth	Mouth shape	25<=L<35	The height of the mouth is between 25 and 35 pixels

Table 5. The experimental results of test .

<table>
<tr><td colspan="8">Result</td></tr>
<tr><td rowspan="3">face</td><td rowspan="3"></td><td>Head coordinates</td><td>(457,319,330,360)</td><td rowspan="3">Fatigue level</td><td rowspan="3">1</td><td rowspan="7">summary</td><td rowspan="7">2</td></tr>
<tr><td>Angular variable</td><td>1</td></tr>
<tr><td>Nod</td><td>1</td></tr>
<tr><td rowspan="2">Mouth</td><td rowspan="2"></td><td>Mouth opening area</td><td>2</td><td rowspan="2">Fatigue level</td><td rowspan="2">0</td></tr>
<tr><td>Mouth state</td><td>normal</td></tr>
<tr><td rowspan="2">Eyes</td><td rowspan="2">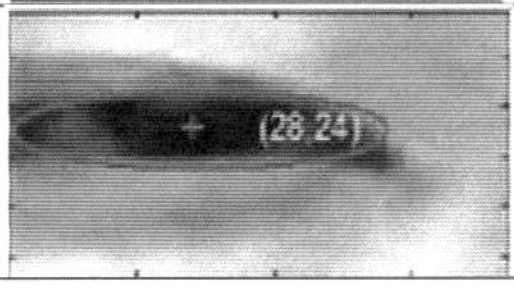</td><td>Pupil length</td><td>16</td><td rowspan="2">Fatigue level</td><td rowspan="2">2</td></tr>
<tr><td>Closed eye percentage</td><td>60%</td></tr>
</table>

3. Experimental results

According to the above-mentioned feature extraction step and the fatigue level determination criterion of multi-feature fusion, the test results are shown in the following table5.

This result can be seen that although the state of the nodding frequency and the mouth is shown as primary fatigue, it can be seen from the closed eye percentage of the eye that the secondary fatigue has been reached, and the fatigue state of the person at this time is secondary fatigue.

4. Conclusion

In this paper, a more diverse feature extraction scheme is adopted, and the fatigue judgment criterion of multi-feature fusion is improved by the proposed fatigue judgment rule. It can be seen from the experimental results that the test can accurately determine whether the tester is in a fatigue driving state, and also judges the fatigue level and further refines the fatigue level.

Acknowledgments

This work is financially supported by Natural Science Foundation of department of science and technology of Fujian Province(No.2017J01510),Foundation for Science and technology of Longyan (No.2017LY91), Science Research Foundation for Overseas Scholars of Xiamen (No.XRS[2016]314-05).

References

[1] Zhou Yunpeng. Research on driver fatigue detection based on facial feature multi-feature fusion[D]. Hunan University, 2015.

[2] Yin Yanhong. Simulation study of driver fatigue based on brain waves and blinking [D]. Tongji University, 2008.

[3] Jiang Shuilang. Research on vision-based driving fatigue detection [D]. Shanghai Jiaotong University, 2008.

[4] Jie Z, Mahmoud M, Stafford-Fraser Q, et al. Analysis of Yawning Behaviour in Spontaneous Expressions of Drowsy Drivers[C]//2018 13th IEEE International Conference on Automatic Face & Gesture Recognition (FG 2018). IEEE, 2018: 571-576.

[5] Ruiz N, Chong E, Rehg J M. Fine-grained Head Pose Estimation without Keypoints[C]//Proceedings of the IEEE Conference on Computer Vision and Pattern Recognition Workshops. 2018: 2074-2083.

Research on target detection of carrier-based aircraft based on deep convolutional neural network

Zhu Xingdong, Tian Shaobing, Fan Jiali, Wang Zheng, Liu Tianqing
Naval Aviation University, Yantai, Shandong

745157824@qq.com

Abstract. Aiming at the problem that the aircraft carrier deck carrier is dense and easy to block, the carrier aircraft target is difficult to detect, and the detection effect is easily affected by the illumination condition and target scale. Based on the YOLO v3 target detection model, the loss function weighted and linear attenuation NMS strategy is proposed. The experimental results show that the proposed improved method can automatically and comprehensively extract the target characteristics of the carrier aircraft. Compared with the original YOLO v3 detection model, the improved model has a good detection effect on the occlusion aircraft target, the detection accuracy and speed can meet the actual needs, and the adaptability is strong under different illumination conditions and target scales, and the model is highly robust.

Keywords. carrier-based aircraft detection; convolutional neural network; YOLO v3 model

1. Introduction

At present, computer vision technology is developing rapidly, and image target detection has been widely used in the military field. Vigorously developing image recognition technology is of vital importance to improving our combat effectiveness and the safety of our personnel equipment. Aircraft carrier deck target detection is of great significance for ensuring the safety of deck aircraft equipment and improving the efficiency of aircraft transportation [1].

Due to the relatively small deck space of the aircraft carrier, the area of the aircraft usually has a large occlusion area. The occlusion of a large area often causes a serious loss of the target image information to the aircraft target, and the target cannot be accurately detected. Aiming at the problem that YOLO v3 target detection algorithm has low precision and difficult to detect occlusion target in detecting carrier aircraft targets, a loss function weighting and linear attenuation NMS（Non-Maximum Suppression） improvement strategy is proposed to detect the carrier aircraft carrier target on the carrier deck[2,3,4]. Experiments prove that this method Good results have been achieved in occluding carrier aircraft target detection.

2. Improvement method

2.1. Loss function weighting

If the true value of each cell is p^* and the predicted value is $\hat{p}^*$, the loss can be expressed as the difference between the predicted value and the true value, ie $\hat{p}^* - p^*$. In the YOLO v3 training process, the loss is the mean square error between the predicted and true values of all cells. The loss function l consists of three parts, namely the bounding box loss $l_{coordinate}$, the confidence loss $l_{confidence}$ and the classification loss l_{class}, which can be expressed as

$$l = l_{coordinate} + l_{confidence} + l_{class} \tag{1}$$

The bounding box loss consists of the error of the central coordinate and the error of the bounding box width and height, which can be described as

$$\begin{aligned} l_{confidence} = \sum_{i=0}^{S^2}\sum_{j=0}^{B} \Pr_{obj}(i,j)\left[(\hat{x}_i - x_i)^2 + (\hat{y}_i - y_i)^2\right] + \\ \sum_{i=0}^{S^2}\sum_{j=0}^{B} \Pr_{obj}(i,j)\left[\left(\sqrt{\hat{w}_i} - \sqrt{w_i}\right)^2 + \left(\sqrt{\hat{h}_i} - \sqrt{h_i}\right)^2\right] \end{aligned} \tag{2}$$

Where $(\hat{x}_i, \hat{y}_i)$ denotes the coordinates of the prediction boxes; $\hat{w}_i$ and $\hat{h}_i$ c denote the width and height of the prediction boxes respectively; (x, y) denotes the coordinates of the calibration boxes, w_i and h_i denote the width and height, respectively; $\Pr_{obj}(i, j)$ denotes the presence of the target in cell i, and The j-th anchor box is responsible for prediction.
The calculation formulas for confidence loss and classification loss can be described as

$$l_{confidence} = \sum_{i=0}^{S^2}\sum_{j=0}^{B} \Pr_{obj}(i,j)\left(C - \hat{C}_i\right)^2 + \sum_{i=0}^{S^2}\sum_{j=0}^{B} \Pr_{noobj}(i,j)\left[C - \hat{C}_i\right], \tag{3}$$

$$l_{class} = \sum_{i=0}^{S^2}\sum_{j=0}^{B} \Pr_{obj}(i,j) \sum_{c \in classes} \left[p_i(c) - \hat{p}_i(c)\right]^2, \tag{4}$$

Where $\hat{C}_i$ is the confidence of the prediction box; C_i is the confidence of the calibration box. The classification labels are all-hot encoded, $\hat{p}_i(c)$ is the encoded prediction category, $p_i(c)$ is the encoded real category; c is the current detection target category; classes represent all target categories in the data set; $\Pr_{noobj}(i, j)$ indicates that there is no target in the cell i.
If the loss is calculated directly using (1), then the weights of the three types of losses are all 1. The weighted loss function a is

$$l_w = \left(\alpha_{xy} l_{xy} + \alpha_{wh} l_{wh}\right) + \left(\beta_{obj} lconf\_obj + \beta_{no\_obj} l_{conf\_noobj}\right) + \lambda_{class} l_{class} \tag{5}$$

Where α_{xy} is the bounding box center coordinate loss weight, α_{wh} is the bounding box width and height loss weight, β_{obj} is the confidence loss weight of the target in the cell, β_{no_obj} is the confidence loss weight of the target in the cell, λ_{class} is the classification Loss weight.

Figure 1. Two mutually obscured aircraft.

IOP Conf. Series: Materials Science and Engineering **853** (2020) 012057 doi:10.1088/1757-899X/853/1/012057

2.2. Linear attenuation NMS algorithm

As shown in Figure 1, the two aircraft parked on the ship's surface, the A plane in the red box, the confidence score of the prediction box is 0.8, and the blue box is the B plane. The confidence score of the prediction box is 0.6, B. The aircraft severely obscures the A aircraft, and the crossover ratio of the predicted frames of the two aircrafts A and B is greater than 0.5. When using the NMS algorithm to process the excess prediction box, the IOU（Intersection-Over-Union） of the A and B aircraft prediction frames is higher than 0.5 (the YOLO v3 setting threshold is 0.5), and the A aircraft with a higher confidence score is retained, and the confidence score of the prediction frame of the B aircraft is set to 0, causing the B aircraft to be undetectable. This paper attempts to solve the problem that the occlusion aircraft cannot be accurately detected by NMS in YOLO v3 by using the linear attenuation NMS algorithm. When the IOU I_{bi}^{M} is higher than the suppression threshold N, the confidence score $P(class_i)$ in equation is linearly smoothed, and the optimized NMS algorithm is

$$P^{*}_{class_i} = \begin{cases} P_{class_i}, & I_{bi}^{M} < N \\ P_{class_i}\left(1 - I_{bi}^{M}\right), & I_{bi}^{M} \geq N \end{cases} \tag{6}$$

Where $P^{*}_{class_i}$ is the confidence score after linear smoothing.

Optimized NMS algorithm processing flow chart, as shown in Figure 2.

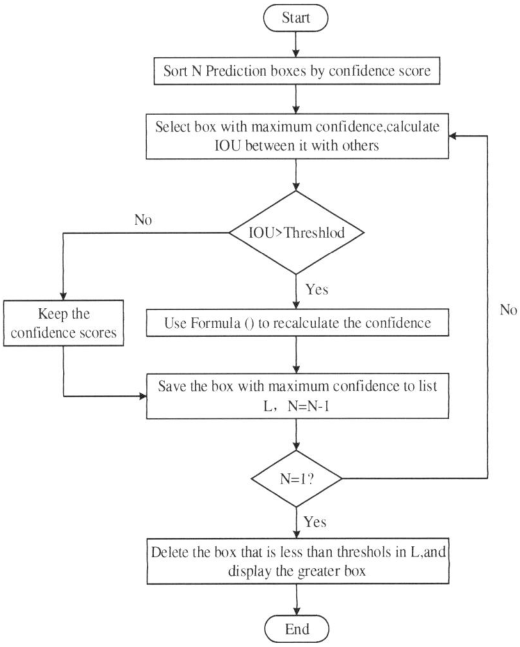

Figure 2. Flow chart of NMS processing

This method adopts the IOU value linear attenuation confidence score method to avoid accidentally deleting the prediction frame of the occlusion target and improving the detection ability of the occlusion target.

3. Experiment and analysis

3.1. Dataset

The data set allocation is shown in Table 1.

Table 1. Experimental dataset

Category	Train	Test	Sum
Wing folding	350		
Wing expansion	366	100	1200
Wing folding + Wing expansion	384		

3.2. Experimental results and analysis

Comparison of detection performance of two models of the same picture.. The parameters are shown in Table 2, and the detection effect is shown in Figure 3.

Table 2. Detection performance of two algorithms of the same picture

Detection algorithm	Detection	False detection	Missed inspection	Recall/%	Precison/%	Detecting a single picture time /s
Original YOLO v3	351	59	81	81.25	85.61	0.059
Improved YOLO v3	387	36	45	89.58	91.49	0.067

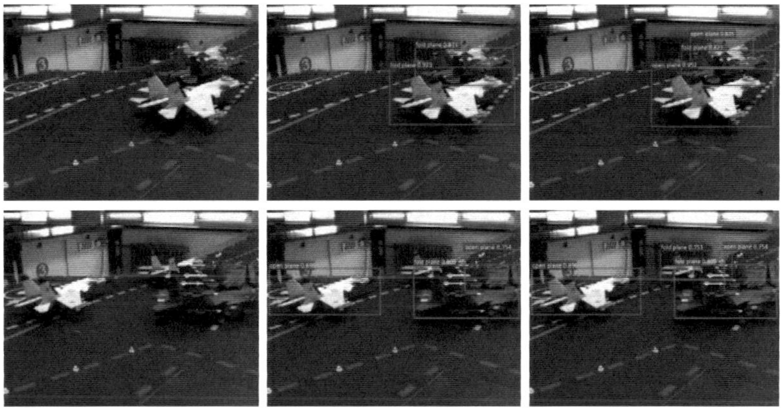

(a) Original images (b) Original YOLO v3 (c) Improved YOLO v3

Figure 3. Compared with the original image, the detection effect of the algorithm before and after the improvement

Figures 3(b) and (c) show the original YOLO v3 algorithm and the improved algorithm proposed in this paper. Combined with Table 2 and Figure 3, it can be concluded that the improved algorithm is higher than the original YOLO v3 in terms of accuracy and recall. The algorithm also has a good detection effect on targets with a large degree of occlusion. Compared with the original YOLO v3 algorithm, the proposed algorithm improves the recall rate by about 8%, the accuracy rate by about 6%, and the detection speed to meet the real-time requirements.

Table 3. Different algorithms deal with the detection effect of the same picture

Detection method	Precison/%	Detecting a single picture time /s
Improved YOLO v3	91.49	0.067
YOLO v3	85.61	0.059
SSD	87.34	0.074
Faster R-CNN	93.58	0.116

Different algorithms deal with the detection effect of the same picture. The SSD, Faster R-CNN, the original YOLO v3 algorithm and the improved method of this paper are used to detect the images in the test set. The accuracy and detection rate of the test are shown in the table. It can be seen that the proposed method has better performance in terms of detection speed and accuracy, and the accuracy is slightly worse than the Faster R-CNN algorithm. Considering this, the method is an effective and feasible carrier-based machine detection method. The test results are shown in Table 3.
Improve YOLO v3 detection effect under different illumination. From the above 100 test set test results, the pictures with different illumination conditions are selected for comparison, and the test results are shown in FIG. 4. In the figure, (a) is the detection effect with strong illumination, and (b) is the detection effect with weak illumination. It can be seen that for different illumination conditions, the improved model can effectively detect the aircraft and has high robustness.

(a) Strong light

(b) Weak light

Figure 4. Detection effect under different lighting conditions

4. Conclusion

In this paper, based on the original YOLO v3 detection model, the loss function weighting and linear attenuation NMS strategy is added to solve the problem of aircraft carrier occlusion aircraft target detection accuracy and difficulty. It has good detection effect on different illumination conditions, and the robustness of the model is high. The detection accuracy and real-time performance can meet the actual needs.

Acknowledgments
Thanks to the teachers and classmates for their guidance and help in the thesis. I hope that students who have the same research direction as me can communicate with each other.

References

[1] Davis B J, Kaszeta R W, Chambers R D, et al. Machine vision tracking of carrier-deck assets for improved launch safety[J]. 2013.
[2] Kharchenko V , Chyrka I . Detection of Airplanes on the Ground Using YOLO Neural Network[C]// IEEE 17th International Conference on Mathematical Methods in Electromagnetic Theory (MMET)2018:294-297.
[3] Buric M., Pobar M., and Ivasic-Kos, M. 2019. "Adapting YOLO network for Ball and Player Detection", In ICPRAM 2019.
[4] Bodla N, Singh B, Chellappa R, et al. Soft-NMS--Improving Object Detection With One Line of Code[C]//Proceedings of the IEEE International Conference on Computer Vision.2017: 5561-5569.

AUTHOR INDEX

Ahmad, Ijaz 388, 397
Ahmad, Irfan 242
Ahmad, Shahab 388, 397
Ahmed, Shihab Uddin 309
Akbar, Zeeshan Ali 234
Bashir, Muhammad Asad 234
Boxia, He 178
Cai, Mengyi 1
Cao, Baoshi 277
Chaohui, Sun 93
Chen, Jifeng 135
Chen, Ruiguo 20, 28
Chen, Xiaoxuan 354
Chengning, Zhang 108
Chong, Wang 7
Chuntao, Zhang 108
Dawei, Liang 7
Diao, Zhaowei 150
Du, Le 54
Du, Zhe 302
Duan, Shengcai 284
Ebihara, Soichiro 322
Fan, Jianping 330
Fang, Shi 86
Fen, Yan 93
Fengqing, Du 204
Fu, Jian 171
Gao, Qi 114
Ghaddar, Alia 342
Ghosn, Mohammad Houssein 342
Gu, Yikun 277
Guo, Ziming 284
Han, Yan 142
Hassan, Shaheer ul 178
He, ZhengYang 315
Hengxu, Zhang 86
Hori, Noriyuki 322
Hu, Bo 100, 171
Hu, Jiaquan 135
Hu, Xianhua 267
Hu, Xingchen 354
Hua, Guowu 135
Huang, Chun-Chieh 372
Huang, Jincai 354
Iqbal, Muhammad Amjad 260
Islam, Md. Monirul 309
Javeed, Danish 388, 397
Ji, Zhongxiao 150
Jiali, Fan 408
Jiang, Luo 379
JianWen, Gan 7
Jiaxiang, Wei 7
Jin, Guangyin 354
Jin, Haiqiang 14, 123, 187, 193
Jin, Minghe 277
Jiye, Wang 365
Junaeid, Sanzid 309
Kang, Peng 20, 28
khan, owais zarin 242
Khayyam, Umar 178
Kun, Wang 93
Lei, Yabin 114
Li, Chunyan 100, 171
Li, Haoyang 210
Li, Hao 223
Li, Huan 272
Li, Jie 223
Li, Lei 128
Li, Manli 77
Li, Mushui 171
Li, Ziqi 223
Liang, Yali 114
Liang, Yu 198
Liao, Zheng 100
Liaquat, Muwahida 242
Lin, Husheng 70, 77
Lini, Li 7
Liu, Hao 128
Liu, Hong 277
Liu, Lu 114
Liu, Mei 150
Liu, Shaohui 128
Liu, Xiaobai 218
Liu, Yan 388, 397
Liu, Yong 302
Lizhi, Zhao 365
Lu, Jianqiang 70
Lu, Jinjun 223
Lu, Jixiang 223
Lumei, Su 156, 403
Luo, MingNan 315
Ma, Jian 330
Ma, Rongrong 128
Malik, Fahad Mumtaz 234, 242
Meixin, Qin 403
Min, Luo 142

Mitton, Nathalie....342
Ni, Ming....77
Niu, Tao....100, 171
Pathik, Bishwajit Banik....309
Qin, He....62
Rehman, Saad Ur....260
Ru, Zhang....365
Rui, Song....142
Ruina, Dang....108
Sarwr, Danish....397
Sha, Hengyu....354
Shamshad, Nadia....397
Shang, Liqun....252
Shao, Fei....218
Shao, Xiang....36
Shaobing, Tian....408
Shaotong, Bai....45
Shi, Dan....198
Shichang, Zhao....142
Shuo, Zhang....108
Sian, Cong....365
Su, Kuei-Ching....372
Sun, Kui....277
Sun, Qingsong....100, 171
Tang, Wenbo....62
Tian, Hao....20
Tianqing, Liu....408
Tianyu, Fang....86
Tsai, Ming-Tang....372
Tu, Chia-Sheng....372
Uzair, Muhammad....234
Wang, Can....284
Wang, Gang....100
Wang, Ruimiao....100, 171
Wang, Tongtong....14, 123, 187, 193
Wang, Wei....267
Wang, Yanxin....1
Wang, Ying....135
Wei, Wang....296
Wen, Jun....14, 123, 187, 193
Wenli, Yang....142
Wenwei, Zhang....93
Wu, Di....135
Wu, Dongmei....291
Wu, Xinyu....284
Wu, Yifei....150
Wu, Yi....150
Wu, Zhuojun....128
Xia, Guoqing....228
Xiaofei, Zhu....161
Xie, Kaigui....100, 171
Xie, Tairong....267
Xing, Chen....403
Xingdong, Zhu....408
Xinqiang, Chen....156
Xu, Liu....204
Xuan, Wang....142
Yan, Jing....1
Yang, Chenguang....252
Yang, Fan....198
Yang, Fei....150
Yang, Lian....272
Yang, Wang....108
Yang, Wei....1
Yang, Wenxiu....272
Yang, Zhou....77
Yanruo, Hou....86
Yao, Xinyu....171
Yi, Li....93
Yin, Jianlin....114
Yiyan, Wang....379
Yu, Caisheng....296
Yu, HongTao....315
Yuan, Ye....284
Zenan, Chu....296
Zeng, Qinghui....128
Zhang, Guobao....1
Zhang, Hanhua....14, 123
Zhang, Longzhi....291
Zhang, Lufeng....284
Zhang, Shaonan....302
Zhang, Tao....223
Zhang, Xingxing....218
Zhang, Ying....315
Zhang, Yi....315
Zhang, Yuan....252
Zhao, Bo....228
Zhao, Hengyang....1
Zhao, Xilan....114
Zhao, Zhao....260
Zhe, Liu....204
Zheng, HongMing....315
Zheng, Wang....408
Zhengxu, Lu....7
Zhicheng, Huang....156
ZhiYong, Xu....260
Zhou, Ping....20, 28
Zhou, Xia....70, 77
Zhu, Cunchao....354
Zhu, Rui....135
Zhu, Shengyi....187, 193
Zhu, Taiyun....1
Zhu, Xiaofei....54
Zhu, Yuan....28